Organic and Sustainable Agriculture

Organic and Sustainable Agriculture

Editor: Jordan Berg

www.callistoreference.com

Callisto Reference,
118-35 Queens Blvd., Suite 400,
Forest Hills, NY 11375, USA

Visit us on the World Wide Web at:
www.callistoreference.com

ISBN: 978-1-63239-782-9 (Hardback)

The publisher's policy is to use permanent paper from mills that operate a sustainable forestry policy. Furthermore, the publisher ensures that the text paper and cover boards used have met acceptable environmental accreditation standards.

Printed in the United States of America.

Cataloging-in-publication Data

Organic and sustainable agriculture / edited by Jordan Berg.
p. cm.
Includes bibliographical references and index.
ISBN 978-1-63239-782-9
1. Organic farming. 2. Agriculture. 3. Sustainable agriculture. I. Berg, Jordan.

S605.5 .O74 2017
631.584--dc23

Table of Contents

Preface

Organic and Sustainable Agriculture aim to replace the current agricultural practices with more environment-friendly practices. This involves avoiding the use of chemical products and using organic fertilizers such as compost or green manure. This book attempts to understand the multiple branches that fall under the discipline of organic and sustainable agriculture and how such concepts have practical applications while understanding the long- term perspectives of the topics, the book makes an effort and highlighting their impact as a modern tool for the progress of the field. This text will serve as a reference to a broad spectrum of readers including agronomists, botanists, researchers, professionals and students engaged in the field of agriculture at various levels.

This book has been an outcome of determined endeavour from a group of educationists in the field. The primary objective was to involve a broad spectrum of professionals from diverse cultural background involved in the field for developing new researches. The book not only targets students but also scholars pursuing higher research for further enhancement of the theoretical and practical applications of the subject.

It was an honour to edit such a profound book and also a challenging task to compile and examine all the relevant data for accuracy and originality. I wish to acknowledge the efforts of the contributors for submitting such brilliant and diverse chapters in the field and for endlessly working for the completion of the book. Last, but not the least; I thank my family for being a constant source of support in all my research endeavours.

Editor

1

Reconciling Pesticide Reduction with Economic and Environmental Sustainability in Arable Farming

Martin Lechenet[1], Vincent Bretagnolle[2], Christian Bockstaller[3,4], François Boissinot[5], Marie-Sophie Petit[6], Sandrine Petit[1], Nicolas M. Munier-Jolain[1]*

1 Institut National de la Recherche Agronomique, Unité Mixte de Recherche 1347 Agroécologie, Dijon, Côte d'Or, France, **2** Centre d'Etudes Biologiques de Chizé - Centre National de Recherche Scientifique, Beauvoir sur Niort, Deux-Sèvres, France, **3** Institut National de la Recherche Agronomique, Unité de Recherche 1121 Agronomie et Environnement, Colmar, Haut-Rhin, France, **4** Université de Lorraine, Vandœuvre-lès-Nancy, Meurthe-et-Moselle, France, **5** Chambre d'Agriculture des Pays de la Loire, Angers, Maine-et-Loire, France, **6** Chambre Régionale d'Agriculture de Bourgogne, Quetigny, Côte d'Or, France

Abstract

Reducing pesticide use is one of the high-priority targets in the quest for a sustainable agriculture. Until now, most studies dealing with pesticide use reduction have compared a limited number of experimental prototypes. Here we assessed the sustainability of 48 arable cropping systems from two major agricultural regions of France, including conventional, integrated and organic systems, with a wide range of pesticide use intensities and management (crop rotation, soil tillage, cultivars, fertilization, etc.). We assessed cropping system sustainability using a set of economic, environmental and social indicators. We failed to detect any positive correlation between pesticide use intensity and both productivity (when organic farms were excluded) and profitability. In addition, there was no relationship between pesticide use and workload. We found that crop rotation diversity was higher in cropping systems with low pesticide use, which would support the important role of crop rotation diversity in integrated and organic strategies. In comparison to conventional systems, integrated strategies showed a decrease in the use of both pesticides and nitrogen fertilizers, they consumed less energy and were frequently more energy efficient. Integrated systems therefore appeared as the best compromise in sustainability trade-offs. Our results could be used to re-design current cropping systems, by promoting diversified crop rotations and the combination of a wide range of available techniques contributing to pest management.

Editor: Raul Narciso Carvalho Guedes, Federal University of Viçosa, Brazil

Funding: Funding for the study was provided by the French National Research Agency ANR (STRA-08-02 Advherb project) and from Région Bourgogne. The Burgundy farms network was developed in the framework of the Réseau Mixte Technologique "Systèmes de Culture Innovants" and the project "Plus d'Agronomie, Moins d'intrants" initiated by Région Bourgogne. The long term experiment at Dijon-Epoisses was partly funded by the European Network of Excellence ENDURE. The funders had no role in study design, data collection and analysis, decision to publish, or preparation of the manuscript.

Competing Interests: The authors have declared that no competing interests exist.

* E-mail: nicolas.munier-jolain@dijon.inra.fr

Introduction

Reconciling agricultural productivity with other components of sustainability remains one of the greatest challenges for agriculture [1]. A key issue will be to achieve substantial reductions in the level of pesticide use for environmental and health reasons [2,3]. Agriculture in temperate climates is widely dominated by conventional intensive farming systems, with highly specialized crop productions and a heavy reliance on pesticides and mineral fertilizers [4]. However, increasing environmental concerns about intensive farming practices has contributed to the emergence of innovative farming systems, such as organic and integrated farming, typically presented as alternative paths to reduce pesticide use as compared to current conventional systems [5,6,7]. Whether these systems better meet sustainability criteria has been a matter of debate [8,9]. Integrated farming, recently promoted in Europe through the 2009/128/EC European directive [10], is defined as a crop protection management based on Integrated Pest Management (IPM) principles, which emphasizes physical and biological regulation strategies to control pests while reducing the reliance on pesticides [11]. It can be regarded as an intermediate between conventional farming, with high levels of inputs, and organic farming, which prohibits the use of synthetic pesticides and fertilizers. Organic and integrated farming have in common the combined use of management approaches to replace, at least in part, synthetic inputs. However, unlike organic farming which is growing both in Europe (by 40 to 50% between 2003 and 2010 [12]) and in the US (by 270% between 2000 and 2008 [13]), integrated arable crop production is not expanding because it is perceived by farmers as a complex system which is difficult to implement, labour-consuming, and associated with reduced and unpredictable economic profitability [14,15]. As a consequence, the amount of pesticides sprayed has only decreased slightly in Europe (−3.6% from 2000 to 2007 [16]) and in the US (−7.5% from 2000 to 2007 [17]). Moreover, this decrease can be partly attributed to the substitution of older chemistry, applied at high dosage, by new products that are efficient at lower doses, which actually cannot be considered as a reduction of pesticide reliance. In France, the national action plan, ECOPHYTO 2018, which had set a target of a 50% decrease in pesticide use by the year 2018, is currently far from achieving this goal [18].

So far, assessments of cropping system sustainability have compared few – typically two or three – experimental prototypes that represent conventional, organic or integrated strategies

[19,20]. However, this approach fails to capture the diversity within each of these farming strategies. Given the diversity of crop management options within a conventional, an integrated or an organic strategy, which might lead to contrasted performances, the generic value of experimental results ignoring this variability may be argued. We assessed the sustainability of 48 cropping systems located in regions of intensive arable farming and covering a wide range of pesticide use levels and cultivation techniques such as crop rotations, from monoculture to highly diversified crop rotations, soil tillage (e.g. inversion tillage, shallow tillage or direct drilling), fertilization (mineral or organic fertilizers), or weed management (e.g. only based on herbicide use, including mechanical weeding). More details about the cropping system sample are available in the online SI section (Dataset S1). All the studied cropping systems were followed for between three and 12 years, between 1999 and 2012. Eight cropping systems were organic, 30 were based on integrated farming and 10 were conventional (Figure 1). Using eight sustainability indicators to evaluate the performance of the study systems, our aims were: (i) to identify possible conflicts between the reduction of pesticide reliance and other components of sustainability; and, (ii) to assess the potential of organic and integrated strategies for improving agricultural sustainability.

As the performance of a cropping system depends not only on the combination of management options it implements, but also on the local production situation [21] (including biophysical and socio-economic local aspects), we standardized the indicators of performance and pesticide use, using a ratio of the performances of the cropping systems over those of a local reference system. This enabled us to focus solely on the effects of the management strategies on sustainability indicators. The local references were cropping systems selected as representative of the most widespread crops and practices within each production situation. Pesticide use was measured as the Treatment Frequency Index (TFI), which is a commonly used indicator in Europe to estimate the cropping system dependence on pesticides [22]. In our sample, organic cropping systems did not use any pesticides (synthetic or natural) so their relative TFI, expressed as a ratio of the local reference TFI, was zero. Integrated cropping systems displayed TFI values that were on average half (−47%) of the local references (Table S1).

Results

Table S1 presents the mean and standard deviation for each performance indicator according to the management strategy (organic, integrated and conventional). The second tab of Dataset S1 provides performance details for each cropping system of the sample.

Productivity and energy efficiency

Given the primary role of agriculture remains to produce food and other goods, we used an indicator of productivity, expressed as the total yearly amount of energy produced by a cropping system, whatever the crops cultivated (Figure 2a). The productivity of organic cropping systems was below that of their local reference (Figure 3b), ranging from −22% to −76%. For non-organic cropping system, productivity was uncorrelated to relative TFI (Figure 2a and Table 1), with some cropping systems that had a low reliance on pesticides even exceeding the productivity of the local reference. Cropping system productivity may strongly depend on crop type, especially if the whole above-ground biomass is harvested or not. Crops other than grain crops were frequently grown in integrated farming, as they are typically associated with low pesticide requirements and can contribute to weed control in subsequent crops [23]. They typically consist of

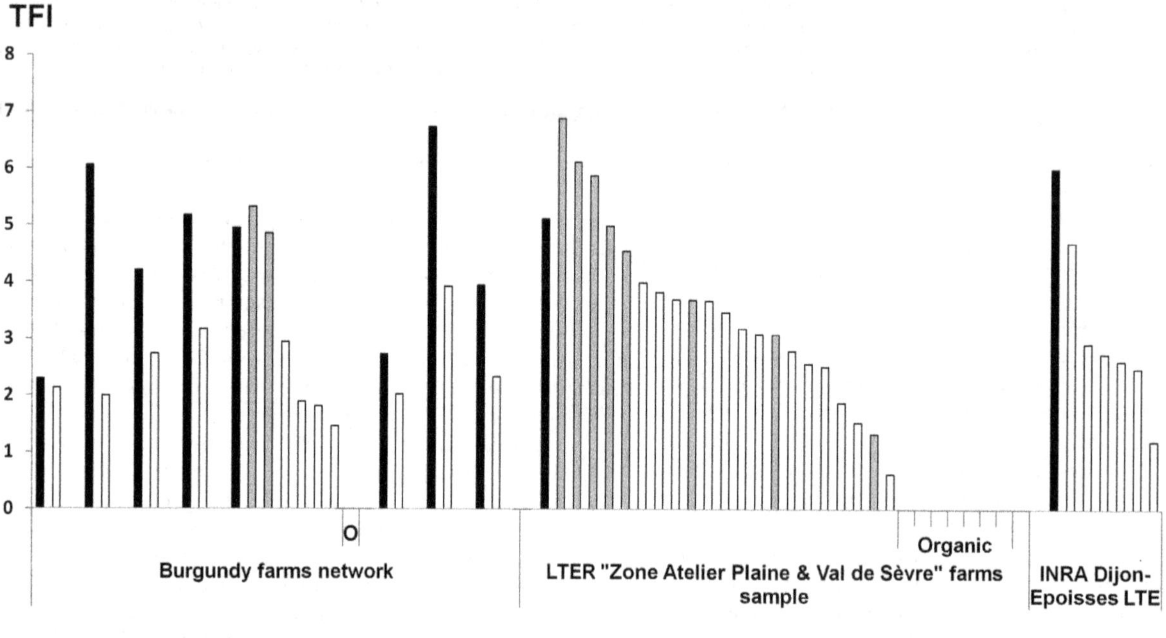

Figure 1. Distribution of the Treatment Frequency Index (TFI) for the studied arable cropping systems. Average TFI for each cropping system composing the study sample. At each site, black bars correspond to the local reference, grey bars to conventional cropping systems and white bars to integrated cropping systems. The sample also includes eight organic cropping systems with TFI = 0 and labelled with "O" or "Organic". Details about the cropping systems are available in Dataset S1.

forage crops, dedicated to livestock feeding with limited energy efficiency, or of crops used for non-food applications. However, distinguishing cropping systems based on grain crops or on crops in which all above-ground biomass is harvested did not change the observed pattern. In systems with grain crops only, productivity was not correlated with relative TFI (Table 1), suggesting that a reduction in pesticide use intensity may not be necessarily translated into a decrease in productivity. The second indicator of energy productivity we used was the energy efficiency of cropping systems, resulting from a ratio between energy output and energy input. It evaluated the ability of a cropping system to convert energy inputs into outputs. Organic cropping systems were significantly less energy efficient than other systems (Figures 2b and 3c, Table 2). Despite their energy consumption being lower (Table 2), notably due to their low reliance on nitrogen fertilizers, it was not sufficient to offset their limited productivity. Energy consumption was negatively correlated with relative TFI in integrated and conventional systems which cultivated only grain crops (Table 1). Energy efficiency was also negatively correlated with relative TFI in these systems, although the relationship was weak and only marginally significant ($r_s = -0.35$, $P = 0.07$). The systems with the highest energy efficiency, whether they included crops with all above-ground biomass harvested or not, were mostly integrated systems (Figures 2b and 3c).

Environmental impact

The environmental impact of cropping systems and their reliance on external inputs were assessed with the indicator I-Pest [24] and with estimates of fuel and nitrogen fertilizer consumption. I-Pest is a predictive indicator that assesses the environmental impacts of pesticide use as the risk of contamination of the air, and surface and ground waters (see Figure S1). As the organic cropping systems composing the sample did not used synthetic or natural pesticide, their cumulated I-Pest was 0. As expected for the rest of the sample, cumulated I-Pest was strongly and positively correlated to relative TFI (Figure 2c and Table 1). Fuel and nitrogen fertilizers together amounted to more than 60% of the total energy inputs for all tested cropping systems. Organic systems consumed more fuel than the rest of the sample (Table 2), with their average consumption exceeding the local references by 17% (Figure 3d). Organic cropping systems had, nonetheless, a lower reliance on N fertilization than the rest of the sample (Figure 3g, Table 2), in line with their lower yield targets and the frequent occurrence of crops with low N requirements used in organic rotations. No relation was detected between fuel consumption and relative TFI in non-organic systems (Figure 2d, Table 1), but a positive correlation was clearly visible between relative TFI and the amount of nitrogen fertilizers applied (Figure 2e).

Economic sustainability and workload

Economic sustainability was assessed by considering (i) the profitability, i.e. the average semi-net margin over a range of ten real price scenarios for agricultural products, fuel and fertilisers, and (ii) the sensitivity of this profitability in a context of price volatility, i.e. the relative standard deviation of the semi-net margin. The range of price scenarios used for the calculations was set to reflect the variability of the economic context over the last decade. Profitability, when averaged over the ten price scenarios was not correlated with relative TFI for integrated and conventional systems (Figure 2f and Table 1), and no significant difference appeared with organic systems (Mann-Whitney test, $P>0.9$). It suggests that low pesticide use would not necessarily result in lower economic return. The strong variability observed within each class (Figure 3f), most notably for integrated cropping systems, confirmed that strategies to reduce pesticide use could even lead to an increase in profitability. As integrated cropping systems were, in contrast to organic systems, evaluated with a conventional price reference, the most profitable integrated systems were able to efficiently reduce their production costs. No relation was detected between the sensitivity to price volatility and relative TFI in conventional and integrated systems (Figure 2g, Table 1). Sensitivity to price volatility was significantly lower in organic cropping systems than in other systems (Table 2), most probably because: (i) they were based on more diversified crop rotations, which spread risks and buffered semi-net margin at the farming system scale; and, (ii) their crop rotations typically included crops with low N demand, that had reduced reliance on N inputs, whose price is directly related to the volatile price of fossil fuels.

The issue of social sustainability was addressed using the 'workload' indicator, which gives emphasis to the potential for bottlenecks where available workforce is a limiting factor at the farm scale (Figure 2h). Workload was calculated for each technical operation but excluded time devoted to transport and crop monitoring. Workload was found not correlated with relative TFI in non-organic cropping systems (Table 1), and no significant difference was found with the organic group (Mann-Whitney test, $P>0.1$), so that reducing pesticide use does not necessarily imply an increased workload. Indeed, in integrated systems, labour requirements ranged from low to high relative values (Fig 2h). The level of workload was, however, related to the type of fertilization, with cropping systems having organic fertilization requiring an average of 13% greater working time, as compared to mineral fertilizer-based cropping systems (Table 2).

Crop diversity

Diversification of crop rotations is often presented as an efficient management tool for controlling pests and to improve agricultural sustainability [25,26]. We used a crop sequence indicator, Isc [27], which estimates the consistency of the crop sequence with regard to the potential of input reduction, by addressing effects of crop rotation on pathogens, pests, weeds, soil structure and nitrogen supply of preceding crops. Even if no significant correlation appeared between Isc and relative TFI (Table 3), organic and integrated cropping systems displayed significantly higher Isc values than conventional systems (Table 2). A negative correlation between Isc and productivity suggests that diversifying crop rotation may reduce cropping system productivity (Table 3), but the Spearman correlation test was no longer significant when organic cropping systems were excluded ($P = 0.07$). We did not detect any significant relationship between energy efficiency and crop diversification, whether organic cropping systems were included or not ($P = 0.44$). No correlation was observed between Isc and semi-net margin, but workload appeared to be lower for systems with higher Isc (Table 3). We found the expected negative correlation between Isc and N fertilization rates, and consequently between Isc and energy consumption (Table 3). We focused therefore more particularly on cropping systems including legume crops, which also displayed higher Isc values than the rest of the sample (Table 2). The role of legume in improving energy efficiency at the cropping system scale was clearly demonstrated by the correlation between the frequency of occurrence of legumes in the crop rotation and the energy efficiency ($r_s = 0.37$, $P<0.05$). The sensitivity to price volatility was negatively correlated with the frequency of occurrence of legumes in the crop rotation ($r_s = -0.33$, $P = 0.02$), but positively correlated with the level of N fertilization ($r_s = 0.49$, $P = 5*10^{-4}$). Fostering exogenous N independence therefore appeared as an efficient way to limit income variability.

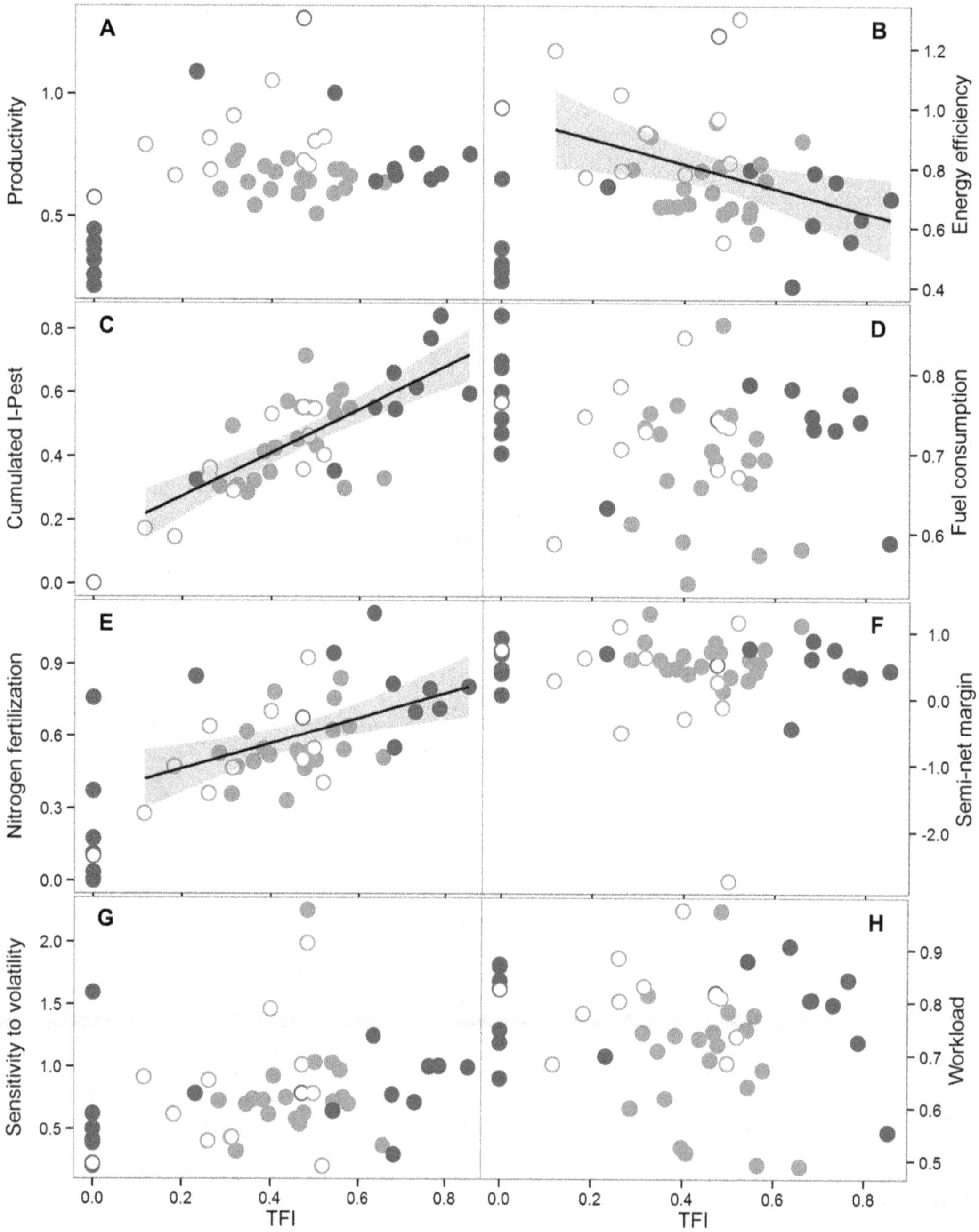

Figure 2. Relationship between sustainability indicators and relative TFI. Cropping system performances according to their relative TFI. Conventional, integrated and organic cropping systems are represented by blue, green and red symbols respectively. Filled symbols correspond to the cropping systems with grain crops only and empty symbols refer to the cropping systems including crops for which the whole above-ground biomass is harvested. Each sustainability indicators is expressed as the natural logarithm of the ratio between the cropping system and the local reference indicators. Linear regressions are represented with their standard error for cumulated I-Pest (Pearson correlation test: $r_p = 0.74$, $P = 5*10^{-8}$), nitrogen fertilization (Pearson correlation test: $r_p = 0.48$, $P = 0.002$), and energy efficiency (Pearson correlation test: $r_p = -0.38$, $P = 0.02$). Performance metric included: a) productivity, b) energy efficiency, c) cumulated I-Pest, d) fuel consumption, e) nitrogen fertilization, f) semi-net margin, g) sensitivity to price volatility, h) workload.

Discussion

This work was aimed at detecting cropping systems able to reconcile low pesticide use and other components of sustainability. Our original multiple dimensions approach, based on a precise description of management practices, was designed to compare and contrast numerous cropping systems from different production situations. This approach, applied at the large-scale, was able to provide generic knowledge about potential trade-offs between the different issues of agricultural sustainability.

Sustainability of integrated and organic farming

Our results show that achieving a low level of pesticide use is possible without triggering negative side effect on any of the

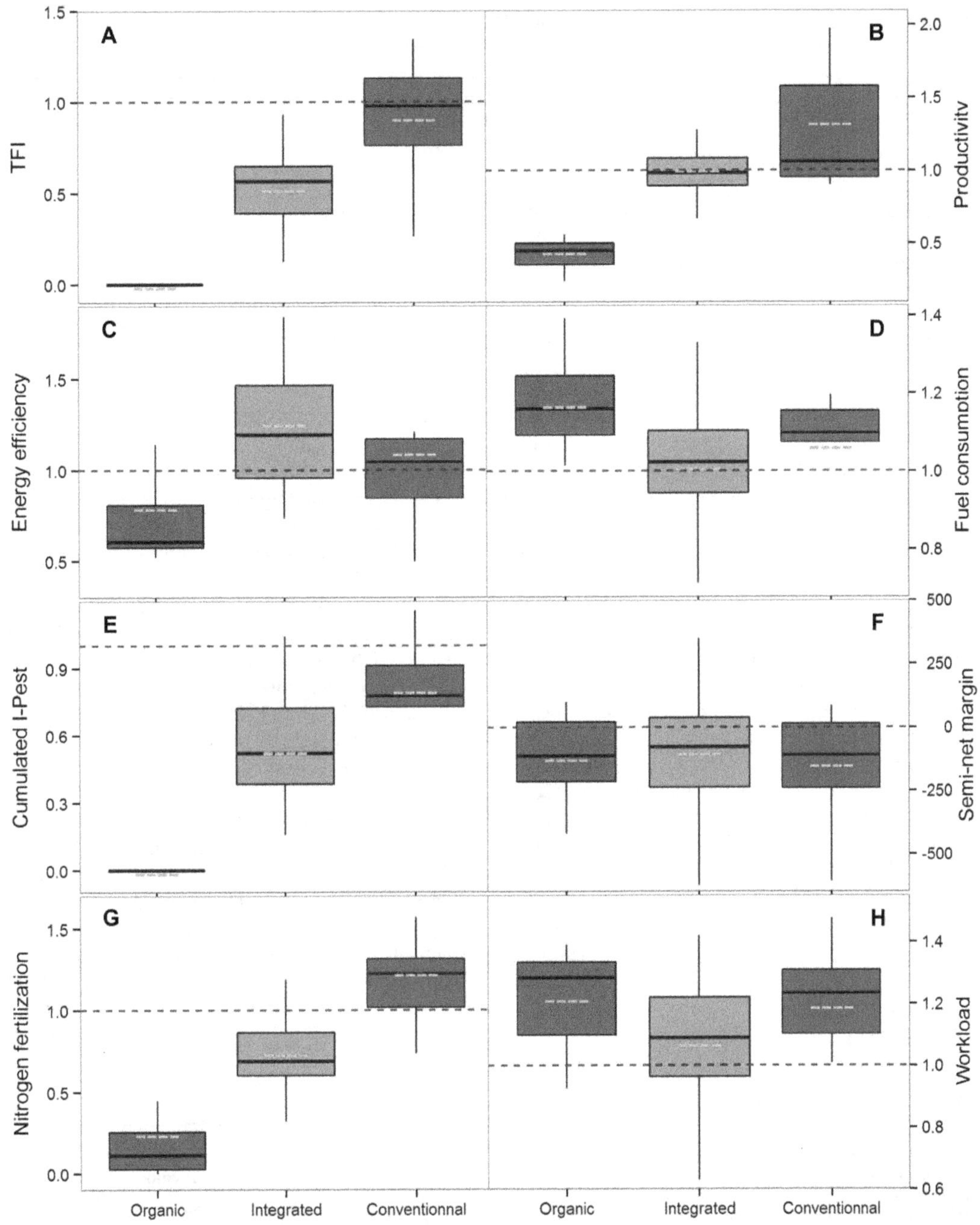

Figure 3. Cropping systems distribution according to sustainability indicators. Performance indicators are expressed as a ratio of the local reference indicator, except for semi-net margin, expressed as a difference with the local reference. Conventional, integrated and organic cropping systems are represented by blue, green and red box plots respectively. The horizontal black bars and grey dashed bars correspond to median and mean values respectively. The horizontal red dashed bar recalls the position of the local references. Outliers are not represented. Performance metrics included: a) Treatment Frequency Index, b) productivity (organic farming: one outlier, v = 0.78; integrated farming: two outliers, v1 = 1.48 and v2 = 1.87), c) energy efficiency (organic farming: one outlier, v = 1.72), d) fuel consumption (integrated farming: one outlier, v = 1.37; conventional farming: two outliers v1 = 0.8 and v2 = 0.88), e) cumulated I-Pest (conventional farming: two outliers v1 = 0.38 and v2 = 0.42), f) semi-net margin, g) nitrogen fertilization (organic farming: one outlier, v = 1.13; integrated farming: two outliers, v1 = 1.32 and v2 = 1.52), h) workload (conventional farming: one outlier v = 0.74).

components of cropping system sustainability we assessed in this study. Integrated cropping systems were not only associated with low pesticide use and low risks for contamination of air and water with pesticide residues; they also displayed lower energy consumption than more intensive cropping systems and are likely to improve energy efficiency without impact on productivity and profitability. Lower pesticide usage in arable cropping systems did not imply a heavier workload, another critical point conditioning strongly the adoption of an innovative strategy.

Organic farming prohibits the use of synthetic pesticides and fertilizers, and this approach is often associated with low nitrogen fertilization, as observed in our sample. In addition to the positive

Table 1. Rank correlation between TFI and sustainability indicators for integrated and conventional cropping systems.

Spearman correlation	Productivity	Productivity (grain crops only)	Energy consumption	Energy consumption (grain crops only)	N fertilization	Pesticides environmental impact
r_s	−0.17 (NS)	0.06 (NS)	0.30 (NS)	0.42	0.51	0.67
P-value	0.3	0.8	0.06	0.03	$9*10^{-4}$	$4*10^{-6}$
Spearman correlation	**Fuel consumption**	**Energy efficiency**	**Semi-net margin**	**Sensitivity to prices volatility**	**Workload**	**Crop Sequence Indicator (Isc)**
r_s	0.06 (NS)	−0.40	−0.05 (NS)	0.22 (NS)	$9*10^{-3}$ (NS)	−0.22
P-value	0.7	0.01	0.8	0.2	0.95	0.2

Spearman rank correlation tests ($\alpha = 0.05$). r_s is the Spearman correlation coefficient. Values of r_s followed by (NS) are not significant.

effects on environmental quality, numerous studies underlined other environmental benefits of organic farming such as effects on pollinator dynamics [28], on landscape floristic composition [29], as well as on soil microbial diversity [5,30]. Here we demonstrated that organic farming does not necessarily affect profitability and workload, and conversely, might strengthen farm financial stability in a variable and unpredictable economic context. Organic cropping systems were however less productive and less energy efficient than integrated systems in our sample. Although highly dependent on crops and production context, productivity in organic farming was already reported as lower than in conventional farming by other comparative studies [31]. The poor land use efficiency associated with organic farming is a key issue in the current land sharing – land sparing debate about the growing competition for land use [32], and notably urban sprawl [33] as well as the necessity to keep natural spaces undisturbed [34,35]. Both aspects – environmental benefits of organic farming and the limited productivity per unit of land – should therefore be considered by decision makers in their incentives for sustainable agriculture.

Crop diversification

Our results support the hypothesis that crop diversification may be an effective means to enhance cropping system performance. At the cropping system scale, crop diversification provides agronomic advantages, such as the regulation of pests, diseases and weeds [36,37,25]. In our sample, the most diversified cropping systems, which displayed the highest values of the crop sequence indicator, Isc, were indeed less dependent on pesticides. Their low environmental impact on water and air quality makes crop diversification an interesting potential pathway for reducing the damage caused by agriculture on natural resources (e.g., biodiversity [38]), as well as on human health (e.g. neurological degenerative disorders [39]). By mitigating the adverse effects of climate variability, crop diversification may also improve system resilience for productivity [40], with the increasing likelihood of extreme weather events requiring farm adaptation [41]. Economic market volatility is an additional source of variation and risk factor for farm economic stability. We found that crop diversification, particularly through the introduction of legumes in the crop rotation, is likely to limit dependence on inputs that have unstable prices. By allowing a decrease in the use of exogenous N fertilizer across a crop rotation, legume cultivation reduces production cost fluctuations and consequently makes the cropping system less sensitive to market volatility. Legumes come with a supplementary advantage [42] in the face of the considerable amount of fossil energy necessary to produce mineral N fertilizers, and we noted a substantial increase in energy efficiency for crop rotations where

Table 2. Significantly different groups for a given performance indicator.

Indicator	Test designation	P-value	Statistic W
Productivity	Difference of productivity between organic cropping systems and the rest of the sample	$2*10^{-8}$	318
Productivity	Difference of productivity between cropping systems including crops with the whole above-ground biomass harvested and the rest of the sample	$5*10^{-4}$	76
Energy efficiency	Difference of energy efficiency between organic cropping systems and the rest of the sample	0.005	258
Energy consumption	Difference of energy consumption between organic cropping systems and the rest of the sample	0.001	272
Fuel consumption	Difference of fuel consumption between organic cropping systems and the rest of the sample	0.02	72
N fertilization rate	Difference of N fertilization rate between organic cropping systems and the rest of the sample	$2*10^{-4}$	285
Sensitivity to price volatility	Difference of sensitivity to price volatility between organic cropping systems and the rest of the sample	0.01	250
Workload	Difference of workload between cropping systems based on organic fertilization and the rest of the sample	0.045	189
Crop sequence indicator	Difference of Isc between organic cropping systems and conventional cropping systems	0.01	69
Crop sequence indicator	Difference of Isc between integrated cropping systems and conventional cropping systems	0.001	255
Crop sequence indicator	Difference of Isc between cropping systems including legumes and the rest of the sample	$6*10^{-6}$	63

Mann-Whitney tests ($\alpha = 0.05$). All P-values are below 0.05, indicating that the differences between means of the sub-samples are significant for the corresponding indicators.

Table 3. Rank correlation between Crop Sequence Indicator Isc and sustainability indicators.

Spearman correlation	Productivity	Energy consumption	N fertilization	Pesticides environmental impact	Fuel consumption
r_s	−0.35	−0.42	−0.41	−0.39	*−0.07 (NS)*
P-value	0.01	0.003	0.004	0.006	*0.7*
Spearman correlation	**Energy efficiency**	**Semi-net margin**	**Sensitivity to price volatility**	**Workload**	
r_s	*−0.05 (NS)*	*0.10 (NS)*	*−0.23 (NS)*	−0.29	
P-value	*0.7*	*0.5*	*0.1*	0.04	

Spearman correlation tests ($\alpha = 0.05$). r_s is the Spearman correlation coefficient. Values of r_s followed by (NS) are not significant.

legume crops are more frequent. The most part of legumes introduced as diversification crops are however forage crops, and livestock production is commonly considered more energy consuming than plant production [43]. Conversely, the use of farmyard manure may contribute to reduce mineral fertilizers reliance for grain and forage crop production. The necessity of (i) integrating these situation-dependent parameters into energy balancing calculations, and, (ii) evaluating other environmental indicators [44] will be critical for the assessment of livestock production as a management option for enhancing agricultural sustainability.

A key agronomical advantage of crop diversification is related to the management of weed resistance to herbicides. Crop diversification is an efficient means to alternate herbicide modes of action and to introduce diversified measures of weed control, allowing changing selection pressure on weed communities and thus maintaining a sensitive weed population (i.e. maintaining high herbicide efficiency) [45].

Our results demonstrate a negative correlation between the Isc value and workload. We can nevertheless assume that diversifying crop rotations increases cropping system complexity and time devoted to field observations. Another aspect is that crop diversification may lead to a more evenly distributed workload over the seasons. Crops diversity implies a greater diversity in sowing and harvest periods, which are both times of peak labour that strongly influence task organisation and farmer decision making [15]. By reducing the amplitude of these peaks in labour, crop diversification could contribute to ensuring greater farmer decisional flexibility at the farm scale.

Beyond technical and organizational issues at the farm level, diversifying crop production as a component of an integrated strategy at regional or national scale would inevitably lead to important changes in production volumes, as well as markedly changing agricultural sectors within each production basin. It would definitely require an adaptation in the organisation of the whole agricultural sector and the development of new local markets. These economic and social lock-ins are rightly highlighted as the main limiting constraints hindering crop diversification [46]. However, by creating a particular economic sub-context, niche markets can be attractive and able to support innovation. Promoting such niche markets, for integrated farming development, would be the first step along an accelerating cycle of improvement based on mutually positive feed-backs between production and outlets.

Materials and Methods

In all cases, the field studies did not involve endangered or protected species.

For future permissions about the private farm network of Burgundy, please contact Marie-Sophie Petit (co-author of the research article, Chambre Régionale d'Agriculture de Bourgogne) and Sandrine Petit (co-author of the research article, INRA).

For future permissions about the private farms survey carried out on the LTER "Zone Atelier Plaine & Val de Sèvre", please contact Nicolas Munier-Jolain (corresponding author, INRA).

Study areas

The main objective of this study was to highlight potential conflicts between pesticide use and a set of sustainability indicators, so the cropping systems we consider were selected to maximize the contrast across the range of possible pesticide use intensities. The sample of cropping systems we used originates from:

(i) A long term experiment conducted since 2000 at the INRA Dijon-Epoisses farm in Bretenière (Burgundy, eastern France; 47°20′N, 5°2′E) in order to assess Integrated Weed Management-based cropping systems [15,26]. Seven cropping systems were tested between 2000 and 2012 including different combinations of technical levers likely to reduce pesticide reliance.

(ii) An experimental network (bringing together 14 cropping systems) monitored (1) by the local agricultural extension services and coordinated by the Chambre Régionale d'Agriculture de Bourgogne, and (2) by the INRA de Dijon. This network involved contrasting private farms of the Burgundy region, and was developed to test feasibility of innovative cropping systems with reduced pesticide use in a realistic context.

(iii) A survey of private farms carried out in 2010 on the LTER "Zone Atelier Plaine & Val de Sèvre" [47] located in the Poitou-Charentes region (450 km^2 study area in western France), and set up to explore a diversity of pesticide reliance, including organic farming, conventional intensive systems, and intermediate, IPM-based systems. Twenty nine varied cropping systems were surveyed in this area.

Cropping systems classification

Details of cropping systems, including crop sequences, performances, and detailed crop management operations are made available in the Dataset S1 and S2. Cropping systems were considered as conventional, integrated or organic according to the following rules. Cropping systems complying with the organic farming specifications were treated as 'organic'. Other systems were considered as 'integrated', when they were based either on diversified crop rotations including unusual alternative crops for the production situation (i.e. not present in the local reference crop rotation), or when crop management included at least one non-chemical management approach that contributed to the control of pests, diseases or weeds. These included for instance biocontrol, mechanical weeding and false seed bed techniques. Systems that were not classified as 'organic' or 'integrated' were classified as 'conventional'.

Local reference definition

For each of the 48 systems, a local cropping system reference was selected to reflect the most widespread crop rotation and associated technical management, as well as the typical agricultural performance in the production situation. Using this local cropping system reference made it possible to distinguish the effects of agronomic strategies from the effects of the production situation (soil, climate, economic and social context) when assessing the various components of sustainability of each cropping system. The Dijon-Epoisses experiment included a reference standard system that follows recommendations of local extension services [26], and which was used as the local reference. For each farm of the network across Burgundy, the local reference was defined as the cropping system implemented within the farm before the set-up of the alternative cropping system, even though crop management sequences were slightly updated according to expert appraisal to match with current standards (e.g. active ingredients allowed). For the Zone Atelier "Plaine et Val de Sèvre", local expert knowledge was used to select one system from the survey, with a standard crop rotation for the area, and a crop management representative of local practices. This system was then used as the local reference for all the remaining surveyed systems of the area.

Assessment of sustainability

The assessment of sustainability at the cropping system scale was based on a range of indicators covering economic, environmental and social issues. The Treatment Frequency Index (TFI) [22] estimates the number of registered doses applied, for each pesticide, per hectare and per crop season. Averaged over the cropping system, this indicator summarizes the level of dependence on pesticides, which should be distinguished from the environmental impact of pesticide use. This indicator is calculated for each pesticide application according to the following formula:

$$TFI = \frac{Application\ rate\ \times\ Treated\ surface\ area}{Registered\ dose\ \times\ Plot\ surface\ area}$$

The application rate and the registered dose were both expressed for a given commercial product (which possibly contains several active ingredients). The recommended application dose depends obviously on the treated crop and on the targeted pest. Here we defined the registered dose as the lowest application dose which is recommended for a given crop. The TFI for a given crop season was then calculated as the sum of the TFI for each pesticide application performed during this crop season. Productivity was evaluated as the amount of energy harvested yearly. This approach allowed the comparison of different crop rotations that included crops with different yielding potentials and different energy content. For each crop, yields were transformed into the energy metric using their Lower Heating Value (LHV) [48], which corresponds to the amount of energy released per unit of mass by the combustion of the harvested biomass. Energy consumption was estimated from the conversion of inputs into energy according to the Dia'terre reference database [48]. Dia'terre is an assessment tool developed by the Agency for the Environment and Energy Management (ADEME) in the framework of the French Plan for Energy Performance (PPE) to evaluate a carbon-energy balance at the farm scale. The reference database used to design this assessment tool provides energy values for indirect energy consumption associated with the production of farming inputs. For instance, the calculation of the energy cost associated with the production of nitrogen fertilizers integrates the energy necessary from raw material production (e.g. Haber–Bosch process) through materials processing, manufacture and distribution. We used the reference energy cost provided by the carbon calculator Dia'terre to compute the energy balances of the cropping systems. In this way, the inputs necessary for crop production were converted into energy, using the energy cost of fertilizers, pesticides, seeds, water spread for irrigation, fuel consumed by the equipment and the amount of steel necessary to manufacture this equipment, i.e. energy cost of mechanization (see Dataset S3). The energy requirements for preparing farmyard manure are farm-specific and very difficult to quantify precisely. A simplification was consequently required: following previous studies based on energy balancing methods in crop production [49], the energy equivalent of farmyard manure was equated with that of the mineral fertilizers they substituted (using a substitution value related to the fertilizing efficiency of manure). Energy efficiency was computed from the ratio between productivity and energy consumption. For assessing the economic productivity, the gross product derived from the direct conversion of crops yields into economic values. The 'semi-net' margin was calculated as the gross product per hectare from which we subtracted the input costs (fertilizers, pesticides, seeds, fuel, water and mechanisation). This 'semi-net' margin assessed the system profitability without taking into account subsidies or incentives. The sensitivity to price volatility was defined as the relative standard deviation of the semi-net margin calculated over ten contrasting real price scenarios selected between 2000 and 2010, and thus measured the ability of a cropping system to generate a stable income in a variable economic context. The ten scenarios integrated the prices of crops but also the prices of volatile inputs such as fertilizers or fuel. Each price scenario was defined at a given moment between 2000 and 2010, and it therefore reflected the correlations between the prices of crop products and inputs. This approach notably made it possible to integrate better the effects coming from crop diversity (proportion of cereal crops, oil crops or protein crops) on cropping system profitability and economic stability. Fuel consumption and workload were estimated according to in-field cropping operations only, without considering fuel and time consumed for farm-to-field transports, or extra-workload dedicated to equipment maintenance or field observations. The size, fuel requirements and working output of the various equipment types were standardized for all cropping systems, and defined from a national database [50], consistent with the aim of evaluating management strategies, and of ignoring the potential effects of the equipment specifications (See Dataset S4 for the details of the equipment used for the calculations).

Pesticide environmental impact was expressed as cumulated I-Pest [24]. This indicator measures the risk associated with pesticide application for three compartments of the environment, namely the air, the surface water and the groundwater. This risk indicator, ranging from 0 to 1 (maximum risk), and calculated for each active substance application, is based on: (i) field inherent sensitivity to pesticide transfer toward these three compartments; (ii) characteristics of the active substance (e.g. ecotoxicity, mobility, half-life); and, (iii) information about the conditions of the spraying operation (e.g. amount of active substances employed, canopy cover at the date of treatment) in order to calculate three impact factors, one for each compartment. I-Pest index is obtained using fuzzy decision trees that allow the aggregation of these three impact factors into one synthetic indicator. The diagram presented in Figure S1 illustrates how this indicator of pesticide environmental impact was computed for each pesticide active substance that was sprayed within the field.

The crop sequence indicator Isc [27] is used as an additional indicator to quantify the agronomic effects of crop diversification. Isc ranges on a qualitative scale between 0 and 10 (best value) and is calculated as shown in the following equation:

$$Isc = kp \times kr \times kd$$

Isc is based on the assessment of the effects of the previous crop on the current crop (kp), with respect to the development of pathogens, pests and weeds, to soil structure and nitrogen supply. kp, ranging between 1 and 6, was assessed for 470 couples crop/previous crop. kp is corrected by two factors taking into account the crop frequency (kr ranging between 0.3 and 1.2) and the crop rotation whole diversity (kd ranging between 1.0 and 1.4). Isc yields respectively 0.5 for wheat monoculture, 3.3 for a rape/wheat rotation, 5.1 for a rape/wheat/barley rotation, and 7.6 for a maize/wheat/sunflower/spring barley rotation.

Computation of sustainability indicators at the cropping system level

As a first step each indicator was calculated for each cropping operation composing our database (Dataset S2). These values of indicators were summed over the crop season year, and then averaged across years and across plots, each plot being considered as a replicate of a given cropping system. Each indicator was therefore calculated at the cropping system level, integrating (i) the different crops composing the crop sequence, (ii) the variability of crop production related with the inter-annual climatic variability, and (iii) the possible variation in plot properties.

All sustainability indicators were expressed per hectare and per year. For distinguishing specifically the effects of the management strategy on cropping system sustainability from the effects of the production situation, each indicator computed for a given cropping system was then expressed as a ratio (or as a distance in the case of semi-net margin) between the system indicator and the local reference indicator. To increase the quality of the graphs drawing the relationship between sustainability indicators and pesticide use, values of assessment indicators were translated into natural logarithm (Figure 2), which reduced the visual effect of extreme values.

Statistical analyses

Spearman and Pearson correlations were estimated using the 'rcorr' correlation matrix function in the *Hmisc* package of R v2.15.0 [51]. The difference between the means of two subsamples for a given indicator was tested with a non-parametric Mann-Whitney test ('wilcox.test' function with two samples) in the *stats* package of R v2.15.0.

Supporting Information

Figure S1 Simplified description of the assessment process of pesticide environmental impact in the I-Pest model.

Dataset S1 Cropping systems details. A.xlsx file describing the cropping systems of the studied sample (e.g. crop rotation, tillage and weed management strategies). This file also provides information about the local reference associated with the evaluation of each cropping system. The second tab provides the respective performances of each cropping system described in the first tab.

Dataset S2 Cropping operations database. A.xlsx file which provides the details of all cropping operations carried out in each cropping system: type of cropping operation, date (when recorded), application rates (for pesticides, fertilizers, seeds and irrigation) and proportion of the plot surface targeted.

Dataset S3 Energy balancing database. A.xlsx file with two sheets. The first sheet provides energy cost values for inputs: pesticides active substances, fuel, fertilizers, irrigation water and seeds. The second sheet includes the Lower Heating Values (LHV) for usual crops, that is to say the energy contained in one mass unit of crop harvested.

Dataset S4 Standard equipment characteristics. A.xlsx file describing the technical characteristics of the standard equipment we associated with each cropping operation. Details include the purchase price, the payback period and the maintenance cost to calculate the mechanization costs, but also the equipment size and weight, the working output, the fuel consumption rate and the energy cost value.

Table S1 Means and standard deviations for the range of performance indicators according to the management strategy. A.xlsx file summarizing and comparing the performances of organic, integrated and conventional cropping systems which compose the study sample. Significant difference between groups was tested with a Mann-Whitney test.

Acknowledgments

We thank A. Villard, C. Vivier and M. Geloen for their contribution to the experimental network in Burgundy, and D. Meunier, P. Farcy, P. Chamoy for technical assistance. We particularly want to thank D. Bohan for the precious advice on style and the language corrections.

Author Contributions

Conceived and designed the experiments: NMJ VB CB SP. Performed the experiments: ML FB. Analyzed the data: ML FB. Contributed reagents/materials/analysis tools: CB MSP. Wrote the paper: NMJ ML. Discussed the results and commented on the manuscript: NMJ ML VB CB SP MSP FB.

References

1. Foley JA, Ramankutty N, Brauman KA, Cassidy ES, Gerber JS, et al. (2011) Solutions for a cultivated planet. Nature 478: 337–342.
2. Pimentel D (1995) Amounts of pesticides reaching target pests: environmental impacts and ethics. Journal of Agricultural and Environmental Ethics 8: 17–29.

3. Richardson M (1998) Pesticides-friend or foe? Water science and technology 37: 19–25.
4. Tilman D, Cassman KG, Matson PA, Naylor R, Polasky S (2002) Agricultural sustainability and intensive production practices. Nature 418: 671–677.
5. Maeder P, Fliessbach A, Dubois D, Gunst L, Fried P, et al. (2002) Soil fertility and biodiversity in organic farming. Science 296: 1694–1697.
6. Holland JM, Frampton GK, Cilgi T, Wratten SD (1994) Arable acronyms analysed - a review of integrated arable farming systems research in Western Europe. Annals of applied biology 125: 399–438.
7. Ferron P, Deguine JP (2005) Crop protection, biological control, habitat management and integrated farming. A review. Agronomy for Sustainable Development 25: 17–24.
8. Trewavas A (2001) Urban myths of organic farming. Nature 410: 409–410.
9. Pimentel D, Hepperly P, Hanson J, Douds D, Seidel R (2005) Environmental, energetic, and economic comparisons of organic and conventional farming systems. Bioscience 55: 573.
10. Directive 2009/128/EC of the European Parliament and of the Council (2009) Official journal of the European Union. Available: http://eur-lex.europa.eu/LexUriServ/LexUriServ.do?uri = OJ:L:2009:309:0071:0086:en:PDF Accessed 2013 Sep 10.
11. Munier-Jolain N, Dongmo A (2010) Evaluation de la faisabilité technique de systèmes de Protection Intégrée en termes de fonctionnement d'exploitation et d'organisation du travail. Comment adapter les solutions aux conditions locales? Innovations Agronomiques 8: 57–67.
12. European commission (2010) Eurostat Agriculture online database. Available: http://epp.eurostat.ec.europa.eu/portal/page/portal/agriculture/farm_structure/database. Accessed 2013 Jul 19.
13. USDA Economic Research Service (2010) U.S. certified organic farmland acreage, livestock number, and farm operations. Available: http://www.ers.usda.gov/data-products/organic-production.aspx "\l ".UiWVOX8QO89. Accessed 2013 Jul 22.
14. Bastiaans L, Paolini R, Baumann DT (2008) Focus on ecological weed management: what is hindering adoption? Weed Research 48: 481–491.
15. Pardo G, Riravololona M, Munier-Jolain N (2010) Using a farming system model to evaluate cropping system prototypes: Are labour constraints and economic performances hampering the adoption of Integrated Weed Management? European Journal of Agronomy 33: 24–32.
16. Food and Agriculture Organization of the United Nations (FAO) (2013) FAOSTAT Resources: Pesticides Use. Available: http://faostat.fao.org/site/424/DesktopDefault.aspx?PageID = 424#ancor. Accessed 2013 Jul 25.
17. U.S. Environmental Protection Agency (EPA) (2011) Pesticides industry sales and usage: 2006 and 2007 market estimates. Washington, D.C.: U.S. Environmental Protection Agency. Available: www.epa.gov/opp00001/pestsales/07pestsales/market_estimates2007.pdf. Accessed 2013 Jul 25.
18. Ministère de l'Agriculture, de l'Agro-alimentaire et de la Forêt (2012) Note de suivi du plan Ecophyto 2018: tendances de 2008 à 2011 du recours aux produits phytopharmaceutiques. Available: http://agriculture.gouv.fr/IMG/pdf/121009_Note_de_suivi_2012_cle0a995a.pdf. Accessed 2013 Jul 26.
19. Reganold JP, Glover JD, Andrews PK, Hinman HR (2001) Sustainability of three apple production systems. Nature 410: 926–930.
20. Deike S, Pallutt B, Christen O (2008) Investigations on the energy efficiency of organic and integrated farming with specific emphasis on pesticide use intensity. European Journal of Agronomy 28: 461–470.
21. Aubertot JN, Robin MH (2013) Injury Profile SIMulator, a qualitative aggregative modelling framework to predict crop injury profile as a function of cropping practices, and the abiotic and biotic environment. I. Conceptual bases. PLoS one 8: e73202.
22. OECD (2001) Environmental Indicators for Agriculture, Volume 3: Methods and Results. Available: www.oecd.org/tad/sustainable-agriculture/40680869.pdf. Accessed 2014 Feb 26.
23. Meiss H, Mediene S, Waldhardt R, Caneill J, Bretagnolle V, et al. (2010) Perennial lucerne affects weed community trajectories in grain crop rotations. Weed Research 50: 331–340.
24. Van der Werf H, Zimmer C (1998) An indicator of pesticide environmental impact based on a fuzzy expert system. Chemosphere 36: 2225–2249.
25. Davis AS, Hill JD, Chase CA, Johanns AM, Liebman M (2012) Increasing cropping system diversity balances productivity, profitability and environmental health. PloS one 7: e47149.
26. Chikowo R, Faloya V, Petit S, Munier-Jolain NM (2009) Integrated Weed Management systems allow reduced reliance on herbicides and long-term weed control. Agriculture, Ecosystems and Environment 132: 237–242.
27. Bockstaller C, Girardin P (2000) Using a crop sequence indicator to evaluate crop rotations. 3rd International Crop Science Congress 2000 ICSC, Hambourg, 17–22 August 2000, p. 195.
28. Andersson GKS, Rundlöf M, Smith HG (2012) Organic Farming Improves Pollination Success in Strawberries. PloS one 7: e31599.
29. Aavik T, Liira J (2010) Quantifying the effect of organic farming, field boundary type and landscape structure on the vegetation of field boundaries. Agriculture, Ecosystems and Environment 135: 178–186.
30. Li R, Khafipour E, Krause DO, Entz MH, de Kievit TR, et al. (2012) Pyrosequencing Reveals the Influence of Organic and Conventional Farming Systems on Bacterial Communities. PloS one 7: e51897.
31. Seufert V, Ramankutty N, Foley JA (2012) Comparing the yields of organic and conventional agriculture. Nature 485: 229–232.
32. Foley JA, Defries R, Asner GP, Barford C, Bonan G, et al. (2005) Global consequences of land use. Science 309: 570–574.
33. Theobald DM (2001) Land-use dynamics beyond the American urban fringe. Geographical Review 91: 544.
34. Phalan B, Onial M, Balmford A, Green R (2011) Reconciling food production and biodiversity conservation: land sharing and land sparing compared. Science 333: 1289–1291.
35. Hulme MF, Vickery JA, Green RE, Phalan B, Chamberlain DE, et al. (2013) Conserving the birds of Uganda's banana-coffee arc: land sparing and land sharing compared. PLoS One 8: e54597.
36. Altieri MA, Nicholls CI, Ponti L (2009) Crop diversification strategies for pest regulation in IPM systems. Integrated pest management Cambridge University Press, Cambridge, UK. pp. 116–130.
37. Krupinsky J, Bailey K, McMullen M, Gossen B, Turkington T (2002) Managing plant disease risk in diversified cropping systems. Agronomy Journal 94: 198–209.
38. Beketov MA, Kefford BJ, Schäfer RB, Liess M (2013) Pesticides reduce regional biodiversity of stream invertebrates. Proceedings of the National Academy of Sciences 110: 11039–11043.
39. Ascherio A, Chen H, Weisskopf MG, O'Reilly E, McCullough ML, et al. (2006) Pesticide exposure and risk for Parkinson's disease. Annals of Neurology 60: 197–203.
40. Di Falco S, Chavas JP (2008) Rainfall shocks, resilience, and the effects of crop biodiversity on agroecosystem productivity. Land Economics 84: 83–96.
41. Reidsma P, Ewert F, Lansink AO, Leemans R (2010) Adaptation to climate change and climate variability in European agriculture: The importance of farm level responses. European Journal of Agronomy 32: 91–102.
42. Nemecek T, von Richthofen JS, Dubois G, Casta P, Charles R, et al. (2008) Environmental impacts of introducing grain legumes into European crop rotations. European Journal of Agronomy 28: 380–393.
43. Pimentel D, Pimentel M (2003) Sustainability of meat-based and plant-based diets and the environment. The American Journal of Clinical Nutrition 78: 660S–663S.
44. Halberg N, van der Werf HMG, Basset-Mens C, Dalgaard R, de Boer IJM (2005) Environmental assessment tools for the evaluation and improvement of European livestock production systems. Livestock Production Science 96: 33–50.
45. Beckie HJ (2009) Herbicide Resistance in Weeds: Influence of Farm Practices. Prairie Soils and Crops 2:3.
46. Meynard JM, Messéan A, Charlier A, Charrier F, Farès M, et al. (2013) Freins et leviers à la diversification des cultures. Etudes au niveau des exploitations agricoles et des filières. Synthèse du rapport d'étude, INRA. Available: http://inra.dam.front.pad.brainsonic.com/ressources/afile/223799-6afe9-resource-etude-diversification-des-cultures-synthese.html. Accessed 2013 Mar 15.
47. Centre d'Etudes Biologiques de Chizé (2009) Zone Atelier «Plaine & Val de Sèvre». Available: http://www.zaplainevaldesevre.fr. Accessed 2013 Jun 5.
48. Agence de l'Environnement et de la Maîtrise de l'Energie (ADEME) (2011) Guide des valeurs Dia'terre. Version du référentiel 1.13. Available: http://www2.ademe.fr/servlet/KBaseShow?sort = -1&cid = 96&m = 3&catid = 24390. Accessed 2013 Jul 12.
49. Hülsbergen KJ, Feil B, Biermann S, Rathke GW, Kalk WD, et al. (2001) A method of energy balancing in crop production and its application in a long-term fertilizer trial. Agriculture, Ecosystems and Environment 86: 303–321.
50. Bureau de Coordination du Machinisme Agricole (BCMA) (2012) Simcoguide online decision tool. Available: http://simcoguide.pardessuslahaie.net/#accueil. Accessed 2013 Apr 12.
51. R Development Core Team (2012). R: A language and environment for statistical computing (R Foundation for Statistical Computing, Vienna, Austria).

2

Decreased Functional Diversity and Biological Pest Control in Conventional Compared to Organic Crop Fields

Jochen Krauss[1,2]*, Iris Gallenberger[2], Ingolf Steffan-Dewenter[1,2]

1 Department of Animal Ecology and Tropical Biology, University of Würzburg, Biocentre, Würzburg, Germany, **2** Population Ecology Group, Department of Animal Ecology I, University of Bayreuth, Bayreuth, Germany

Abstract

Organic farming is one of the most successful agri-environmental schemes, as humans benefit from high quality food, farmers from higher prices for their products and it often successfully protects biodiversity. However there is little knowledge if organic farming also increases ecosystem services like pest control. We assessed 30 triticale fields (15 organic vs. 15 conventional) and recorded vascular plants, pollinators, aphids and their predators. Further, five conventional fields which were treated with insecticides were compared with 10 non-treated conventional fields. Organic fields had five times higher plant species richness and about twenty times higher pollinator species richness compared to conventional fields. Abundance of pollinators was even more than one-hundred times higher on organic fields. In contrast, the abundance of cereal aphids was five times lower in organic fields, while predator abundances were three times higher and predator-prey ratios twenty times higher in organic fields, indicating a significantly higher potential for biological pest control in organic fields. Insecticide treatment in conventional fields had only a short-term effect on aphid densities while later in the season aphid abundances were even higher and predator abundances lower in treated compared to untreated conventional fields. Our data indicate that insecticide treatment kept aphid predators at low abundances throughout the season, thereby significantly reducing top-down control of aphid populations. Plant and pollinator species richness as well as predator abundances and predator-prey ratios were higher at field edges compared to field centres, highlighting the importance of field edges for ecosystem services. In conclusion organic farming increases biodiversity, including important functional groups like plants, pollinators and predators which enhance natural pest control. Preventative insecticide application in conventional fields has only short-term effects on aphid densities but long-term negative effects on biological pest control. Therefore conventional farmers should restrict insecticide applications to situations where thresholds for pest densities are reached.

Editor: Andrew Hector, University of Zurich, Switzerland

Funding: I.S.-D. and J.K. acknowledge project funding from the European Community's Seventh Framework Programme (FP7/2007–2013) under grant agreement no. 226852, Scales Project (http://www.scales-project.net). The publication was funded by the German Research Foundation (DFG) and the University of Wuerzburg in the funding programme Open Access Publishing. The funder had no role in study design, data collection and analysis, decision to publish, or preparation of the manuscript.

Competing Interests: The authors have declared that no competing interests exist.

* E-mail: j.krauss@uni-wuerzburg.de

Introduction

Ecosystem services like pollination and pest control are essential benefits for farmers throughout the world [1–3]. Pollinators enhance crop production for many cash crops like fruits and vegetables [4] and biological pest control is an important ecosystem service for crops [5,6]. In the last century agricultural intensification caused significant biodiversity loss in most agroecosystems, underlying the need for restoration and conservation schemes in agroecosystems [5,7]. Biodiversity and ecosystem services might be protected with agri-environmental schemes, where farmers get subsidies, partly to produce ecological benefits. Some of these schemes have been criticised, due to their low success in protecting biodiversity [8], while other schemes were successful [9,10].

One important agri-environmental scheme is organic farming, where synthetic fertilisation and pesticide treatments are not applied, while both are common in conventional farming systems. Organic farming might decrease the biomass of the crop by 25% [11], but increases the diversity of most functional species groups [6,12–18], but see [19] for an exception. Particularly, organic farming enhances guilds relevant for ecosystem services like pollinators [20] and predators [21]. Studies focusing in parallel on species diversity of different functional groups and ecosystem services are rare. In a recent review on pest control in organic and conventional farms, the authors call for additional studies on the relationship of biodiversity and pest control [6]. In this context, field edges and field centres often contain different species communities, with higher diversities, abundances and ecosystem service provision at the edges compared to centres [5,18]. It is therefore necessary to consider in field studies edges and centres separately.

Aphids are major insect pests on cereals and can cause massive yield loss [22,23]. An application of systemic insecticides in conventional fields is therefore a common practice [24–26]. Unexpectedly, the application of insecticides was not a common praxis in our study region. Therefore we had the chance to compare conventional fields with and without insecticide application. As aphid predators might be similarly reduced by insecticides as aphids, and as aphid population growth rates are very high [27,28], it is plausible that top-down control could be reduced in fields with insecticide treatment [26].

Most studies comparing cereals of organic vs. conventional farming systems were conducted in wheat fields [13,14,20] with some studies on barley and oat fields [29]. Beside the field scale some studies focused on field margins [30], on effects of organic farming at the landscape scale [31] or use a farm scale approach [32]. As far as we know, studies on triticale were not performed at any spatial scale. Triticale is a cereal emerged from crossing and backcrossing of wheat (*Triticum aestivum* L.) and rye (*Secale cereale* L.) and is mainly used in low-input systems as animal feed. Its importance might grow because of its potential role in biofuel production [33,34]. The worldwide production in the year 2009 was 15,040,432 t and therefore comparable with rye 17,856,568 t and oat 23,032,118 t, but far below maize 817,110,509 and wheat 681,915,838 (FAOSTAT 2009: http://faostat.fao.org/).

In this study we compared conventional and organic fields of the cereal triticale, distinguishing between effects of field edges and centre of the fields, and considering the diversity of functional groups including plants and pollinators as well as densities of aphids and predators, and predator-prey ratios. We further tested the effect of insecticide treatment on aphids and their predators in conventional fields.

We tested the following hypotheses:

1. Vascular plant and pollinator diversity is enhanced in organic compared to conventional farming.
2. Higher predator abundance (pest control) leads to reduced aphid abundance in organic compared to conventional farming systems.
3. Field edges are more species rich and contain higher abundances than field centres.
4. In conventional fields sprayed with insecticides herbivore abundances recover faster than predator abundances.

Materials and Methods

Study region and study sites

A total of 30 (15 organic, 15 conventional) winter triticale fields were selected as study sites in the vicinity of Bayreuth (49°56′53″N, 11°34′42″E) located in the region Upper Franconia (South Germany, Bavaria) (Fig. 1a). The study sites were within an area of approximately 300 km^2 and the minimum distance between the studied triticale fields was 500 m. Upper Franconia is characterised by relatively heterogeneous landscapes, rich in forests (40.4% of the land area) and agricultural land (47.3%) including arable land (69.1%), grassland (30.5%) and permanent crops (0.4%) (Bayerisches Landesamt für Statistik und Datenverarbeitung 2004; http://www.statistik.bayern.de). The average temperature in the study region (Bayreuth: 1971–2000) is approximately 8.2°C with an average rainfall of 724 mm per year (http://www.klimadiagramme.de/Deutschland).

The 15 conventional fields, we investigated, were treated with agrochemicals like herbicides, inorganic fertilisers and growth regulators (for detailed information see Table 1). For cereal aphid control five of the 15 conventional fields were sprayed preventatively with the insecticides Karate®Zeon (Syngenta) with 75 ml/ha and Pirimor (Syngenta) with 100 g/ha. Karate®Zeon is a contact insecticide against sucking and chewing herbivores and contains lambda-cyhalothrin, a pyrethroid, as active agent. Pirimor comprises the cabarmate Primicarb and, similar to Karate®Zeon influences the nervous system and leads to paralysis and mortality, but Pirimor is specific against aphids [35] and (Syngenta product information http://www.syngenta.de/).

By contrast the 15 organic fields were cultivated under the European Union regulation (EEC) N° 2092/91 based on a prohibition of inorganic fertilisers and pesticide application.

Forest directly adjacent to the 30 study fields was recorded to test if forest act as a source habitat for species [36,37]. However, the proportion of forest surrounding the fields had no significant effect on species richness and abundance (all $p>0.2$) and was therefore excluded from all statistical models and is not presented in the results.

Data collection

Study design. We established 10 study plots on each of the 30 triticale fields. Five plots were located at the edge (0.5 m away from the outer field border) and 5 plots in the centre of the fields (Fig. 1b). The study plots had an area of 2 m^2 and were arranged in a row along the edge or centre every 10 m. Depending on the field size there was a distance of 20 to 80 m between the study plots at the edge and those in the centre of the fields (Fig. 1b).

Vascular plants and pollinators. Vascular plant species richness was recorded once in a random sequence between the 5th and the 27th June 2008 in the 10 plots on all 30 triticale fields. Most arable wild plants were determined directly in the fields; unknown species were taken into laboratory for subsequent identification. The vegetation cover was estimated by vertical projection of non crop plant elements on the ground. For statistical analyses we used the total number of plant species at the edge and in the centre of the field, while the vegetation cover is the mean of the 5 plots (field edge or field centre).

Pollinators were recorded between 10th of June and the18th of July 2008 in 50 m×2 m transect corridors, separately conducted for field edges and field centres. The walks were repeated three times at different days and lasted approximately 10 minutes for each transect corridor. All flower-visiting insects of the families Apidae, Syrphidae and Lepidoptera were recorded. Unknown species were netted and taken into laboratory for subsequent identification. For statistical analyses we used the total number of species and the summed number of individuals of the three walks to calculate abundances (separately for field edges and field centres).

Cereal aphids and their natural enemies. In 2008 cereal aphids and their natural enemies were recorded in four surveys at different days in the period between the 10th of June and the 19th of July. On each of the 10 study plots per triticale field (5 edge, 5 centre) 10 sweeps with a net were carried out to count cereal aphids and aphidophagous predators in the nets. We focused our study on three cereal aphid species which are known as pests in European agroecosystems: *Sitobion avenae* (Linnaeus), *Metopolophium dirhodum* (Walker) and *Rhopalosiphum padi* (Fabricius) (Hemiptera, Aphididae) [21,23]. We also recorded specialised aphidophagous predators, including all larvae and adults of the ladybirds *Coccinella septempunctata* (L.) and *Prophylea quatuordecimpunctata* (L.) (Coccinelidae, Coleoptera), lacewing larvae (Chrysopidae) and hoverfly larvae (Syrphidae). These stenophagous predators are known to contribute effectively to cereal aphid control [23,38]. For statistical analyses we used the summed individual numbers of the four surveys, whereby each survey contains the sum of the five study plots (either at the field edge or the field centres). The

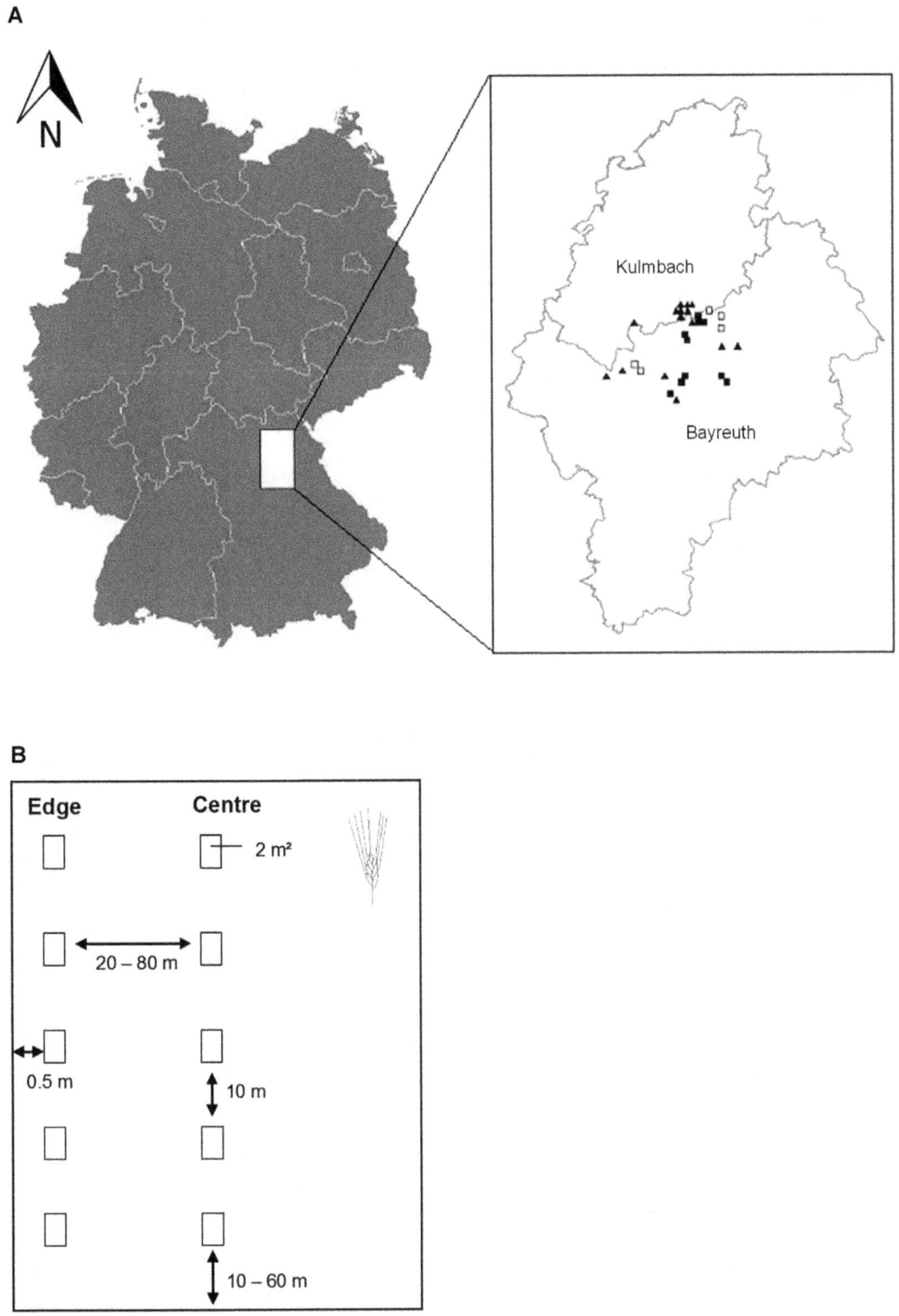

Figure 1. Location of study region in southern Germany and used sampling scheme. (a) The 30 triticale fields were located in the northern part of Bavaria in the districts of Bayreuth and Kulmbach (Symbols: ■ conventional, not treated; □ conventional treated with insecticides; ▲ organic farming system). (b) Sampling scheme in each triticale field. Five study plots were surveyed at the edge and five in the centre of each field.

predator-prey ratio was calculated by dividing the number of (aphidophagous) predators by the number of prey (cereal aphids). To assess the temporal dynamics of aphids and their predators on conventional fields with and without insecticide application we also tested aphid and predator abundances for each survey separately. Temporal dynamics were not considered for pollinators, as they occurred in too low densities throughout the season.

Table 1. Application of agrochemicals on conventional triticale fields in the region of Oberfranken (x = treated).

Field number	Field name (location)	Inorganic fertiliser	Herbicide	Insecticide	Fungicide	Growth regulator
1	Eichelberg 1	X	X		X	
2	Eichelberg 2	X	X		X	
3	Geigenreuth 1	X	X			
4	Geigenreuth 2	X	X			
5	Mistelbach	X	X		X	
6	Obergräfenthal 1	X	X			X
7	Obergräfenthal 2	X	X			X
8	Obergräfenthal 3	X				
9	Unterkonnersreuth 1	X	X			X
10	Unterkonnersreuth 2	X	X			
11	Crottendorf	X	X	X	X	
12	Eschen	X	X	X	X	
13	Ramsenthal	X	X	X	X	
14	Schaitz	X	X	X	X	
15	Windhof	X	X	X	X	

Statistical analyses

The statistical analyses were performed using the software R 2.9.1 for Windows [39]. Linear mixed effects models were calculated with the sequence of explanatory variables being (i) farming system (conventional/organic), (ii) field position (edge/centre), and (iii) the interaction between farming system and field position. Edge and centre plots or transects were nested within fields by using field identity as random effect [40]. The four time steps in the insecticide treatment analyses were performed separately apart from one comparison between time step 1 and time step 4 in aphid abundances. In this specific case we added time step as fixed effect with explanatory variable interactions and further nested it as random effect within field identity. The response variables plant species richness, pollinator species richness, pollinator abundance and predator abundance were $\log_{10}$ (c+0.1) and aphid individuals $\log_{10}$ - transformed to meet the assumptions of normality and homoscedasticity in the statistical models. Vegetation cover and predator-prey ratios were not transformed. Pearson correlations were used to show relations between response variables (Table 2). Means ± one standard error are presented throughout the text.

Results

Vascular plants and pollinators

In the vascular plant surveys 55 weed species were found in conventional and 114 species in organic fields (in total 122 species, see Material S1, Appendix A). The species richness and the vegetation cover of non-crop species were significantly higher in organic fields compared to conventional fields. Further, field edges showed consistently higher species richness and vegetation cover compared to field centres (Fig. 2a; Table 3; vegetation cover: conventional/edge = 2.4±0.7%, conventional/centre = 1.1±0.6%; organic/edge = 30.7±2.4%; organic/centre = 16.3±2.3%). The interaction terms between the explanatory variables was also significant, indicating that the difference between edges and centres is more pronounced in organic fields.

In total 31 species and 3113 individuals of potential pollinators were recorded (Material S1, Appendix B). Species richness and

Table 2. Pearson correlations (r-values) between response variables of the 60 study locations in 30 fields.

	Vegetation cover	Pollinator species richness	Pollinator abundance	Aphid abundance	Predator abundance	Predator-prey ratio
Plant species richness	0.72****	0.73****	0.77****	−0.52****	0.60****	0.53****
Vegetation cover		0.77****	0.83****	−0.65****	0.56****	0.79****
Pollinator species richness			0.98****	−0.75****	0.49***	0.60****
Pollinator abundance				−0.75****	0.52****	0.66****
Aphid abundance					−0.21 ns	−0.56****
Predator abundance						0.59****

Variables are transformed (see statistical analyses).
Significance levels:
****$P<0.0001$;
***$P<0.001$;
n.s. = not significant.

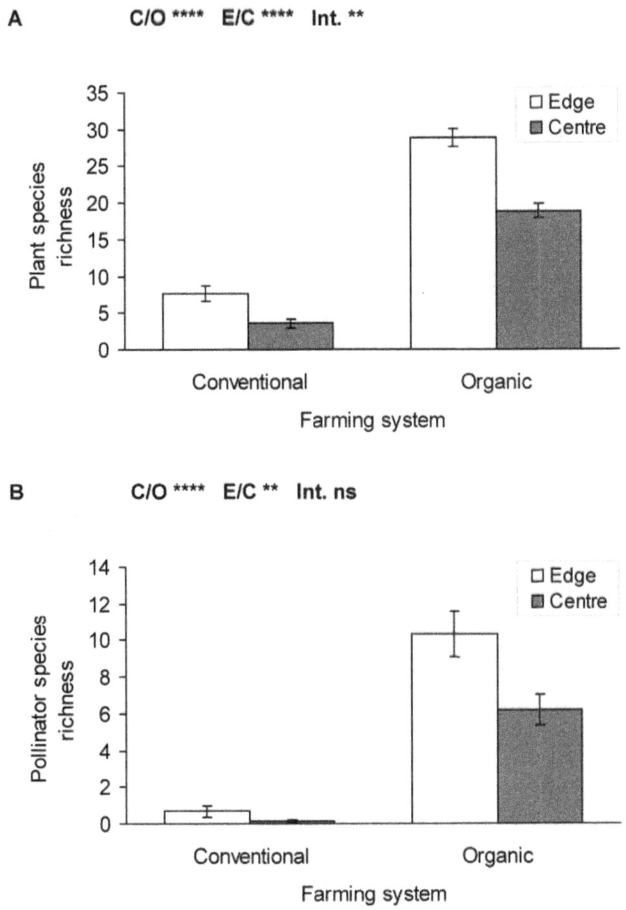

Figure 2. Species richness of (a) vascular plants and (b) pollinators in conventional and organic triticale field edges and centres (mean ± se). C/O: conventional/organic fields, E/C: edge/centre in field, Int: Interaction term. **** p≤0.0001, ** p≤0.01, ns p>0.1. Statistics see Table 3.

abundance of pollinators showed consistently similar patterns as vascular plants because they were highly correlated (Table 2). Pollinator species richness and abundance were significantly higher in organic compared to conventional fields, and higher at edges compared to field centres (Fig. 2b; Table 3; pollinator abundance: conventional/edge = 1.5±0.8 individuals, conventional/centre = 0.1±0.1 individuals; organic/edge = 167.6±45.7 individuals; organic/centre = 38.3±9.8 individuals). Thereby pollinator species richness with a total of only 5 recorded species in all 15 conventional fields was substantially lower than the 31 species found in the 15 organic fields.

The interaction term (conventional/organic vs. edge/centre) was not significant for pollinator species richness, but was just below the significance level for pollinator abundance (Table 3), indicating that abundance differences in organic fields between edge and centre were slightly more pronounced than for species richness. The edges to centre differences for pollinators are very small in conventional fields, due to low individual numbers at both field locations.

Cereal aphids and their natural enemies

A total of 8835 aphid individuals were collected in the 30 triticale fields. Altogether, almost five times more aphids were recorded in the 15 conventional fields (7296) than on the 15 organic fields (1539). *Sitobion avenae* (Fabricius) was the most dominant species (90%), followed by *Rhopalosiphum padi* (L.) (5.4%) and *Metopolophium dirhodum* (Walker) (4.6%). Aphid abundance was significantly higher in conventional compared to organic fields (Fig. 3a, Table 3), contrasting the patterns of pollinator and vascular plant abundance, which were lower in conventional compared to organic fields. Field edges in conventional fields contained much higher aphid abundances than field centres, whereas the low aphid numbers in organic fields did not differ between edges and centres (Fig. 3a, Table 3).

The abundance of aphid predators, in contrast to their prey, was higher in organic compared to conventional fields. Further, the field edges had consistently higher abundances than field centres. The interaction was not significant (Fig. 3b, Table 3), indicating that organic and conventional fields have similarly two to three times higher predator abundances at field edges compared

Table 3. Mixed effects model statistics of the seven response variables with the explanatory variables farming system, field location and the interaction of farming system and field location.

	df		Plant species richness	Vegetation cover	Pollinator species richness	Pollinator abundance	Aphid abundance	Predator abundance	Predator-prey ratio
Farming system (conventional/organic)	1,28	*F*	96.99	86.01	134.32	180.16	91.04	12.90	25.60
		P	<0.0001	<0.0001	<0.0001	<0.0001	<0.0001	0.001	<0.0001
			↓	↓	↓	↓	↑	↓	↓
Field location (edge/centre)	1,28	*F*	71.13	66.27	10.88	35.45	13.35	26.04	11.33
		P	<0.0001	<0.0001	0.003	<0.0001	0.001	<0.0001	0.002
			↑	↑	↑	↑	↑	↑	↑
Framing system×Field location	1,28	*F*	7.89	46.44	<0.01	5.13	7.63	0.44	9.91
		P	0.009	<0.0001	0.971	0.032	0.010	0.515	0.004

Field identity was used as a random effect. Response variables were transformed (see statistical analyses). Mean and SE are shown in Fig. 2 and Fig. 3 or in the text. ↑ = conventional or edge is higher, ↓ = conventional or edge is lower.

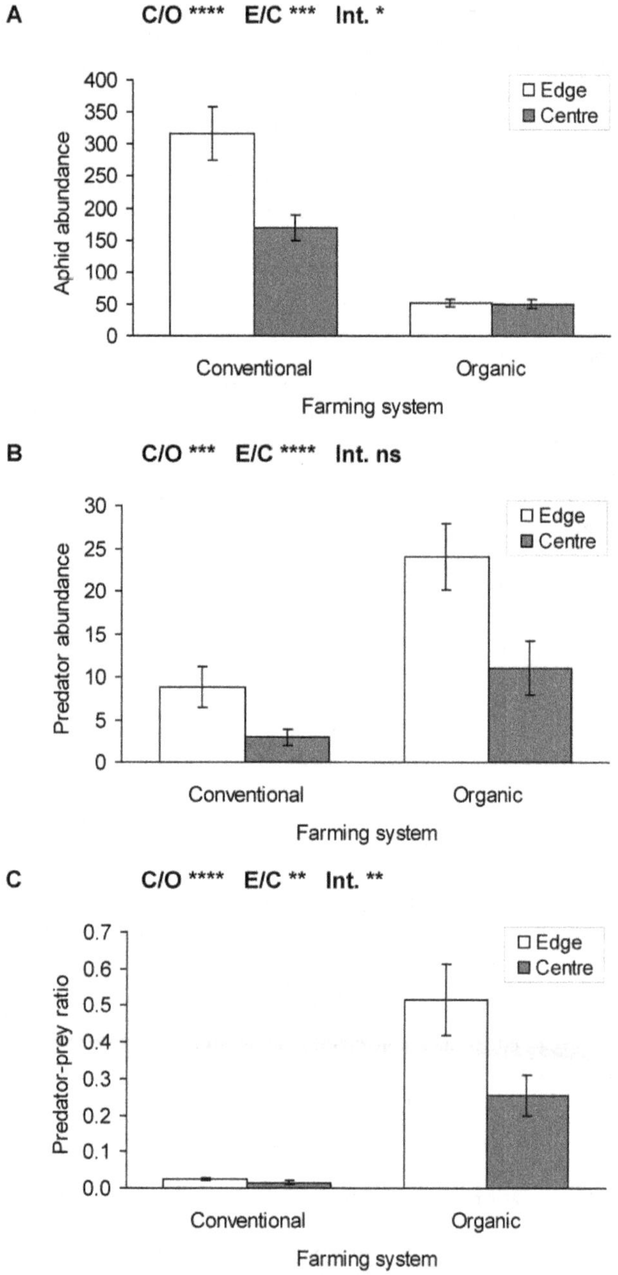

Figure 3. Abundances of (a) aphids, (b) aphid-predators and (c) predator-prey ratio in conventional and organic triticale field edges and centres (mean ± se). C/O: conventional/organic fields, E/C: edge/centre in field, Int: Interaction term. **** p≤0.0001, *** p≤0.001, ** p≤0.01, * p≤0.05, ns p>0.1. Statistics see Table 3.

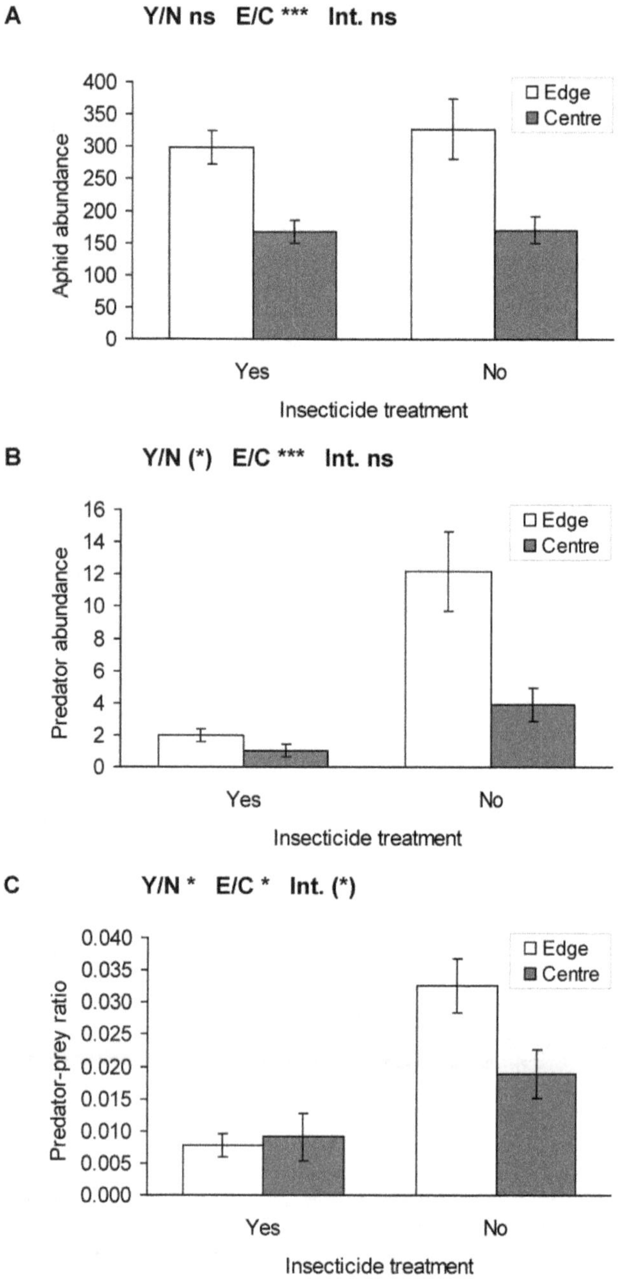

Figure 4. Abundances of (a) aphids, (b) aphid-predators and (c) predator-prey ratio in insecticide treated and non-treated conventional triticale field edges and centres (mean ± se). Y/N: treated/not treated fields, E/C: edge/centre in field, Int: Interaction term. *** p≤0.001, * p≤0.05, (*) p≤0.1, ns p>0.1. Statistics see Table 4.

to field centres. Due to the higher predator, but lower aphid abundances in organic fields, the predator-prey ratio in organic fields was 21 times higher at field edges and 16 times higher in field centres compared to conventional fields (Fig. 3c, Table 3).

Insecticide treatment on conventional farms

Five of the 15 conventional field sites were treated with insecticides before the surveys started. However the aphid abundances were not significantly different between sprayed and unsprayed field sites, while higher abundances were detected at field edges compared to field centres (Fig. 4a, Table 4). In contrast, the aphid predators were more abundant in not sprayed fields, being also higher at edges compared to field centres (Fig. 4b, Table 4). Thus, the predator-prey ratio was significantly higher in not sprayed conventional fields compared to insecticide treated fields (Fig. 4c, Table 4).

To find the potential mechanism behind these patterns we analysed the temporal dynamics of aphids and their predators during the 4 surveys. Aphid abundance was significantly reduced after insecticide application only in the first survey, but afterwards rapidly increased. In the last survey there was even a trend for higher aphid abundances in sprayed compared to unsprayed

Table 4. Mixed effects model statistics for aphid abundance, predator abundance and predator-prey ratio with the explanatory variables insecticide treatment, field location and the interaction of insecticide treatment and field location.

	df		Aphid abundance	Predator abundance	Predator-prey ratio
Insecticide treatment (yes/no)	1,13	*F*	<0.01	4.58	5.16
		P	0.992	0.052	0.041
				↓	↓
Field location (edge/centre)	1,13	*F*	24.01	21.47	5.68
		P	0.0003	0.0005	0.033
			↑	↑	↑
Insecticide treatment×Field location	1,13	*F*	<0.01	1.37	3.67
		P	0.945	0.263	0.078

Field identity was used as a random effect. Response variables were transformed (see statistical analyses). Mean and SE are shown in Fig. 4.
↑ = insecticide treated (yes) or edge is higher, ↓ = insecticide treated (yes) or edge is lower.

conventional fields (Fig. 5a, Table 5). In a mixed effect model including time step 1 and time step 4 in one model (see statistical analyses), the interaction term between insecticide treatment×time step was highly significant ($F_{1,13} = 20.98$, $P<0.001$), providing evidence that the effect of insecticide treatment was not constant across the sampling dates, with lower aphid abundances in the treated fields at the first survey (first time step), but higher aphid abundances in the treated fields at the last survey. The generally low abundance of predators at the first two surveys was not affected by insecticide application. However the third and fourth surveys show significantly higher abundances for predators in non-treated fields, indicating a time delayed response to the insecticide treatment (Fig. 5b, Table 5). During the four surveys aphid abundances were consistently higher at field edges, while predator abundances were higher at field edges only at the last two surveys (Fig. 5, Table 5).

Discussion

Our results show that organic farming increases biodiversity of vascular plants and pollinators, as well as vegetation cover and pollinator abundances. In addition the abundances of aphidophagous predators were enhanced, allowing a better top down control of aphids, which had clearly lower abundances in organic compared to conventional triticale fields. Insecticide spraying in conventional fields did decrease aphid abundances, but only for a short time period. After two weeks the insecticide effect was gone and at the end of the season aphid abundances were even higher in sprayed fields compared to not sprayed fields. In contrast, abundances of aphid-predators remained low in insecticide sprayed fields throughout the study period, but did increase their abundance in not sprayed fields.

We show that vascular plant species richness and vegetation cover are higher in organic compared to conventional fields, as similarly shown for other study systems [13,15,41,42]. In general most functional species groups have higher species diversities and abundances in organic fields [6,12]. Pollinator diversity and abundances in organic wheat fields in Germany were also enhanced [20], but not in organic tomato and watermelon fields in the USA [43]. In our organic triticale fields we found more species and higher abundances of pollinators, which can provide important ecosystem services for wild plants. These enhanced pollinator numbers might be also linked to the correlation between pollinator diversity and plant diversity in the triticale fields. Such correlations have been also reported for other study systems [20,44].

We assume that pest control is enhanced in organic compared to conventional fields because of a higher predator-prey ratio, a free ecosystem service which needs further exploration [6,45]. Cereal aphids are disastrous pests across the world, and their control can be time-consuming and cost intensive for farmers. The aphid species *R. padi* alone has the potential to decrease the yield of barley by 52% [46]. In our organic triticale fields the predator-

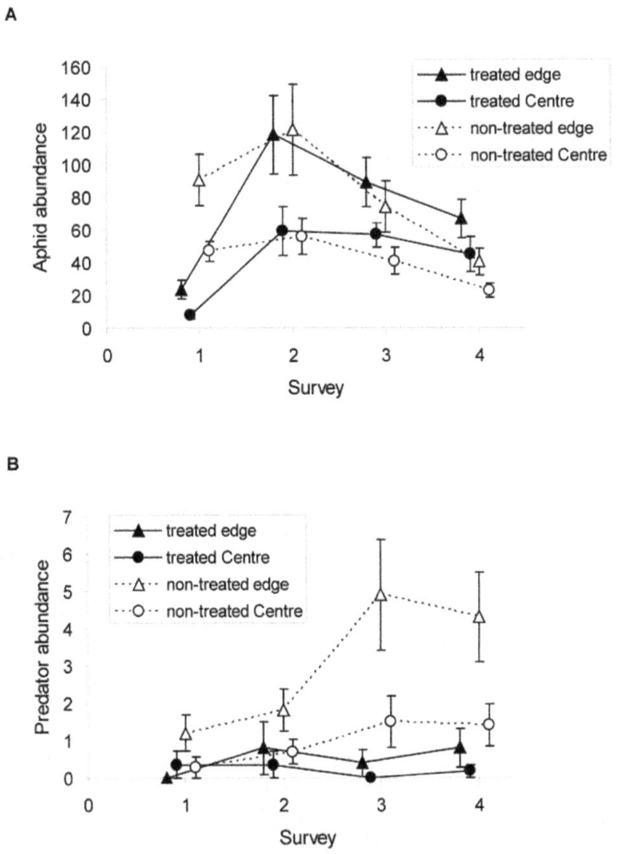

Figure 5. Temporal dynamics of (a) aphid abundance and (b) predator abundance in insecticide treated and non-treated conventional triticale field edges and centres (mean ±se). Statistics see Table 5.

Table 5. Mixed effects model statistics for temporal changes in aphid abundance and predator abundance with the explanatory variables insecticide treatment, flied location and the interaction of insecticide treatment and field location.

	df		Aphid Survey 1	Aphid Survey 2	Aphid Survey 3	Aphid Survey 4
Insecticide treatment (yes/no)	1,13	*F*	22.40	<0.01	1.92	4.02
		P	0.0004	0.989	0.190	0.066
			↓			↑
Field location (edge/centre)	1,13	*F*	54.31	9.51	13.51	30.54
		P	<0.0001	0.009	0.003	0.0001
			↑	↑	↑	↑
Insecticide treatment×Field location	1,13	*F*	8.07	0.42	0.50	0.49
		P	0.014	0.528	0.492	0.493
	df		**Predator Survey 1**	**Predator Survey 2**	**Predator Survey 3**	**Predator Survey 4**
Insecticide treatment (yes/no)	1,13	*F*	1.41	1.38	6.80	7.25
		P	0.257	0.261	0.022	0.019
					↓	↓
Field location (edge/centre)	1,13	*F*	1.81	2.21	11.85	20.82
		P	0.202	0.161	0.004	0.0005
					↑	↑
Insecticide treatment×Field location	1,13	*F*	3.75	0.68	1.43	2.42
		P	0.075	0.424	0.253	0.144

Field identity was used as a random effect. Response variables were transformed (see statistical analyses). Mean and SE are shown in Fig. 5.
↑ = insecticide treated (yes) or edge is higher, ↓ = insecticide treated (yes) or edge is lower.

prey ratio was almost 20 times higher compared to the conventional fields. However, this enhanced pest control was not only evident for organic vs. conventional fields, but also within conventional fields where we found higher predator-prey ratios in insecticide untreated fields compared to insecticide treated fields. This seems surprising, as the insecticide was sprayed to reduce cereal aphid abundances. However we show that this aphid reduction works only for a short time period. Afterwards aphid abundances increased rapidly, which is typical for aphid phenologies [26]. Most aphid predators are insects and could have been simultaneously poisoned after insecticide applications. The used insecticides, Karate®Zeon and Pirimor, were described on the one side as specific for sucking or chewing herbivores, or even as specific for aphids (Pirimor), but on the other side as harmful for several aphid predators like ladybirds, syrphids and green lacewings (Syngenta product information: http://www.syngenta.de/). More herbivore specific insecticides could essentially reduce negative effects of insecticide application on beneficial insects. Aphid predators usually occur in low densities at the beginning of the season and increase their abundances following the availability of their aphid prey [28,47]. However in the sprayed conventional triticale fields the aphid predators did not increase their abundances, beside an increase in aphid abundances, during the study period. In contrast aphid predator abundance increased in the not sprayed conventional fields. Possible explanations for the low abundances of predators at the end of the study period in the sprayed fields are that (1) the insecticides were systemic and prevented development of aphid predators or that (2) colonisation of aphid predators was delayed due to low aphid abundances directly after the insecticide treatment. In a pan European study it was shown that the predation rate on aphids in cereal fields declined, when farmers increased the amount of applied insecticides [26].

Due to the low number of pollinators in conventional triticale fields, we could not test if pollinators showed similar negative responses to insecticide treatment. A recent study indicates that wild bee species are negatively affected by insecticide treatments in agricultural systems, at least when two insecticide treatments in the growing season were conducted [48]. However, the low densities of pollinators in conventional triticale fields without insecticide application suggest that enhanced diversity of pollinators in organic fields is mainly related to higher floral resource availability.

Apart from differences between triticale fields, we also showed that field edges within fields are more important for species diversity and ecosystem services than field centres. Field edges contained higher plant and pollinator diversity as well as higher predator abundances and higher predator-prey ratios. Previous studies also reported that field edges contain higher diversities of plants, spiders, beetles and pollinators [15,49]. These edge effects might be caused by lower farming intensity at field edges or spillover from adjacent habitats [50]. Field edges therefore contribute essentially to species diversity and ecosystem services in cereal fields.

We conclude that organic farming contributes to the maintenance of biodiversity in agricultural systems. Organic farming also enhances species groups that provide ecosystem services with benefits for farmers due to better top-down control of pest species. The preventative insecticide application in conventional fields had significant direct costs in terms of material and labour with no long-term benefit for aphid control and negative effects on natural antagonists. As all triticale fields had relatively low aphid abundances we assume that the aphid abundances in cereals in our study year 2008 were below an economic injury level. We therefore conclude that the application of insecticides without a priori monitoring of aphid abundances and below critical thresholds increases direct management costs for farmers and indirect costs due to reduced ecosystem services like an effective biological pest control.

Supporting Information

Material S1 APPENDIX A. Plant species recorded in organic and conventional fields. APPENDIX B. Pollinator species recorded in organic and conventional fields.

Acknowledgments

We thank Andrea Holzschuh, Andy Hector and two anonymous reviewers for helpful comments on the manuscript; Pedro Gerstberger for help with plant species identification; the farmers for access to their triticale fields and information about their farming practices and the Landwirtschaftsamt Bayreuth for general information about agriculture in the study region.

Author Contributions

Conceived and designed the experiments: ISD JK. Performed the experiments: IG. Analyzed the data: JK IG. Wrote the paper: JK ISD IG.

References

1. Costanza R, d'Arge R, de Groot R, Farber S, Grasso M, et al. (1997) The value of the world's ecosystem services and natural capital. Nature 387: 253–260.
2. Sandhu HS, Wratten SD, Cullen R, Case B (2008) The future of farming: The value of ecosystem services in conventional and organic arable land. An experimental approach. Ecol Econ 64: 835–848.
3. Crowder DW, Northfield TD, Strand MR, Snyder WE (2010) Organic agriculture promotes evenness and natural pest control. Nature 466: 109–112.
4. Klein AM, Vaissière BE, Cane JH, Steffan-Dewenter I, Cunningham SA, et al. (2007) Importance of pollinators in changing landscapes for world crops. Proc R Soc B-Biol Sci 274: 303–313.
5. Tscharntke T, Klein AM, Kruess A, Steffan-Dewenter I, Thies C (2005) Landscape perspectives on agricultural intensification and biodiversity - ecosystem service management. Ecol Lett 8: 857–874.
6. Letourneau DK, Bothwell SG (2008) Comparison of organic and conventional farms: challenging ecologists to make biodiversity functional. Front Ecol Environ 6: 430–438.
7. Krebs JR, Wilson JD, Bradbury RB, Siriwardena GM (1999) The second silent spring? Nature 400: 611–612.
8. Kleijn D, Baquero RA, Clough Y, Diaz M, De Esteban J, et al. (2006) Mixed biodiversity benefits of agri-environment schemes in five European countries. Ecol Lett 9: 243–254.
9. Albrecht M, Duelli P, Müller CB, Kleijn D, Schmid B (2007) The Swiss agri-environment scheme enhances pollinator diversity and plant reproductive success in nearby intensively managed farmland. J Appl Ecol 44: 813–822.
10. Dallimer M, Gaston KJ, Skinner AMJ, Hanley N, Acs S, et al. (2010) Field-level bird abundances are enhanced by landscape-scale agri-environment scheme uptake. Biol Lett 6: 643–646.
11. Hald AB (1999) Weed vegetation wild flora. Of long established organic versus conventional cereal fields in Denmark. Ann Appl Biol 134: 307–314.
12. Bengtsson J, Ahnström J, Weibull AC (2005) The effects of organic agriculture on biodiversity and abundance: a meta-analysis. J Appl Ecol 42: 261–269.
13. Hole DG, Perkins AJ, Wilson JD, Alexander IH, Grice PV, et al. (2005) Does organic farming benefit biodiversity? Biol Conserv 122: 113–130.
14. Roschewitz I, Gabriel D, Tscharntke T, Thies C (2005a) The effects of landscape complexity on arable weed species diversity in organic and conventional farming. J Appl Ecol 42: 873–882.
15. Clough Y, Holzschuh A, Gabriel D, Purtauf T, Kruess A, et al. (2007) Alpha and beta diversity of arthropods and plants in organically and conventionally managed wheat fields. J Appl Ecol 44: 804–812.
16. Gabriel D, Tscharntke T (2007) Insect pollinated plants benefit from organic farming. Agric Ecosyst Environ 118: 43–48.
17. Hiltbrunner J, Scherrer C, Streit B, Jeanneret P, Zihlmann U, et al. (2008) Long-term weed community dynamics in Swiss organic and integrated farming systems. Weed Res 48: 360–369.
18. Romero A, Chamorro L, Sans FX (2008) Weed diversity in crop edges and inner fields of organic and conventional dryland winter cereal crops in NE Spain. Agric Ecosyst Environ 124: 97–104.
19. Purtauf T, Roschewitz I, Dauber J, Thies C, Tscharntke T, Wolters V (2005) Landscape context of organic and conventional farms: influences on carabid beetle diversity. Agric Ecosyst Environ 108: 165–174.
20. Holzschuh A, Steffan-Dewenter I, Kleijn D, Tscharntke T (2007) Diversity of flower-visiting bees in cereal fields: effects of farming system, landscape composition and regional context. J Appl Ecol 44: 41–49.
21. Roschewitz I, Hücker M, Tscharntke T, Thies C (2005b) The influence of landscape context and farming practices on parasitism of cereal aphids. Agric Ecosyst Environ 108: 218–227.
22. Kieckhefer RW, Gellner JL, Riedell WE (1995) Evaluation of the aphid-day standard as a predictor of yield loss caused by cereal aphids. Agron J 87: 785–788.
23. Schmidt MH, Lauer A, Purtauf T, Thies C, Schaefer M, Tscharntke T (2003) Relative importance of predators and parasitoids for cereal aphid control. Proc R Soc B-Biol Sci 270: 1905–1909.
24. Banks JE, Stark JD (2004) Aphid response to vegetation diversity and insecticide applications. Agric Ecosyst Environ 103: 595–599.
25. MacFadyen S, Gibson R, Raso L, Sint D, Traugott M, et al. (2009) Parasitoid control of aphids in organic and conventional farming systems. Agric Ecosyst Environ 133: 14–18.
26. Geiger F, Bengtsson J, Berendse F, Weisser WW, Emmerson M, et al. (2010) Persistent negative effects of pesticides on biodiversity and biological control potential on European farmland. Basic Appl Ecol 11: 97–105.
27. Snyder WE, Ives AR (2003) Interactions between specialist and generalist natural enemies: parasitoids, predators, and pea aphid biocontrol. Ecology 84: 91–107.
28. Härri SH, Krauss J, Müller CB (2008) Natural enemies act faster than endophytic fungi in population control of cereal aphids. J Anim Ecol 77: 605–611.
29. Öberg S, Ekbom B, Bommarco R (2007) Influence of habitat type and surrounding landscape on spider diversity in Swedish agroecosystems. Agric Ecosyst Environ 122: 211–219.
30. Rundlöf M, Smith HG (2006) The effects of organic farming on butterfly diversity depends on landscape context. J Appl Ecol 43: 1121–1127.
31. Holzschuh A, Steffan-Dewenter I, Tscharntke T (2008) Agricultural landscapes with organic crops support higher pollinator diversity. Oikos 117: 354–361.
32. Gibson RH, Pearce S, Morris RJ, Symondson WOC, Memmott J (2007) Plant diversity and land use under organic and conventional agriculture: a whole-farm approach. J Appl Ecol 44: 792–803.
33. Jorgensen JR, Deleuran LC, Wollenweber B (2007) Prospects of whole grain crops of wheat, rye and triticale under different fertilizer regimes for energy production. Biomass and Bioenergy 31: 308–317.
34. Oettler G (2005) The fortune of a botanical curiosity – triticale: past, present and future. J Agr Sci 143: 329–346.
35. Whare DM, Whitacre GW (2004) The pesticide book. Meister Pro Information Resources, Willoughby Ohio.
36. Hradetzky R, Kromp B (1997) Spatial distribution of flying insects in an organic rye field and an adjacent hedge and forest edge. Biol Agric Hort 15: 353–357.
37. Devlaeminck R, Bossuyt B, Hermy M (2005) Seed dispersal from a forest into adjacent cropland. Agric Ecosyst Environ 107: 57–64.
38. Freier B, Triltsch H, Möwes M, Moll E (2007) The potential of predators in natural control of aphids in wheat: Results of an ten-year field study in two German landscapes. Bio Control 52: 775–788.
39. R Development Core Team (2004) R: a language and environment for statistical computing. Foundation for Statistical Computing, Vienna; URL: http://www.R-project.org.
40. Pinheiro JC, Bates DM (2004) Mixed effects models in S and S-Plus. Springer, USA.
41. Fuller RJ, Norton LR, Feber RE, Johnson PJ, Chamberlain DE, et al. (2005) Benefits of organic farming to biodiversity vary among taxa. Biol Lett 1: 431–434.
42. Boutin C, Baril A, Martin PA (2008) Plant diversity in crop field and woody hedgerows of organic and conventional farms in contrasting landscapes. Agric Ecosyst Environ 123: 185–193.
43. Winfree R, Williams NM, Gaines H, Ascher JS, Kremen C (2008) Wild bee pollinators provide the majority of crop visitation across land-use gradients in New Jersey and Pennsylvania, USA. J Appl Ecol 45: 793–802.
44. Krauss J, Steffan-Dewenter I, Tscharntke T (2003) How does landscape context contribute to effects of habitat fragmentation on diversity and population density of butterflies? J Biogeogr 30: 889–900.
45. Zehnder G, Gurr GM, Kühne S, Wade MR, Wratten SD, et al. (2007) Arthropod pest management in organic crops. Annu Rev Entomol 52: 57–80.
46. Östman Ö, Ekbom B, Bengtsson J (2003) Yield increase attributable to aphid predation by ground-living polyphagous natural enemies in spring barley in Sweden. Ecol Econ 45: 149–158.
47. Holland JM, Thomas SR (1997) Quantifying the impact of polyphagous invertebrate predators in controlling cereal aphids and in preventing wheat yield and quality reductions. Ann Appl Biol 131: 375–397.
48. Brittain CA, Vighi M, Bommarco R, Settele J, Potts SG (2010) Impacts of a pesticide on pollinator species richness at different spatial scales. Basic Appl Ecol 11: 106–115.
49. Kiss J, Penksza K, Tóth F, Kádár F (1997) Evaluation of fields and field margins in nature production capacity with special regard to plant protection. Agric Ecosyst Environ 63: 227–232.
50. Rand TA, Tylianakis JM, Tscharntke T (2006) Spillover edge effects: the dispersal of agriculturally subsidized insect natural enemies into adjacent natural habitats. Ecol Lett 9: 603–614.

3

Conserving the Birds of Uganda's Banana-Coffee Arc: Land Sparing and Land Sharing Compared

Mark F. Hulme[1]*, Juliet A. Vickery[2], Rhys E. Green[2,3], Ben Phalan[3], Dan E. Chamberlain[4], Derek E. Pomeroy[5], Dianah Nalwanga[6], David Mushabe[6], Raymond Katebaka[5], Simon Bolwig[7], Philip W. Atkinson[1]

1 British Trust for Ornithology, Thetford, Norfolk, United Kingdom, **2** The Royal Society for the Protection of Birds, Sandy, Bedfordshire, United Kingdom, **3** Department of Zoology, University of Cambridge, Cambridge, United Kingdom, **4** Dipartimento di Biologiá Animale e dell'Uomo, University of Turin, Turin, Italy, **5** Department of Biological Sciences, Makerere University, Kampala, Uganda, **6** NatureUganda, Kampala, Uganda, **7** Department of Management Engineering, Technical University of Denmark, Copenhagen, Denmark

Abstract

Reconciling the aims of feeding an ever more demanding human population and conserving biodiversity is a difficult challenge. Here, we explore potential solutions by assessing whether land sparing (farming for high yield, potentially enabling the protection of non-farmland habitat), land sharing (lower yielding farming with more biodiversity within farmland) or a mixed strategy would result in better bird conservation outcomes for a specified level of agricultural production. We surveyed forest and farmland study areas in southern Uganda, measuring the population density of 256 bird species and agricultural yield: food energy and gross income. Parametric non-linear functions relating density to yield were fitted. Species were identified as "winners" (total population size always at least as great with agriculture present as without it) or "losers" (total population sometimes or always reduced with agriculture present) for a range of targets for total agricultural production. For each target we determined whether each species would be predicted to have a higher total population with land sparing, land sharing or with any intermediate level of sparing at an intermediate yield. We found that most species were expected to have their highest total populations with land sparing, particularly loser species and species with small global range sizes. Hence, more species would benefit from high-yield farming if used as part of a strategy to reduce forest loss than from low-yield farming and land sharing, as has been found in Ghana and India in a previous study. We caution against advocacy for high-yield farming alone as a means to deliver land sparing if it is done without strong protection for natural habitats, other ecosystem services and social welfare. Instead, we suggest that conservationists explore how conservation and agricultural policies can be better integrated to deliver land sparing by, for example, combining land-use planning and agronomic support for small farmers.

Editor: Dorian Q. Fuller, University College London, United Kingdom

Funding: This work was supported by grants from The Darwin Initiative (14-032), the Cambridge Conservation Initiative (CCI 05/10/006) and The Leverhulme Trust (F/01 503/B). The funders had no role in study design, data collection and analysis, decision to publish, or preparation of the manuscript.

Competing Interests: The authors have declared that no competing interests exist.

* E-mail: mark.hulme@bto.org

Introduction

Increases in human population and per capita consumption are likely to lead to greatly increased agricultural demand over at least the next 40 years [1], which could lead to further habitat destruction, loss of ecosystem services, ecosystem simplification and species loss [2]. This has raised the question of how food production and biodiversity conservation can best be reconciled [3,4,5]. In temperate regions, and in some cases in the tropics, much emphasis has been placed on agri-environment and certification schemes to encourage wildlife-friendly farming, or land sharing, where lower-yield farming enables a high biodiversity to be maintained within farmed landscapes [6,7,8]. Land sharing has been championed by many in conservation practice and research (e.g. [9–11]) because wildlife-friendly farmland typically supports higher species richness and more species of the natural or semi-natural habitat it replaced than does intensively-managed farmland [12,13]. However, some studies have cast doubt on the effectiveness of land sharing initiatives, in both temperate and tropical areas [14–16]. This is both because wildlife-friendly farmland often offers a poor substitute habitat, particularly for the most sensitive species, and because it often entails a yield penalty and thus requires a greater area to produce any given amount of food (but see criticism of this interpretation of the evidence in [17]).

An alternative proposal to land sharing is land sparing, where agricultural land is farmed to produce a high yield of crops. This requires a smaller area of land than would be needed to grow the same total production target by lower-yielding methods. If this spared land is maintained or restored as natural habitat, then species associated with natural habitats are expected to benefit [6]. Future realised agricultural yield is likely to have a strong effect on the amount of land demanded by a growing and increasingly affluent population in the developing world [18,19] so the land sparing approach appears to be a strategy worth considering. There is a range of possible intermediate strategies but, for typical species assemblages, attempting to combine land sparing and land sharing benefits fewer species than adopting the better of the two pure strategies [4,6]. The pertinent question for policy-makers and conservationists is the extent to which conservation resources should be allocated towards preventing habitat loss, relative to ameliorating the negative impacts of intensification.

Explicit comparisons between expected biodiversity outcomes under land sparing and land sharing approaches are rare and more are needed from a wider range of locations in order to inform the debate on this issue [20]. A study on the responses of bird and tree species to varying agricultural yields in forested regions in Ghana and India found that many species were expected to have larger total population sizes with high yield farming combined with land sparing so that more forest was retained, than with low yield farming and land sharing [4]. This was particularly the case for species expected to have smaller total populations with, than without farming present (losers). In this paper, we perform a similar analysis of data on population densities of birds as a function of yield in the form of income and food energy collected across a large region in southern Uganda which includes both forest and agriculture. We use comparable methods to those used in Ghana and India [4] to assess whether the conclusions from those studies also apply to birds in Uganda.

Materials and Methods

Ethics Statement

This research was conducted through the NGO NatureUganda which has a MoU allowing research to be carried out in the majority of forest sites which were managed by the National Forestry Authority and the Uganda Wildlife Authority. Permission to access privately owned land (all farmland sites and one forest site) was given by the land owners. Some forest sites were part of the Mabira Forest Reserve, managed by the National Forestry Authority.

Study Area

The study area lies within the banana-coffee farming system in the Lake Victoria crescent, southern Uganda, a farming landscape covering more than 50,000 km^2. The area is one of high human population density and good access to infrastructure and markets. There are two wet seasons per year and annual rainfall is between 1000–1500 mm, making it one of the wetter regions of the country [21,22]. Major land uses include perennial crops, mainly banana and coffee, but there is an increasing shift towards cultivation of annual crops, largely in response to emerging disease and pest issues associated with traditional coffee and banana production. Land under agriculture increased by 11.4% between 1975 and 2000 in the wider region [23] and deforestation trends have been high across Africa in recent years [24]. Whilst it is likely that some forest fragments in Uganda have been isolated within savanna for hundreds or thousands of years other patches will have been part of much larger areas of forest before extensive forest clearance, particularly in the twentieth century, for timber, agriculture and as a measure against sleeping sickness [25].

The farmland study was conducted at 22 sites, each consisting of a 1 km x 1 km square, selected to represent a broad range of agricultural land uses from small-scale mixed holders to large-scale monoculture plantations. Population density for southern Ugands was derived from the 2002 Uganda National Census (URL: www.ubos.org) and was used as a surrogate for cultivation intensity with sites selected across a population gradient. Forest sites were selected from native forest patches within the farmed landscape described above. Forest sites were limited by the availability of patches of sufficient size and thirty forest patches of at least 1 km^2 in area were identified from the Biomass Map of Uganda [26]. Each of these sites was visited in November 2007 in order to determine (i) whether the forest patch still existed (ii) the extent of degradation and (iii) whether there were any access problems. Ten forest sites that had large clear-felled areas for cultivation or charcoal burning, and that therefore had open canopies (all sites <50% canopy cover), were excluded from the study. Of the remaining 20 forest sites selected for the bird surveys, one was partially deforested between the first visit and the commencement of bird surveys and was therefore also excluded, leaving 19 sites (Table 1). A map of site locations is given in Figure 1.

Bird Surveys

At each site a folded line transect of 2 km in length was followed, beginning at a random location and following paths and tracks where this was necessary to avoid trampling crops and for ease of access. Point counts [27] were located at 200 m intervals along each transect, totalling 10 per site for farmland but often fewer for forest, depending on the size of the forest patch. After a preliminary visit to the forest sites, low bird activity was apparent within a short time of arrival of the observer at the survey point. This is thought to have been caused by the noise generated by moving towards the point through forest vegetation and the lack of habituation to people compared to farmland birds; this effect did not appear to occur in the more open habitat on farmland. As a result, a 2-minute settling period was used before the 10-minute bird recording period began in forest, but no settling period was considered necessary in farmland. A comparison between forest points with and without a settling period indicated that this difference in methods might have made a difference to two species observed in both forest and farmland, but that the relationship was weak and would have made, at most, a very small difference to the results of the analysis presented here (Text S1, Table S1). Birds were recorded during the 10-minute survey period and each record assigned to one of three distance bands (<25 m, 25–50 m, >50 m) according to the distance from the point to the location of the bird when it was first detected. Distances were estimated by eye, but with regular checks against directly measured distances. Birds first seen when in flight were recorded separately. Farmland points were visited five times between February 2006 and January 2007, with intervals of at least six weeks between visits. Forest points were visited twice between February and April 2008.

Habitat Surveys

Five 1 km parallel transects were arranged from east to west across each farmland site, separated by 200 m. Between February and June 2006 the length of each transect passing through different vegetation and crop types was measured using a tape measure. Vegetation types recorded were: cultivated, fallow, woodlot, homestead (building and yard where people and domestic animals reside), road, managed pasture, unmanaged pasture, school/market place, kraal, garden, natural vegetation and whether or not this was forest. The list of crops used for yield calculation is given in Table 2. The proportion of total land covered by each crop was estimated as its proportion of the total length of transect.

Estimation of Crop Yield for Farm Income and Food Energy Measures

To evaluate the potential performance of land sparing and land sharing we need to model the total population size of species whilst achieving a given fixed level of agricultural production (the

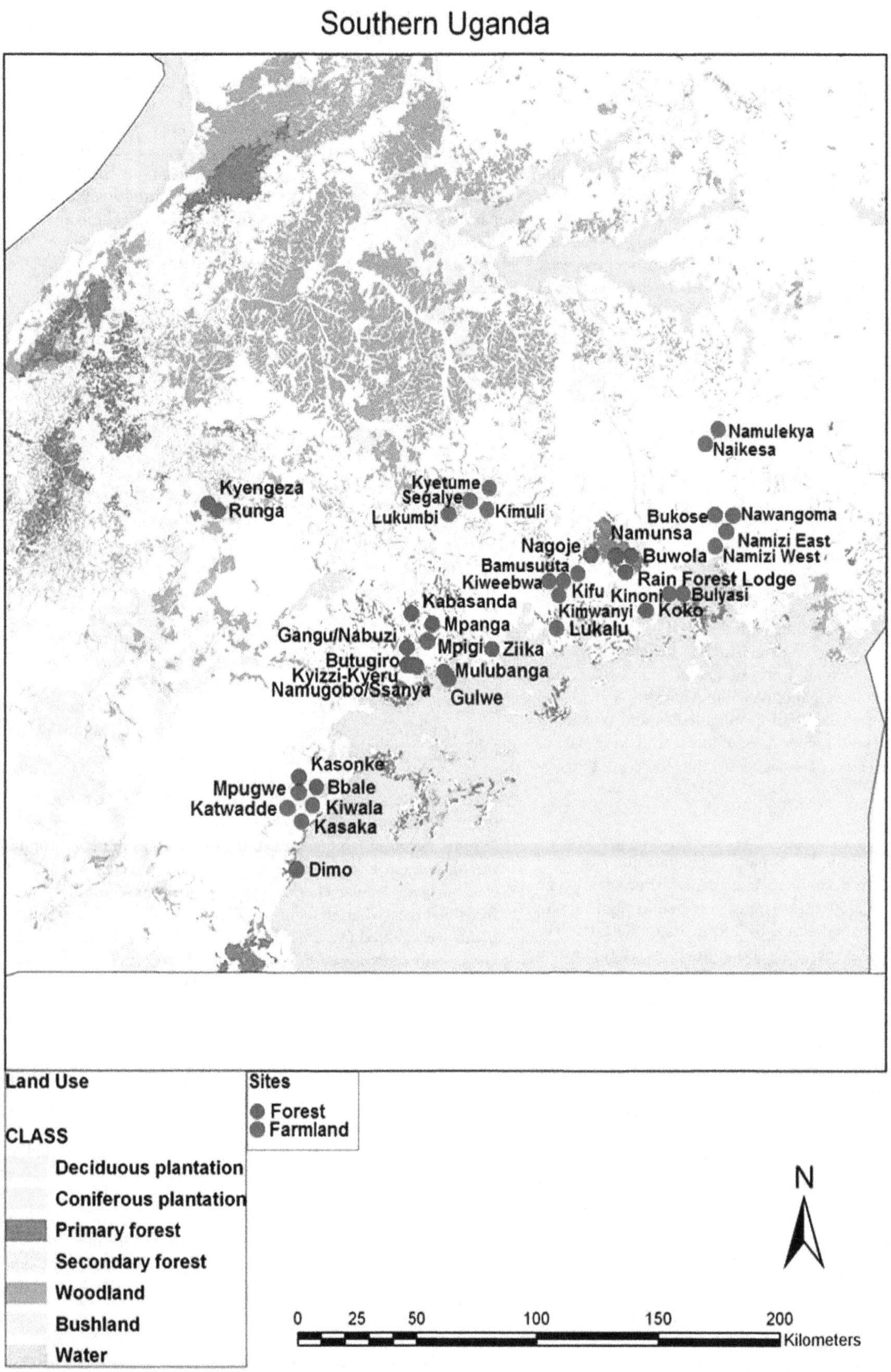

Figure 1. Location of study sites in Uganda. Forest sites are denoted with blue circles, farmland sites with red circles.

Table 1. Sites surveyed, area of forest sites, effort (number of point counts conducted) and annual yield per hectare (GJ ha^{-1} $year^{-1}$ and US$ ha^{-1} half $year^{-1}$).

Site	Habitat	Area of forest (ha)	Forest perimeter (km)	Effort (Point counts conducted)	Food Energy (GJ ha^{-1} $year^{-1}$)	Income (US$ ha^{-1} $year^{-1}$)
Bamusuuta	Farmland			50	8.736	431.26
Bukose	Farmland			50	13.172	555.26
Bulyasi	Farmland			50	5.43	510.92
Kasaala	Farmland			50	10.296	526.68
Katwadde	Farmland			50	7.342	414.98
Kifu	Farmland			40	11.13	312.9
Kimuli	Farmland			50	3.614	374.42
Kimwanyi	Farmland			50	8.958	370.02
Kinoni	Farmland			50	4.416	448.48
Kiwaala	Farmland			50	6.488	338.3
Kiweebwa	Farmland			50	8.346	330
Kyetume	Farmland			50	2.554	340.2
Lukalu	Farmland			50	4.784	483.14
Lukumbi	Farmland			50	8.696	546.84
Mpigi	Farmland			50	6.276	591.8
Mpugwe	Farmland			50	12.124	477.24
Naikesa	Farmland			50	17.516	583.1
Namizi East	Farmland			50	10.256	556.34
Namizi West	Farmland			50	14.128	589.98
Namulekya	Farmland			50	22.84	579.48
Nawangoma	Farmland			50	24	667.68
Segalye	Farmland			50	2.63	407.74
Bbale	Forest	5.27	40.21	16	0	0
Butugiro	Forest	10.22	89.85	20	0	0
Buwola	Forest	70.78	26.96	20	0	0
Dimo	Forest	17.19	93.23	20	0	0
Gangu/Nabuzi	Forest	19.77	97.32	14	0	0
Gulwe	Forest	21.74	71.01	6	0	0
Kabasanda	Forest	24.62	128.15	20	0	0
Kasonke	Forest	1.91	13.32	12	0	0
Koko	Forest	14.7	114.1	12	0	0
Kyengeza	Forest	27.05	129.73	8	0	0
Kyizzi-Kyeru	Forest	10.27	82.01	16	0	0
Mpanga	Forest	17.69	104.99	20	0	0
Mulubanga	Forest	27.26	90.39	7	0	0
Nagoje	Forest	35.83	75.07	20	0	0
Namugobo/Ssanya	Forest	11.6	78.79	20	0	0
Namunsa	Forest	67.77	55.53	20	0	0
Rain Forest Lodge	Forest	36.78	77.27	20	0	0
Runga	Forest	5.71	28.01	6	0	0
Ziika	Forest	9.99	111.7	8	0	0

production target) at a range of yields of food energy or farm income per unit area of the farmed landscape. This section describes how we estimated food energy and income yield from surveys.

Biomass Yield of Agricultural Products

We estimated the biomass yield of crops in each of our study areas as an intermediate step in obtaining yields in terms of food energy and farm income. The conversion of biomass yield to food energy and income yield is described in later sections. Direct

Table 2. Energy of edible mass and % inedible refuse (skins, husks, stalks *etc.*) for crops for which yield and area data were available.

Crop	Refuse (%)	GJ tonne-1
Banana	36	3.71
Bean	0	13.93
Cabbage	20	1.03
Cassava	14	6.67
Coffee	0	0
Eggplant	19	1.01
Groundnut	0	23.74
Maize	0	15.27
Millet	0	15.82
Pineapple	49	2.09
Pumpkin	30	1.09
Rice	0	14.98
Simsim	0	23.97
Sorghum	0	14.18
Soy bean	0	18.66
Spinach	28	0.97
Sugarcane	52	2.11
Sweet potato	28	3.59
Tea	0	0
Tomato	9	0.75
Vanilla	0	0
Yam	14	4.94

measurements of biomass yield were impractical because of the large number of study areas and their large size. Instead, reported yields from farmer interviews were used. Ten farmers at each site were interviewed in 2007 and 2008 about their cropping practices. For each site a list was drawn up, in conjunction with a local leader, of about 20 households. Those farmers with less than 50% of their land in production were removed from the list. The remaining famers were categorised according to the three most common crop types grown in the square kilometre. Up to 10 farmers were selected to ensure each of the major crop types in the area were included with farms broadly typical of the area. Six individual farmers/households were selected at random from the list of 10 farmers for interviewing. Within the selected households, the head, spouse or any other knowledgeable person was the target respondent. The crop yield information they reported was collected separately for two 6-month growing periods: April - August and September - March. The local units given were converted to tonnes. Biomass yield per unit area was calculated for each farmer, crop and growing season in tonnes ha^{-1}. A number of very high yields were reported. For example, the highest recorded maize yield in Season 2 of 62 tonnes ha^{-1} was 15.5 times the next highest recorded yield for that season of 4 tonnes ha^{-1}, so we assumed that the higher value was inaccurately reported. For crops where outliers of this kind were present (banana, cassava, maize, tomato and sweet potato) values above the 95th percentile were excluded for both seasons. The yield per hectare for each crop at a site was taken to be the sum of the average yields for the two growing seasons. Pasture was a very small proportion of overall land use (1.41% of land under managed pasture, 2.41% under unmanaged pasture) and livestock productivity was not included in agricultural yield calculations.

Farm Income

For estimating the yield for food and non-food products monetary currencies are an appropriate measure. The potential income generated per unit area of the farmed sites was calculated, by multiplying the biomass yield per hectare per year, calculated as described above, by the local market price per unit weight of each crop, which was obtained as a mean from market surveys in both seasons and for all clusters with farmed sites. The mean farm income for each crop at each site was then calculated by multiplying the mean value of a crop per hectare by the area of that crop at the site, separately for each of the two seasons. Incomes were then summed across all crops and divided by the total area of the site to give total income per unit area per season, which was converted to US$ ha^{-1} at 2007 exchange rates of 1 US$ = 1690 Ugandan Shillings (URL: WWW.OANDA.com: accessed on 01/11/2011). The values for the two seasons were then summed. Mean farm incomes were therefore estimates of the potential income per hectare of the whole farmed landscape per year which might have been derived from the crops grown, regardless of whether they were sold, bartered or consumed by the farmers, their families or livestock. Farm income for forest sites was assumed to be zero.

Food Energy Yield

Food energy, unlike income, is not affected by market fluctuations. However, it is not as appropriate for products which have a high monetary value but low food energy, such as coffee and vanilla. Hence, we use both food energy and income yield in our analyses to check whether conclusions about the responses of bird densities to yield are robust to the choice of measures, neither of which is perfect. The amount of food energy contained in each crop per unit biomass harvested was assessed using values obtained from the literature for the energy content per unit weight of processed crop and the average proportion by weight of the harvested crop which is discarded as inedible refuse during the preparation of the crop for consumption, such as skin and husks. The values obtained for the crops present on transects are shown in Table 2. Values for most raw crops which occurred on the farmland transects were obtained from the United States Department of Agriculture (USDA) National Nutrient Database for Standard Reference (URL: http://www.nal.usda.gov/fnic/foodcomp/search/: accessed on 01/11/2011). The details of the data and methods used to calculate the energy content of various raw and processed foodstuffs are found in the project documentation [28]. For coffee the energy value derived from black coffee was assumed to be negligible so coffee was not considered to contribute to food energy yield. Similarly vanilla, a flavouring which is used in very small quantities, contributes negligibly to energy intake and was not considered in energy yield calculations. For sugar cane the proportion of refuse was determined from the percentage of fibrous bagasse and liquid juice and percentage of sugar in the juice [29], and the energy content calculated using USDA data for raw sugar. The energy value of edible food per unit of harvested biomass was then multiplied by the biomass yield per hectare minus the refuse and the area of each crop at each site to give total food energy production per site per season. Energy production values were then summed across all crops and divided by the total area of the site to give total food energy yield per unit area per season, expressed as GJ ha^{-1}. The values for the two seasons were then added together.Food energy yields were therefore estimates of the food energy which might have been derived from the crops grown per hectare of the whole farmed landscape per year, regardless of whether they were sold, bartered or consumed by the farmers, their families or livestock. Food energy yields for forest sites were assumed to be zero.

Data Analysis

Species densities. Records of birds in flight and over 50 m from the point were discarded. Some aerial species, such as swallows and martins, were only seen in flight and are therefore not included. Eligible records were pooled across survey points for each site and each species and the number of point counts at each site was recorded and is shown in Table 1. Twenty point counts were performed at most forest sites (10 points visited twice) and 50 counts for most farmland sites (10 points visited five times). Effort and area for each site are shown in Table 1. Distance version 6.0 [30] was used to estimate detection probabilities for each species. Since there were only two distance bands, limiting the number of key parameters available for use to one, the half normal function with no adjustment terms was used for all species. For each species, point habitat type (forest or farmland), was included as a covariate if at least 20 individuals had been recorded in both habitats and this model was chosen if the AIC value was lower than that for the model with no covariate. For three species the habitat model had the lowest AIC but failed to converge so the model that ignored habitat type was used. For another species the habitat model had the lowest AIC but the variance estimation was invalid so the non-habitat model was used. Due to the low number of residual degrees of freedom goodness of fit tests were not possible so models were accepted based on visual determination of the plausibility of detection functions and the variance of the overall density estimate. Density values were estimated for each site with a detection function by habitat where habitat was selected as a covariate [31].

Too few registrations were available for some species to estimate a detection function from only the data for that species. These species were each assigned a detectability group depending on our assessments of their diet, habitat stratum and activity level. Species were classified as carnivore, frugivore, granivore, insectivore or omnivore and further classified as to which stratum of vegetation they usually inhabit by allocating each to one of five classes: 1) canopy or in sub-canopy of forest, or in canopy of large trees in other habitats, perch high in the canopy, 2) lower or middle layers of vegetation in forest or other habitats with dense tree cover 3) bushes or small trees, usually in open habitats 4) on the ground in open areas 5) low vegetation, often heard rather than seen. Species' activity was classified as usually active (e.g., sunbirds) or usually static (e.g., kingfishers). Some groups were further aggregated to achieve an adequate sample size. Using these classifications 26 groups were formed which included all species recorded and which had sufficient observations to calculate detection functions. Density values were estimated by stratifying by species and estimating density by site. Habitat was included in some of these group models as a covariate using the same model selection process as for individual species above.

Fitting of density-yield curves. Parametric models were fitted to relate bird density to yield, as described previously (see supplementary online material in [4]). The method is summarised here. Univariate parametric Poisson regression models were fitted for each species by a maximum-likelihood (M-L) method. The dependent variable was population density, determined using Distance, with each of the two measures of yield being the independent variable. The following two alternative model formulations were used:

Model A

$$n/v = exp(b_0 + b_1(x^{\alpha}))$$

Model B

$$n/v = exp(b_0 + b_1(x^{\alpha}) + b_2(x^{2\alpha}))$$

where n is the number of individuals of that species recorded, $v = (a \times e)$, where a is the effective detection area per survey point from Distance and e is effort (number of point counts conducted at the site). Note that a differed between farmland and forest sites for those species for which a habitat-specific detection function was used in Distance. The variable x represents yield (in either GJ ha^{-1} year^{-1} or income ha^{-1} year^{-1}), and b_0, b_1, b_2 and α are constants estimated from the data. The value of α was constrained to be positive and not to exceed 4.6. This maximum value of α was used because for species with high α, the likelihood of the data was usually approximately constant with increasing α beyond this value, making a precise M-L model impossible to identify. However, the shape of the functions determined by the models with high α varied little as α was changed. These model formulations were selected because they give a wide range of shapes of curves. In particular, the M-L Model B curves were often hump-shaped, but with an asymmetrical shape. This asymmetry was visible in plots of density against yield for many species and was well described by the inclusion of the shape parameter α. For each square, the expected density under either Model A or Model B was calculated for a given set of parameter values and multiplied by the value of v for that square to give the expected number of individuals for that square. The natural logarithm of the Poisson probability of the observed number of individuals for the square, given the expected number under the model, was then obtained and summed across all squares to give the log-likelihood of the data. This log-likehood was then maximised to give M-L values of b_0, b_1, and α, for Model A and b_0, b_1, b_2 and α for Model B. Under Model B, the best-fitting hump-shaped functions sometimes had a high peak density value in a gap between groups of sites in the distribution of the yield variable. For some species, this peak density was much larger than the observed density at any site: sometimes thousands of times larger. We considered such models to be unrealistic and therefore constrained the model parameter values to give peak densities no greater than 1.5 times the maximum observed density. The maximised log-likelihoods were multiplied by -2 to give the residual deviances for models A and B. If the residual deviance for Model B was more than 3.84 (X^2 with 1 degree of freedom for $P = 0.05$) lower than that for Model A then Model B was selected. Otherwise Model A was selected for reasons of parsimony. For species which were only observed in forest sites no model was fitted and a simple step function was assumed with the only non-zero density value density at zero yield in forest.

Densities for all species were also estimated for forest and farmland by $\Sigma n/\Sigma v$, with summation across all sites in each habitat and n and v as defined above.

Model of population size of a species in relation to yield and production target. We used a model developed previously [6], and used in Ghana and India [4], in which the expected total population of a species within a region is given by adding its total population in forest to its total population on farmed land. The expected population in forest was calculated as the product of the area of forest and the density of the species in forest obtained from its density-yield function. For a scenario in which the whole site was covered by forest, the total population is given by the density in forest, taken from the fitted density-yield curve, multiplied by the total area of the region. This is referred to as the baseline population and all other calculated total population sizes are expressed as proportions of the baseline population. The total population of the species on farmed land was calculated as the product of the area of farmed land in the region under a given farming scenario and the population density derived from the fitted density-yield function, given the yield assumed in that scenario. The impact of land allocation strategies on species populations was assessed at a defined level of production of food energy or income, referred to as the production target [6]. The production target can be produced at any yield per unit area of farmed land within a range defined by minimum and maximum permissible yields. The minimum permissible yield is that obtained by dividing the production target by the total cultivable area of the site. At lower yields than this, the amount of energy or income produced would be less than the production target. A maximum permissible yield is assumed set by maximum feasible levels of crop production, here designated as 1.25 times the maximum yield observed for the set of farmland sites we studied. This multiplier is arbitrary, but it has been shown previously that conclusions are robust to variation in the multiplier [4]. Within the permissible range of yields, the area of farmed land is obtained by dividing the production target by the assumed yield. The area of forest is assumed to be the remaining area of the region that is not required for farming and can therefore be obtained by subtracting the area of farmed land needed to grow the production target from the total area of the region. Hence, for a given production target, areas of farmed land and forest and the population density of the species on farmed land can be calculated for all yields within the permissible range. The total population of the species in the region can be calculated from these areas and the population densities, as described above. When this is done for all yields in the permissible range, the yield at which the highest total population occurs can be obtained. This is referred to as the optimal yield on farmed land for the species, conditional on the production target.

Determining best farming strategy for biodiversity. At a given production target and for species whose optimal yield was the lowest permissible yield, land sharing with low-yield farming would be the strategy under which those species would have the largest total population, which we call the best strategy. For species whose optimal yield was the highest permissible yield, land sparing with high-yield farming would be the best strategy. For species whose optimal yield was neither the lowest permissible yield nor the highest permissible yield, an intermediate yield would be best. For a given production target, species were classified as doing best with land sharing and low yields, land sparing and high yields or some intermediate strategy. At each production target all bird species were also classified as winners or losers in relation to agriculture according to whether their total population size would be higher or lower than the baseline if there was any farmed land within the study region. Winners were those species for which the total population size in the province was always equal to, or larger than, the baseline population, regardless of the yield of farming within the permissible range. Losers were those species with total populations lower than the baseline population at some (or all) permissible yields. Winners would be expected always to have more favourable conservation status than the average state for the distant past because their population is higher than the baseline population, regardless of production target and yield. Losers are species whose total population could potentially fall below the baseline as a result of agriculture, and therefore their conservation status is more sensitive to choices made about land allocation to farming at different yields. Our definition of losers includes both species which always decline as agricultural production increases, and others which have higher populations than the baseline at some yields, but lower populations at others. Figure 2 shows example density-yield curves for winners and losers. We calculated the optimal strategy for each species at production targets ranging from that equivalent to producing a single unit (1 GJ or $1 US) of output per hectare over the entire region, to the equivalent of farming the entire region at the maximum value of observed yield in any of our study sites.

Global range size. We compared the proportion of species that were winners and losers and with different optimal strategies between species with large global range sizes and those with smaller global range sizes, which are often those of greater conservation concern. Range sizes were obtained as the Extent of Occurrence (EOO) given by the World Bird Database (URL: http://www.birdlife.org/datazone/species/search: accessed on 27/10/2011). Species with an EOO of greater than 3,000,000 km^2 were classified as having a large range size with those below having a small range size (see [4]). In the 26 cases where no extent of occurrence was for given breeding range other sources were referred to [32,33]. This resulted in 91 species having small ranges and 165 species having large ranges.

Sample sizes. Since many species of greater conservation concern are likely to be found at low densities we wished to avoid biasing our conclusions by removing those species with low sample sizes. However, such species are likely to have less precisely estimated density-yield functions and this might have undue influence on the frequencies of different types of density-yield curves. In order to gauge the effect of retaining or excluding rare species, we compared our conclusions when species with fewer than 30 records were excluded with conclusions based upon all species.

Results

The mean population density of each species in forest and farmland is shown in Table S2 and coefficients of the fitted density-yield functions are given in Table S3 for both the food energy and income models.

The numbers of winner and loser species with each of three categories of optimal yield (high yield, intermediate yield and low yield) were plotted against production target. We included all species when drawing conclusions based upon Figure 3, Figure 4, Figure 5 and Figure 6, since for both measures of yield removing species with 30 or fewer records did not alter the results markedly (Figures 3 and 5). More species were losers than were winners. Land sparing gave higher total populations of more of the loser species than did land sharing at all production targets, but the proportion of loser species doing best with land sparing increased as the production target increased. Land sharing gave higher total populations of more of the winner species than did land sparing at all production targets. The proportion of winner species doing best with land sharing increased as the production target increased. Ignoring whether species were winners or losers, there was an overall majority of species that would benefit from land sparing compared with species that would benefit from land sharing (Figures 3A and 5A). Intermediate yields were best for a relatively small proportion of species for both winners and losers and that proportion decreased markedly with increasing production target. The proportion of loser species was higher for species with small than large global ranges (Figures 4 and 6). For both losers and winners the proportion of species doing best with land sparing

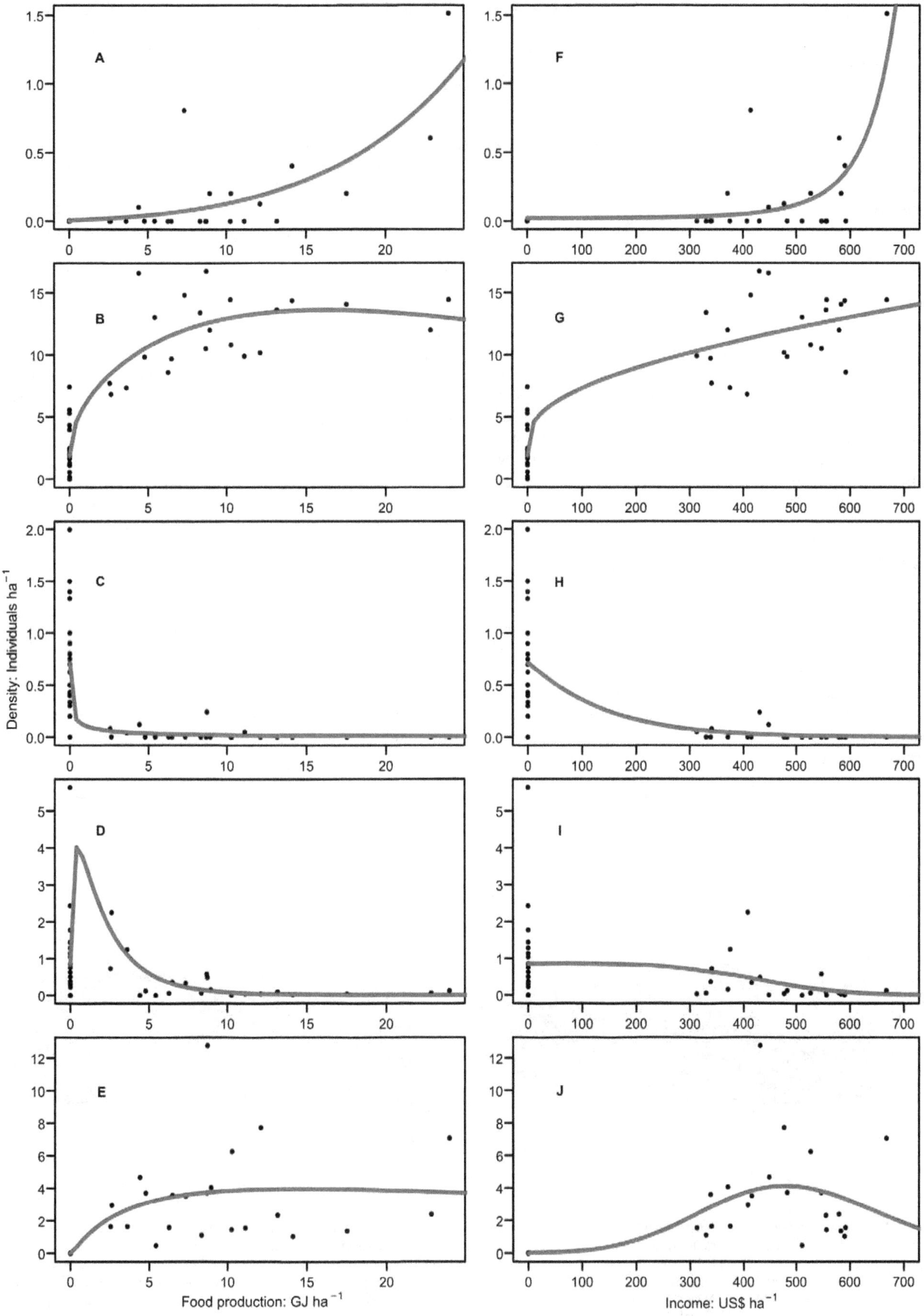

A
B
C
D
E
F
G
H
I
J
Density: Individuals ha^{-1}
Food production: GJ ha^{-1}
Income: US$ ha^{-1}

Figure 2. Examples of species with different types of fitted density-yield functions. (A, F) Cattle Egret *Bubulcus ibis*, which at all production targets is a winner for which land sparing is the best strategy. (B, G) Common Bulbul *Pycnonotus barbatus*, a winner for which land sharing is always the best strategy. (C, H) Black-necked Weaver *Ploceus nigricollis*, a loser for which land sparing is always the best strategy. (D, I) Splendid-glossy Starling *Lamprotornis splendidus*, a loser for which the best strategy depends on the production target. (E, J) Black-headed Weaver *Ploceus melanocephalus*, a winner for which the best strategy depends on the production target.

rather than land sharing was higher for small range than large range species (Figures 4 and 6).

Discussion

Our results are remarkably consistent with those for birds and trees in Ghana and India [4], where land sparing with high-yield farming was estimated to give the largest total population size for most species of both groups, this finding being robust to the use of two different measures of yield. This was especially the case when species expected to have lower total populations than the baseline when agriculture is present (losers) are considered, but was also the case overall. Indeed, the Uganda models predict an even greater proportion of bird species would benefit from land sparing in comparison to land sharing than in Ghana and India. The greater preponderance of species that benefit from land sparing among losers than winners is considered to be of relevance to conservation because we think it likely that winner species tolerant of open habitats on farmland are likely to have larger populations under current conditions than at most times during their evolutionary history [34]. Hence, we expect loser species to be at higher current and future risk of adverse conservation status than winners, and therefore the response of losers to farming yield to be of greater conservation significance than that of winners. The greater representation of losers amongst species with smaller global ranges is also consistent with results for both birds and trees in Ghana and India. Small geographical range size is currently the single best predictor of threat of extinction in terrestrial bird species [35], so this information on the degree to which species in the two range size groups tolerate farming is of relevance to their future conservation. Although all species included in this analysis are classified globally as Least Concern by the IUCN red list (URL: http://www.iucnredlist.org/: accessed on 27/10/2011), deforestation and agricultural encroachment are continuing and likely to affect some of these species in future. Species tend to have much more restricted ranges than is indicated by their Extent of Occurrence [35]. Whilst the yields presented here were based, of necessity, on interviews with farmers who relied mostly on memory, the similarities between this and previous studies suggest that the yield data was sound and represented a true gradient of production intensity.

The forest fragments in our survey were mostly small and bird species that prefer forest interiors might be absent or at low density. A targeted survey of the largest and best quality forest sites in the region (probably only possible in Mabira and Dimmo) might produce higher population densities in forests (with zero agricultural yield) for these interior species. This would make our findings about the predominance of loser species for which land sparing is optimal conservative. Recent research in Uganda has suggested that forest birds move among forest fragments to a greater extent than was previously thought [25], so isolated small forests are still likely to have conservation value. Recent deforestration trends in Africa are such that they predict substantial forest loss over the next 50 years [24] and it is also possible that there is a lag effect in the response of bird density to recent deforestation. This might cause density-yield relationships to change somewhat over time.

Of 256 species 10 were Palearctic migrants, wintering in Africa between September and May [22], totalling 31 registrations (Table S2) of which Wood Warbler *Phylloscopus sibilatrix* was detected in forest only and Willow Warbler *Phyllosopus trochilus* was detected in forest and farmland with the remainder in farmland only, consistent with expected habitat requirements [22]. This indicates that the forest surveys were sufficient to register use of forest habitat by migrants between February and April. The period

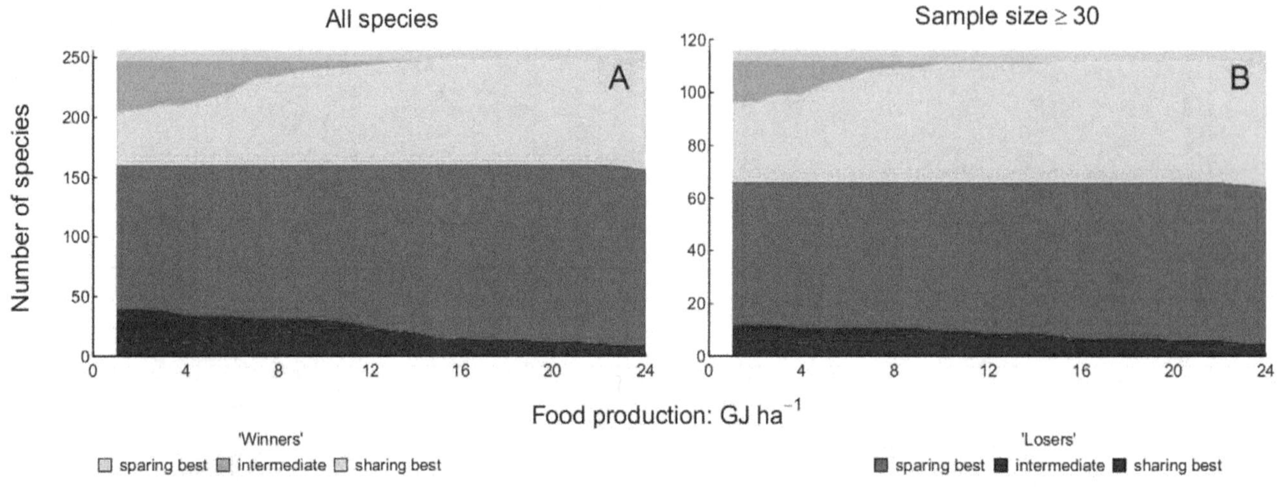

Figure 3. Winners and losers with food energy production targets by sample size. Number of species which have larger total populations with than without agriculture (winners: light colours) and those with smaller total populations (losers: dark colours) in relation to the production target for food energy. Species which have their largest total populations with the highest energy yield and land sparing (red/pink) those with largest populations with lowest permissible energy yield (land sharing: dark/light blue) and those benefitting most from intermediate yield (dark/light purple) are shown separately Maximum permissible yield was 30 GJ ha^{-1} year^{-1}, 1.25 times the maximum observed yield. A is for all species, B is for species with a sample size of 30 individuals or greater.

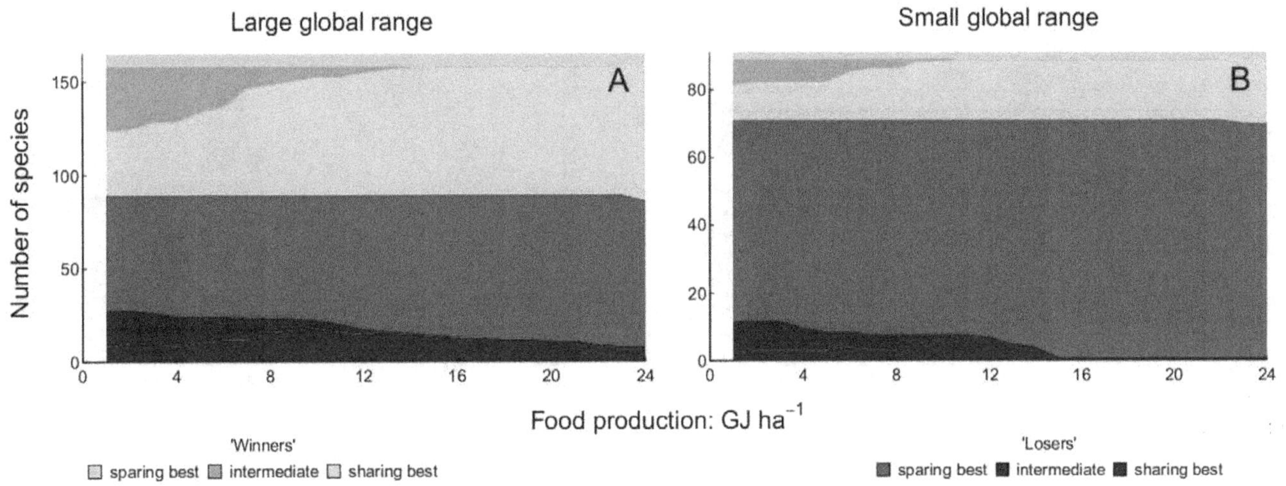

Figure 4. Winners and losers with food energy production targets by range size. Number of species which have larger total populations with than without agriculture (winners: light colours) and those with smaller total populations (losers: dark colours) in relation to the production target for food energy. Species which have their largest total populations with the highest energy yield and land sparing (red/pink) those with largest populations with lowest permissible energy yield (land sharing: dark/light blue) and those benefitting most from intermediate yield (dark/light purple) are shown separately. Conventions are as for Figure 3. A is for species with a large global range, B is for species with a small global range.

during which migrants are on breeding grounds will not have affected the density-yield relationships for those occurring in one habitat only. Partial intra-African migrants totalled 10 species, all of which are recorded in southern Uganda during the period of the forest surveys [22]. Two were observed only in forest and five only in farmland, Broad-billed Roller *Eurystomus glaucurus* and White-throated Bee-eater *Merops albicollis* were detected only rarely in forest (4 out of 22 registrations and 23 out of 484 registrations respectively), all of which is consistent with the expected habitat requirements of these nine species [22], so density-yield relationships are unlikely to have been biased. Red-chested Cuckoo *Cuculus solitaries* was recorded 24 times in forest and 27 times in farmland but the fitted model (Table S3) is consistent with what might be expected of this forest generalist [22]. Certain species might have seasonally-variable detectability due to changes in behaviour, vocalisations or vegetation but surveys were conducted throughout the year in farmland so maximising the chance that relative occurrence at each site will have been recorded. We have no reason to suspect that, other than the movements of potential migrants discussed above, species change their habitat use at particular times during the year [22].

Our results suggest that, at least within this tropical forested landscape, bird conservation would be best served by maintaining as much natural habitat as possible. This could benefit forest specialist species with small ranges which were observed only in our forest sites, such as, Weyns's Weaver *Ploceus weynsi* and Joyful

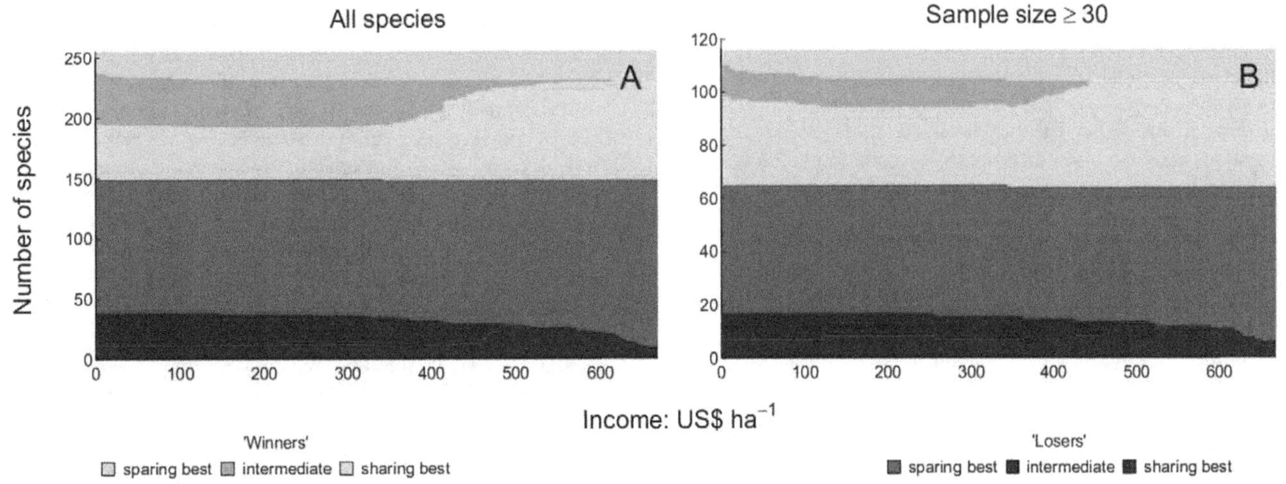

Figure 5. Winners and losers with gross income production targets by sample size. Number of species which have larger total populations with than without agriculture (winners: light colours) and those with smaller total populations (losers: dark colours) in relation to the production target for income. Species which have their largest total populations with the highest income and land sparing (red/pink) those with largest populations with lowest permissible income (land sharing: dark/light blue) and those benefitting most from intermediate yield (dark/light purple) are shown separately. Maximum permissible income was 835 US$ ha^{-1} year^{-1}, 1.25 times the maximum observed yield. A is for all species, B is for species with a sample size of 30 individuals or greater.

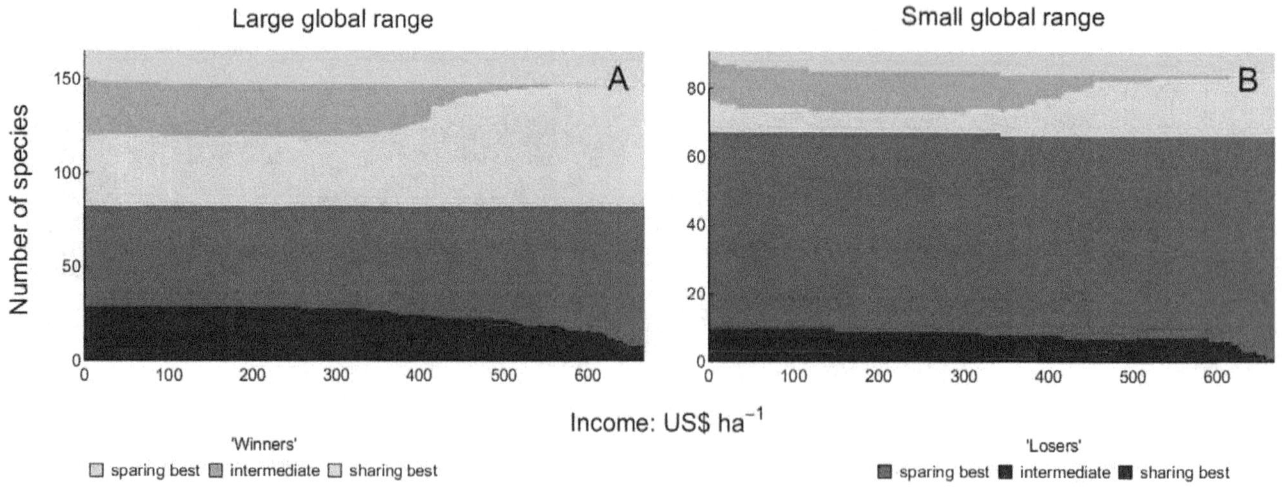

Figure 6. Winners and losers with gross income production targets by range size. Number of species which have larger total populations with than without agriculture (winners: light colours) and those with smaller total populations (losers: dark colours) in relation to the production target for income. Species which have their largest total populations with the highest income yield and land sparing (red/pink) those with largest populations with lowest permissible income yield (land sharing: dark/light blue) and those benefitting most from intermediate yield (dark/light purple) are shown separately. Conventions are as in Figure 5. A is for species with a large global range, B is for species with a small global range.

Greenbul *Chlorocichla laetissima* as well as, potentially, forest specialists with larger ranges such as Yellow-spotted Barbet *Buccanodon duchaillui*. In our study sites small range, more generalist species, such as Black-necked Weaver *Ploceus nigricollis* and Black-headed Weaver *Ploceus melanocephalus*, which both occurred on farmland (Figure 2), may, potentially, benefit from land sharing. There are examples, particularly from temperate regions (e.g. [36]) of some bird species being dependent on farmland habitat. Recently, dependency on low-intensity farming has also been claimed for some globally-threatened bird species in developing countries [37]. Any land sparing initiative aiming to increase yields on farmland should avoid doing so on farmland important for such species.

There has been criticism of the conclusions from Ghana and India [4] for paying insufficient attention to social and ecological complexities [38]. However, the approach taken here is not intended to provide detailed prescriptions for future landscape change, nor to address all of the complex issues involved in land-use change. Instead, it aims to test the widespread assumption that encouragement of low-yielding farmlands is necessarily the best option for conservation, using species-level and yield data for birds in Uganda that are more detailed than have been collected previously.

There are several aspects which the model described here does not take into account, such as social complexities and ecosystem services. To address these aspects will require further information on trade-offs and synergies between socio-economic and service outcomes and the biodiversity outcomes we have described, including quantification of the reliance of agriculture on ecosystem services [34,39,40]. Studies to collect such information should address the flaws in sampling design, inappropriate metrics, and/or failure to measure biodiversity baselines that have undermined the conclusions of many previous studies [34,41].

A further important concern is to identify the social and governance contexts in which increasing yields might be effective as part of a strategy to protect natural habitats. There is good evidence that yield-increasing technologies can increase rather than decrease habitat conversion at local scales [11,42], and at larger scales the evidence for sparing without any explicit policies to deliver it is weak [43,44]. However, land sharing interventions are also often ineffective in practice [14,45], and it seems premature to dismiss land sparing as a strategy when policy interventions specifically designed to achieve it have not yet been designed and tested. To help ensure that decision-makers, whether government bodies or local community leaders, take biodiversity into account it is imperative to integrate biodiversity conservation into policies and decision frameworks for resource production and consumption [44,46,47].

Conclusion

Despite the close agreement between our results and those from Ghana and India [4], there are reasons to remain cautious about generalising our conclusion that land sparing has greater potential biodiversity benefits than land sharing. Studies are needed in more regions. In addition, further work is needed to understand how our conclusions might be affected by the inclusion of other objectives (such as social objectives), the spatial configuration of land uses, and the social or political feasibility of implementing particular strategies. However, we can draw some firm conclusions. None of the farming systems we examined in the banana-coffee arc around Lake Victoria are a substitute for relatively intact forests. We suggest that conservationists should avoid the promotion of low-yield farming where that is likely to result in further expansion into forests, unless a quantitative study on likely impacts on species' populations indicates that this will be beneficial. Instead, we suggest that they explore the potential of linked policies to deliver land sparing, for example by directing development aid towards small farmers to increase yields on existing farmland, within a land-use planning framework (at regional or community level) which limits expansion of farmland into forests.

Supporting Information

Table S1 Mean difference in total bird registrations and species richness between preliminary and main survey visits. N = number of site pairs (sites where a species was absent on both visits are not included).

Table S2 Population densities (individuals ha^{-1}) of all bird species, estimated for forest and farmland by sum(n)/sum(v) across all sites in each habitat, where n is the number of individuals of that species recorded, $v = (a \times e)$, where a is the effective detection area per survey point from Distance and e is effort (number of point counts conducted at the site). Note that a differed between farmland and forest sites for those species for which a habitat-specific detection function was used in program Distance.

Table S3 Maximum-likelihood estimates of the coefficients for density-yield models for each species. Density is expressed as individuals ha^{-1}, yield in food energy, GJ ha^{-1} $year^{-1}$ and gross income, US$ ha^{-1} $year^{-1}$. Where species were observed only in forest b_0 was set at the natural logarithm of the calculated density in forest and zeroes are given for the other model parameters. Scientific names are given in Table S2.

Text S1 The effect of a settling period on the number of individuals seen in the ten minute point count period at forest sites.

Acknowledgments

Many thanks to the farmers who gave us permission to access the land. We would also like to thank Ibrahim Ssenfuma for assisting with fieldwork on the forest sites.

Author Contributions

Contributed major revisions and editorial input leading to final draft: REG RK DN DM BP JAV PWA DEP DEC. Conceived and designed the experiments: DN DM PWA DEP JAV DEC SB. Performed the experiments: DN DM RK DEC. Analyzed the data: MFH REG. Contributed reagents/materials/analysis tools: REG BP. Wrote the paper: MFH.

References

1. Godfray HCJ, Beddington JR, Crute IR, Haddad L, Lawrence D, et al. (2010) Food security: the challenge of feeding 9 billion people. Science 327: 812–818. doi: 10.1126/science.1185383.
2. Tilman D, Fargione J, Wolff B, D'Antonio C, Dobson A, et al. (2001) Forecasting agriculturally driven global environmental change. Science 292: 281–284. doi:10.1126/science.1057544.
3. Fischer J, Brosi B, Daily GC, Ehrlich PR, Goldman R, et al. (2008) Should agricultural policies encourage land sparing or wildlife-friendly farming? Front Ecol Environ 6: 380–385. doi:10.1890/070019.
4. Phalan B, Onial M, Balmford A, Green RE (2011a) Reconciling food production and biodiversity conservation: Land sharing and land sparing compared. Science 333: 1289–1291. doi: 10.1126/science.1208742.
5. Foley JA, Ramankutty N, Brauman KA, Cassidy ES, Gerber JS, et al. (2011) Solutions for a cultivated planet. Nature 478: 337–342. doi:10.1038/nature10452.
6. Green RE, Cornell SJ, Scharlemann JPW, Balmford A (2005) Farming and the fate of wild nature. Science 307: 550–555. doi:10.1126/science.1106049.
7. Matson PA, Vitousek PM (2006) Agricultural intensification: Will land spared from farming be land spared for nature? Conserv Biol 20: 709. doi:10.1111/j.1523-1739.2006.00442.x.
8. Perfecto I, Vandermeer J, Mas A, Soto Pinto L (2005) Biodiversity, yield, and shade coffee certification. Ecol Econ 54: 435–446. doi:http://dx.doi.org/10.1016/j.ecolecon.2004.10.009.
9. Daily GC, Ehrlich PR, Sanchez-Azofeifa GA (2001) Countryside biogeography: use of human-dominated habitats by the avifauna of southern Costa Rica. Ecol Appl 11: 1–13. doi: 10.1890/1051-0761(2001)011[0001:CBUOHD]2.0.CO;2.
10. Perfecto I, Vandermeer J (2008) Biodiversity conservation in tropical agroecosystems: A new conservation paradigm. Ann N Y Acad Sci 1134: 173–200. doi:10.1196/annals.1439.011.
11. Perfecto I, Vandermeer J (2010) The agroecological matrix as alternative to the landsparing/agriculture intensification model. PNAS 107: 5786–5791. doi:10.1073/pnas.0905455107.
12. Bhagwat SA, Willis KJ, Birks HJB, Whittaker RJ (2008) Agroforestry: A refuge for tropical biodiversity? Trends Ecol Evol 23: 261–267. doi:10.1016/j.tree.2008.01.005.
13. Ranganathan J, Daniels RJR, Chandran MDS, Ehrlich PR, Daily GC (2008) Sustaining biodiversity in ancient tropical countryside PNAS 105: 17852–17854. doi: 10.1073_pnas.0808874105.
14. Davey CM, Vickery JA, Boatman ND, Chamberlain DE, Parry HR, et al. (2010) Assessing the impact of Entry Level Stewardship on lowland farmland birds in England. Ibis 152: 459–474. doi: 10.1111/j.1474-919X.2009.01001.x.
15. Edwards DP, Hodgson JA, Hamer KC, Mitchell SL, Ahmad AH, et al. (2010) Wildlife-friendly oil palm plantations fail to protect biodiversity effectively. Conserv Lett 3: 236–242. doi:10.1111/j.1755-263X.2010.00107.x.
16. Waltert M, Bobo KS, Kaupa S, Montoya ML, Nsanyi MS, et al. (2011) Assessing conservation values: Biodiversity and endemicity in tropical land use systems. PLoS ONE 6: e16238. doi:10.1371/journal.pone.0016238.
17. Tscharntke T, Clough Y, Wanger TC, Jackson L, Motzke I, et al. (2012) Global food security, biodiversity conservation and the future of agricultural intensification. Biol Cons 51: 53–59. doi:10.1016/j.biocon.2012.01.068.
18. Balmford A, Green RE, Scharlemann JPW (2005) Sparing land for nature: Exploring the potential impact of changes in agricultural yield on the area needed for crop production. Glob Change Biol 11: 1594–1605. doi:10.1111/j.1365-2486.2005.001035.x.
19. Tilman D, Balzer C, Hill J, Befort BL (2011) Global food demand and the sustainable intensification of agriculture. PNAS 108: 20260–20264. doi:10.1073/pnas.1116437108.
20. Godfray HCJ (2011) Food and biodiversity. Science 333: 1231–1232. doi:10.1126/science.1211815.
21. NEMA (2002) State of the environment report for Uganda. National Environmental Management Authority (NEMA): Kampala, Uganda.
22. Carswell M, Pomeroy DE, Reynolds J, Tushabe H (2005) Bird Atlas of Uganda. British Ornithologists' Club: Tring.
23. Brink AB, Eva HD (2009) Monitoring 25 years of land cover change dynamics in Africa: A sample based remote sensing approach. Appl Geogr 29: 501–512. doi: 10.1016/j.apgeog.2008.10.004.
24. Barnes RFW (2008) Deforestation trends in tropical Africa. Afr J Ecol 28: 161–173. doi: 10.1111/j.1365-2028.1990.tb01150.x.
25. Dranzoa C, Williams C, Pomeroy DE (2011) Birds of isolated small forests in Uganda. Scopus 31: 1–8.
26. Uganda Forest Department (1999) The Uganda Forest Biodiversity Database. Version 1.0. Uganda Forest Department: Kampala.
27. Bibby CJ, Burgess ND, Hill DA, Mustoe S (2000) Bird Census Techniques. Academic Press: London.
28. USDA (2011) Composition of Foods Raw, Processed, Prepared USDA National Nutrient Database for Standard Reference, Release 24 September 2011. Beltsville, Maryland. Available: http://www.ars.usda.gov/SP2UserFiles/Place/12354500/Data/SR24/sr24_doc.pdf.
29. Speedy AW, Seward L, Langton N, Du Pleiss J, Dlamini B (1991) A comparison of sugar cane juice and maize as energy sources in diets for growing pigs with equal supply of essential amino acids. Livestock Research for Rural Development 3: 1. Available: http://www.cipav.org.co/lrrd/lrrd3/1/speedy.htm.
30. Thomas L, Buckland ST, Rexstad EA, Laake JL, Strindberg S, et al. (2010) Distance software: design and analysis of distance sampling surveys for estimating population size. J Appl Ecol 47: 5–14. doi:10.1111/j.1365-2664.2009.01737.x.
31. Buckland ST, Anderson DR, Burnham KP, Laake JL, Borchers DL, et al. (2001) Introduction to Distance Sampling. Oxford Univ. Press: Oxford.
32. Urban EK, Keith S, Fry CH, Woodcock M (1993–2004) The Birds of Africa Vol I-VII. Princeton University Press: Princeton.
33. Snow DW, Perrins CM (1998) The Birds of the Western Palearctic. Oxford University Press: Oxford.
34. Phalan B, Balmford A, Green RE, Scharlemann JPW (2011) Minimising the harm to biodiversity of producing more food globally. Food Policy 36: S62. doi:10.1016/j.foodpol.2010.11.008.
35. Harris G, Pimm SL (2008) Range size and extinction risk in forest birds. Conserv Biol 22 : 163–171. doi:10.1111/j.1523-1739.2007.00798.x.
36. Fuller RJ, Gregory RD, Gibbons DW, Marchant JH, Wilson JD, et al. (1995) Population declines and range contractions among lowland farmland birds in Britain. Conserv Biol 9: 1425–1441. doi:10.1046/j.1523-1739.1995.09061425.x.
37. Wright HL, Lake IR, Dolman PM (2011) Agriculture – a key element for conservation in the developing world. Conserv Lett 5: 11–19. doi:10.1111/j.1755-263X.2011.00208.x.
38. Fischer J, Batáry P, Bawa KS, Brussaard L, Chappell MJ, et al. (2011) Conservation: limits of land sparing. Science 334: 593–595. doi:10.1126/science.334.6056.593-a.
39. Vandermeer J, Perfecto I, Philpott S (2010) Ecological complexity and pest control in organic coffee production: uncovering an autonomous ecosystem service. BioScience 60: 527–537. doi:http://dx.doi.org/10.1525/bio.2010.60.7.8.
40. Thies C, Haenke S, Scherber C, Bengtsson J, Bommarco R, et al. (2011) The relationship between agricultural intensification and biological control: experimental tests across Europe. Ecol Appl 21: 2187–2196. doi:http://dx.doi.org/10.1890/10-0929.1.

41. Phalan B, Onial M, Balmford A, Green RE (2011) Conservation–Response. Science 334: 594–595.
42. Angelsen A (2010) Policies for reduced deforestation and their impact on agricultural production. PNAS 107: 19639–19644. doi:10.1073/pnas.0912014107.
43. Ewers RM, Scharlemann JPW, Balmford A, Green RE (2009) Do increases in agricultural yield spare land for nature? Glob Change Biol 15: 1716–1726. doi:10.1111/j.1365-2486.2009.01849.x.
44. Rudel TK, Schneider L, Uriarte M, Turner II BL, DeFries R, et al. (2009) Agricultural intensification and changes in cultivated areas, 1970–2005. PNAS 106: 20675–20680. doi:10.1073/pnas.0812540106.
45. Kleijn D, Baquero RA, Clough Y, Dıaz M, De Esteban J, et al. (2006) Mixed biodiversity benefits of agri-environment schemes in five European countries. Ecol Lett 9: 243–254. doi:10.1111/j.1461-0248.2005.00869.x.
46. Rands MRW, Adams WM, Bennun L, Butchart SHM, Clements A, et al. (2010) Biodiversity conservation: Challenges beyond 2010. Science 329: 1298–1303. doi:10.1126/science.1189138J.
47. Balmford A, Phalan B, Green RE (2012) What conservationists need to know about farming. Proc R Soc Lond B Biol Sci 279: 2705–2713. doi:10.1098/rspb.2012.0515.

4

Fish Product Mislabelling: Failings of Traceability in the Production Chain and Implications for Illegal, Unreported and Unregulated (IUU) Fishing

Sarah J. Helyar[1¤*], Hywel ap D Lloyd[1¤], Mark de Bruyn[1], Jonathan Leake[2], Niall Bennett[3], Gary R. Carvalho[1]

1 Molecular Ecology and Fisheries Genetics Laboratory, Bangor University, Bangor, Wales, United Kingdom, **2** Sunday Times, London, United Kingdom, **3** Greenpeace UK, London, United Kingdom

Abstract

Increasing consumer demand for seafood, combined with concern over the health of our oceans, has led to many initiatives aimed at tackling destructive fishing practices and promoting the sustainability of fisheries. An important global threat to sustainable fisheries is Illegal, Unreported and Unregulated (IUU) fishing, and there is now an increased emphasis on the use of trade measures to prevent IUU-sourced fish and fish products from entering the international market. Initiatives encompass new legislation in the European Union requiring the inclusion of species names on catch labels throughout the distribution chain. Such certification measures do not, however, guarantee accuracy of species designation. Using two DNA-based methods to compare species descriptions with molecular ID, we examined 386 samples of white fish, or products labelled as primarily containing white fish, from major UK supermarket chains. Species specific real-time PCR probes were used for cod (*Gadus morhua*) and haddock (*Melanogrammus aeglefinus*) to provide a highly sensitive and species-specific test for the major species of white fish sold in the UK. Additionally, fish-specific primers were used to sequence the forensically validated barcoding gene, mitochondrial cytochrome oxidase I (COI). Overall levels of congruence between product label and genetic species identification were high, with 94.34% of samples correctly labelled, though a significant proportion in terms of potential volume, were mislabelled. Substitution was usually for a cheaper alternative and, in one case, extended to a tropical species. To our knowledge, this is the first published study encompassing a large-scale assessment of UK retailers, and if representative, indicates a potentially significant incidence of incorrect product designation.

Editor: Konstantinos I. Stergiou, Aristotle University of Thessaloniki, Greece

Funding: This study was jointly funded by Greenpeace and The Sunday Times. The funder Greenpeace provided support in the form of salary for author NB, but did not have any additional role in the study design, data collection and analysis, decision to publish, or preparation of the manuscript. The funder The Sunday Times provided support in the form of salary for author JL, but did not have any additional role in the study design, data collection and analysis, decision to publish, or preparation of the manuscript. The specific roles of these authors are articulated in the 'author contributions' section.

Competing Interests: The authors have the following interests: This study was jointly funded by Greenpeace and The Sunday Times. Co-author Jonathan Leake is employed by The Sunday Times. Co-author Niall Bennett is employed by Greenpeace. There are no patents, products in development or marketed products t o declare.

* E-mail: sarah.helyar@matis.is

¤ Current address: Food Safety, Environment & Genetics, Matís, Reykjavík, Iceland

Introduction

In recent years, concerns about the health of the oceans and the effects of over-exploitation of fisheries have increased. Consumer demand for seafood is growing with the contribution of fish to the average annual diet reaching a record of 18.8 kg per person per year in 2011 [1], as compared to 17.1 Kg in 2008 [2]. This is partly due to an increase in the range of species consumed, and an increase in aquaculture. Fish products were worth a record $217.5 billion in 2010, up over 9% from 2009, and these trends are expected to continue. The increasing demand for fish highlights the need for the sustainable management of aquatic resources; 87.3% of world fish stocks are classed as overexploited, depleted or recovering: a number which continues to increase [1], with 29.9% of stocks classed as overexploited and unlikely to meet the targets of the Johannesburg Plan of Implementation to restore them to a level that can produce maximum sustainable yield by 2015 [3].

A major threat for the sustainable management of these valuable resources is Illegal, Unreported and Unregulated (IUU) fishing. Current estimates suggest that globally up to 25% of fisheries catches fall within IUU practices [4–6], identifying it as the single largest threat to achieving sustainability. Both the FAO [7] and the European Union [8] have placed increasing emphasis on the use of trade measures to prevent IUU-sourced fish and fish products from entering international trade. One component of this increased regulation has required the inclusion of binomial species nomenclature on catch labels throughout the distribution chain [9].

In addition to top down pressure for improved labelling and traceability of fish products, many consumers are increasingly aware of nutritional and environmental issues regarding fisheries, leading to shifts in attitude regarding acceptable species, catch location and catch methods [10]. In parallel, due to globalization of the industry, consumers are encountering an increasing number of fish species and/or an escalation in common names applied to the same species. Such drivers have led to a greater demand for informative labelling, including the use of 'eco-labelling'. Although labelling to provide additional ecological information about a product is often voluntary, the FAO recognised that it could contribute to improved fisheries management and convened a Technical Consultation in 1998, which resulted in their Guidelines for the Eco-labelling of Fish and Fishery Products from Marine Capture Fisheries [11]. Informative labelling is particularly important for processed items because any recognizable external morphological features are typically removed, leaving consumers reliant on product labelling for content information. However, it has been argued that any such labelling scheme, whether voluntary or legislated, requires policing in order to prevent misuse and fraud [12].The mislabelling of a fish product may be unintentional if, for example, species that are morphologically similar are caught together, such as in many tropical or coral reef fisheries [13–17]. Alternatively, mislabelling may not be accidental, such as where product substitutions are from species that do not occur in the same ocean [18–20], or for lesser value species [21,22]. However, whether intentional or not, the outcome can be serious for management and sustainability targets. In addition to the direct impacts of depletion from IUU fishing, substitutions and misidentification that occur before fish are landed will inflate the inaccuracies in catch and forecast statistics.

Several recent studies of mislabelling have been undertaken in Europe [19,22,23], yielding rates of mislabelling of up to 32% [19]. Most mislabelled products have originated from small-scale retailers and convenience food outlets (e.g. fish and chip shops) but the major supermarkets have not hitherto been thoroughly investigated. Supermarket chains account for 72% of the total fish retail market in the UK (excluding canned products) [24]. If comparable rates of mislabelling occur in supermarket products it is thereby likely to have a substantial impact on efforts to manage the respective fisheries sustainably. It is therefore necessary to establish to what extent mislabelling of fish products occurs in the major retailers of the fish food supply chain, which is addressed in this study.

The current study uses two DNA-based methods to identify the species of origin for 386 samples collected from major supermarket chains around the UK. Species-specific real-time PCR probes [25] for cod (*Gadus morhua*), and haddock (*Melanogrammus aeglefinus*) were used to provide a highly sensitive test for the major species of white fish sold in British supermarkets. Additionally, DNA barcoding [26] using fish-specific COI primers [27] was employed. The COI mitochondrial gene has been validated for forensic species identification [28] to determine its reproducibility and limitations by testing its ability to provide accurate results under a variety of conditions. To our knowledge, the current findings represent the first large-scale assessment of fish product authentication across major UK supermarket retailers.

Materials and Methods

Sample collection

386 samples of processed white fish, ranging from fillets to fish fingers and fish cakes, were collected from six leading supermarket chains, at multiple locations across England, Scotland and Wales (Table S1in File S1). Approximately 20 mg of tissue was taken from the centre of each product to ensure minimal DNA damage from production, processing, or contamination. These were placed into numbered tubes filled with 96% ethanol. Sample details including the place and date of purchase, species designation, and eco-labelling were entered into a database linked to photographs of the packaging. Sample identities were not disclosed until completion of molecular genetic analyses, when molecular and sample IDs were cross-referenced.

Molecular methods

DNA was extracted with the E-Z 96 Tissue DNA kit (Omega-biotek), then quantified with a Nanodrop 1000 (Thermo Scientific), and standardised to either 5 ng/µL or 2 ng/µL depending on original concentration. Real-time PCR assays were carried out on all samples on an Applied Biosystems 7700 real-time sequence detection system. The 25 µL reactions contained 200 nM of each of the two species specific probes (see Table 1), 300 nM of the GAD-F and GAD-R primers (Taylor *et al.* 2002), 9.163 µl 2X Taqman Universal PCR Master Mix (UNG+ROX and passive reference) (Applied Biosystems), 15 ng of DNA, and (depending on DNA concentration) either 10.417 or 6.917 µL PCR grade H_2O (Sigma). Reactions were run in optical 96-well reaction plates using optical adhesive covers (Applied Biosystems). Plates were analysed under real-time conditions on two dye layers with eight 'no template controls' (NTCs) per 96-well plate, and 2 positive controls for each of the two target species. The assay was run using the default cycling conditions [25].

In addition to the real-time PCR, all samples were sequenced for approximately 655 bp from the 5′ region of the COI gene from mitochondrial DNA using primers developed by Ward [27]. Tests were run with all combinations of the four available primers, but the combination of FishF1/FishR2 produced consistently good PCR products in the species tested, and was therefore used throughout (see Table 1). PCRs were carried out in 30 µL reactions containing 15 µL of 2 x PCR Mastermix (containing 0.75 U of *Taq* polymerase (buffered at pH 8.5), 400 µM each dNTP, 3 mM $MgCl_2$ (Promega)), 9 µL PCR grade H_2O (Sigma), 15 pmol each primer, and 3.0 µL of DNA template. The PCRs consisted of a denaturation step of 2 min at 95°C followed by 35 cycles of 30 seconds at 94°C, 30 seconds at 54°C, and 1 min at 72°C, followed by a final extension of 10 min at 72°C and then held at 4°C. PCR products were visualized on 1.2% agarose gels. If a single clear band was produced, PCR products were sent to GATC (Germany, http://www.gatc-biotech.com) for sequencing. DNA from 48 samples was re-extracted as independent replicates of real-time PCR and sequencing, including all samples where molecular data contradicted species designations, and an additional randomly chosen 33 samples to test repeatability of DNA-based species ID.

Species identification

Real-time PCR. The results were analysed using the Sequence Detection Software version 1.71 (Applied Biosystems). The ΔRn values for each cycle and dye layer were then exported to MS Excel and additional manual processing was carried out. First, the mean and standard deviation of the endpoint (PCR cycle 40) ΔRn values of the NTCs were calculated for each dye layer. z*M-values (z*M = M+(3.89xSD)+C) were then calculated where M = mean of the NTC ΔRn, SD is the standard deviation of the NTC ΔRn and 3.89 is the one tailed *Z*-value for the 99.999% confidence interval, C is a constant (0.3) introduced to overcome the slight increase in fluorescence of samples above the NTC fluorescence due to spectral bleeding between dye layers. Samples which had ΔRn values larger than the value of z*M were considered to have a fluorescence significantly greater than the NTCs, and therefore to be positive reactions.

Table 1. list of all primers used.

	Sequence 5′-3′	Reporter	Quencher
COD P	CTTTTTACCTCTAAATGTGGGAGG	-	-
HAD P	CTTTCTTCCTTTAAACGTTGGAGG	-	-
GAD-F	GCAATCGAGTYGTATCYCTWCAAGGAT	FAM	Non-fluorescent
GAD-R	CACAAATGRGCYCCTCTWCTTGC	TET	Non-fluorescent
FishF1	TCAACCAACCACAAAGACATTGGCAC	-	-
FishR2	ACTTCAGGGTGACCGAAGAATCAGAA	-	-

COD P, HAD P, GAD-F, and GAD-R were used in the real-time-PCR, and FishF1 and FISHR2 were used for the sequencing PCRs.

COI sequencing. Successfully sequenced COI amplicons were manually checked and edited to remove ambiguous base calling in BioEdit (Ibis Biosciences). Sequences were tested against the Barcode of Life database (BOLD) [29]. In addition, reference sequences for all species genetically identified and all species indicated on sample packaging, were downloaded from BOLD and aligned with the sample sequences in Clustal X [30], the Neighbour-joining tree was constructed in MEGA5 [31] with 1000 bootstrap replicates.

Results

For consistency, all samples are referred to by the labelled species unless otherwise stated. Of 386 samples, 371 (97.4%) produced DNA of sufficient quality for further analysis. Label designations indicated primarily cod (179), haddock (155) and pollock (32).

Real-time-PCR. All samples labelled as hake or Alaskan pollack showed negative results for probes designed to identify cod and haddock. The sample labelled as whiting was positive for cod. For the samples labelled as haddock (155), the haddock probe was positive in 134 samples (86.5%), while the cod probe gave a positive result for 6 samples, both probes were amplified in 7 samples (inconclusive result) and neither were amplified in 8 samples (negative). All cod labelled as originating from the Pacific were negative for both the cod and haddock probes. Out of the Atlantic cod samples (57), the cod specific probe amplified in 47 samples (82.5%), both probes were positive in 3 samples (inconclusive result) and neither in 7 samples (negative). For the cod samples which did not indicate a catch location (102), the cod specific probe was positive in 80 samples, the haddock specific probe was positive in 2 samples, both amplified in 8 samples (inconclusive result) and neither in 12 samples (negative). Real-time-PCR results are presented in Table 2.

COI sequencing. All sequence data has been submitted to NCBI, under accession numbers KJ614671 to KJ615069 (Table S2 in File S1). 48 samples have two sequences listed as these samples were re-extracted as independent replicates to ensure the repeatability of the methods.

The majority of sequences were identified with a sequence identity greater than 99.5% in the BOLD database, with sequences from two samples falling below this threshold. Additionally, two samples could not be matched unambiguously due to 100% sequence identity at COI at the taxon-pairs involved. The sequence data matched with *Gadus chalcogramma/G. finnmarchica* (Alaskan and Norwegian Pollock respectively; previously *Theragra* sp.), or *Gadus macrocephalus* and *Gadus ogac* (Pacific and Greenland cod respectively): these are both instances where the (sub-) species designation is debatable (see Discussion).

Of 179 samples labelled as cod, 57 were specified as Atlantic cod (*Gadus morhua*) and 20 as Pacific cod (*Gadus macrocephalus*), while for the remainder (102) there was no specification for either the species or catch area. In total, 9 (5.03%) of these cod samples were not verified as cod by DNA data, including 1 (0.56%) identified as *Melanogrammus aeglefinus* (haddock), and 2 (1.11%) highly processed samples that were found to have a mixed species composition (see Table 2: #1892; *G. morhua/G. chalcogramma* and #1886; *G. morhua/M. aeglefinus*). From the 57 samples labelled specifically as Atlantic cod, 51 had congruent label and DNA-based designations, while 6 (10.5%) were genetically identified as Pacific cod (*G. macrocephalus*).

155 samples were labelled as haddock (*M. aeglefinus*). Of these, 146 generated a molecular ID in agreement with labelling (5.81% mislabelled), with 6 (3.87%) identified as *G. morhua* (Atlantic cod), 1 (0.65%) as *G. macrocephalus* (Pacific cod) and 2 (1.29%) exhibited a mixed species composition (see Table 2: #1452 and #1847; *G. morhua/M. aeglefinus*).

In addition, one of the four hake (labelled as *Merluccius capensis*) samples was identified as *Merluccius paradoxus* (cape hake), one whiting (*Merlangius merlangus*) sample was identified as *Micromesistius poutassou* (blue whiting), and one Alaskan Pollack was also found to contain the Vietnamese catfish *Pangasius hypophthalmus*. Overall, our survey indicated a rate of mislabelling of 5.66%. All samples and results are presented in Tables S1 and S2 in File S1, with detailed results of the mislabelled samples in Table 2 and Table S3 in File S1. Sequence similarity with all reference samples is demonstrated in Figure 1, and the details of the reference sequences used are in Table S4 in File S1.

Discussion

Our study represents, to our knowledge, the largest published survey to date of mislabelling within the fish products sold by UK supermarkets. Samples were taken of products from leading brands and supermarket "own brands" from 6 major supermarket chains across the UK. Previous studies have examined the food retail sector and found high rates of mislabelling, particularly in restaurants and fast-food outlets [22,32]. Within our study of supermarket-sourced samples the overall inconsistency between product label and genetic species identification was 5.66%. This is considerably lower than observed in other sectors: 25% within mixed sectors [22]; 25% within markets and restaurants [32]; 32% within fishmongers [19]. Nevertheless, if our data are representative of overall trends, with over 4 billion fish products consumed (C. Roberts, unpublished data) the incidence of mislabelling could exceed 200 million products annually in the UK alone. This level of misinformation raises considerable concern in terms of consumer information and protection. It also presents substantial

Table 2. Summary of all mislabelled samples.

Identification Code	Species reported (type)	Area of Catch	real-time PCR	First sequence identity	Second sequence identity
1415	Cod (breaded fillet)	Atlantic	Negative	*Gadus macrocephalus*	*Gadus macrocephalus*
1426	Cod (breaded fillet)	Atlantic	Negative	*Gadus macrocephalus*	*Gadus macrocephalus*
1446	Cod (breaded fillet)	Atlantic	Negative	*Gadus macrocephalus*	*Gadus macrocephalus*
1747	Cod (precooked meal)	Atlantic	Negative	*Gadus macrocephalus*	*Gadus macrocephalus*
1889	Cod (precooked meal)	Atlantic	Negative	*Gadus macrocephalus*	*Gadus macrocephalus*
1975	Cod (breaded fillet)	Atlantic	Negative	*Gadus macrocephalus*	*Gadus macrocephalus*
1886	Cod (fish cakes)	NA	Inconclusive	*Gadus morhua*	*Melanogrammus aeglefinus*
1765	Cod (fish cakes)	NA	*Melanogrammus aeglefinus*	*Melanogrammus aeglefinus*	*Melanogrammus aeglefinus*
1892	Cod (fish fingers)	NA	*Gadus morhua*	*Gadus chalcogrammus*	*Gadus morhua*
1470	Haddock (precooked meal)	Atlantic	*Gadus morhua*	*Gadus morhua*	*Gadus morhua*
1812	Haddock (fish cakes)	Atlantic	*Gadus morhua*	*Gadus morhua*	*Gadus morhua*
1888	Haddock (precooked meal)	Atlantic	*Gadus morhua*	*Gadus morhua*	*Gadus morhua*
1977	Haddock (breaded fillet)	Atlantic	*Gadus morhua*	*Gadus morhua*	*Gadus morhua*
1989	Haddock (precooked meal)	Atlantic	*Gadus morhua*	*Gadus morhua*	*Gadus morhua*
1868	Haddock (precooked meal)	Atlantic	*Gadus morhua*	*Gadus morhua*	*Gadus morhua*
1851	Haddock (precooked meal)	Atlantic	Negative	*Gadus macrocephalus*	*Gadus macrocephalus*
1452	Haddock (fish cakes)	Atlantic	Inconclusive	*Gadus morhua*	*Melanogrammus aeglefinus*
1847	Haddock (fish cakes)	Atlantic	Inconclusive	*Gadus morhua*	*Melanogrammus aeglefinus*
1763	Alaskan Pollack (fish cakes)	Pacific	Negative	*Pangasius hypophthalamus*	*Gadus chalcogrammus*
1813	Hake *(M. capensis)* (breaded fillet)	NA	Negative	*Merluccius paradoxus*	*Merluccius paradoxus*
1848	Whiting (precooked meal)	NA	Inconclusive	*Micromesistius poutassou*	*Micromesistius poutassou*

NA: Not available from packaging. Negative: neither of the real-time PCR probes amplified. Inconclusive: both real-time PCR probes amplified. First and second sequence identities are the result of independent DNA extractions and sequencing (see methods for details).

challenges for the sustainable management of the respective fisheries.

Genetic identification of products was carried out with species specific real-time PCR, and by matching sample COI sequences with those of known species in the BOLD database with high (≥ 99.5%) sequence identity [33]. Such independent testing yields a high degree of certainty to the identifications, as more than 98% of species pairs have shown greater than 2% COI sequence divergence [34]. The BOLD database was used in preference to the nucleotide sequence database in GenBank (www.ncbi.nlm.nih.gov/), to ensure that the queried sequences were matched to taxonomically-validated specimens. Of all the sequences submitted, only two returned a match with less than 99.5% identity. Both of these were from highly processed samples (one labelled as cod, the other as haddock), and also returned inconclusive results for the real-time-PCR (both cod and haddock probes amplified). Both sequences were genetically identified as *M. aeglefinus* (haddock), although with relatively low sequence similarity (99.49% and 98.6%). For both of these sequences, the next closest match was *G. morhua*, rather than the next closest relative of haddock, *Merlangius merlangus* (see Figure 1), supporting the conclusion that the DNA amplified was a mix of more than one species, and therefore that these products had a mixed species composition.

Ambiguous results occurred when a sample matched with both Alaskan and Norwegian pollock (*Gadus (= Theragra) chalcogramma* and *G. finnmarchica*, respectively), or with Pacific and Greenland cod (*Gadus macrocephalus* and *G. ogac*, respectively), because congeners have 100% sequence identity at COI. However, in the case of Pacific and Greenland cod, catches of *G. ogac* are thought to be extremely low, and currently only of local importance. The total reported catch for this stock from 2009–2011 was 586 metric tons [35], while for the same three years, the total reported catch for *G. macrocephalus* was 1,165,420 metric tons. Greenland cod is also no longer considered a separate species, but is now classed as a subspecies of Pacific cod, *G. macrocephalus* [36,37]. In the case of the Pollack species, *G. finnmarchica* was identified from a few samples from the northern tip of Norway [38] and recent molecular evidence has shown it to be indistinct from the Alaskan Pollock (*G. chalcogramma*) [39–41].

From all samples labelled as Atlantic cod, the majority of those found to be mislabelled were genetically identified as Pacific cod. This category of mislabelling could not originate at the pre-landing stage; as is evident from their common names; these species are harvested from different oceans. The implication, therefore, is that intentional mislabelling has occurred at a later stage in the supply chain. The incentive could be to supply products that mirror the preferences of the buying public, and so presumably fetch a higher price. This class of mislabelling may have little direct impact on the Atlantic cod stocks but it may influence efforts to sustainably manage stocks of Pacific cod. More importantly perhaps for this particular case of mislabelling is the issue of consumer misinformation and protection as it indicates that at some point in the supply chain there appears to be either negligence or a wilfully fraudulent attempt to provide inaccurate product information. Such instances erode consumer confidence and can undermine trust in product labelling, including any associated eco-labels.

Samples labelled as *M. aeglefinus* (haddock) show a different pattern of mislabelling. The majority of mislabelled products were

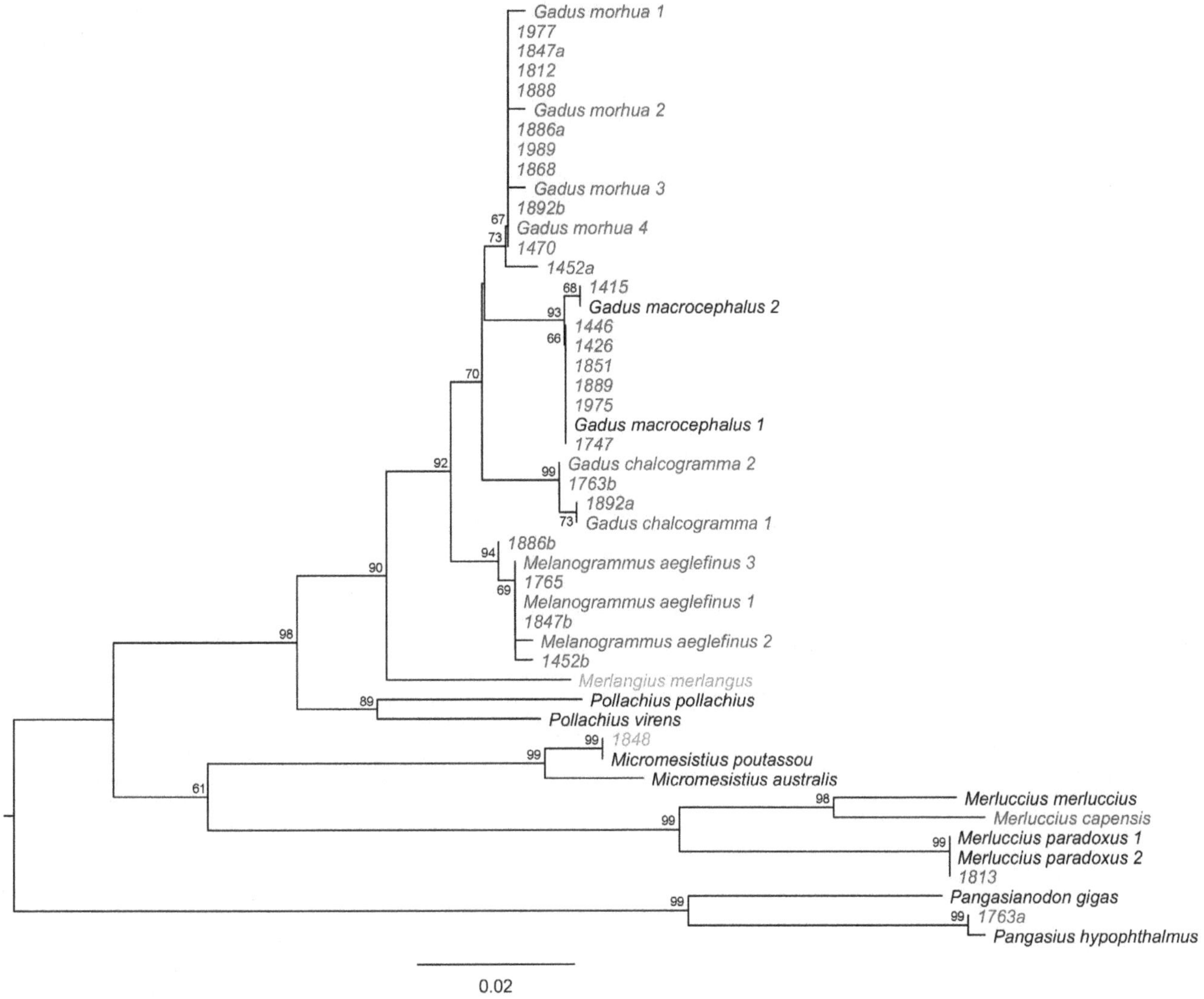

Figure 1. Neighbour-joining tree showing all mislabelled samples together with representative reference sequences taken from BOLD. Reference sequences are colour coded according to species and samples tested are colour coded according to the species stated on the packaging. Samples that have two sequences are labelled a and b.

identified as *G. morhua* (Atlantic cod). Haddock and Atlantic cod are frequently caught together in a mixed fishery and have similar market values, with cod slightly more valuable on average. As a result there is minimal direct benefit to intermediaries in the production chain to encourage such mislabelling. Alternatively, it has been suggested that such mislabelling may arise by an accidental consequence of the mixed fishery [23]. However, while we accept such possibility, mislabelling undeniably benefits the primary producer. Mislabelling *G. morhua* (Atlantic cod) as *M. aeglefinus* (haddock) enables fishermen to land undersized or over quota Atlantic cod and so profit from fish that should currently be discarded. Irrespective of the underlying cause, if the mislabelling occurs before the fish are landed (for example, if filleted and frozen at sea), such IUU activities will likely exceed catch quotas (TAQ) for a major North Atlantic fishery. The rate of mislabelling (3.87%) is comparatively low compared to other recent studies [22]. However, if we extrapolate such incidence to the TAQ for 2011, it represents an additional 2188 tonnes of Atlantic cod (or an excess 2.9% of the Atlantic cod TAQ for 2011) being landed and recorded as haddock.

In addition to the mislabelling of cod and haddock presented here, other mislabelling instances were found. One highly processed (fish cake) sample labelled as containing Alaskan Pollack (*G. chalcogramma*) was found to also contain *Pangasius hypophthalmus*. *P. hypophthalmus*, or Vietnamese catfish, is a freshwater species from Southeast Asia, legally described in the UK as Basa, Panga(s), *Pangasius*, River cobbler or any of these combined with 'catfish' [42]. Without performing a quantitative test for the presence of *P. hypophthalmus*, we were unable to estimate the relative quantities of the 2 species in this product (made of minced fish). It was therefore not possible to determine whether this reflected inadvertent contamination through inadequate cleaning of the production line between products, or deliberate substitution of a cheaper product. In either case it is unlikely to significantly affect catch data or to contribute to IUU. However, this accidental or fraudulent behaviour is a serious issue for consumer misinformation and trust, given the concerns over potentially increased contaminant levels in *Pangasius* species (such as mercury) [43], which may result in avoidance by some consumer groups.

Four hake samples were tested and one, labelled as *M. capensis*, was identified as *M. paradoxus* (25% mislabelled, although the low sample size requires caution). Historically, hake has been assessed as a single species, as separation of catches has not always been possible [44,45]. However species-specific assessments are now being conducted. The shallow water *M. capensis* stock is above sustainable levels, with catches below maximum sustainable levels and is certified by the Marine Stewardship Council (MSC). The deep-water *M. paradoxus* stock is below precautionary levels, and a rebuilding plan is in place [46]. The mislabelling of this species, whether intentional or not, at a rate even well below that observed here is a cause for serious concern, as such a practice would compromise restoration of *M. paradoxus* to sustainable levels.

One noteworthy pattern to emerge is the variation in amount of mislabelling found among the different levels of processing: within the fresh/frozen fillets (n = 84) no mislabelling was identified; in battered/breaded fillets (n = 84), fish fingers (n = 31), and pre-cooked meals (n = 128), the respective mislabelling rates were 7.14%, 6.45% and 5.47% respectively. In fishcakes (n = 44), which are composed of minced fish, mislabelling rates of 13.6% were identified. However, these data are insufficient to identify where in this production chain, pre- or post-landing, yield higher rates of illegal activity. Targeted sampling at discrete stages across the supply chain is required: from on-board during sample catch to final retailer outlet. Alternatively it may be an inadvertent consequence of the particular processing activity, such as inadequate cleaning of processing machinery. Huxley-Jones et al. [23] found lower levels of mislabelling in processed products, such as fish fingers, than filleted products, and suggested that this may be due to greater economic gains associated with the mislabelling of fillets. In contrast, our study included more diverse forms of processing (from fresh fillets through fish fingers and precooked meals to fish cakes consisting of minced fish), and has demonstrated a clear pattern of mislabelling, from zero in unprocessed fish fillets to the highest levels of mislabelling in the most highly-processed category.

The main trends highlighted here have been the substitution of *G. morhua* (Atlantic cod) with *G. macrocephalus* (Pacific cod) in primarily filleted products, and the substitution of *M. aeglefinus* (haddock) with *G. morhua* (Atlantic cod) in precooked meals and fish cakes. Both aspects of mislabelling have a detrimental effect on *G. morhua*: substitution with *G. microcephalus* creates an erroneous impression of the abundance of the former, undermining work carried out by seafood awareness campaigns such as Seafood Watch and the Marine Stewardship Council, to educate consumers and provide tools for informed purchasing decisions. However, cod is one of the species for which there are now sufficient genomic resources to move beyond species identification and allow traceability to population level [47]. Testing by regulatory and certifying bodies would improve consumer confidence in products that are proven to fulfil claims of having been sourced from sustainably harvested stocks. In addition, as suggested here, if the substitution of *M. aeglefinus* with *G. morhua* is occurring at sea, the implications of such IUU activity would compromise the recovery of these heavily exploited species.

Previous studies have reported relatively high rates of mislabelling of seafood products globally [13,48], in Europe [19,22,49], and South Africa [21,50]. However, many studies have focused on smaller convenience food outlets and/or restaurants. Actions such as increasing media attention, the importance of consumer confidence in the fisheries sector and revised EU legislation [51,52] will collectively highlight and tackle mislabelling practices. Nevertheless, only genetic testing across the supply chain can assess the scale and likely key stages of highest risk. It also appears increasingly likely that such practices are more frequent at the more highly processed end of the market, where opportunities for detection and/or levels of discrimination are reduced. As witnessed recently in the wake of the horsemeat scandal across Europe, the complexities of the modern food production chain demand close scrutiny at all stages to ensure authenticity and compliance. A forensic framework of genetic testing using validated reference databases [47,53–55] is expected to provide an increasingly effective approach for detection, prosecution and ultimate deterrence of illegal activity. Such actions are likely not only to protect policy compliant end-users and the wider fishing industry but importantly also enhance prospects for achieving sustainability of exploited marine resources.

Supporting Information

File S1 Supporting Information. Table S1. Summary of sampling effort for samples which produced a DNA of sufficient quality for testing. Samples are recorded by reported species and are split by supermarket (own brand/other brand items). **Table S2.** Genetic analyses for all samples. The sample identification number, product labelling (reported species and catch location), processing level* and results from the real time PCR and COI sequencing are given for each sample. Second sequence identity is the result of new DNA extraction and sequencing for mislabelled, ambiguous, and control samples. Genbank accession numbers are provided for each sequence. * classification of processing level (1: fresh or frozen fillets, 2: battered or breaded fillets, 3: fish fingers, 4: pre-cooked meals, 5: fishcakes). **Table S3.** Additional data for the mislabelled samples. Query sample details, including species labelled on packaging, COI sequence and the Genbank accession number are given. Reference species is the closest sequence match in the BOLD database, together with the catch location, BOLD ID number and Genbank accession number of the reference sample. Sequence similarity (% identity) between sample and reference is shown (* indicates sequence matches lower than 99.5%). Sequence similarity to the next closest match is also shown. **Table S4.** BOLD and Genbank identifiers for the reference sequences used in Figure 1.

Acknowledgments

Help with sampling was provided by Amy Sherborne, Willie McGee, and Charlotte Hunt-Grubbe. Martin Taylor, Delphine Lallias and Wendy Grail all provided technical advice. We would also like to thank the two reviewers for their insightful comments, which have greatly improved the clarity.

Author Contributions

Conceived and designed the experiments: SJH GRC JL. Performed the experiments: HapDL MdB. Analyzed the data: SJH HapDL. Wrote the paper: SJH HapDL GRC MdB JL NB.

References

1. FAO (2012) The State of the World's Fisheries and Aquaculture. Rome: Food and Agriculture Organization of the United Nations.
2. FAO (2011) The State of the World's Fisheries and Aquaculture. Rome: Food and Agriculture Organization of the United Nations.

3. WSSD (2002) United Nations Plan of Implementation of the World Summit on Sustainable Development A/CONF.199/20.
4. Pauly D, Christensen V, Guénette S, Pitcher TJ, Sumaila UR, et al. (2002) Towards sustainability in world fisheries. Nature 418: 689–695.
5. MRAG (2008) the Global Extent of Illegal Fishing. London: MRAG. Available: http://www.mrag.co.uk/Documents/ExtentGlobalIllegalFishing.pdf .Accessed: 20 September 2013.
6. Agnew DJ, Pearce J, Pramod G, Peatman T, Watson R, et al. (2009) Estimating the Worldwide Extent of Illegal Fishing. PLoS ONE 4(2): e4570.
7. FAO. (2001) International Plan of Action to prevent, deter and eliminate illegal, unreported and unregulated fishing. Rome, FAO. 24p. Available: http://www.fao.org/fishery/ipoa-iuu/en. Accessed: April 2012.
8. European Union (2008) Regulation (EC) No 1005/2008.
9. European Union (2009) Regulation (EC) No 1224/2009.
10. Potts T, Brennan R, Pita C, Lowrie G (2011) Sustainable Seafood and Eco-labelling: The Marine Stewardship Council, UK Consumers, and Fishing Industry Perspectives. SAMS Report: 270–211 Scottish Association for Marine Science, Oban. 78 pp. ISBN: 0-9529089-2-1.
11. FAO (2009) Guidelines for the ecolabelling of fish and fishery products from marine capture fisheries. Revision 1. Rome Available: http://www.fao.org/docrep/012/i1119t/i1119t00.htm. Accessed: April 2012.
12. Stokstad E (2010) To fight illegal fishing, forensic DNA gets local. Science 330: 1468–1469.
13. Marko PB, Lee SC, Rice AM, Gramling JM, Fitzhenry TM, et al. (2004). Mislabelling of a depleted reef fish. Nature 430: 309–310.
14. Ardura A, Pola IG, Ginuino I, Gomes V, Garcia-Vazquez E (2010) Application of barcoding to Amazonian commercial fish labelling. Food Research International 43: 1549–1552.
15. Crego-Prieto V, Campo D, Perez J, Garcia-Vazquez E (2010) Mislabelling in megrims: implications for conservation. In:Tools for Identifying Biodiversity Progress and Problems (Proceedings of the International Congress, Paris, eds. EUT Edizioni Universita di Trieste, Trieste, pp. 315–322.
16. Iglesias SP, Toulhoat L, Sellos DY (2010) Taxonomic confusion and market mislabelling of threatened skates: important consequences for their conservation status. Aquatic Conservation 20(3):319–333.
17. Gold JR, Voelker G, Renshaw MA (2011) Phylogenetic relationships of tropical western Atlantic snappers in subfamily Lutjaninae (Lutjanidae: Perciformes) inferred from mitochondrial DNA sequences. Biological Journal of the Linnaean Society 102(4):915–929.
18. Barbuto M, Galimberti A, Ferri E, Labra M, Malandra R, et al. (2010) DNA barcoding reveals fraudulent substitutions in shark seafood products: the Italian case of "palombo" (*Mustelus* spp.). Food Research International 43: 376–381.
19. Filonzi L, Chiesa S, Vaghi M, Marzano FN (2010) Molecular barcoding reveals mislabelling of commercial fish products in Italy. Food Research International 43: 1383–1388.
20. Armani A, Castigliego L, Tinacci L, Gianfaldoni D, Guidi A (2011) Molecular characterization of icefish, (*Salangidae* family), using direct sequencing of mitochondrial *cytochrome b* gene. Food Control 22: 888–895.
21. von der Heyden S, Barendse J, Seebregts AJ, Matthee CA (2010) Misleading the masses: detection of mislabelled and substituted frozen fish products in South Africa. ICES Journal of Marine Science 67: 176–185.
22. Miller DM, Mariani S (2010) Smoke, mirrors and mislabelled cod: poor transparency in the European seafood industry. Frontiers in Ecology and the Environment 8: 517–521.
23. Huxley-Jones E, Shaw JLA, Fletcher C, Parnell J, Watts PC (2012) Use of DNA Barcoding to Reveal Species Composition of Convenience Seafood. Conservation Biology 26: 367–371.
24. FAO (2004) Fishery and Aquaculture Country Profile UK. Available: ftp://ftp.fao.org/FI/DOCUMENT/fcp/en/FI_CP_UK.pdf. Accessed: April 2012.
25. Taylor MI, Fox C, Rico I, Rico C (2002) Species-specific TaqMan probes for simultaneous identification of (*Gadus morhua* L.), haddock (*Melanogrammus aeglefinus* L.) and whiting (*Merlangius merlangus* L.). Molecular Ecology Notes 2(4): 599–601.
26. Hebert PDN, Ratnasingham S, DeWaard JR (2003) Barcoding animal life: cytochrome c oxidase subunit 1 divergences among closely related species. Proceedings of the Royal Society B: Biological Sciences 270: S96–9.
27. Ward RD, Zemlak TS, Innes BH, Last PR, Hebert PDN (2005) DNA barcoding Australia's fish species. Philosophical Transactions of the Royal Society B 360: 1847–57.
28. Dawnay N, Ogden R, McEwing R, Carvalho GR, Thorpe RS (2007) Validation of the barcoding gene COI for use in forensic genetic species identification. Forensic Science International 173: 1–6.
29. Ratnasingham S, Hebert PDN (2007) BOLD: The Barcode of Life Data System (www.barcodinglife.org) Molecular Ecology Notes 7(3): 355–364.
30. Higgins DG, Sharp PM (1988). CLUSTAL: a package for performing multiple sequence alignment on a microcomputer. Gene 73: 237–244.
31. Tamura K, Peterson D, Peterson N, Stecher G, Nei M, Kumar S (2011) MEGA5: Molecular Evolutionary Genetics Analysis using Maximum Likelihood, Evolutionary Distance, and Maximum Parsimony Methods. Molecular Biology and Evolution 28: 2731–2739.
32. Wong EHK, Hanner RH (2008). DNA barcoding detects market substitution in North American seafood. Food Research International 41(8): 828–837.
33. Ward RD, Holmes BH (2007) An analysis of nucleotide and amino acid variability in the barcode region of cytochrome c oxidase I (cox1) in fishes. Molecular Ecology Notes 7(6): 899–907.
34. Hebert PDN, Ratnasingham S, de Waard JR (2003). Barcoding animal life: cytochrome c oxidase subunit 1 divergences among closely related species. Proceedings of the Royal Society London Series B 270: S96–99.
35. FAO (2012) Species Fact Sheet. Available: http://www.fao.org/fishery/species/2219/en. Accessed: 12 April 2012.
36. Carr SM, Kivlichan DS, Pepin P, Crutcher DC (1999) Molecular systematics of gadid fishes: Implications for the biogeographic origins of Pacific species. Canadian Journal of Zoology 77: 19–26.
37. Coulson MW, Marshall HD, Pepin P, Carr SM (2006) Mitochondrial genomics of gadine fishes: implications for taxonomy and biogeographic origins from whole-genome data sets. Genome 49 (9): 1115–1130.
38. FAO (1990) FAO species catalogue. Vol.10. Gadiform fishes of the world (Order Gadiformes). An annotated and illustrated catalogue of cods, hakes, grenadiers and other gadiform fishes known to date. FAO Fisheries Synopsis S125 Vol.10.
39. Ursvik A, Breines R, Christiansen JS, Fevolden SE, Coucheron DH, et al. (2007) A mitogenomic approach to the taxonomy of pollocks: *Theragra chalcogramma* and *T. finnmarchica* represent one single species. BMC Evolutionary Biology 7 1): 86.
40. Byrkjedal I, Rees DJ, Christiansen JS, Fevolden SE (2008) The taxonomic status of *Theragra finnmarchica* Koefoed, 1956 (Teleostei: Gadidae): perspectives from morphological and molecular data. Journal of Fish Biology 73 5): 1183–1200.
41. Carr SM, Marshall HD (2008) Phylogeographic analysis of complete mtDNA genomes from walleye pollock (*Gadus chalcogrammus* Pallas, 1811) shows an ancient origin of genetic biodiversity. Mitochondrial DNA 19 6): 490–496.
42. Fish Labelling (England) Regulations (2010) Statutory instrument No. 420 FOOD, ENGLAND.
43. Ferrantelli V, Giangrosso G, Cicero A, Naccari C, Macaluso A, et al. (2012) Evaluation of mercury levels in Pangasius and Cod fillets traded in Sicily (Italy) Food Additives & Contaminants 29 (7): 1046–1051.
44. Butterworth DS, Rademeyer RA (2005) Sustainable management initiatives for the Southern African hake fisheries over recent years. Bulletin of Marine Science 76(2): 287–319.
45. Johnsen E, Kathena J (2012) A robust method for generating separate catch time-series for each of the hake species caught in the Namibian trawl fishery. African Journal of Marine Science 34(1): 43–53.
46. Rademeyer RA, Butterworth DS, Plaganyi EE (2008) Assessment of the South African hake resource taking its two-species nature into account. African Journal of Marine Science 30(2): 263–290.
47. Nielsen E, Cariani A, Mac Aoidh E, Maes G, Milano I, et al. (2012) Gene-associated markers provide tools for tackling illegal fishing and false eco-certification. Nature Communications 3:851.
48. Logan CA, Alter SE, Haupt AJ, Tomalty K, Palumbi SR (2008) An impediment to consumer choice: overfished species are sold as Pacific red snapper. Biological Conservation 141: 1591–1599.
49. Pepe T, Trotta M, Di Marco I, Anastasio A, Bautista JM, Cortesi ML (2007) Fish species identification in surimi-based products. Journal of Agricultural and Food Chemistry 55(9): 3681–3685.
50. Cawthorn DM, Steinman HA, Witthuhn RC (2012) DNA barcoding reveals a high incidence of fish species misrepresentation and substitution on the South African market. Food Research International 46(1): 30–40.
51. EC (European Commission) (2000) European Council Regulation No 104/2000 of 17 December 1999 on the common organization of the markets in fishery and aqua-culture products. Official Journal of the European Communities, L17, 22–52.
52. EC (European Commission) (2001) Commission Regulation (EC) No 2065/2001 of 22 October 2001 laying down detailed rules for the application of Council Regulation (EC) No 104/2000 as regards informing consumers about fishery and aquaculture products. Official Journal of the European Communities, L278, 6–8.
53. Glover KA (2010) Forensic identification of fish farm escapees: the Norwegian experience. Aquaculture Environment Interactions 1: 1–10.
54. Glover KA, Haug T, Oien N, Walloe L, Lindblom L, et al. (2012) The Norwegian minke whale DNA register: a data base monitoring commercial harvest and trade of whale products. Fish and Fisheries 13: 313–332.
55. Ogden R (2008) Fisheries forensics: the use of DNA tools for improving compliance, traceability and enforcement in the fishing industry. Fish and Fisheries 9(SI): 462–472.

5

Organic vs. Conventional Grassland Management: Do ^{15}N and ^{13}C Isotopic Signatures of Hay and Soil Samples Differ?

Valentin H. Klaus[1]*, Norbert Hölzel[1], Daniel Prati[2], Barbara Schmitt[2], Ingo Schöning[3], Marion Schrumpf[3], Markus Fischer[2], Till Kleinebecker[1]

1 University of Münster, Institute of Landscape Ecology, Münster, Germany, **2** University of Bern, Institute of Plant Sciences, Bern, Switzerland, **3** Max-Planck-Institute for Biogeochemistry, Jena, Germany

Abstract

Distinguishing organic and conventional products is a major issue of food security and authenticity. Previous studies successfully used stable isotopes to separate organic and conventional products, but up to now, this approach was not tested for organic grassland hay and soil. Moreover, isotopic abundances could be a powerful tool to elucidate differences in ecosystem functioning and driving mechanisms of element cycling in organic and conventional management systems. Here, we studied the $\delta^{15}N$ and $\delta^{13}C$ isotopic composition of soil and hay samples of 21 organic and 34 conventional grasslands in two German regions. We also used $\Delta\delta^{15}N$ ($\delta^{15}N$ plant - $\delta^{15}N$ soil) to characterize nitrogen dynamics. In order to detect temporal trends, isotopic abundances in organic grasslands were related to the time since certification. Furthermore, discriminant analysis was used to test whether the respective management type can be deduced from observed isotopic abundances. Isotopic analyses revealed no significant differences in $\delta^{13}C$ in hay and $\delta^{15}N$ in both soil and hay between management types, but showed that $\delta^{13}C$ abundances were significantly lower in soil of organic compared to conventional grasslands. $\Delta\delta^{15}N$ values implied that management types did not substantially differ in nitrogen cycling. Only $\delta^{13}C$ in soil and hay showed significant negative relationships with the time since certification. Thus, our result suggest that organic grasslands suffered less from drought stress compared to conventional grasslands most likely due to a benefit of higher plant species richness, as previously shown by manipulative biodiversity experiments. Finally, it was possible to correctly classify about two third of the samples according to their management using isotopic abundances in soil and hay. However, as more than half of the organic samples were incorrectly classified, we infer that more research is needed to improve this approach before it can be efficiently used in practice.

Editor: Shuijin Hu, North Carolina State University, United States of America

Funding: The work has been funded by the DFG Priority Program 1374 "Infrastructure-Biodiversity-Exploratories" (FI 1246/6-1, FI1246/9-1, HO 3830/2-2, SCHR 1181/2-1). The funders had no role in study design, data collection and analysis, decision to publish, or preparation of the manuscript.

Competing Interests: The authors have declared that no competing interests exist.

* E-mail: v.klaus@uni-muenster.de

Introduction

Distinguishing organic and conventional products is a major issue of food security and authenticity and much research on method development has been conducted to tackle this issue [1]. Stable isotope analysis was proven to give important insight in ecosystem functioning and was successfully used to detect differences between organic and conventional agriculture [2,3,4]. Since Nakano et al. [5] proposed the use of natural abundances of stable isotopes to separate organic and conventional products, several studies tested this approach successfully for fruits, vegetables and other plant products [1,6,7,8,9,10] as well as for beef [11,12] and milk [13], but not for grassland hay or soil samples. Differences among organic and conventional plant products were mostly attributed to differences in $\delta^{15}N$ isotopic signatures of applied fertilizers [14], because organic farming abandons the use of synthetic mineral fertilizers. While such conventional (synthetic) N sources exhibit $\delta^{15}N$ values close to 0‰, organic N sources such as cattle dung or slurry are strongly enriched in $\delta^{15}N$ [15]. Consequently, organic farming products are mostly enriched in $\delta^{15}N$ compared to conventional ones due to the replacement of synthetic N sources by organic fertilizers. However, in nature $\delta^{15}N$ abundances in plants are affected by a multiplicity of factors such as type and degree of mycorrhization, the chemical type of N-compounds taken up or further soil characteristics, which can be barely separated from each other [16,17].

Similarly, $\delta^{13}C$ in plant and soil are also of broad ecological interest [18]. In C3 plants, which represent Central European grassland vegetation, $\delta^{13}C$ abundances in biomass are first of all affected by water availability and drought stress, but show also significant interactions with nutrient availability and fertilization [2,19]. Additionally, $\delta^{13}C$ values are related to a different contribution of CO_2 from soil respiration to plant photosynthesis y [6,20] and thus contain valuable ecological information related to agricultural management. Furthermore, was shown to be related to functional aspects of plant communities [18].

Although grasslands play a central role in the production of organic meat and dairy products [21], and proportions of organic grasslands have increased significantly during the last decade [22], stable isotope analysis was so far not used to distinguish between soils and yield (hay) of organically and conventionally managed grasslands. As organic fertilizers can even in grasslands lead to higher $\delta^{15}N$ values in soil and vegetation [23], this might give the ability to classify organic and conventional plant products using isotopic abundances [7]. Moreover, isotopic abundances are related to important ecosystem processes affecting nutrient cycling and balances [24] and thus bear the potential to elucidate possible differences in ecosystem functioning of organic vs. conventional grasslands, which are otherwise difficult to detect.

Here, we studied 21 organic and 34 conventional grasslands in two German regions and analyzed $\delta^{13}C$ and $\delta^{15}N$ of soil and hay (plant biomass). Furthermore, $\Delta\delta^{15}N$ values ($\delta^{15}N$ plant - $\delta^{15}N$ soil) were calculated to estimate differences in nitrogen dynamics. We also assessed the time since organic certification to test for temporal trends. In detail, we analyzed whether (a) differences in isotopic abundances among organic and conventional grasslands exist and whether there are (b) significant trends in isotopic composition with time since conversion to organic management. Additionally (c), we used discriminant analysis to deduce the respective management type from the isotopic composition of hay and/or soil samples.

Methods

Ethics statement

Field work permits were given by the responsible state environmental offices of Baden-Württemberg, Thüringen, and Brandenburg (according to § 72 BbgNatSchG).

Study design

We studied agriculturally used permanent grasslands in two regions in Germany which belong to the *Biodiversity Exploratories* project [25]: (I) *Hainich-Dün* in Thuringia in central Germany situated in and around the National Park Hainich and (II) the Biosphere Reserve *Schwäbische Alb* in Baden-Württemberg in south-west Germany. In grasslands of both regions Cambisols occur, while in the Schwäbsiche Alb Leptosols and in Hainich-Dün Stagnosols and Vertisols can also be found. Grassland types could be categorized as pastures, meadows and mown pastures [25]. To get information on land use for each grassland, farmers and land owners were annually questioned about the amount and type of fertilizer (kg N ha^{-1}) from 2006 to 2010 [26]. We chose organic and conventional grasslands from a randomly selected dataset of 50 plots in each region. Organic management of grasslands abandons pesticides and synthetic fertilizers, restricts livestock density and the use of organic fertilizers from animal husbandry to a maximum of 170 kg $N*ha^{-1}*a^{-1}$ (European Union, 2008). Accordingly, study plots can be distinguished in two sub-sets: organic plots which are managed according to an official organic farming certificate [27] and uncertified (conventional) plots, where management goes against certification criteria. Please note that all unfertilized but not certified grasslands were excluded from the analysis. Finally, we used 17 conventional plots per study region as well as 17 organic plots at Hanich-Dün and 4 organic plots at the Schwäbische Alb for comparison. Duration of organic management of the grasslands differs from 3 up to 20 years. Although we have no detailed data on the management prior conversion to organic farming, it seems to be likely that at least some of the organic grasslands were already previously managed at a low to medium intensity. The proportion of legumes varied widely among study plots (from 0.0 to 60.5%) but not among organic and conventional management (data not shown). C4 plants are generally no regular component of Central European grassland vegetation.

Field work and chemical analyses

Soil sampling was conducted in early May 2011. On each plot mixed samples of 14 soil cores from 0 to 10 cm depth were collected using a split tube sampler with a diameter of 5 cm. Cores were taken along two 20 m transects at each plot. Roots were removed from the samples in the field and soil samples were air-dried, sieved to <2 mm and ground. For δ13C analyses, soil samples were weighted into tin capsules and treated with sulphurous acid inside the capsules to remove carbonates. Samples were dried again at 70°C before combustion in an oxygen stream using an elemental analyser (NA 1110, CE Instruments, Milan, Italy). Evolved CO2 was analysed using an isotope ratio mass spectrometer (IRMS; Delta C or DELTA+XL, Thermo Finnigan MAT, Bremen, Germany). For biomass sampling, we harvested aboveground community biomass in four quadrates of 0.25 m^2 from mid-May to mid-June 2011 in both regions simultaneously. Temporary fences ensured that no mowing or grazing took place before yield was sampled. Plant material was dried for 48 h at 80°C and ground to fine powder for lab analyses. Both isotopic abundances of biomass samples and $\delta^{15}N$ of soil samples were determined by mass spectrometry (Finnigan MAT DeltaPlus with Carlo Erba Elementar Analysator with ConFlo II Interface). Isotope ratios are given in per mille (‰), whereby $\delta^{13}C$ is relative to the international reference standard v-PDB using NBS19 and $\delta^{15}N$ relative to AIR-N_2 [28].

The nearby climate station in *Hainich-Dün*, located in Schönstedt (193 m A.S.L.), revealed that the precipitation prior to sampling (April and May 2011) was 70% lower than the mean of the same month from 2003–2010 (33.8 mm instead of 116.2 mm) [29]. At the *Schwäbische Alb*, the next climate station is situated in Münsingen-Rietheim (732 m A.S.L). Compared to the same month in the period 2005–2010, precipitation in April and May 2011 was 20% lower (153.8 mm instead of 191.5 mm) [30].

Data analysis

Multiple analysis of variance (ANOVA) models were used to examine differences among management types (organic vs. conventional) while also accounting for farm or land owner ($n = 7$), soil type ($n = 4$), study region ($n = 2$) and grasslands type (meadow, pasture, mown pasture) including all two-way interactions with management type. Therefore, the lm() and step() functions in R for stepwise reduction of explanatory variables were used. Obtaining model results by using the anova() output assured that the order of variables entering the model had no effect on significance gained for the respective variable, because all other variables were taken into account previously as co-variables. Model assumptions were checked using diagnostic plots function. Similarly, also applying the lm() function analysis of co-variance (ANCOVA) models were calculated to relate the isotopic composition to the time since certification in all organic plots including farm, soil type and study region. Soil texture (proportions of clay, silt and sand) was also incorporated in lm() models but removed for final analysis as they did not add to explained variance. Additionally, we employed the qda() function from the MASS package [31] to perform Quadratic Discriminant Analysis for the classification of organic and conventional grasslands based on ^{15}N and ^{13}C isotopic values in soil and biomass samples. To ensure normal distribution of variables log10 transformation was applied prior to all analyses where necessary. All statistical tests were performed with R [32].

Results

Farmer's questionnaires revealed that organic grasslands received average fertilizer applications of 21.6 ($\pm$37.0) kg N a^{-1}, while conventional grasslands received more than the threefold amount: 72.3 ($\pm$41.5) kg N a^{-1}. However, both management systems show wide variation in fertilization intensity at the plot level. Organic plots received only organic fertilizers, as prescribed by organic management guidelines, whereas conventional plots received mostly mineral fertilizers (75% of total fertilizer N).

Grasslands of both management types overlap widely in ^{13}C and ^{15}N isotopic abundances in soil and hay (Figure 1). Nevertheless, organic grasslands were characterized by significantly lower $\delta^{13}C$ abundances in soil (Figure 2a). Meanwhile, $\delta^{13}C$ in hay, $\delta^{15}N$ in hay and soil and $\Delta\delta^{15}N$ did not differ between organic and conventional grasslands but only among farms, study regions and/or grassland types (Table 1). In none of the models the soil type was a significant predictor of ^{13}C and ^{15}N isotopic signals. Although this does not mean that isotopic signals are independent of further soil characteristics, it, nevertheless, underlines the comparability of the selected plots. Generally, analyses explained only 6 to 34% of the variance (Table 1).

When related to the time since certification, ^{13}C in hay and soil showed significant negative relationships, while the explained variance of the models increased considerably compared to previous models (Table 2). This was most significant for $\delta^{13}C$ in soil (Figure 2b). Neither $\delta^{15}N$ in soil or hay nor $\Delta\delta^{15}N$ showed significant relationships with the time since certification (Table 2).

Due to significant effects of study region on some of the isotopic abundances (Table 3), quadratic discriminant analysis (QDA) was carried out using data which was centered (standardized) according to the regional mean. Using QDA, 60% of the samples could be correctly classified as organic or conventional using $\delta^{13}C$ and $\delta^{15}N$ isotopic information of soils, while 70% could be correctly classified using the isotopic information of the hay, assumingly due to higher variation in isotopic values of hay compared to soil samples. Combining both, increased correct classifications only slightly up to 73%. In all cases analyses mismatched organic samples to a higher degree than conventional samples and classified significantly more organic samples wrongly as conventional ones (Table 3).

Discussion

Isotopic abundances of ^{15}N and ^{13}C can give insight in ecosystem functioning and were often shown to be a useful tool to separate between organic and conventional products [1,6]. In the case of organic management in grasslands, this is only partly true for $\delta^{15}N$ and δ ^{13}C in soil and hay.

^{15}N in organic grasslands

Organic grasslands received only organic fertilizers and were characterized by significantly lower fertilization intensity, in line with Klaus et al. [33]. Nevertheless, this difference in fertilization regime did not imprint in the $\delta^{15}N$ signal of soil and hay samples. While $\delta^{15}N$ abundances of conventional synthetic fertilizers vary between −2 and 2‰, organic fertilizers such as cattle dung or slurry are strongly enriched in $\delta^{15}N$ (5 to 35‰) [15]. While experimental studies revealed an effect of organic compared to synthetic fertilization on the $\delta^{15}N$ isotopic composition of grassland soils and vegetation [23], it was likewise shown that both organic but also mineral fertilizers can lead to increasing $\delta^{15}N$ vales in soil and vegetation probably due to increased microbial activity and subsequent losses of ^{15}N depleted N from the system [17]. Furthermore, conventional management includes both the use of organic and mineral fertilizers and farmers may considerably change respective proportions among years [26]. In our study, the naturally diverse abiotic conditions such as strong variation in soil properties and land-use history among plots might have additionally impeded clear patterns of $\delta^{15}N$ in organic and conventional grasslands. This is supported by findings from Wrage et al. [34] giving poor to missing associations among N balances and $\delta^{15}N$ in soil and vegetation due to spatially heterogeneous conditions and short-term changes in stocking densities in pastures. Moreover, as only a certain proportion of the N stored

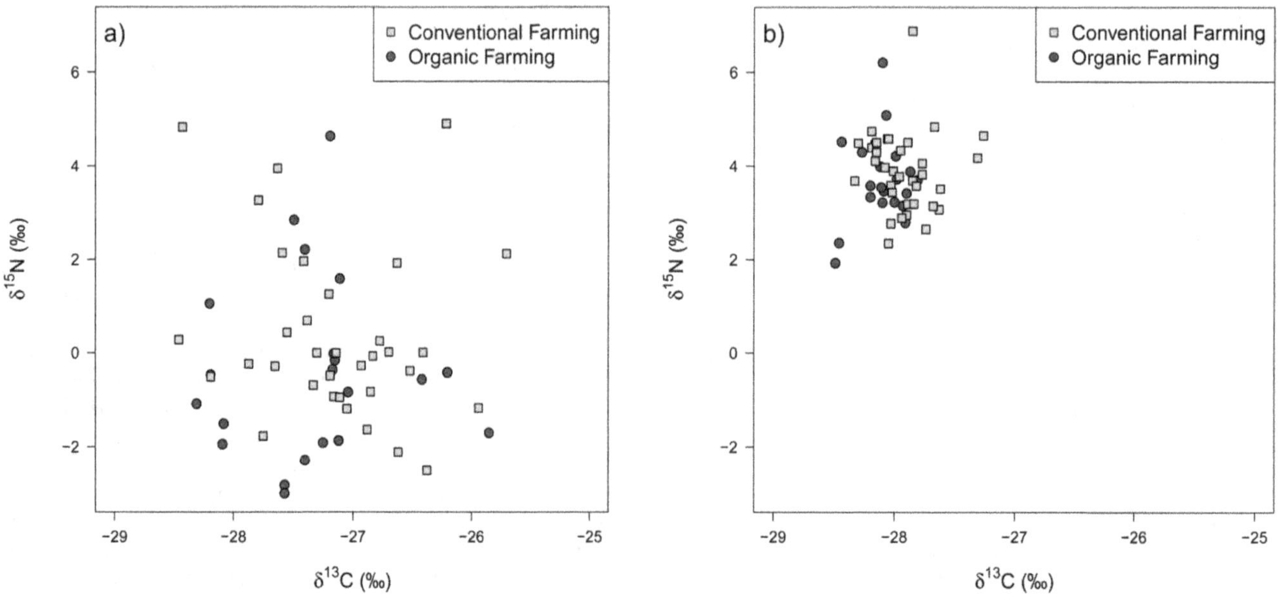

Figure 1. $\delta^{15}N$ and $\delta^{13}C$ composition of a) hay and b) soil samples of organic and conventional grasslands.

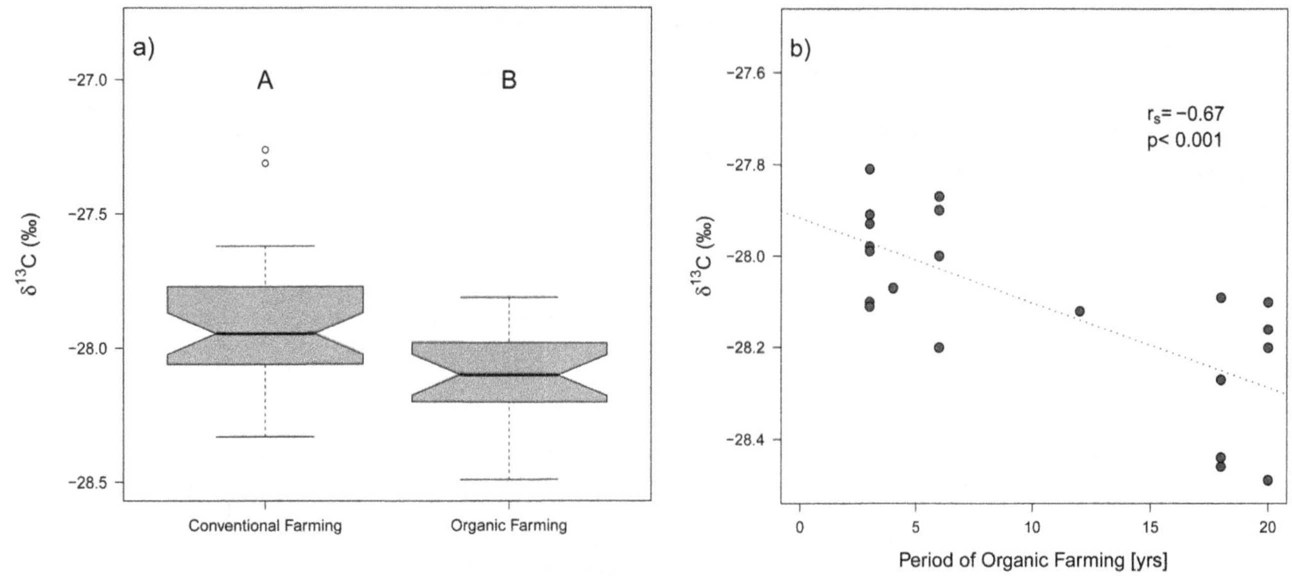

Figure 2. $\delta^{13}C$ abundances in soil of a) organic vs. conventional grasslands and b) organic grasslands in relation to the time since certification ($r_s = -0.70$; $p<0.001$). Letters indicate significant group differences according to ANOVA analyses (for details see Table 1 & 2).

in biomass and soil originates from fertilization, biologically fixed N (e.g. through legumes) have additionally diluted isotopic signals of fertilizer applications [35].

Missing differences in $\Delta\delta^{15}N$ between organic and conventional plots suggest that the studied grasslands do not substantially differ in nitrogen cycling [24]. Instead, the respective farm (or land owner) was responsible for most of the variation explained in $\Delta\delta^{15}N$ indicating that nitrogen cycling is strongly driven by individual decisions and practice at the farm level. Because plant species richness was shown decease $\Delta\delta^{15}N$ in plant and soil assumedly via nitrogen partitioning [35], missing or poor differences in plant diversity between organic and conventional grasslands [e.g. 33,36,37] could be another reason why $\Delta\delta^{15}N$ did not respond to organic grassland management. Contrary, in our study organic grasslands had higher mean plant species richness compared to conventional ones (t-test; $p<0.01$; data not shown), apparently not leading to significant differences in $\Delta\delta^{15}N$.

Haas et al. [21] suggest that lower fertilization intensity under organic management is associated with fewer losses of N to neighbouring habitats or to the ground water. However in our case, even after up to 20 years of organic farming ^{15}N isotopic abundances revealed no significant indication for fewer losses of N from the studied organic grasslands.

^{13}C in organic grasslands

We found significant differences in $\delta^{13}C$ abundances between organic and conventional soils and trends in decreasing $\delta^{13}C$ abundances with time since certification in both organic soil and hay samples. Missing statistical significances of soil types in the analyses suggested that this is not only caused by differences in water availability among study plots. However, clear differences in $\delta^{13}C$ in hay might only get apparent in dry years, because $\delta^{13}C$ isotopic patterns can differ considerably among seasons [38]. Thus, $\delta^{13}C$ abundances in soils are probably a more reliable indicator for water supply for plants, because they integrate $\delta^{13}C$ signals over a longer period of time, while $\delta^{13}C$ in hay strongly depends on the photosynthetic activity and stomatal conductance influenced by current weather conditions [2,38]. However, it is questionable whether differences in $\delta^{13}C$ in hay can get more pronounced than in the studied year as prior to sampling the weather was particularly dry.

It was shown that vegetation $\delta^{13}C$ decrease with plant diversity in grassland experiments due to facilitation and/or complementarity among plant species [39,40]. Assumedly, we also detected such a pattern of a positive feedback of higher plant diversity under organic management causing less water stress in organic grasslands. However, as mentioned above, organic and conventional grasslands

Table 1. Summary of multiple ANOVA models of isotopic abundances (no interaction with management was significant) ($n = 55$). "Grassland type" = pasture, meadow or mown pasture.[a]

	Adj. R^2	Effect of organic management	Farm	Region	Grassland type	Soil type
δ13C soil	**0.13****	**negative****	ns	ns	ns	ns
δ13C hay	**0.34*****	ns	*	***	ns	ns
δ15N soil	**0.06***	ns	ns	*	ns	ns
δ15N hay	**0.24***	ns	*	ns	*	ns
Δδ15N	**0.20***	ns	**	ns	*	ns

[a]Significance levels: *** = $p<0.001$; ** = $0.001<p<0.01$; * = $0.01<p<0.05$.

Table 2. Summary of ANCOVA models of isotopic abundances testing for relationships with the time since certification (only organic plots, $n=21$).

	Adj. R^2	Effect of time since certification	Farm	Region	Grassland type	Soil type
δ13C soil	**0.49*** **	**negative*** **	ns	ns	ns	ns
δ13C hay	**0.73** **	**negative** **	*	ns	ns	ns
δ15N soil	ns	ns	ns	ns	ns	ns
δ15N hay	ns	ns	ns	ns	ns	ns
Δδ15N	ns	ns	ns	ns	ns	ns

[a]Significance levels: *** $=p<0.001$; ** $=0.001<p<0.01$; * $=0.01<p<0.05$.
"Grassland type" = pasture, meadow or mown pasture.[a]

do not always differ in plant diversity and more research is needed to give full evidence for reduced water stress under higher plant diversity in established permanent grasslands. Furthermore, changes in $\delta^{13}C$ with time since certification cannot be traced back to changes in plant diversity, because diversity was stable over time (data not shown).

There are several other mechanisms, which could have also occurred, but cannot be easily separated from the whole complex of factors influencing isotopic abundances. Georgi et al. [6] found organic biomass (vegetables) to be depleted in $\delta^{13}C$. They suggest higher soil respiration rates in organic soils due to higher microbial activity might cause a lowering of the $\delta^{13}C$ value of the soil CO_2 pool available to plants. As soil density fractions can have different ^{13}C signals [41], differences in the distribution of soil carbon among density fractions could also have also led to isotopic patterns.

Other reasons for lower $\delta^{13}C$ values in organic grasslands could be direct or indirect effects of (former) fertilization. As organic farms use less often corn (a C4 plant) to feed their livestock, conventional slurry is thus enriched in ^{13}C and may have led to higher $\delta^{13}C$ values in soil organic matter in conventional grasslands [42]. As the "conventional" organic matter deceases after time since certification, this effect is likely to have caused decreasing $\delta^{13}C$ in soil with time since certification. Similarly, it was shown that high N availability can lead to higher $\delta^{13}C$ in plant biomass mostly due to structural changes in plant tissue further tightening drought conditions [43]. However, differences in $\delta^{13}C$ values of vegetables observed by Georgi et al. [6] were independent of optimal or reduced N supply. Finally, we can't rule out that further factors such as the local rainfall distribution also influenced observed $\delta^{13}C$ abundances.

Classification of organic and conventional samples

Although Rapisarda et al. [6] showed that approximately 90% of fruit samples could be correctly categorized as organic or conventional products using isotopic abundances, classification of grassland samples revealed rather weak results. Only one third of organic and conventional samples were correctly classified and a high proportion especially of organic samples were mismatched. This can be largely explained by missing differences in especially $\delta^{15}N$ abundances. These were significantly higher in organic fruits used in the study by Rapisarda et al. [6]. Thus, we have to conclude that separating organic and conventional soil and hay samples is barely possible using only $\delta^{15}N$ and $\delta^{13}C$ abundances.

Table 3. Results of quadratic discriminant analysis (QDA) of management types (organic vs. conventional) deduced from regionally standardized $\delta^{15}N$ and $\delta^{13}C$ isotopic abundances of soil and/or hay samples of grasslands.

Soil samples		**Classified**			
		organic	conventional	total	correct
Origin	organic	7	14	21	33%
	conventional	8	26	34	76%
				total	**60%**
Hay samples		**Classified**			
		organic	conventional	total	correct
Origin	organic	9	12	21	43%
	conventional	4	30	34	88%
				total	**71%**
Soil & hay samples		**Classified**			
		organic	conventional	total	correct
Origin	organic	10	11	21	48%
	conventional	4	30	34	88%
				total	**73%**

This is especially true, if we would also include uncertified but unfertilized grasslands, which occur frequently [33,44]. However, including further stable isotopes of S, O and H or other indicative substances might offer additional possibilities to improve the classification of organic vs. conventional samples from permanent grasslands [11,12].

Acknowledgments

We thank Harald Strauβ and Artur Fugman for friendly isotopic analyses and Svenja Agethen and Max Moenikes for help during laboratory work. Willi A. Brand, Heike Geilmann, Jessica Schäfer and Theresa Klötzing are acknowledged for ^{13}C analyses of soil samples. Furthermore, our thank goes to the managers of the exploratories, Swen Renner, Sonja Gockel annd Kerstin Wiesner for their work in maintaining the plot and project infrastructure; Simone Pfeiffer and Christiane Fischer giving support through the central office, Michael Owonibi for managing the central data base, and Markus Fischer, Eduard Linsenmair, Dominik Hessenmöller, Jens Nieschulze, Daniel Prati, Ingo Schöning, François Buscot, Ernst-Detlef Schulze, Wolfgang W. Weisser and the late Elisabeth Kalko for their role in setting up the Biodiversity Exploratories project.

Author Contributions

Conceived and designed the experiments: NH DP TK IS MS MF. Performed the experiments: VHK BS IS MS. Analyzed the data: VHK. Contributed reagents/materials/analysis tools: IS TK. Wrote the paper: VHK TK NH DP MF BS MS IS.

References

1. Camin F, Perini M, Bontempo L, Fabroni S, Faedi W, et al. (2011) Potential isotopic and chemical markers for characterizing organic fruits. Food Chemistry 125: 1072–1082.
2. Adams MA, Grierson PF (2001) Stable isotopes at natural abundance in terrestrial plant ecology and ecophysiology: An update. Plant Biology 3: 299–310.
3. Franke BM, Gremaud G, Hadorn R, Kreuzer M (2005) Geographic origin of meat - elements of an analytical approachto its authentication. European Food Research and Technology 221: 493–503.
4. Schwertl M, Auerswald K, Schäufele R, Schnyder H (2005) Carbon and nitrogen stable isotope composition of cattle hair: ecological fingerprints of production systems? Agriculture, Ecosystems and Environment 109: 153–165.
5. Nakano A, Uehara Y, Yamauchi A, (2003) Effect of organic and inorganic fertigation on yields, δ15N values, and δ13C values of tomato (Lycopersicon esculentum Mill. cv. Saturn). Plant and Soil 255: 343–349.
6. Georgi M, Voerkelius S, Rossmann A, Grassmann J, Schnitzler WH (2005) Multielement isotope ratios of vegetables from integrated and organic production. Plant and Soil 275: 93–100.
7. Rapisarda P, Calabretta ML, Romano G, Intrigliolo F (2005) Nitrogen metabolism components as a tool to discriminate between organic and conventional citrus fruits. Journal of Agricultural and Food Chemistry 53: 2664–2669.
8. Bateman AS, Kelly SD, Woolfe M (2007) Nitrogen Isotope Composition of Organically and Conventionally Grown Crops. Journal of Agricultural and Food Chemistry 55: 2664–2670.
9. del Amor FM, Navarro N, Aparicio PM (2008) Isotopic Discrimination as a Tool for Organic Farming Certification in Sweet Pepper. Journal of Envionmental Quality 37: 182–185.
10. Camin F, Moschella A, Miselli F, Parisi B, Versini G, et al. (2007) Evaluation of markers for the traceability of potato tubers grown in an organic versus conventional regime. Journal of the Science of Food and Agriculture 87: 1330–1336.
11. Boner M, Förstel H (2004) Stable isotope variation as a tool to trace the authenticity of beef. Analytical & Bioanalytical Chemistry 378: 301–310.
12. Schmidt O, Quilter JM, Bahar B, Moloney AP, Scrimgeour CM, et al. (2005) Inferring the origin and dietary history of beef from C, N and S stable isotope ratio analysis. Food Chemistry 91: 545–549.
13. Molkentin J (2009) Authentication of Organic Milk Using δ13C and the r-Linolenic Acid Content of Milk Fat. Journal of Agricultural and Food Chemistry 2009 57: 785–790.
14. Bateman AS, Kelly SD, Jickells TD (2005) Nitrogen Isotope Relationships between Crops and Fertilizer:Implications for Using Nitrogen Isotope Analysis as an Indicator of Agricultural Regime. Journal of Agricultural and Food Chemistry 53, 5760–5765.
15. Bateman AS, Kelly SD (2007) Fertilizer nitrogen isotope signatures. Isotopes in Environmental and Health Studies 43: 237–247.
16. Hobbie EA, Högberg P (2012) Nitrogen isotopes link mycorrhizal fungi and plants to nitrogen dynamics. New Phytologist 196: 367–382.
17. Kleinebecker T, et al. (submitted) 15N natural abundances reveal plant diversity effects on nitrogen dynamics in permanent grassland ecosystems. Submitted
18. de Bello F, Buchmann N, Casals P, Leps J, Sebastia MT (2009) Relating plant species and functional diversity to community δ^{13}C in NE Spain pastures. Agriculture, Ecosystems and Environment 131: 303–307.
19. Högberg P, Johannisson C, Hallgren J-E (1993) Studies of 13Cin the foliage reveal interactions between nutrients and water in forest fertilization experiments. Plant and Soil 152: 207–214.
20. Šantrůčková H, Bird MI, Lloyd J (2000) Microbial processes and carbon-isotope fractionation in tropical and temperate grassland soils. Functional Ecology 14: 108–114.
21. Haas G, Wetterich F, Köpke U (2001) Comparing intensive, extensified and organic grassland farming in southern Germany by process life cycle assessment. Agriculture, Ecosystems and Environment 83: 43–53.
22. Schaack D, Iller S, Würtenberger E (2010) AMI-Marktbilanz Öko-Landbau 2010, Bonn: Agrarmarkt Informationsgesellschaft mbH.
23. Watzka M, Buchgraber K, Wanek W (2006) Natural N-15 abundance of plants and soils under different management practices in a montane grassland. Soil Biology & Biochemistry 38: 1564–1576.
24. Kahmen A, Wanek W, Buchmann N (2008) Foliar delta ^{15}N values characterize soil N cycling and reflect nitrate or ammonium preference of plants along a temperate grassland gradient. Oecologia 156: 861–870.
25. Fischer M, Bossdorf O, Gockel S, Hansel F, Hemp A, et al. (2010) Implementing large-scale and long-term functional biodiversity research: The Biodiversity Exploratories. Basic and Applied Ecology 11: 473–485.
26. Bluethgen N, Dormann CF, Prati D, Klaus VH, Kleinebecker T, et al. (2012) A quantitative index of land-use intensity in grasslands: Integrating mowing, grazing and fertilization. Basic and Applied Ecology 13: 207–220.
27. European Union (2008) Commission Regulations (EC) No 889/2008, Brussels.
28. Werner RA, Brand WA (2001) Referencing strategies and techniques in stable isotope ratio analysis. Rapid Communications in Mass Spectrometry 15: 501–519.
29. LUFTGEIST website. Available: http://www.luftgeist.de. Accessed 2013 Jun 3.
30. Wetterstation Riethein Lichse website. Available: http://www.wetterstation-rietheim-lichse.de/html/archiv.html. Accessed 2013 Jun 3.
31. Venables WN, Ripley BD (2002) Modern Applied Statistics with S. Fourth Edition. New York: Springer.
32. R Development Core Team (2011) R: A language and environment for statistical computing. R Foundation for Statistical Computing, Vienna.
33. Klaus VH, Kleinebecker T, Prati D, Fischer M, Alt F, et al. (2013) Does organic grassland farming benefit plant and arthropod diversity at the expense of yield and soil fertility? Agriculture, Ecosystems and Environment 177: 1–9.
34. Wrage N, Küchenmeister F, Isselstein J (2011) Isotopic composition of soil, plant or cattle hair no suitable indicator of nitrogen balances in permanent pasture. Nutrient Cycling in Agroecosystems 90: 189–199.
35. Gubsch M, Roscher C, Gleixner G, Habekost M, Lipowsky A, et al. (2011) Foliar and soil delta 15N values reveal increased nitrogen partitioning among species in diverse grassland communities. Plant Cell and Environment 34: 895–908.
36. Hole DG, Perkins AJ, Wilson JD, Alexander IH, Grice PV, et al. (2005) Does organic farming benefit biodiversity? Biological Conservation 122: 113–130.
37. Batáry P, Báldi A, Sárospataki M, Kohler F, Verhulst J, et al. (2010) Effect of conservation management on bees and insect-pollinated grassland plant communities in three European countries. Agriculture, Ecosystems and Environment 136: 35–39.
38. Neilson R, Hamilton D, Wishart J, Marriott CA, Boag B, et al. (1998) Stable isotope natural abundances of soil, plants and soil invertebrates in an upland pasture. Soil Biology and Biochemistry 30: 1773–1782.
39. Caldeira MC, Ryel RJ, Lawton JH, Pereira JS (2001) Mechanisms of positive biodiversity-production relationships: insights provided by δ13C analysis in experimental Mediterranean grassland plots. Ecology Letters 4: 439–443.
40. Jumpponen A, Mulder CPH, Huss-Danell K, Högberg P (2005) Winners and losers in herbaceous plant communities: insights from foliar carbon isotope composition in monocultures and mixtures. Journal of Ecology 93: 1136–1147.
41. Baisden WT, Amundson R, Cook AC, Brenner DL (2002) Turnover and storage of C and N in five density fractions from California annual grassland surface soils. Global Biogeochemical Cycles 16: 64–71.
42. Bol R, Moering J, Preedy N, Glaser B (2004) Short-term sequestration of slurry-derived carbon into particle size fractions of a temperate grassland soil. Isotopes in Environmental and Health Studies 40: 81–87.
43. Shangguan ZP, Shao MA, Dyckmans J (2000) Nitrogen nutrition and water stress effects on leaf photosynthetic gas exchange and water use efficiency in winter wheat. Environmental and Experimental Botany 44: 141–149.
44. Socher S, Prati D, Müller J, Klaus VH, Hölzel N, et al. (2012) Direct and productivity-mediated indirect effects of fertilization, mowing and grazing intensities on grassland plant species richness. Journal of Ecology 100: 1391–1399.

Spatial Covariance between Aesthetic Value & Other Ecosystem Services

Stefano Casalegno*, Richard Inger, Caitlin DeSilvey, Kevin J. Gaston

Environment and Sustainability Institute, University of Exeter, Penryn, Cornwall, United Kingdom

Abstract

Mapping the spatial distribution of ecosystem goods and services represents a burgeoning field of research, although how different services covary with one another remains poorly understood. This is particularly true for the covariation of supporting, provisioning and regulating services with cultural services (the non-material benefits people gain from nature). This is largely because of challenges associated with the spatially specific quantification of cultural ecosystem services. We propose an innovative approach for evaluating a cultural service, the perceived aesthetic value of ecosystems, by quantifying geo-tagged digital photographs uploaded to social media resources. Our analysis proceeds from the premise that images will be captured by greater numbers of people in areas that are more highly valued for their aesthetic attributes. This approach was applied in Cornwall, UK, to carry out a spatial analysis of the covariation between ecosystem services: soil carbon stocks, agricultural production, and aesthetic value. Our findings suggest that online geo-tagged images provide an effective metric for mapping a key component of cultural ecosystem services. They also highlight the non-stationarity in the spatial relationships between patterns of ecosystem services.

Editor: Tobias Preis, University of Warwick, United Kingdom

Funding: This work was supported by the University of Exeter. The funders had no role in study design, data collection and analysis, decision to publish, or preparation of the manuscript

Competing Interests: The authors have declared that no competing interests exist.

* E-mail: stefano@casalegno.net

Introduction

Key to the successful maintenance and management of environmental resources is an understanding of their spatial distribution. Recognition of the significance of, and threats to, ecosystem services (the benefits that humans gain from ecosystems), has thus been associated with growth in attempts to map their patterns of variation and covariation [1–4]. In many instances, however, this remains challenging. Of the four main categories of services (supporting, provisioning, regulating and cultural [5]), this is particularly true of cultural services: the non-material benefits people gain from ecosystems (including aesthetic, recreational and spiritual benefits). Therefore, most studies have tended to focus on provisioning and regulating services, rather than cultural ones [6,7]. Indeed, whilst there has been substantial improvement in understanding of the fundamental importance of cultural services, their effective integration into the application of ecosystem service frameworks has been limited by the challenges of quantifying, valuing and mapping them [8].

Efforts to quantify and map cultural services have concentrated foremost on estimates of the relative numbers of recreational visitors to particular areas [1,2,9,10]. Other measures include the number of tourist attractions, tax value of summer cottages, number of reported sightings of rare species [11], tourist expenditure [12], accessibility to natural areas [13], days spent fishing [14] and indices combining multiple such variables [15–17]. When compared, these measures have, in the main, been found to be weakly correlated with spatial variation in other ecosystem services, which has significant implications for identifying and managing priority areas for their maintenance [1,2,11]. However, although these data are useful, a much broader portfolio of metrics and proxies for cultural services, and an understanding of how these covary with measures of other ecosystem services, is urgently required.

Quantification of the aesthetic value that people place on different parts of the landscape represents an innovative development in the mapping of cultural services. One potential measure of aesthetic value can be found in the spatial distribution of photographs of the natural environment that people post online, working from the premise that areas more highly valued for their aesthetic attributes will generate 'hotspots' of activity. Particularly useful are Internet platforms which specifically facilitate posting of geo-tagged digital images, and which are populated by an increasing number of users worldwide. Here we exploit one such resource, the "Panoramio" web platform (www.panoramio.com), to provide a valuable additional metric of cultural ecosystem services (aesthetic value), and document its relationship with two other ecosystem goods and services: a provisioning service (agricultural production) and a supporting/regulating service (carbon stocks in soil).

Materials and Methods

The study was carried out in Cornwall, UK, as a regional approach is considered most appropriate for the analysis of cultural activity [18]: tourism and agriculture are the largest components of the economy of this particular region [19]; and both residents and visitors have extensive (albeit not universal) access to digital cameras and the internet [20–21] that are required for the use of posted geo-tagged images.

Aesthetic value (cultural service)

Panoramio hosts photos of "places of the world", with a particular focus on images of landscapes, natural features (such as woodlands) and animals in their natural environment [22]. Images that have as their central subject people, machines, vehicles or the interiors of structures, or that depict public events such as fairs or concerts, are excluded from the platform [22]. The semantic content of Panoramio's images makes it better adapted to measure the perceived aesthetic value of ecosystems than other geo-tagged web platforms, that do not focus on landscape and environment. The number of individuals per unit area (1 km^2) uploading photographs to Google Earth via the Panoramio web platform was used as our measure of aesthetic value. This measure is more appropriate than the total number of photographs uploaded in each area, which reflects the level of activity of individual photographers rather than the overall value placed on a site by visitors.

Soil carbon (supporting and regulating service)

Data on carbon storage in soil (1 km^2 resolution map) were obtained from the European Commission Joint Research Centre [23,24]. These data are especially accurate for England, as detailed ground survey verification has been carried out.

Agricultural production (provisioning service)

Following Anderson et al. [1] and Eigenbrod et al. [25], we calculated an overall measure of agricultural production by summing the gross margins for all major crops/livestock. As inputs we used agricultural census data [26] at ward level, the CORINE land cover map [27], and gross margin estimates [28]. Agricultural production was expressed in units of £ per ha, and processed at 100 m resolution and then resampled at 1 km resolution. We improved on the original methodology by computing an averaged agricultural value from 2000 to 2005 (instead of using one year of data); differentiating gross margins according to lowlands, disadvantaged and severely disadvantaged areas; and using more precise input land cover data to achieve a higher resolution.

Analysis

Ecosystem service data were normalised to a 0–100 scale for comparison. We tested potential bias of aesthetic value by population density and by coastal/non-coastal locations. Cornish population densities were obtained from the 2011 census [29]. We selected 55 centres including the main towns (population greater than 3,000) and the populations coincident with hotspots of aesthetic value (>40 photographers per grid cell: the upper 99th quantile).

Spatially autocorrelated data violate the assumption of sample independence for a traditional test of significance. Therefore, we tested the presence of spatial autocorrelation using Moran's I coefficient [30] and quantified correlations between ecosystem services using the CRH-method [31,32], which correct correlation statistics for spatial autocorrelation [33–35]. Because of the distribution of our data we rank transformed inputs as outlined in Zar (2007) [36] correctly to apply the CRH method [33,34]. To determine the influence of spatial extent on results, we also divided Cornwall into six separate zones: coastal, Lizard peninsula, west, north-east, south-east and central Cornwall (Figure 1). We performed comparison tests for the overall study area and within separate zones, and corrected results for multiple test significance by applying Benjamin-Hochberg corrections on p-values [37].

The aesthetic and agricultural production surfaces were generated using Bash-Awk and GRASS [38]. Statistical analyses were performed in R [39] using "SpatialPack" and "raster" libraries for spatial related statistics.

Results

A total of 113,686 photographs were uploaded by 15,413 users in Cornwall from 2005 to 2011; 9,632 photographers in coastal, 1,414 in west, 1,411 in north-east, 1,385 in central, 1,227 in south-east and 344 in the Lizard peninsula (Figure 1). Hotspots of aesthetic value (35 of 3,843 grid cells; Figure 2) were all located in coastal areas: seven were in coastal towns (population>3000), 17 close to sparsely populated settlements (<3000 inhabitants), and 11 in unpopulated areas (beaches or touristic coastal sites).

There was a negative correlation between population density and aesthetic value (CRH correlation = −0.56, n = 55, *p-value<0,001*).

Soil carbon storage was highest in four main areas, located in the west and in three inland parts of Cornwall (Figure 1). The zonal statistics showed low carbon storage in coastal areas and highest in west and north-east Cornwall (Figure 1). Agricultural production was highest in west Cornwall and the Lizard peninsula, and lowest in coastal zones and in the north-east region.

We found positive spatial autocorrelation in all ecosystem service layers (Figure 3): soil carbon had the highest while aesthetic value had more dispersed spatial patterns. Compared to the overall study area, regional zones such as Coastal, North-East and West Cornwall had weaker patterns of spatial autocorrelation. Across the whole region, agricultural production was negatively correlated with both aesthetic value and soil carbon storage, and the latter two were themselves weakly negatively correlated (Table 1). However, within zones the relationships were quite variable in strength (Table 1).

Discussion

Most mapping of geographic variation in ecosystem services, and analysis of the patterns of covariation between services, has been conducted by ecologists and conservation biologists [6]. This has tended to result in a heavy reliance on conventional data sources with an established geographically explicit component. This is particularly the case for cultural ecosystem services, for which measures have in consequence been quite restricted, and for which novel approaches will need to be explored. Here we present one such approach, using data derived from social media to capture variation in the value that people place on different parts of the landscape.

The emerging field of computational social science exploits the capacity to collect and analyse data about human interactions on an unprecedented scale [40]. The potential for online digital data sets to provide valuable information for scientific studies has been well recognised [41] and applied to diverse disciplines [42–47], but to our knowledge they have not previously been employed in the context of ecosystem services. The research shares its approach with diverse computational social sciences applications [42–44,46–47] which aim to identify relationships between people's behaviour online and real world quantities.

We find substantial variation in our measure of aesthetic value across the study region. The peaks are distinct from the centres of human population, contrasting with the findings of previous studies that have measured recreational usage [1,25], but reflecting a priori expectation.

Our findings lend further support to the general conclusion that spatial variation in cultural services tends to be poorly or

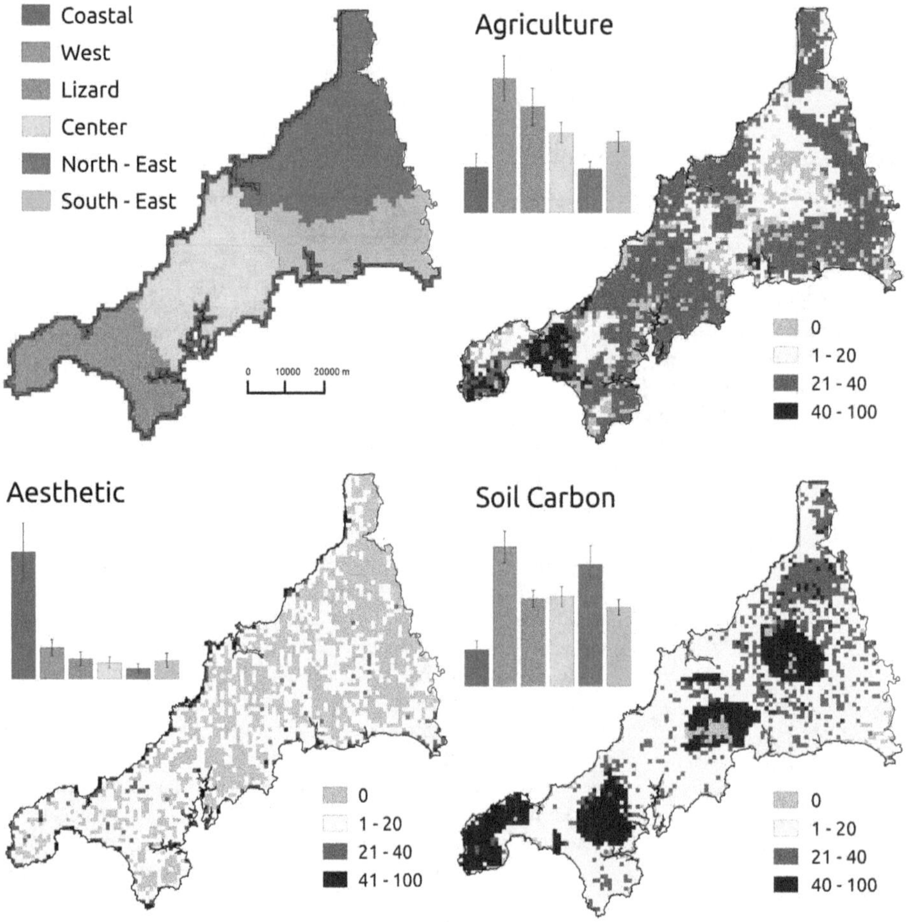

Figure 1. Study area and ecosystem services distribution. Geographical zonation of Cornwall (upper left), the distribution of agriculture, aesthetics and soil carbon (other maps; variation scaled from 0–100), and the mean value of each ecosystem service within each geographical zone (histograms).

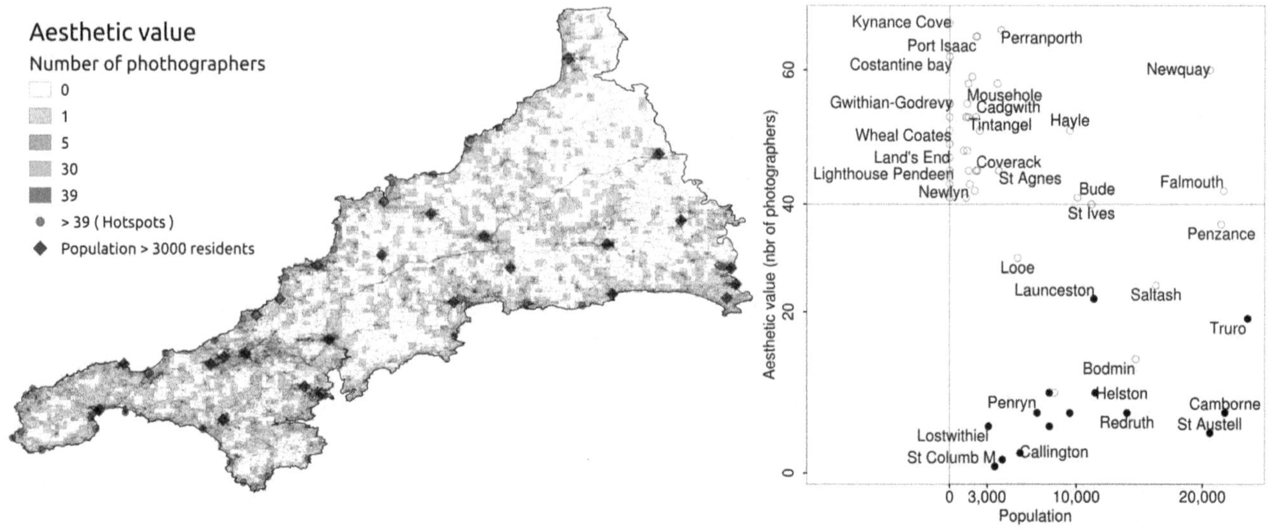

Figure 2. Aesthetic value and population of Cornwall. Aesthetic value map of Cornwall (left) and spatial covariance with the most populated places in Cornwall (right). Population data: Office for National Statistics, mid year estimates 2010. White dots: coastal locations; black dots: inland locations (not all location labels are shown).

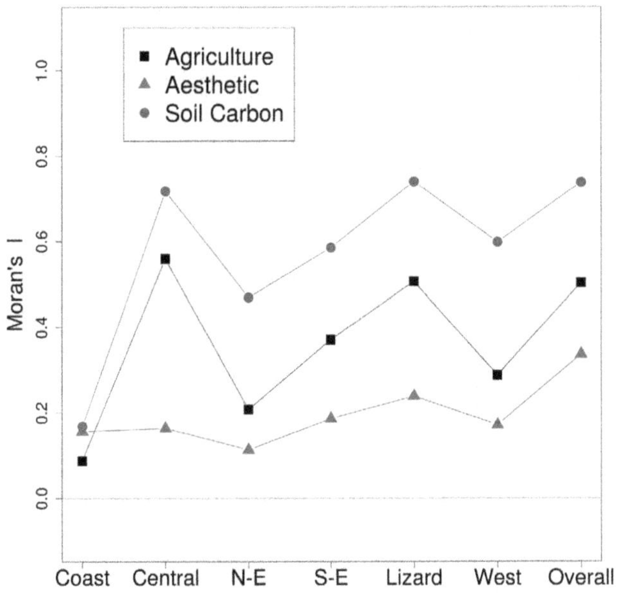

Figure 3. Test of spatial autocorrelation for agriculture, aesthetics and soil carbon data in Cornwall and within each geographical zone.

negatively correlated with that in many other ecosystem services [1,2,11], albeit we consider only single supporting/regulating and provisioning services. This would seem to highlight the need both to explore and test a variety of measures of cultural services, and to identify some such measures that are particularly well-founded and robust. Any failure to do so will likely serve to hinder attempts to ensure that cultural services get fuller consideration when planning for the maintenance of ecosystem service provision both at present and in the future.

We also find that the extent over which patterns of covariation in ecosystem services are determined can have a marked influence on the outcome, with these patterns typically becoming much more variable among smaller areas. Such an outcome has previously been documented [1,3] but over yet larger extents than those examined here. This will make generalizing about the relationships between different services challenging, and raises the spectre that observed relationships between variation in different ecosystems services is highly context-specific. Indeed, one can see an analogous situation arising to that which has developed in the context of attempts to understand spatial relationships between patterns of species richness and environmental variables, in which two schools of thought exist, a prevailing one that continues largely to ignore such complexities and another which focuses on the challenges of statistically handling non-stationary processes [48–50].

Table 1. CRH correlation test between different ecosystem services across Cornwall, and within different regions thereof.

	Coastal	Central	North-E	South-E	Lizard	West	Overall
Agriculture vs Aesthetics							
n	602	419	160	872	1,167	623	3,843
CRH	−0.02	−0.26*	−0.07	−0.07*	−0.12	−0.23*	−0.16*
Agriculture vs Soil Carbon							
n	308	417	160	836	1,154	603	3,478
CRH	0.10	−0.52**	−0.03	−0.07*	−0.45***	−0.16***	−0.22***
Soil Carbon vs Aesthetics							
n	308	417	160	836	1,154	603	3,478
CRH	0.04	0.06	−0.13	0.03	0.02	0.08	−0.11***

Multiple significance tests adjusted using Benjamin-Hochberg's (Benjamin et al. 1995) corrections: *p-value<0.05; ** p-value<0.01; *** p-value<0.001.

Author Contributions

Conceived and designed the experiments: SC KJG. Performed the experiments: SC. Analyzed the data: SC RI KJG. Contributed reagents/materials/analysis tools: SC. Wrote the paper: SC RI CD KJG.

References

1. Anderson BJ, Armsworth PR, Eigenbrod F, Thomas CD, Gillings S, et al. (2009) Spatial covariance between biodiversity and other ecosystem service priorities. J. Appl. Ecol. 46: 888–896. doi:10.1111/j.1365-2664.2009.01666.x.
2. Eigenbrod F, Armsworth PR, Anderson BJ, Heinemeyer A, Gillings S, et al. (2010) Error propagation associated with benefits transfer-based mapping of ecosystem services. Biol. Conserv.143: 2487–2493. doi:10.1016/j.biocon.2010.06.015.
3. Holland RA, Eigenbrod F, Armsworth PR, Anderson BJ, Thomas CD, et al. (2011) The influence of temporal variation on relationships between ecosystem services. Biodivers. Conserv. 20: 3285–3294. doi:10.1007/s10531-011-0113-1.
4. Egoh B, Reyers B, Rouget M, Bode M, Richardson D (2009) Spatial congruence between biodiversity and ecosystem services in South Africa. Biological Conservation 142: 553–562. doi:10.1016/j.biocon.2008.11.009.
5. Millennium Ecosystem Assessment (2005) Ecosystems and human well-being: synthesis. Reid WV, editor. Washington DC:Island Press. 137 p.
6. Seppelt R, Dormann CF, Eppink FV, Lautenbach S, Schmidt S (2011) A quantitative review of ecosystem service studies, approaches, shortcomings and the road ahead. J. Appl. Ecol.48: 630–636. doi:10.1111/j.1365-2664.2010.01952.x.
7. Martın-Lopez B, Iniesta-Arandia I, Garcıa-Llorente M, Palomo I, Casado-Arzuaga I, et al. (2012) Uncovering ecosystem service bundles through social preferences. PLoS ONE 7(6): e38970. doi:10.1371/journal.pone.0038970.
8. Daniel TC, Muhar A, Arnberger A, Aznar O, Boyd JW, et al. (2012) Contributions of cultural services to the ecosystem services agenda. P. Natl. Acad. Sci. USA 109: 8812–8819. doi:10.1073/pnas.1114773109.
9. Gimona A, Horst D (2007). Mapping hotspots of multiple landscape functions, a case study on farmland afforestation in Scotland. Landscape Ecol.22: 1255–1264. doi:10.1007/s10980-007-9105-7.
10. Hill GW, Courtney PR (2006) Demand analysis projections for recreational visits to countryside woodlands in Great Britain. Forestry 79: 185–200. doi:10.1093/forestry/cpl005.
11. Raudsepp-Hearne C, Peterson GD, Bennett EM (2010) Ecosystem service bundles for analyzing tradeoffs in diverse landscapes. P. Natl. Acad. Sci.USA 107: 5242–5247.
12. O'Farrell PJ, De Lange WJ, Le Maitre DC, Reyers B, Blignaut JN, et al. (2011) The possibilities and pitfalls presented by a pragmatic approach to ecosystem service valuation in an arid biodiversity hotspot. J. Arid Environ.75: 612–623. doi:10.1016/j.jaridenv.2011.01.005.
13. Chan KM, Shaw MR, Cameron DR, Underwood EC, Daily GC (2006) Conservation planning for ecosystem services. PLoS Biology 4: 2138–2152.
14. Chan KM, Hoshizaki L, Klinkenberg B (2011) Ecosystem services in conservation planning, targeted benefits vs. co-benefits or costs? PloS ONE 6: e24378. doi:10.1371/journal.pone.0024378.
15. Joyce K, Sutton S (2009) A method for automatic generation of the recreation opportunity spectrum in New Zealand. Appl Geogr 29: 409–418. doi:10.1016/j.apgeog.2008.11.006.
16. Maes J, Paracchini ML, Zulian G (2011) A European assessment of the provision of ecosystem services, towards an atlas of ecosystem services. Tech. rep. European Commission Joint Research Centre. doi:10.2788/63557.
17. Norton L, Inwood H, Crowe A, Baker A (2012) Trialling a method to quantify the cultural services of the English landscape using countryside survey data. Land Use Policy 29: 449–455. doi:10.1016/j.landusepol.2011.09.002.
18. Hein L, Vankoppen K, Degroot R, Vanierland E (2006) Spatial scales, stakeholders and the valuation of ecosystem services. Ecol Econ 57: 209–228. doi:10.1016/j.ecolecon.2005.04.005.
19. Cornwall Council (2009) Economic evidence review. Appendix 1. Green paper (economic priorities and strategic intent). Community Intelligence Chief

Executives Department, Technical report. Available: http://www.cornwall.gov.uk/idoc.ashx?docid = d82fc54b-1d94-4c3e-ae92-b8a445bdef55&version = -1 Accessed 2012 July 3.
20. International Telecommunications Unions website (2012) Percentage of individuals using the internet 2000–2010. Available: http://www.itu.int/ITU-D/ict/statistics/material/excel/2010/IndividualsUsingInternet_00-10.xls.Accessed 2012 July 3.
21. International Telecommunications Unions website (2012) Key global telecom indicators for the world telecommunication service sector. Available: http://www.itu.int/ITU-D/ict/statistics/at_glance/keytelecom.html. Accessed 2012 July 3.
22. Panoramio website (2013) Panoramio acceptance policy for Google Earth and Google Maps. Available: http://www.panoramio.com/help/acceptance_policy. Accessed 2013 May 3.
23. European Commission Joint Research Centre website (2012) The map of organic carbon content in topsoil in Europe: version 1.2. Available: http://eusoils.jrc.ec.europa.eu/ESDB_Archive/octop/octop_data.html. Accessed 2012 July 3.
24. Jones RJA, Hiederer R, Rusco E, Loveland PJ, Montanarella L (2005) Estimating organic carbon in the soils of Europe for policy support. European J. Soil Sci.56: 655–671.
25. Eigenbrod F, Anderson BJ, Armsworth PR, Heinemeyer A, Jackson SF, et al. (2009) Ecosystem service benefits of contrasting conservation strategies in a human-dominated region. Proc. R. Soc. B.276: 2903–11. doi:10.1098/rspb.2009.0528.
26. Departement for Environment, Food and Rural Affairs website (2012) June survey of agriculture 2000. Available: http://www.defra.gov.uk/statistics/foodfarm/landuselivestock/junesurvey/junesurveyresults/. Accessed 2012 July 3.
27. European Environment Agency website (2012) Corine land cover 2000 raster data. Available: http://www.eea.europa.eu/data-and-maps/data/corine-land-cover-2000-raster-1 .Accessed 2012 July 3.
28. Farm Management Handbook(2005) Agricultural Economics Unit. Exeter, IK:University of Exeter. 117 p.
29. Cornwall Council website (2012) Office for national statistics mid year estimates 2010: lower layer super output areas. Available: http://www.cornwall.gov.uk/default.aspx?page = 26436. Accessed July 2012 July 3.
30. Moran PAP (1950) Notes on continuous stochastic phenomena. *Biometrika* 37:17–23.
31. Clifford P, Richardson S, Hemon D (1989) Assessing the significance of the correlation between two spatial processes. *Biometrics* 45: 123–134.
32. Dutilleul P (1993) Modifying the t test for assessing the correlation between two spatial processes. *Biometrics* 49: 305–314.
33. Thomson JD, Weiblen BA, Thomson BA, Alfaro S, Legendre P (1996) Untangling multiple factors in spatial distributions: Lilies, Gophers, and Rocks. *Ecology* 77: 1698–1715.
34. Carroll C, Zielinski W, Noss R (1999) Using presence-absence data to build and test spatial habitat models for the fisher in the Klamath region, U.S.A. *Cons. Biol.* 13: 1344–1359.
35. Dormann CF, McPherson JM, Araujo MB, Bivand R, Bolliger J, et al. (2007) Methods to account for spatial autocorrelation in the analysis of species distributional data: a review. *Ecography* 30: 609–628. doi: 10.1111/j.2007.0906-7590.05171.x.
36. Zar JH (2007) Biostatistical analysis. Upper Saddle River, NJ:Prentice Hall .960 p.
37. Benjamini Y, Hochberg Y (1995) Controlling the false discovery rate: a practical and powerful approach to multiple testing. *J. Roy. Statist. Soc. Ser. B* 57: 289–300.
38. GRASS development team (2012) Geographic Resources Analysis Support System (GRASS) Software. v.6.4.2. Open Source Geospatial Foundation Project. Available: http://grass.osgeo.org. Accessed 2013 Jun 7.
39. R Core Team (2013) R: A Language and Environment for Statistical Computing. R foundation for statistical computing. Available: http://www.R-project.org. Accessed 2013 Jun 7.
40. Lazer D, Pentland A, Adamic L, Aral S, Barabási AL, et al. (2009) Computational social science. *Science* 323: 721–723 doi: 10.1126/science.1167742.
41. Bainbridge WS (2007) The scientific research potential of virtual worlds. *Science* 317 472–476 doi:10.1126/science.1146930.
42. Brownstein CA, Brownstein JS, Williams DS, Wicks P, Heywood JA (2009) The power of social networking in medicine. Nat. biotechnol. 27: 888–90. doi:10.1038/nbt1009-888.
43. Malizia A, Bellucci A, Diaz P, Aedo I, Levialdi S (2011) eStorys, A visual storyboard system supporting back-channel communication for emergencies. J. Visual Lang.Comput. 22: 150–169. doi:10.1016/j.jvlc.2010.12.003.
44. Signorini A, Segre AM, Polgreen PM (2011) The use of Twitter to track levels of disease activity and public concern in the U.S. during the influenza A H1N1 pandemic. *PLoS ONE* 6: e19467. doi:10.1371/journal.pone.0019467.
45. Gonzalez MC, Hidalgo CA, Barbasi AL (2008) Understanding individual human mobility patterns. *Nature* 453: 779–782. doi:10.1038/nature06958.
46. Preis T, Moat HS, Stanley HE (2013) Quantifying trading behavior in financial markets using google trends. *Sci. Rep.*3: 1684. doi:10.1038/srep01684.
47. Preis T, Moat HS, Stanley HE, Bishop SR (2012) Quantifying the advantage of looking forward. *Sci. Rep.* 2: 350. doi:10.1038/srep00350.
48. Casalegno S, Amatulli G, Bastrup-Birk A, Durrant TH, Pekkarinen A (2011) Modelling and mapping the suitability of European forest formations at 1-km resolution. *Eur. J. For. Res.* 130: 971–981. doi:10.1007/s10342-011-0480-x.
49. Bradford AH (2012) Eight (and a half) deadly sins of spatial analysis. *J. Biogeogr.* 39: 1–9. doi:10.1111/j.1365-2699.2011.02637.x.
50. Miller JA (2012) Species distribution models: Spatial autocorrelation and non-stationarity. *Progress in Phys. Geogr.* 36: 681–692.

7

The Impact of Organic Farming on Quality of Tomatoes Is Associated to Increased Oxidative Stress during Fruit Development

Aurelice B. Oliveira[1], Carlos F. H. Moura[2], Enéas Gomes-Filho[1], Claudia A. Marco[3], Laurent Urban[4], Maria Raquel A. Miranda[1]*

1 Universidade Federal do Ceará, Depto. Bioquímica e Biologia Molecular, Fortaleza-CE, Brazil, **2** Embrapa Agroindustria Tropical, Fortaleza-CE, Brazil, **3** Universidade Federal do Ceará, Campus do Cariri, Av. Tenente Raimundo Rocha s/n - Cidade Universitária, Juazeiro do Norte-CE, Brazil, **4** Université d'Avignon et des Pays de Vaucluse, Campus Agroparc, Avignon, France

Abstract

This study was conducted with the objective of testing the hypothesis that tomato fruits from organic farming accumulate more nutritional compounds, such as phenolics and vitamin C as a consequence of the stressing conditions associated with farming system. Growth was reduced in fruits from organic farming while titratable acidity, the soluble solids content and the concentrations in vitamin C were respectively +29%, +57% and +55% higher at the stage of commercial maturity. At that time, the total phenolic content was +139% higher than in the fruits from conventional farming which seems consistent with the more than two times higher activity of phenylalanine ammonia lyase (PAL) we observed throughout fruit development in fruits from organic farming. Cell membrane lipid peroxidation (LPO) degree was 60% higher in organic tomatoes. SOD activity was also dramatically higher in the fruits from organic farming. Taken together, our observations suggest that tomato fruits from organic farming experienced stressing conditions that resulted in oxidative stress and the accumulation of higher concentrations of soluble solids as sugars and other compounds contributing to fruit nutritional quality such as vitamin C and phenolic compounds.

Editor: Hany A. El-Shemy, Cairo University, Egypt

Funding: Funding was provided by BNB-Fundeci, CAPES-REUNI, and CNPq/INCT – Frutos Tropicais. The funders had no role in study design, data collection and analysis, decision to publish, or preparation of the manuscript.

Competing Interests: The authors have declared that no competing interests exist.

* E-mail: rmiranda@ufc.br

Introduction

The consumption of fruit and vegetables has been associated with lower risk of chronic human health problems like cardiovascular diseases, cancer, hypertension and diabetes type two due to their high contents in dietary bioactive compounds, the so-called phytochemicals, endowed with protective properties [1;2;3;4;5]. Until recently, the health benefits of fruits and vegetables have been attributed to the antioxidant properties of the phytochemicals they provide. However, nowadays, the routine consumption of antioxidant supplements has become highly controversial as studies have demonstrated that they can actually be harmful [6]. Besides the antioxidant theory, there are now alternative theories about the way phytochemicals induce protective mechanisms in consumers. For instance, it has been shown that a large range of secondary metabolites in fruit and vegetables as phenolic compounds [7] act as elicitors that activate Nrf2, a transcription factor that binds to the antioxidant response element in the promoter region of genes coding for enzymes involved in protective mechanisms [4]. Supplementation of processed tomato products, containing lycopene, has shown to lower biomarkers of oxidative stress and carcinogenesis in healthy and type II diabetic patients, and prostate cancer patients, respectively. The mechanisms of action involve protection of plasma lipoproteins, lymphocyte DNA and serum proteins against oxidative damage, and anticarcinogenic effects, including reduction of prostate-specific antigen, up-regulation of connexin expression and overall decrease in prostate tumor aggressiveness [5].

As stressed in a recent survey of literature, environmental factors represent a powerful lever to increase the concentrations in phytochemicals [8]. Among all the factors that seem effective in enhancing the concentrations in phytochemicals in fruits and vegetables, stress emerges as especially promising. This makes sense considering that all stresses, either biotic or abiotic, are conducive to oxidative stress in plants [9] and that oxidative signaling controls synthesis and accumulation of secondary metabolites [10;11;12]. Thereby, it may be hypothesized that cropping systems that allow plants to undergo (moderate) stress such as organic farming result in products with higher concentrations in phytochemicals resulting of low mineral availability and, therefore of the diversion of carbon skeletons from protein synthesis [13]. Indeed, several studies have demonstrated that fruits and vegetables from organic farming generally are endowed with enhanced nutritional properties [14;15;16;17]. A recent comparative study showed that organic tomato juice has a higher phenolic content and hydrophilic antioxidant activity when compared to conventional tomato juice [17]. Organic tomatoes from Felicia, Izabella and Paola varieties had higher vitamin C and carotenoid contents which were more pronounced when expressed on fresh matter than on dry matter [14]. Organic strawberries present higher antioxidant concentrations and have been shown to inhibit the proliferation of human colon (HT29) and breast (MCF-7) cancer cells more effectively than conventional ones [15].

The hypothesis that oxidative stress is involved in enhanced concentrations in phytochemicals of fruits and vegetables from organic farming has rarely been tested to our knowledge [18]. The objective of this paper is to contribute to fulfilling this gap by comparing not only the concentrations in compounds contributing to quality in fruits from organic and conventional farming, but also by measuring indicators of oxidative stress, namely the activities of antioxidant enzymes, superoxide dismutase (SOD), ascorbate peroxidase (APX) and catalase (CAT), the concentration in ascorbate (AsA) and cell membrane lipid peroxidation (LPO) degree. The study was conducted on tomatoes which are climacteric fruits, representing a relevant source of vitamins C and E and other phytochemicals such as carotenoids and polyphenols [19]. We focused in this study on phenolic compounds and the activity of phenylalanine ammonia-lyase (PAL) because this enzyme controls a rate-determining step of the biosynthetic pathway of phenolic compounds in plants and is well-known to be induced by environmental stress [20;21;22;23].

Materials and Methods

Fruit Material

Organic and conventional tomato fruits (*Solanum lycopersicum* cv. Débora) cultivated organically and conventionally were obtained from local producers of the district of Crato-Ceará State at northeastern part of Brazil (7°14′03″S and 39°24′34″W). The organic and conventional farms were within 1.5 km from each other, therefore with similar environmental conditions and plants were planted in rows 1 m apart with 0.4 m between plants. The soil was representative of this region of Brazil and classified as humic yellowish latosol with medium phosphate and potassium levels, medium-textured clay and 0–20 cm top layer with pH (H_2O) 6.2; cation exchange capacity of (cmolc.dm^{-3}) Ca^{2+}2.0; Mg^{2+}0.8; K^{+}0.2; Al^{3+}0.0; P (mg.dm^{-3}) 0.6 and organic matter (g.kg^{-1}) 2.6. As reported by the growers, the organic cultivation system used a compost of animal manure (10 t.ha^{-1}), legume cocktail and sugar cane bagasse incorporated into soil just before sowing; and every 10 days, 0.5% Bordo mix (a mixture of copper (II) sulfate ($CuSO_4$) and slaked lime) was used preventively as a fungicide. In the conventional system, the pesticide FASTAC®100 was applied at 0.01% when needed and inorganic fertilizer was used as recommended by the Brazilian agricultural development department at the following rates: 100 kg.ha^{-1} for nitrogen, 400 kg.ha^{-1} for P_20_5 and 80 kg.ha^{-1} for K_20.

Organic and conventional tomatoes were evaluated at three developmental stages: immature (green colored), physiologically mature (breaker), and at the harvesting stage (red). Fruits were harvested manually from 30 plants for each system, then washed in tap water and carefully selected to insure good uniformity in maturity and size. After cleansing and selection, fruits from each treatment were divided into samples made up of four replicates each consisting of twelve fruits. Fruit pericarp was ground and homogenized using a domestic food processor and then stored at −20°C for further analysis.

Fruit Quality Parameters

Weight was measured on a semi-analytical scale (TECNAL, São Paulo-Brazil) as whole fruits were individually weighed and results expressed in grams (g). Size measurements were made with a handheld pachymeter (ZAAS Precision, São Paulo-Brazil) and expressed in centimeters (cm). Soluble solids (SS) content was determined by refractometry as described by AOAC [24] using a digital refractometer (ATAGO N1, Kirkland-USA) with automatic temperature compensation. The results were expressed in °Brix (concentration of sucrose w/w). The pH was measured using an automatic pHmeter (Labmeter PHS-3B, São Paulo-Brazil) as recommended by AOAC (24). Titrable acidity (TA) was evaluated following AOAC (24) by using an automatic titrator (Mettler-Toledo DL12, Columbus-USA). Results were expressed as % of the predominant acid for each species.

Table 1. Changes in quality parameters and chlorophyll content of tomatoes cultivated organically (OG) and conventionally (CV).

Tomato			
Parameters	**Stage**	**OG**	**CV**
Weight (g)	Immature	67.10±12.40 Bb	103.02±15.53 Ab
	Mature	84.88±26.40 Ba	131.52±22.29 Aa
	Ripe	75.15±7.24 Bab	124.93±41.85 Aa
Width (cm)	Immature	4.68±0.33 Bab	5.16±0.51 Aa
	Mature	5.14±0.82 Aa	5.58±0.55 Aa
	Ripe	4.20±0.37 Bb	5.46±0.83 Aa
pH	Immature	4.36±0.06 Ba	4.53±0.07 Aa
	Mature	4.46±0.06 Aa	4.43±0.04 Aa
	Ripe	4.39±0.10 Aa	4.50±0.07 Aa
TA (% citric acid)	Immature	0.33±0.01 Ab	0.28±0.02 Ba
	Mature	0.25±0.00 Ac	0.28±0.03 Aa
	Ripe	0.36±0.00 Aa	0.28±0.00 Ba
SS (°Brix)	Immature	4.80±0.17 Aa	4.17±0.15 Aa
	Mature	5.03±1.06 Aab	4.20±0.10 Aa
	Ripe	6.00±0.00 Aa	3.83±0.51 Ba
Total phenolics (mg GAE.kg^{-1})	Immature	308.5±3.04 Ab	249.1±5.65 Aa
	Mature	508.3±1.51 Aa	299.8±2.39 Ba
	Ripe	556.5±5.40 Aa	232.5±0.62 Ba
Anthocyanins (mg. kg^{-1})	Immature	5.1±0.10 Ba	8.0±0.19 Aa
	Mature	2.5±0.05 Ba	9.0±0.16 Aa
	Ripe	3.6±0.09 Ba	9.9±0.11 Aa
Yellow Flavonoids (mg. kg^{-1})	Immature	27.8±0.15 Bb	37.4±0.33 Aa
	Mature	26.1±0.33 Bb	33.3±0.43 Aab
	Ripe	43.7±0.49 Aa	25.7±0.33 Bb
Total Vitamin C (mg.kg^{-1})	Immature	134.1±0.20 Ac	89.4±0.05 Bb
	Mature	220.5±0.12 Ab	175.3±0.20 Ba
	Ripe	264.7±0.40 Aa	170.9±0.16 Ba
Relative chlorophyll content		40.18±7.20 A	40.29±5.20 A

Mean values in the same column followed by the same small letter did not differ significantly between the developmental stages; by Tukey's test ($p \geq 0.05$).
Mean values in the same line followed by the same CAPITAL letter did not differ significantly between the cultural systems, by Tukey's test ($p \geq 0.05$).

Table 2. Metabolism parameters evaluated during the development of tomato cultivated organically (OG) and conventionally (CV).

Tomato			
Parameters	**Stage**	**OG**	**CV**
Lipid Peroxidation (nmol MDA g^{-1} FW)	Immature	16.93±1.54 Aab	9.84±0.90 Ba
	Mature	14.09±2.71 Ab	7.20±2.18 Aa
	Ripe	19.24±0.09 Aa	8.06±0.65 Ba
PAL Activity (μmol cinnamic ac. g^{-1} mg^{-1} P)	Immature	6.72±0.88 Ab	2.54±0.24 Ba
	Mature	8.22±0.67 Ab	4.06±0.24 Ba
	Ripe	11.43±0.71 Aa	4.79±0.89 Ba
Antioxidant Activity (μM Trolox g^{-1} FW)	Immature	98.72±38.65 Aa	98.18±30.42 Aa
	Mature	143.54±44.52 Aa	161.23±6.15 Aa
	Ripe	128.34±22.89 Aa	136.28±57.54 Aa
APX Activity (μmol H_2O_2 g^{-1} min^{-1} mg^{-1} P)	Immature	143.54±44.52 Aa	161.23±6.15 Aa
	Mature	128.34±22.89 Aa	136.28±57.54 Aa
	Ripe	1.01±0.11 Aab	0.98±0.11 Aa
CAT Activity (μmol H_2O_2 g^{-1} min^{-1} mg^{-1} P)	Immature	4.27±0.20 Aa	4.78±5.22 Ab
	Mature	3.90±1.63 Aa	8.28±4.41 Aab
	Ripe	5.07±1.67 Ba	16.46±6.05 Aa
SOD Activity (UA g^{-1} mg^{-1} P)	Immature	104.95±21.15 Aab	42.16±6.90 Ba
	Mature	77.05±21.02 Ab	47.38±6.50 Aa
	Ripe	121.76±8.33 Aa	22.27±10.08 Ba

Mean values in the same column followed by the same SMALL letter did not differ significantly between the developmental stages; by Tukey's test ($p \geq 0.05$).
Mean values in the same line followed by the same CAPITAL letter did not differ significantly between the cultural systems, by Tukey's test ($p \geq 0.05$).

Antioxidants

The total antioxidant activity (TAA) was determined using the ABTS method as described by Re and others [25], which measures the ability of lipophilic and hydrophilic antioxidants to quench a 2,2′-azino-bis 3-ethylbenzthiazoline-6-sulphonic acid ($ABTS^{*+}$, Sigma) radical cation. Before the colorimetric assay, the samples were subjected to a procedure of extraction in 50% methanol and 70% acetone as described by Larrauri and others [26]. The radical solution was formed using 7 mM $ABTS^{*+}$ and 140 mM potassium persulfate, incubated and protected from light for 16 h. Once the radical was formed, the reaction was started by adding 30 μL of extract in 3 mL of radical solution. Absorbance was measured (734 nm) after 6 min, and the decrease in absorption was used to calculate the TAA. A calibration curve was prepared and different Trolox (Sigma) concentrations (standard trolox solutions ranging from 100 to 2000 μM) were also evaluated against the radical. Antioxidant activity was expressed as Trolox equivalent antioxidant capacity (TEAC), μmol Trolox. g^{-1} FW (fresh weight).

The total phenolic content was measured by a colorimetric assay using Folin–Ciocalteu reagent (Sigma) as described by Obanda and Owuor [27]. Before the colorimetric assay, the samples were subjected to a procedure of extraction in 50% methanol and 70% acetone as described by Larrauri and others [26]. Absorbance was measured at 700 nm, gallic acid (Acros Organics) was used as the standard and results were expressed as galic acid equivalents (GAE) $mg.kg^{-1}$ FW.

Anthocyanins and yellow flavonoids were extracted and determined as described by Francis [28]. The absorbance of the filtrate was measured at 535 nm and at 374 nm for total anthocyanin and for yellow flavonoid content using absorption coefficients of 98.2 and 76.6, respectively. The results were expressed as mg. kg^{-1} FW.

Total vitamin C was determined by titration with Tillman solution (0.02% 2,6 dichloro-indophenol, DFI from Sigma) described by Strohecker and Henning [29]. The results were expressed as mg. kg^{-1} FW.

Enzymes

Samples of two grams of pulp were homogenized in an ice-cold extraction buffer (100 mM potassium-phosphate buffer pH 7.0+0.1 mM EDTA). The homogenate was filtered through a muslin cloth and centrifuged at 3300×*g* for 40 min. The supernatant fraction was used as a crude extract for antioxidant enzyme activity assays. All the procedures were performed at 4°C. The total protein content was determined according to Bradford [30].

Catalase (CAT, EC 1.11.1.6) activity was measured according to Beers and Sizer [31]. The decrease in H_2O_2 (Merck) was monitored through absorbance at 240 nm and quantified by its molar extinction coefficient (36 M^{-1} cm^{-1}). The results were expressed as μmol H_2O_2. min^{-1}. mg^{-1} P (protein).

Ascorbate peroxidase (APX, EC 1.11.1.1) activity was assayed according to Nakano and Asada [32]. The reaction was started by adding ascorbic acid and ascorbate oxidation was measured through absorbance at 290 nm. Enzyme activity was measured using the molar extinction coefficient for ascorbate (2.8 mM. cm^{-1}) and the results expressed in μmol H_2O_2. min^{-1}. mg^{-1} P, taking into account that 1 mol of ascorbate is required for the reduction of 1 mol H_2O_2.

Superoxide dismutase (SOD, EC 1.15.1.1) activity was determined by spectrophotometry, based on the inhibition of the photochemical reduction of nitroblue tetrazolium chloride (NBT, Sigma) [33]. The absorbance was measured at 560 nm and one unit of SOD activity (UA) was defined as the amount of enzyme required to cause a 50% reduction in the NBT photo-reduction rate. Thus, results were expressed as UA. mg^{-1} P.

Phenylalanine ammonia-lyase (PAL, EC 4.3.1.24) activity was assayed with samples extracted by a modified version of the method developed by Mori and others [34]. Pulp samples (1 g) were homogenized for 3 min at 4°C with (2 mL) 0.1 M Tris-HCl buffer solution (pH 8.0), 1 mM EDTA and 0.5 g PVP, and then centrifuged at 5000×*g* for 20 min. The supernatant was the extract used to determine the PAL activity which was assayed using an assay modified from that of El-Shora [21]. The reaction mixture contained 100 mM Tris-HCl buffer (pH 8.4), 40 mM L-phenylalanine and 100 μL of enzyme to a total volume of 880 μL. The reaction was stopped by adding 6 M HCl and the A_{290} of the clear solution was measured. The PAL activity was expressed as μmol of trans-cinnamic acid. mg^{-1} P.

Lipid Peroxidation Degree

The thiobarbituric acid reactive substances (TBARS) assay measures lipid peroxidation degree through determination of hydroperoxides and aldehydes such as malondialdehyde (MDA). MDA reacts with thiobarbituric acid (TBA, 1:2) to form a fluorescent adduct MDA-$(TBA)_2$ using a modified version of the method developed by Zhu and others [35]. Pulp samples (0.5 g) were homogenized in 5 mL of 0.1% trichloroacetic acid (TCA) and centrifuged at 3300×*g* at 4°C for 20 min. The supernatant (750 μL) was collected and added to 3 mL 0.5% TBA in 20% TCA and incubated at 95°C for 30 min. Following incubation, the tubes were immediately cooled in ice bath and centrifuged at 3000×*g* for 10 min. Absorbance at 532 nm was measured and corrections were made for unspecific turbidity by subtracting the absorbance at 600 nm. TBARS are expressed as MDA equivalents, calculated using an extinction coefficient of 155 $mmol.cm^{-1}$ and expressed as nmol. g^{-1} FW.

Leaf Relative Chlorophyll Content

The relative levels of total chlorophyll were estimated with a portable chlorophyll meter (SPAD-502 Minolta, Osaka, Japan). Measurements were performed on four of the youngest fully expanded leaves on five different plants and results were expressed as SPAD values.

Statistical Analysis

All results are expressed as means ± standard error (SE). Analysis of variance, followed by multiple comparisons of means was performed using SISVAR version 5.1. Means were compared using Tukey's test at $\alpha = 0.05$.

Results and Discussion

Fruits from Organic Farming were More Stressed than Fruits from Conventional Growing Systems

The quality parameters of organic and conventional tomatoes were evaluated throughout the period of maturation of fruits (Table 1). There were substantial differences in growth between the fruits of the two growing systems compared. Mass and size were about 40% higher in fruits from conventional growing systems than in fruits from organic farming. Such differences could originate either from differences in nitrogen availability or from limitations to growth imposed by the more stressing conditions prevailing in organic farming [13]. The idea that stressing conditions impact negatively fruit size and mass is a very common one. Zushi and others [36], for instance, observed that salt-stressed tomatoes had a shorter developmental period with lower mass and accelerated ripening.

The hypothesis of lower nitrogen availability in organic farming cannot be totally excluded from our observations even though one would have expected substantial differences in leaf chlorophyll content, which was clearly not the case (Table 1). All the same, there are ample reasons to consider that, in our trial, conditions were more stressing in organic farming than in conventional growing systems. Superoxide dismutase (SOD) activity was significantly higher in organic tomatoes, especially at the harvesting stage (+90%), but not the activities of APX and catalase (Table 2). SOD scavenges radical superoxide by catalyzing its conversion to H_2O_2, which subsequently is neutralized by CAT or APX. It may, thus be hypothesized that in fruits from organic farming there was an increase in concentration in H_2O_2. This could explain the ca. 60% higher LPO degree (Table 2) and the up to 140% higher PAL activity (Table 2) we observed. The higher concentration in ascorbate (+57% at the stage of harvesting maturity) seems to confirm that oxidative stress was higher in fruits from organic farming than in fruits from conventional growing systems (Table 1). A higher concentration in ascorbate in fruits from organic farming seems consistent with other observations like the ones made by Chassy and others [37] on tomatoes.

The Metabolism of Phenolic Compounds was Stimulated in the Fruits from Organic Farming when Compared to Fruits from Conventional Growing Systems

Phenylalanine ammonia lyase (PAL) activity increased slightly during the development of tomatoes and was significantly higher (up to 140%) in organic fruits (Table 2). The total phenolic content differed greatly between cropping systems. Conventional tomato fruits presented lower and constant total phenol concentrations during their development (~250 mg GAE. kg^{-1}), while there was an increase from 308 to 556 mg $GAE.kg^{-1}$ in organic fruits. PAL is a key enzyme in both plant development and defense [22]. It catalyzes the conversion of L-phenylalanine into transcinnamate, the initial committed step of the multi-branched phenylpropanoid pathway in higher plants. This step is known to be a rate-limiting one of the biosynthetic pathway of phenolic compounds; which explains why the concentration in total phenolic compounds was eventually higher during maturation of tomato fruits from organic farming (Table 1). Considering the stimulating role stress exerts on PAL activity [22;38;39], it is tempting to link the increase in PAL activity and the subsequent increase in total phenolics we observed in fruits from organic farming with the indicators of oxidative stress we measured. Several other studies have demonstrated that fruits and vegetables from organic farming are usually richer in phenolic compounds than those from conventional growing systems [18;20;23]. What is novel in our study is that our observations seem to confirm the hypothesis of a link between phenolic metabolism and oxidative stress.

Interestingly, yellow flavonoids and anthocyanins did not follow the pattern of total phenolics (Table 1). For instance, the concentration in yellow flavonoids was +70% higher in organic fruits when compared to fruits from conventional growing system, but only at the harvesting stage, which is consistent with similar observations by Mitchell and others [40]. The concentration in anthocyanins was lower in the fruits from organic farming at all three stages of fruit development. These discrepancies indicate that organic farming had the effect of modifying the levels of transcripts or the activities of enzymes controlling intermediary steps of the biosynthetic pathway of phenolic compounds. In spite of the changes in antioxidants, the total antioxidant activity was not significantly different among the organic and conventional tomatoes (Table 2).

Our Observations Only Partly Support the Growth Differentiation Balance Hypothesis

It has often been hypothesized that high carbohydrate supply, that would result of low nitrogen availability and, therefore of the diversion of carbon skeletons from protein synthesis, is favorable to the synthesis of secondary metabolites through its positive influence on precursor availability [41]. But then, ecological theories, such as the much-debated Growth Differentiation Balance Hypothesis (GDBH) predict that at the highest levels of mineral resource availability, plants would decrease their relative investment in the so-called differentiation processes, including the secondary metabolism [42,43,44]. Indeed, our observations apparently support the GDBH. In conventional fruits, that correspond to the highest availability of mineral resources as mass and size data would suggest, the concentration in total phenolics was the lowest, in contrary of those found for fruits from organic farming (Table 1). As suggested by observations made on Citrus fruits [45], the primary/secondary metabolisms interaction/competition theory may apply to whole plants which have to continuously arbitrate between growth and defense, but is probably not relevant for storage organs like fruits where secondary metabolites accumulate in response to a development programme aiming, among others, at advertising its nutritional status to potential seed disseminators.

Conclusions

Our work clearly demonstrates that tomato fruits from organic farming have indeed a smaller size and mass than fruits from conventional growing systems, but also a substantially better quality in terms of concentrations in soluble solids and phytochemicals such as vitamin C and total phenolic compounds. Until recently, the focus has been mainly on yield rather than on gustative and micronutritional quality of fresh plant products. This might be all right for staple food, but, as far as fruits and vegetables are concerned, it may be argued that gustative and micronutritional quality matter more than energy supply. Our observations suggest that, at least for fruit and vegetable production, growers should not systematically try to reduce stress to maximize yield and fruit size, but should accept a certain level of stress as that imposed by organic farming with the objective of improving certain aspects of product quality. More research is needed in the future to better understand the links between stress and oxidative stress, on one side, and oxidative stress and secondary metabolism in fruits, on the other side. Also the physiological mechanisms behind the positive effect of organic farming on fruit quality will require additional studies to be conducted.

Author Contributions

Conceived and designed the experiments: MRAM LU EG-F. Performed the experiments: ABO CFHM CAM. Analyzed the data: ABO MRAM. Contributed reagents/materials/analysis tools: MRAM EG-F. Wrote the paper: ABO MRAM.

References

1. Manach C, Scalbert A, Morand C, Remesy C, Jimenez L (2004) Polyphenols: food sources and bioavailability. Am J Clin Nutr 79: 727–7472.
2. Chun OK, Kim DO (2004) Consideration on equivalent chemicals in total phenolic assay of chlorogenic acid-rich plums. Food Res. Int. 37: 337–342.
3. Kuskoski M, Asuero A, Troncoso A (2005) Aplicación de diversos métodos químicos para determinar actividad antioxidant em pulpa de frutos. Ciênc Tecnol Aliment 25(4): 726–732.
4. Surh YJ, Na HK (2008) NF-κB and Nrf2 as prime molecular targets for chemoprevention and cytoprotection with anti-inflammatory and antioxidant phytochemicals. Genes & Nut 2: 313–317.
5. Basu A, Imrham V (2007) Tomatoes versus lycopene in oxidative stress and carcinogenesis: conclusions from clinical trials. European Journal of Clinical Nutrition, vol 61: 295–303.
6. Bjelakovic G, Nikolova D, Gluud LL, Simonetti RG, Gluud C (2007) Mortality in randomized trials of antioxidant supplements for primary and secondary prevention systematic review and meta-analysis. JAMA: J Am Med Assoc 297: 842–857.
7. Foyer CH, Noctor G (2009) Redox regulation in photosynthetic organisms: signaling, acclimation, and practical implications. Antioxid. Redox Signaling 11: 861–905.
8. Poiroux-Gonord F, Bidel LPR, Fanciullino AL, Gautier H, Lauri-Lopez F, et al. (2010) Health benefits of vitamins and secondary metabolites of fruits and vegetables and prospects to increase their concentrations by agronomic approaches. J Agric Food Chem 58: 12065–12082.
9. Grassmann J, Hippeli S, Elstner EF (2002) Plant's defence and its benefits for animals and medicine: role of phenolics and terpenoids in avoiding oxygen stress. Plant Physiol Biochem 40: 471–478.
10. Bouvier F, Backhaus RA, Camara B (1998) Induction and control of chromoplast-specific carotenoid genes by oxidative stress. J Biol Chem 273: 30651–30659.
11. Fujita M, Fujita Y, Noutoshi Y, Takahashi F, Narusaka Y, et al. (2006) Crosstalk between abiotic and biotic stress responses: a current view from the points of convergence in the stress signaling networks. Curr Opinion Plant Biol 9: 436–442.
12. Kuntz M, Chen HC, Simkin AJ, Romer S, Shipton CA, et al. (1998) Upregulation of two ripening-related genes from a nonclimacteric plant (pepper) in a transgenic climacteric plant (tomato). Plant J 13: 351–361.
13. Urban L, Gonord F, Berti L, Sallanon H, Lauri-Lopez F (2008) Effet de l'agriculture biologique sur la valeur-santé des fruits et des légumes : réflexions et études en cours sur clémentine et tomate. Colloque DinABio. Montpellier.
14. Caris-Veyrat C, Amiot MJ, Tyssandier V, Grasselly D, Buret M, et al. (2004) Influence of organic versus conventional agricultural practice on the antioxidant microconstituent content of tomatoes and derived purees; consequences on antioxidant plasma status in humans. J Agric Food Chem 52: 6503–6509.
15. Olsson ME, Andersson S, Oredsson S, Berglund RH, Gustavsson KE (2006) Antioxidant levels and inhibition of cancer cell proliferation in vitro by extracts from organically and conventionally cultivated strawberries. J Agric Food Chem 54: 1248–1255.
16. Luthria D, Singh AP, Wilson T, Vorsa N, Banuelos GS, et al. (2010) Influence of conventional and organic agricultural practices on the phenolic content in eggplant pulp: plant-to-plant variation. Food Chem 121: 406–411.
17. Vallverdu-Queralt A, Medina-Remon A, Casals-Ribes I (2012) Is there any difference between the phenolic content of organic and conventional tomato juices? Food chem. 130(1): 222–227.
18. Jin P, Wang SY, Wang CY, Zheng Y (2011) Effect of cultural system and storage temperature on antioxidant capacity and phenolic compounds in strawberries. Food Chem 124: 262–270.
19. Lima VLAG, Melo EA, Maciel MIS, Prazeres FG, Musser RS, et al. (2005) Total phenolic and carotenoid contents in acerola genotypes harvested at three ripening stages. Food Chem. 90: 565–568.
20. Hafiz IM, Hawa ZEJ, Asmah R (2011) Effects of nitrogen fertilization on synthesis of primary and secondary metabolites in three varieties of Kacip Fatimah (*Labisia pumila* B.). Int J Mol Sci 12(8): 5238–5254.
21. El-Shora HM (2002) Properties of phenylalanine ammonia-lyase from Marrow cotyledons. Plant Sci 162: 1–7.
22. Chang A, Lim M, Lee S, Robb EJ, Nazar RN (2008) Tomato phenylalanine ammonia-lyase gene family, highly redundant but strongly underutilized. J Biol Chem 283(48): 33591–33601.
23. El-Mergawi RA, Al-Redhaiman K (2010) Effect of organic and conventional production practices on antioxidant activity, antioxidant constituents and nutritional value of tomatoes and carrots in Saudi Arabia markets. J Food, Agric Environ 8: 253–258.
24. Association of Official Analytical Chemistry (2005) Official methods of Analysis of the Association of Official Analytical Chemistry. 15th ed. Washington: AOAC.
25. Re R, Pellegrini N, Proteggente A, Pannala A, Yang M, et al. (1999) Antioxidant activity applying an improved ABTS radical cation decolorization assay. Free Radical Biol Med 26(9/10): 1231–1237.
26. Larrauri JA, Rupérez P, Saura-Calixto F (1997) Effect of drying temperature on the stability of polyphenols and antioxidant activity of red grape pomace peels. J Agric Food Chem 45: 1390–1393.
27. Obanda M, Owuor PO, Taylor SJ (1997) Flavonol composition and caffeine content of green leaf as quality potential indicators of Kenyan black teas. J Sci Food Agric 74: 209–215.

28. Francis FJ (1982) Analysis of anthocyanins. In: MARKAKIS, P. Anthocyanins as food color. Academic Press, New York, 181–207.
29. Strohecker R, Henning HM (1967) Analisis de vitaminas: métodos comprobados. Madrid: Ed. Paz Montalvo, 428 p.
30. Bradford MM (1976) A rapid and sensitive method for the quantification of microgram quantities of protein utilizing the principle of protein-dye binding. Anal Biochem 722: 248–254.
31. Beers-Jr RF, Sizer IW (1952) A spectrophotometric method for measuring the breakdown of hydrogen peroxide by catalase. J Biol Chem 195: 133–140.
32. Nakano Y, Asada K (1981) Hydrogen peroxide is scavenged by ascorbate-specific peroxidases in spinach chloroplast. Plant Cell Physiol 22: 867–880.
33. Giannopolitis CN, Ries SK (1977) Superoxide dismutase. I. Occurrence in higher plants. Plant Physiol 59: 309–314.
34. Mori T, Sakurai M, Sakuta M (2001) Effects of conditional medium on activities of PAL, CHS, DAHP synthases (DS-Co and Ds-Mn) and anthocyanin production in suspension cultures of *Fragaria ananassa*. Plant Sci 160: 355–360.
35. Zhu S, Sun L, Liu M, Zhou J (2008) Effect of nitric oxide on reactive oxygen species and antioxidant enzymes in kiwifruit during storage. J Sci Food Agric 88: 2324–2331.
36. Zushi K, Matsuzoe N, Kitano M (2009) Developmental and tissue-specific changes in oxidative parameters and antioxidant systems in tomato fruits grown under salt stress. Sci Hortic 122: 362–368.
37. Chassy AW, Bui L, Renaud ENC, Horn MV, Mitchell A (2006) Three-year comparison of the content of antioxidant, microconstituents and several quality characteristics in organic and conventional managed tomatoes and bell peppers. J Agric Food Chem 54: 8244–8252.
38. Beno-Moualem D, Prusky D (2000) Early events during quiescent infection development by *Colletotrichum gloeosporioides* in unripe avocado fruits. Phytopathology 90: 553–559.
39. Guo J, Wang MH (2010) Ultraviolet A-specific induction of anthocyanin biosynthesis and PAL expression in tomato (*Solanum lycopersicum* L.). Plant Growth Regul 62: 1–8.
40. Mitchell AE, Hong YJ, Koh E, Barret DM, Bryant DE, et al. (2007) Ten-year comparison of the influence of organic and conventional crop management practices on the content of flavonoids in tomatoes. J Agric Food Chem 55: 6154–6159.
41. Cunningham FX (2002) Regulation of carotenoid synthesis and accumulation in plants. Pure Appl. Chem. 74: 1409–1417.
42. Lorio PL (1986) Growth differentiation balance - A basis for understanding southern pine-beetle tree interactions. For Ecol Manage 14: 259–273.
43. Luxmore RJ (1991) A source sink framework for coupling water, carbon, and nutrient dynamics of vegetation. Tree Physiol 9: 267–280.
44. Herms DA, Mattson WJ (1992) The dilemma of plants - To grow or to defend. Q Rev Biol 67: 283–335.
45. Poiroux-Gonord F, Fanciullino AL, Berti L, Urban L (2012) Effect of fruit load on maturity and carotenoid content of clementine (*Citrus clementina* Hort.) fruits. J. Sci Food Agric., 92, 2076–2083.

8

Organic Farming Favours Insect-Pollinated over Non-Insect Pollinated Forbs in Meadows and Wheat Fields

Péter Batáry[1,2]*, Laura Sutcliffe[3], Carsten F. Dormann[4,5], Teja Tscharntke[1]

1 Agroecology, Georg-August University, Göttingen, Germany, **2** MTA-ELTE-MTM Ecology Research Group, Budapest, Hungary, **3** Plant Ecology and Ecosystem Research, Georg-August University, Göttingen, Germany, **4** Computational Landscape Ecology, UFZ Centre for Environmental Research, Leipzig, Germany, **5** Biometry and Environmental System Analysis, University of Freiburg, Freiburg, Germany

Abstract

The aim of this study was to determine the relative effects of landscape-scale management intensity, local management intensity and edge effect on diversity patterns of insect-pollinated vs. non-insect pollinated forbs in meadows and wheat fields. Nine landscapes were selected differing in percent intensively used agricultural area (IAA), each with a pair of organic and conventional winter wheat fields and a pair of organic and conventional meadows. Within fields, forbs were surveyed at the edge and in the interior. Both diversity and cover of forbs were positively affected by organic management in meadows and wheat fields. This effect, however, differed significantly between pollination types for species richness in both agroecosystem types (i.e. wheat fields and meadows) and for cover in meadows. Thus, we show for the first time in a comprehensive analysis that insect-pollinated plants benefit more from organic management than non-insect pollinated plants regardless of agroecosystem type and landscape complexity. These benefits were more pronounced in meadows than wheat fields. Finally, the community composition of insect-pollinated and non-insect-pollinated forbs differed considerably between management types. In summary, our findings in both agroecosystem types indicate that organic management generally supports a higher species richness and cover of insect-pollinated plants, which is likely to be favourable for the density and diversity of bees and other pollinators.

Editor: Bruno Hérault, Cirad, France

Funding: P.B. was supported by the Alexander von Humboldt Foundation and during the preparation of the paper by the Deutsche Forschungsgemeinschaft (DFG BA 4438/1-1). The publication was funded by the University of Göttingen in the funding programme Open Access Publishing. The funders had no role in study design, data collection and analysis, decision to publish, or preparation of the manuscript.

Competing Interests: The authors have declared that no competing interests exist.

* E-mail: pbatary@gmail.com

Introduction

Agricultural intensification is a major driver of biodiversity loss, affecting not only endangered species but increasingly also common and generalist species [1,2]. Agricultural intensification can occur on different spatial scales, from local and landscape to regional scales [3,4,5]. Common types of local-scale intensification are the increased use of agrochemicals and deep ploughing in arable crops, and increased grazing and mowing in grasslands. During the last two decades the importance of landscape-scale intensification has also been recognised, including landscape simplification due to the loss of non-crop natural habitats such as hedges and natural field boundaries.

In order to counteract the negative effects of agricultural intensification, agri-environment schemes (AES) have been initiated in many countries [6]. AES aim to compensate farmers for the potential loss of income when they reduce intensity of production, with expected positive outcomes for biodiversity, ecosystem services and environmental pollution [7,8]. AES measures cover a wide range of approaches, including in some regions the promotion of organic farming. Both organic farming, which explicitly forbids the use of chemical inputs, and other AES measures have been the focus of increasing numbers of studies in recent years, assessing their effectiveness in terms of biodiversity conservation, e.g. [7]. Many studies focus on the importance of AES for species richness and populations of endangered species, see e.g. the syntheses of [4,5,6], whereas relatively few deal with the effects on the functional composition of plant communities, such as pollination type, but see [9,10,11,12]. Dependence on pollinators is an important plant trait facilitating genetic exchange [13]. Animal (mostly insect) pollination is an important ecosystem function, supporting 88% of all plant species across the globe [14]. However, disturbances such as intensive agricultural practices lead to disruptions of plant-pollinator interactions, and plant communities in disturbed environments are thus characterized by higher proportions of wind-pollinated species [15]. Gabriel and Tscharntke [16] showed that organic farming can benefit insect-pollinated arable weeds resulting in a shift in arable weed community structure towards a higher proportion of insect-pollinated species in organic crop fields. Similarly, Power et al. [17] found that insect-pollinated plants benefit more from organic management than non-insect pollinated plants in grasslands. Biesmeijer et al. [18] additionally showed that insect-pollinated plants and wild bees decrease in tandem.

In this study, we compared the species richness and cover of insect-pollinated vs. non-insect pollinated forbs in organic vs. conventional meadows and wheat fields along a landscape-scale management intensity gradient. To the best of our knowledge, there are no published studies on effects of organic vs. conventional management on insect vs. non-insect pollinated

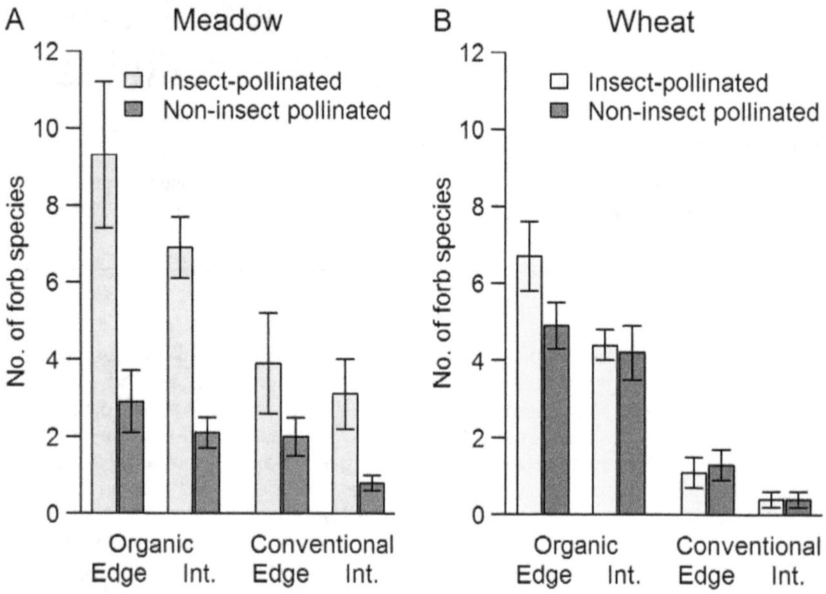

Figure 1. Mean (± SEM) forb richness in organic vs. conventional meadows (A) and in organic vs. conventional wheat fields (B). Data were gathered in edge and interior (Int.) transects of 20 m^2.

plant communities considering both arable fields and grasslands in their landscape context. Previous studies on individual agroecosystem types found a lower percentage of insect-pollinated vs. non-insect pollinated forb species in conventional than in organic management: for wheat fields this was ca. 9.0% lower [16] and for meadows this was ca. 5.5% lower. Thus we hypothesized that organic management favours insect-pollinated over non-insect pollinated forbs in wheat fields [16] and also in meadows [17], but with stronger effects in the more disturbed agroecosystem, i.e. in the wheat fields. In addition to effects of management type, we hypothesized that field edges support more insect-pollinated forbs than field centres due to influences such as lower management intensity and increased light availability [16]. Decreasing landscape-scale management intensity was also hypothesised to enhance insect-pollinated forb species and their abundance through increased propagule pressure from surrounding semi-natural habitats [19].

Methods

Ethics Statement

Permission to access fields and survey vegetation was obtained from all farmers.

Study Area and Field Survey

We selected nine landscapes along a landscape-scale management intensity gradient (percent intensively used agricultural area, i.e. proportion of conventionally managed crop fields and grasslands, IAA: 48–98%) within a 35 km radius of the city of Göttingen, Lower Saxony, Germany (Fig. S1 in Supporting Information). In each landscape, a pair of conventional and organic winter wheat fields and a pair of conventional and organic permanent meadows were selected in close vicinity to each other (within-pair distance of wheat fields (mean ± SEM): 716±185 m; within-pair distance of meadows: 715±185 m; distance between wheat fields and meadows within the same landscape: 1101±109 m; distance between landscapes: 21.1±1.9 km). The two pairs per landscape resulted in 36 fields belonging to 24 farmers (most farmers managed mixed arable and livestock farms).

The study area is characterised by an agricultural mosaic of mostly intensively used arable crops and fertilised meadows, which also contains forest remnants and small fragments of semi-natural habitats such as calcareous grassland, naturally developed fallows, field margin strips and hedges. Around each field, the surrounding landscape was characterized based on official digital thematic maps (ATKIS DTK 50, year 2003) within a circle of 500 m radius using ArcGIS 9.2 (ESRI 2006). Prior calculating the landscape composition for each field, the thematic maps were improved based on field surveys. This distance has previously been found suitable to analyse landscape effects on plants in cereal cropping systems [20,21]. The centre of the 500 m radius buffer was in the mid-point of the rectangle formed by the two transects in each field (see designation of transects below). There were in some cases overlaps of buffers within landscapes, but none between landscapes (Fig. S1). The proportion of intensively used agricultural area (% IAA) in a 500 m radius area did not differ significantly between organic and conventional fields for wheat fields or for meadows (t-test for paired samples, $p>0.15$).

The selected organic and conventional wheat fields received twice as much nitrogen fertiliser as meadows, while conventionally managed fields of both agroecosystems (i.e. meadows and wheat fields) received about four times as much fertiliser than organic ones (organic meadow (mean ± SEM): 29±19 kg N/ha; conventional meadow: 116±30 kg N/ha; organic wheat: 44±22 kg N/ha; conventional wheat: 209±22 kg N/ha). Organic fields were all managed without pesticides and synthetic fertilisers, and organic management had been practiced in all fields were older than 10 years. Conventional meadows were in a few cases treated with herbicides, and all meadows were improved meadows. Conventional meadows were mown nearly twice as frequently (2.8±0.3 times per year) as the organic ones (1.7±0.3 times per year), with the first cut in mid-May. 6 organic and 4 conventional meadows were additionally grazed in the fall (for more details on management and landscape structure see [22]).

Table 1. Results of general linear mixed models testing the effects of landscape composition (intensive agricultural area %), agroecosystem type (meadow vs. wheat field), management (organic vs. conventional), position in field (edge vs. interior) and pollination (insect-pollinated vs. non-insect pollinated) on species richness and percentage cover of forbs in meadows and in wheat fields.

	Variable	df	F	p	effect
Meadow					
Species richness	Landscape	24	1.10	0.306	
	Management	24	13.95	0.001	C<O
	Position in field	24	0.23	0.143	
	Pollination	33	75.65	<0.001	IP> NP
	Landscape×Pollination	33	2.68	0.111	
	Management×Pollination	33	20.68	<0.001	
Cover	Landscape	23	0.41	0.526	
	Management	23	2.41	0.134	
	Position in field	23	0.32	0.577	
	Pollination	33	59.99	<0.001	IP> NP
	Management×Landscape	23	0.79	0.384	
	Landscape×Pollination	33	0.65	0.426	
	Management×Pollination	33	16.58	<0.001	
Wheat field					
Species richness	Landscape	24	3.98	0.057	
	Management	24	118.18	<0.001	C<O
	Position in field	24	7.92	0.010	E>I
	Pollination	33	0.04	0.845	
	Landscape×Pollination	33	1.23	0.275	
	Management×Pollination	33	4.20	0.048	
Cover	Management	25	102.78	<0.001	C<O
	Position in field	25	4.18	0.052	
	Pollination	34	5.46	0.465	
	Management×Pollination	34	1.99	0.168	

df: denominator degrees of freedom. Effect: direction of the significant effect (C: conventional, O: organic; E: edge, I: interior; IP: insect-pollinated, NP: non-insect pollinated).

The yields for wheat fields were (mean ± SEM): organic wheat 44±5 dT/ha; conventional wheat 80±7 dT/ha.

In each field, one edge (in the first wheat row, or in case of meadows next to the edge) and one field interior transect (30 m into the centre and parallel to the edge in both agroecosystems) were surveyed in June 2008. In each transect four 5×1 m plots (288 plots in total) were established, spaced 12 m apart. Field edges were bordered by grassy field margins. Cover of each plant species (%), bare ground (%) and cover of cereal (%; only in wheat fields) was estimated in each plot. Subsequently, relative cover of each species and the total number of plant species (i.e. species richness per 20 m^2) were recorded for each transect. Relative cover (%) per species was calculated by dividing the cover of the given species by total plant cover plus bare ground cover and also wheat cover in case of wheat fields. In the current study we exclusively focus on forbs, hence other plants (mainly grasses) were not included in the analyses.

Statistical Analyses

In order to study the effect of a) landscape-scale management intensity (% IAA), b) management (organic vs. conventional), c) within-field position (edge vs. interior) and d) pollination type (insect vs. non-insect pollinated) on species richness and cover of forbs, we classified all forb species according to their pollination type in two groups: insect vs. non-insect (i.e. self and wind) pollination using the BiolFlor database [23]. Species were classified as insect-pollinated if the database stated that they can be also or exclusively pollinated by insects (many plants that normally are insect pollinated can also be e.g. wind-pollinated). Since forb cover was generally much higher in meadows than in wheat fields, we performed separate statistical analyses for these two agroecosystem types. Additionally, we further classified the forb species in bumblebee-pollinated vs. non-bumblebee pollinated forb species based on the BiolFlor database [23].

First, general linear mixed-effects models with Maximum Likelihood method for nested sampling were used to analyse the effects of the four explanatory variables (a–d) listed above and their two-way interactions on species richness and total relative cover of forbs. The following random factors were used (number of observations –72): landscape (9) and transect (36). Transect as random factor was included to account for the fact that the number of species and cover in each trait group were quantified at the same locations. The same analyses were performed for bumblebee-pollinated vs. non-bumblebee pollinated forb species. Calculations were performed using the nlme package (version 3.1 [24]) for R 2.11.1 [25]. Models were simplified with a stepwise model selection based on AIC by using 'stepAIC' function of MASS package [26].

Second, in order to test whether landscape-scale management intensity (% IAA), local management (organic vs. conventional) and within-field position (edge vs. interior) affect the community composition of forbs, we performed partial redundancy analyses (RDA). To characterise the community composition of forbs we used the relative percentage cover of each species. These analyses were done separately for insect-pollinated and non-insect pollinated forbs in order to investigate whether these factors affect pollination types differently. The species matrix was constrained by the predictor variables landscape-scale management intensity, local management or within-field position, while landscape (9 landscapes) was always included as a conditional variable (to account for nesting). Prior to the analyses, the species matrix was transformed with the Hellinger transformation [27]. This transformation allows the use of ordination methods such as PCA and RDA, which are Euclidean-based, with community composition data (site×species matrix) containing many zeros, i.e. characterised by long gradients. Pseudo-F values with the corresponding p values were calculated by permutation tests based on 999 permutations. Calculations were performed using the vegan package (version 2.0 [28]).

Results

A total of 62 forb species were identified in the meadows, consisting of 40 insect-pollinated (38 in organic, 21 in conventional meadows) and 22 non-insect pollinated (21 in organic, 13 in conventional) species (Table S1). In the wheat fields altogether 57 forbs were identified, consisting of 33 insect-pollinated (29 in organic, 11 in conventional) and 24 non-insect pollinated (22 in organic, 11 in conventional) species (Table S2).

We found a significant positive effect of organic management on species richness of forbs with higher richness in organic than in conventional fields (Table 1). This positive effect, however, was

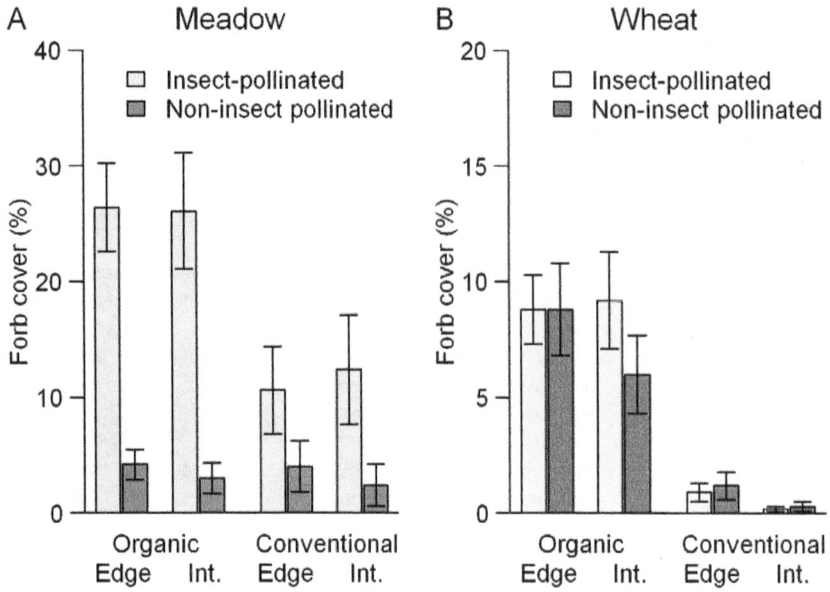

Figure 2. Mean (± SEM) forb cover (%) in organic vs. conventional meadows (A) and in organic vs. conventional wheat fields (B). Data were gathered in edge and interior (Int.) transects of 20 m^2.

more pronounced for insect-pollinated than for non-insect pollinated forbs in both agroecosystems, resulting in an interaction between pollination type and management (Fig. 1a,b). Regarding this interaction, the relative change in the proportion of insect-pollinated vs. non-insect pollinated forb species was higher in wheat fields (28.1% decrease from organic to conventional management) than in the meadows (15.0%) (Table 1; Fig. 1a,b). Finally, species richness of forbs was higher in the edges than in the interiors in organic wheat fields for bumblebee-pollinated vs. non-bumblebee pollinated forb species revealed similar results, with the main exception that we found management×pollination interaction only in case of wheat fields (Table S3).

In wheat fields, the cover of both insect-pollinated and non-insect pollinated forbs was significantly higher in organic than in conventional management (Table 1; Fig. 2a,b). The interaction between pollination type and management on forb cover in meadows indicates that forb cover was enhanced by organic management predominantly in the case of insect-pollinated plants, but not so strongly for wind- or self-pollinated plants.

The partial RDA showed significant effects of management on the community structure of insect-pollinated forbs in both agroecosystems (Table 2; Fig. 3), altering the community composition almost completely. In organic meadows, the community composition was determined by a few characteristic species, such as *Crepis biennis* or *Trifolium spp.*, whereas *Hypochaeris radicata* was more common in conventional meadows (Fig. 3). In wheat fields, some characteristic species (e.g. *Cirsium arvense*, *Matricaria recutita*, *Papaver rhoeas*) occurred much more frequently in organically than in conventionally managed fields (Fig. 3). The community composition of non-insect pollinated forbs was also significantly affected by management, and in meadows transect position had also significant effect (Table 2; Fig. 3).

Discussion

Both diversity and cover of forbs were positively affected by organic management in meadows and wheat fields. This effect differed between pollination types for species richness in both agroecosystem types and for cover in meadows. We show for the first time in a comprehensive analysis (covering arable crops and grasslands paired in the same landscapes) that insect-pollinated plants benefit more from organic management than non-insect pollinated plants, regardless of agroecosystem type and landscape complexity. These benefits were more pronounced in wheat fields than in meadows. The positive organic management effect was also observed for non-insect pollinated plants in both agroecosystems. Additionally, we found that the community composition of insect-pollinated and non-insect-pollinated forb communities differed greatly based on the management type.

Both organic meadows and organic wheat fields contained disproportionally more insect-pollinated forb species than their paired conventional meadows and wheat fields. This supports recent studies showing a higher number of insect-pollinated plant species under organic compared to conventional management [9,10,16,17]. The mechanism behind the difference in sensitivity to organic management could be that both pollinators and their food plants benefit from the absence of pesticide use, e.g. [7,29,30,31,32]. This supports the promotion of organic farming as a means to conserve farmland biodiversity and ecosystem services.

Comparing organic and conventional management, we observed a stronger decrease in the proportion of insect-pollinated vs. non-insect pollinated forb species in wheat fields than in meadows. In wheat fields, soil disturbance by annual ploughing or harrowing is likely to cause larger seed loss and larger extinction probability compared to permanent meadows [33]. In meadows, plant species once established can persist over long time periods also under conventional management. This can lead to less pronounced differences of species richness between the two pollination types in meadows. Finally, the lower light availability at ground level in wheat fields than meadows, may increase the sensitivity of forbs to the negative effects of increased fertilizer and pesticide input involved in conventional agriculture.

Recently, several studies have pointed to a link between declines of pollinators and insect-pollinated plants, e.g. [16,34,35]. Müller et al. [36] analysed the pollen requirements of European bee

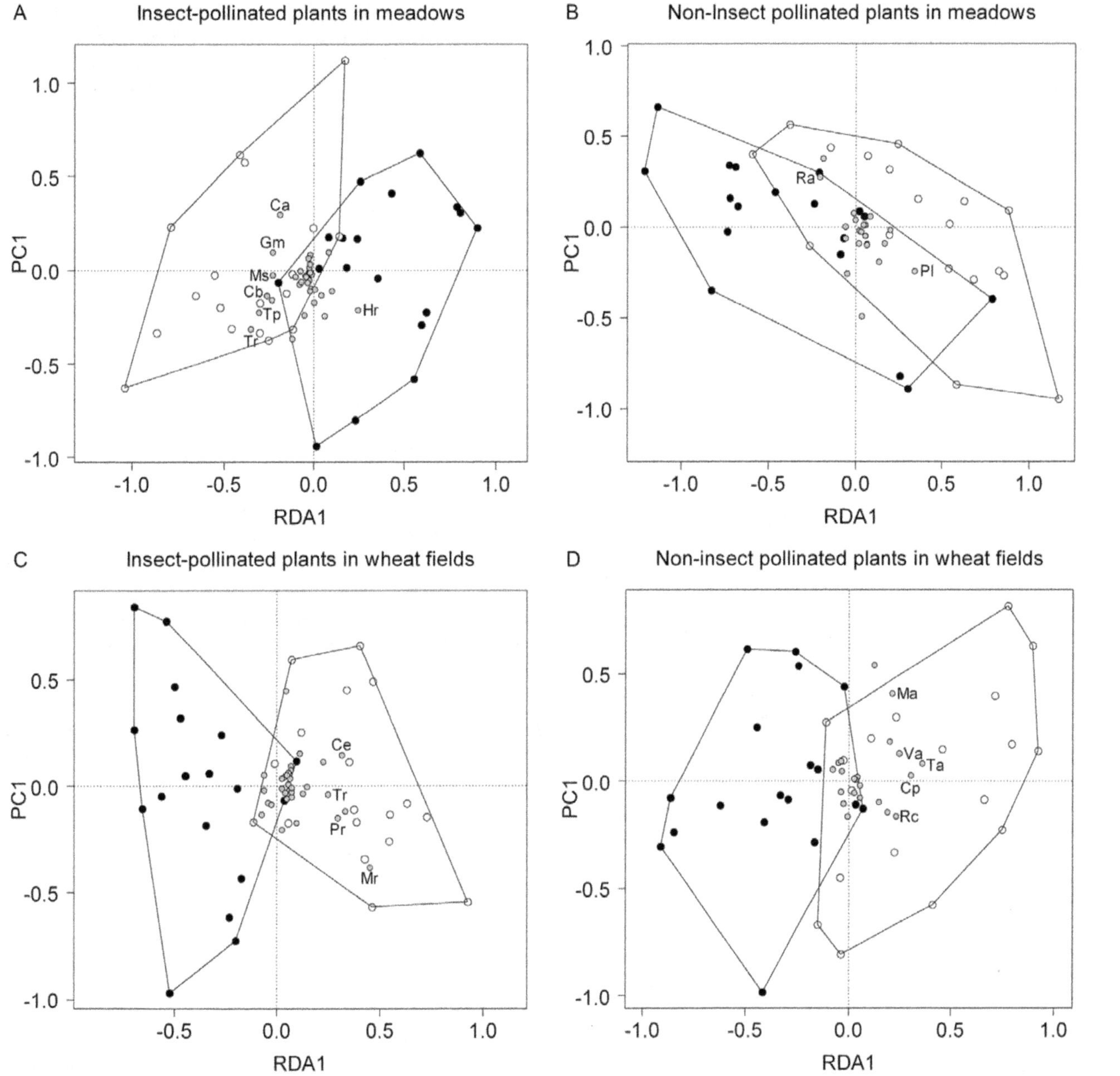

Figure 3. RDA plots for insect-pollinated and non-insect pollinated forbs in meadows (A, B) and wheat fields (C, D). White circles: plant survey transects in organic fields; black circles: plant survey transects in conventional fields; smaller grey circles: forb species with the highest fraction of variance (Ca: *Convolvulus arvensis*; Cb: *Crepis biennis*; Ce: *Cirsium arvense*; Cp: *Capsella bursa-pastoris*; Gm: *Galium mollugo*; Hr: *Hypochaeris radicata*; Ma: *Myosotis arvensis*; Mr: *Matricaria recutita*; Ms: *Medicago sativa*; Pl: *Plantago lanceolata*; Pr: *Papaver rhoeas*; Ra: *Rumex acetosa*; Rc: *Rumex crispus*; Ta: *Thlaspi arvense*; Tp: *Trifolium pratense*; Tr: *Trifolium repens*; Va: *Veronica arvensis*). Minimum convex polygons of the two management types are shown.

species and concluded that recent declines of bee populations are related to the decrease of flower diversity and quantity due to modern agriculture practices. Carvell et al. [37] also reported a national-scale decline in forage availability for bumblebees in the UK during the last century, but moreover showed that changes in abundance of forage plants were greater than those of non-forage plant species. This reflects both qualitative and quantitative decline of foraging resources of bees. These findings suggest that local management effects can cascade up to higher trophic levels including pollinators [22,38]. Nevertheless, organic management can affect the plant-pollinator community from either direction: management enhances plant resources available for pollinators, while increased survival of the pollinator community benefits insect-pollinated plants. Our results support the findings that management-driven changes were stronger for insect-pollinated than non-insect pollinated forbs, which probably has consequences at much larger spatial scales. According to Holzschuh et al. [39], increasing landscape-wide percentage of organic fields results in higher flower resources and bee diversity even outside crop fields.

Furthermore, we found that community composition also differed between conventional and organic fields. Organically managed fields could be characterized by a few typical insect-

Table 2. Results of partial RDA to analyse effects of landscape (intensive agricultural area %), management (organic vs. conventional) and position in field (edge vs. interior) on species composition of insect and non-insect pollinated forbs in meadows and in wheat fields.

	Variable	Variation (%)	pseudo-F	p
Meadow				
Insect-pollinated forbs	Landscape	3.11	1.26	0.239
	Management	9.82	3.96	0.001
	Position in field	2.47	1.00	0.440
Non-insect pollinated forbs	Landscape	2.74	1.07	0.410
	Management	5.33	2.08	0.024
	Position in field	6.34	2.48	0.004
Wheat field				
Insect-pollinated forbs	Landscape	2.91	1.28	0.215
	Management	1.93	5.24	0.001
	Position in field	2.50	1.10	0.362
Non-insect pollinated forbs	Landscape	3.03	1.37	0.164
	Management	8.71	3.94	0.001
	Position in field	1.70	0.77	0.679

Percentage of explained variation, pseudo-F values and p values are given. Denominator degrees of freedom was 24 in all analyses.

pollinated species, such as *Trifolium pratense* in meadows and *Cirsium arvense* in wheat fields, which provide forage for bumblebees [37]. In contrast, in conventionally managed meadows or wheat fields just a mixture of accidentally occurring species was present. This was most likely because conventional fields were impoverished in both insect-pollinated and non-insect pollinated forb species, and therefore the likelihood of occurrence of common species is generally lower.

We did not detect effects of landscape structure on species richness and cover of forbs, which indicates that local management was a more important driver [10,33]. Landscape-scale effects might be more important for mobile pollinators than for their less mobile food resources [39,40], but see [19]. This is also reflected in a recent meta-analysis [5], showing that landscape strongly moderates the response of pollinators to management intensity.

At the smallest scale, i.e. within-field position, we found higher species richness of forbs of both pollination types in the edges than in the interiors. This is most probably because of the less efficient spraying of pesticides and fertilizers, higher light availability close to the borders [16] or mass-effects of higher propagule pressure from adjacent habitats.

In conclusion, our findings in both agroecosystem types (meadows and wheat fields) indicate that organic management supports high species richness and cover of insect-pollinated plants, which is likely to be favourable for the density and diversity of bees and other pollinators [41,42]. These benefits were more pronounced in wheat fields than in meadows. Hence organic management contributes not only to biodiversity conservation but also increases resources for functionally important groups such as pollinators.

Supporting Information

Figure S1 Location of the sample fields.

Table S1 Overview of forb species, their pollination type and frequency in organic and conventional meadows.

Table S2 Overview of forb species, their pollination type and frequency in organic and conventional wheat fields.

Table S3 Results of general linear mixed models bumblebee-pollinated vs. non-bumblebee pollinated forb species analyses.

Acknowledgments

We are indebted to Petra Kubisch for assistance in fieldwork, Jochen Fründ, Bruno Hérault and three anonymous referees for valuable comments and the farmers for access to their fields and information about their farming practices.

Author Contributions

Conceived and designed the experiments: PB TT. Performed the experiments: LS PB. Analyzed the data: PB. Contributed reagents/materials/analysis tools: CFD. Wrote the paper: PB TT CFD LS.

References

1. Gaston KJ (2010) Valuing common species. Science 327: 154–155.
2. Whittingham MJ (2011) The future of agri-environment schemes: biodiversity gains and ecosystem service delivery? J Appl Ecol 48: 509–513.
3. Benton TG, Vickery JA, Wilson JD (2003) Farmland biodiversity: is habitat heterogeneity the key? Trends Ecol Evol 18: 182–188.
4. Smith HG, Öckinger E, Rundlöf M (2010) Biodiversity and the landscape ecology of agri-environment schemes. Asp Appl Biol 100: 225–232.
5. Batáry P, Báldi A, Kleijn D, Tscharntke T (2011) Landscape-moderated biodiversity effects of agri-environmental management – a meta-analysis. Proc R Soc B 278: 1894–1902.
6. Kleijn D, Sutherland WJ (2003) How effective are European agri-environment schemes in conserving and promoting biodiversity? J Appl Ecol 40: 947–969.
7. Kleijn D, Kohler F, Báldi A, Batáry P, Concepción ED, et al. (2009) On the relationship between farmland biodiversity and land-use intensity in Europe. Proc R Soc B 276: 903–909.
8. Flohre A, Fischer C, Aavik T, Bengtsson J, Berendse F, et al. (2011) Agricultural intensification and biodiversity partitioning in European landscapes comparing plants, carabids, and birds. Ecol Appl 21: 1772–1781.
9. Romero A, Chamorro L, Sans FX (2008) Weed diversity in crop edges and inner fields of organic and conventional dryland winter cereal crops in NE Spain. Agric Ecosyst Environ 124: 97–104.
10. Ekroos J, Hyvönen T, Tiainen J, Tiira M (2010) Responses in plant and carabid communities to farming practises in boreal landscapes. Agric Ecosyst Environ 135: 288–293.
11. José-María L, Blanco-Moreno JM, Armengot L, Sans FX (2011) How does agricultural intensification modulate changes in plant community composition? Agric Ecosyst Environ 145: 77–84.
12. Kovács-Hostyánszki A, Kőrösi Á, Orci KM, Batáry P, Báldi A (2011) Set-aside promotes insect and plant diversity in a Central European country. Agric Ecosyst Environ 141: 296–301.
13. Bond WJ (1994) Do mutualisms matter? Assessing the impact of pollinator and disperser disruption on plant extinction. Phil Trans R Soc Lond B 344: 83–90.
14. Ollerton J, Winfree R, Tarrant S (2011) How many flowering plants are pollinated by animals? Oikos 120: 321–326.
15. Regal PJ (1982) Pollination by wind and animals: ecology of geographic patterns. Ann Rev Ecol Syst 13: 497–524.

16. Gabriel D, Tscharntke T (2007) Insect pollinated plants benefit from organic farming. Agric Ecosyst Environ 118: 43–48.
17. Power EF, Kelly DL, Stout JC (2012) Organic farming and landscape structure: effects on insect-pollinated plant diversity in intensively managed grasslands. PLoS ONE 7: e38073.
18. Biesmeijer JC, Roberts SPM, Reemer M, Ohlemüller R, Edwards M, et al. (2006) Parallel declines in pollinators and insect-pollinated plants in Britain and the Netherlands. Science 313: 351–354.
19. Gabriel D, Thies C, Tscharntke T (2005) Local diversity of arable weeds increases with landscape complexity. Perspect Plant Ecol Evol Syst 7: 85–93.
20. Concepción ED, Díaz M, Baquero RA (2008) Effects of landscape complexity on the ecological effectiveness of agri-environment schemes. Landscape Ecol 23: 135–148.
21. Guerrero I, Martínez P, Morales MB, Oñate JJ (2010) Influence of agricultural factors on weed, carabid and bird richness in a Mediterranean cereal cropping system. Agric Ecosyst Environ 138: 103–108.
22. Batáry P, Holzschuh A, Orci KM, Samu F, Tscharntke T (2012) Responses of plant, insect and spider biodiversity to local and landscape scale management intensity in cereal crops and grasslands. Agric Ecosyst Environ 146: 130–136.
23. Klotz F, Kühn I, Durka W (2002) BiolFlor – Eine Datenbank mit biologisch-ökologischen Merkmalen zur Flora von Deutschland. Bundesamt für Naturschutz, Bonn – Bad Godesberg.
24. Pinheiro J, Bates D, DebRoy S, Sarkar D, the R Core team (2011) The nlme package: Linear and nonlinear mixed effects models. R package version 3.1-102. Available: http://cran.r-project.org/src/contrib/Descriptions/nlme.html.
25. R Development Core Team (2011) R: a language and environment for statistical computing. Foundation for Statistical Computing, Version 2.11.1. Vienna, Austria. Available: http://www.R-project.org.
26. Venables WN, Ripley BD (2002) Modern Applied Statistics with S. New York: Springer. 495 p.
27. Legendre P, Gallagher ED (2001) Ecologically meaningful transformation for ordination of species data. Oecologia 129: 271–280.
28. Oksanen J, Blanchet FG, Kindt R, Legendre P, Minchin PR, et al. (2011) The vegan package: Community Ecology Package. R package version 2.0–2. Available: http://cran.R-project.org/package = vegan.
29. Knop E, Kleijn D, Herzog F, Schmid B (2006) Effectiveness of the Swiss agri-environment scheme in promoting biodiversity. J Appl Ecol 43: 120–127.
30. Holzschuh A, Steffan-Dewenter I, Tscharntke T (2008) Agricultural landscapes with organic crops support higher pollinator diversity. Oikos 117: 354–361.
31. Geiger F, Bengtsson J, Berendse F, Weisser WW, Emmerson M, et al. (2010) Persistent negative effects of pesticides on biodiversity and biological control potential on European farmland. Basic Appl Ecol 11: 97–105.
32. Brittain C, Potts SG (2011) The potential impacts of insecticides on the life-history traits of bees and the consequences for pollination. Basic Appl Ecol 12: 321–331.
33. Armengot L, José-María L, Blanco-Moreno JM, Romero-Puente A, Sans FX (2011) Landscape and land-use effects on weed flora in Mediterranean cereal fields. Agric Ecosyst Environ 142: 311–317.
34. Krauss J, Gallenberger I, Steffan-Dewenter I (2011) Decreased functional diversity and biological pest control in conventional compared to organic crop fields. PLoS ONE 6: e19502.
35. Batáry P, Báldi A, Sárospataki M, Kohler F, Verhulst J, et al. (2010) Effect of conservation management on bees and insect-pollinated grassland plant communities in three European countries. Agric Ecosyst Environ 136: 35–39.
36. Müller A, Diener S, Schnyder S, Stutz K, Sedivy C, et al. (2006) Quantitative pollen requirements of solitary bees: implications for bee conservation and the evolution of bee-ower relationships. Biol Conserv 130: 604–615.
37. Carvell C, Roy DB, Smart SM, Pywell RF, Preston CD, et al. (2006) Declines in forage availability for bumblebees at a national scale. Biol Conserv 132: 481–489.
38. Caballero-López B, Blanco-Moreno JM, Pérez N, Pujade-Villar J, Ventura D, et al. (2010) A functional approach to assessing plant–arthropod interaction in winter wheat. Agric Ecosyst Environ 137: 288–293.
39. Holzschuh A, Steffan-Dewenter I, Kleijn D, Tscharntke T (2007) Diversity of flower-visiting bees in cereal fields: effects of farming system, landscape composition and regional context. J Appl Ecol 44: 41–49.
40. Dauber J, Hirsch M, Simmering D, Waldhardt R, Otte A, et al. (2003) Landscape structure as an indicator of biodiversity: matrix effects on species richness. Agric Ecosyst Environ 98: 321–329.
41. Pywell RF, Warman EA, Hulmes L, Hulmes S, Nuttall P, et al. (2006) Effectiveness of new agri-environment schemes in providing foraging resources for bumblebees in intensively farmed landscapes. Biol Conserv 129: 192–206.
42. Potts SG, Woodcock BA, Roberts SPM, Tscheulin T, Pilgrim ES, et al. (2009) Enhancing pollinator biodiversity in intensive grasslands. J Appl Ecol 46: 369–379.

9

Yield and Economic Performance of Organic and Conventional Cotton-Based Farming Systems – Results from a Field Trial in India

Dionys Forster[1], Christian Andres[1]*, Rajeev Verma[2], Christine Zundel[1,3], Monika M. Messmer[4], Paul Mäder[4]

1 International Division, Research Institute of Organic Agriculture (FiBL), Frick, Switzerland, 2 Research Division, bioRe Association, Kasrawad, Madhya Pradesh, India, 3 Ecology Group, Federal Office for Agriculture (FOAG), Bern, Switzerland, 4 Soil Sciences Division, Research Institute of Organic Agriculture (FiBL), Frick, Switzerland

Abstract

The debate on the relative benefits of conventional and organic farming systems has in recent time gained significant interest. So far, global agricultural development has focused on increased productivity rather than on a holistic natural resource management for food security. Thus, developing more sustainable farming practices on a large scale is of utmost importance. However, information concerning the performance of farming systems under organic and conventional management in tropical and subtropical regions is scarce. This study presents agronomic and economic data from the conversion phase (2007–2010) of a farming systems comparison trial on a Vertisol soil in Madhya Pradesh, central India. A cotton-soybean-wheat crop rotation under biodynamic, organic and conventional (with and without Bt cotton) management was investigated. We observed a significant yield gap between organic and conventional farming systems in the 1st crop cycle (cycle 1: 2007–2008) for cotton (−29%) and wheat (−27%), whereas in the 2nd crop cycle (cycle 2: 2009–2010) cotton and wheat yields were similar in all farming systems due to lower yields in the conventional systems. In contrast, organic soybean (a nitrogen fixing leguminous plant) yields were marginally lower than conventional yields (−1% in cycle 1, −11% in cycle 2). Averaged across all crops, conventional farming systems achieved significantly higher gross margins in cycle 1 (+29%), whereas in cycle 2 gross margins in organic farming systems were significantly higher (+25%) due to lower variable production costs but similar yields. Soybean gross margin was significantly higher in the organic system (+11%) across the four harvest years compared to the conventional systems. Our results suggest that organic soybean production is a viable option for smallholder farmers under the prevailing semi-arid conditions in India. Future research needs to elucidate the long-term productivity and profitability, particularly of cotton and wheat, and the ecological impact of the different farming systems.

Editor: Jean-Marc Lacape, CIRAD, France

Funding: Biovision Foundation for Ecological Development, http://www.biovision.ch/; Coop Sustainability Fund, http://www.coop.ch/pb/site/nachhaltigkeit/node/64228018/Len/index.html; Liechtenstein Development Service (LED), http://www.led.li/en/home.html; Swiss Agency for Development and Cooperation (SDC), http://www.sdc.admin.ch/. The funders had no role in study design, data collection and analysis, decision to publish, or preparation of the manuscript.

Competing Interests: The authors have declared that no competing interests exist.

* E-mail: christian.andres@fibl.org

Introduction

The green revolution has brought about a series of technological achievements in agricultural production, particularly in Asia. Worldwide cereal harvests tripled between 1950 and 2000, making it possible to provide enough dietary calories for a world population of six billion by the end of the 20th century [1]. So far, global agricultural development has rather focused on increased productivity than on a more holistic natural resource management for food security and sovereignty. The increase in food production has been accompanied by a multitude of challenges and problems such as the exploitation and deterioration of natural resources, i.e. loss of soil fertility, strong decline of agro-biodiversity, pollution of water [2,3], and health problems associated with the use of synthetic plant protection products [4]. At present, more comprehensive system-oriented approaches are gaining momentum and are expected to better address the difficult issues associated with the complexity of farming systems in different locations and cultures [5].

The concept of organic agriculture builds on the idea of the efficient use of locally available resources as well as the usage of adapted technologies (e.g. soil fertility management, closing of nutrient cycles as far as possible, control of pests and diseases through management and natural antagonists). It is based on a system-oriented approach and can be a promising option for sustainable agricultural intensification in the tropics, as it may offer several potential benefits [6–11] such as: (i) A greater yield stability, especially in risk-prone tropical ecosystems, (ii) higher yields and incomes in traditional farming systems, once they are improved and the adapted technologies are introduced, (iii) an improved soil fertility and long-term sustainability of farming systems, (iv) a reduced dependence of farmers on external inputs, (v) the restoration of degraded or abandoned land, (vi) the access to attractive markets through certified products, and (vii) new partnerships within the whole value chain, as well as a strengthened self-confidence and autonomy of farmers. Critics contend that organic agriculture is associated with low labor productivity and high production risks [1,12–14], as well as high certification costs for smallholders [15]. However, the main criticism reflected in the scientific literature is the claim that organic agriculture is not able to meet the world's growing food demand, as yields are on average 20% to 25% lower than in conventional agriculture [16,17]. It should however be taken into account, that yield deviations among different crops and regions can be substantial depending on system and site characteristics [16,17]. In a meta-analysis by Seufert *et al.* [17] it is shown that yields in organic farming systems with good management practices can nearly match conventional yields, whereas under less favorable conditions they cannot. However, Reganold [18] pointed out that productivity is not the only goal that must be met in order for agriculture to be considered sustainable: The maintenance or enhancement of soil fertility and biodiversity, while minimizing detrimental effects on the environment and the contribution to the well-being of farmers and their communities are equally important as the above mentioned productivity goals. Farming systems comparison trials should thus - besides agronomic determinants - also consider ecological and economic factors over a longer period. These trials are inherently difficult due to the many elements the farming systems are comprised of, thus necessitating holistic research approaches in order to make comparisons possible [19].

Results from various farming systems comparison trials between organic and conventional management have shown, that even though yields may be slightly lower, organic farming systems exhibit several ecological and economic advantages, particularly long-term improvement of soil fertility [20–25]. However, most of the data has been obtained from trials in the temperate zones [20–26]. The little data available under tropical and subtropical conditions [9,27–29] calls for more long-term farming systems comparison trials to provide a better basis for decision making in these regions [17]. To address this issue, the Research Institute of Organic Agriculture (FiBL) has set up three farming systems comparison trials in Kenya, India and Bolivia, thereby encompassing different cropping systems and ethnologies. The main objective of these trials is to collect solid agronomic and socio-economic data on major organic and conventional agricultural production systems in the selected project regions. These trials will contribute to close the existing knowledge gap regarding the estimation of profitability of organic agriculture in developing countries (http://www.systems-comparison.fibl.org/). This paper presents results from cotton-based farming systems in India.

India is the second largest producer (after China) of cotton lint worldwide [30]. Cotton is a very important cash crop for smallholder farmers, but also one of the most exigent crops in terms of agrochemical inputs which are responsible for adverse effects on human health and the environment [27]. Genetically modified (GM) cotton hybrids carrying a gene of *Bacillus thuringiensis* (Bt) for protection against bollworm (*Helicoverpa* spp.) attack, have spread rapidly after their official introduction to India in 2002 [31,32]. By 2012, 7 million farmers cultivating 93% of India's total cotton area had adopted Bt cotton technology [32,33]. This high adoption rate might be attributed to the high pressure caused by cotton bollworms, and associated reductions in pesticide use upon the introduction of Bt cotton technology in India [34,35]. However, the discussion about the impacts of Bt cotton adoption remains highly controversial [36,37]. Giving focus to yields, advocates of Bt cotton claim that the technology has led to an increase in productivity of up to 60% [38–40] and in some cases even "near 100%" [41]. Opponents of Bt cotton on the other hand attributed the yield gains, compared to the pre-Bt period, to other factors. These include (i) the increase of the area under cotton cultivation, (ii) the shift from traditional diploid cotton (*G. arboreum, G. herbaceum* which accounted for 28% of total cotton area in 2000) to tetraploid *G. hirsutum* species [42] and the widespread adoption of hybrid seeds, (iii) the increased use of irrigation facilities, (iv) the introduction of new pesticides with novel action (e.g. Imidacloprid seed treatment), and (v) the increased use of fertilizers in Bt cotton cultivation [43,44]. Critics of Bt crops also stress uncertainties concerning the impact of the technology on human health [45] and on non-target organisms [46], as well as the higher costs of Bt seeds [34,47].

While some argue that GM crops in general can contribute significantly to sustainable development at the global level [33,48], others state that there is no scientific support for this claim [49]. Considering economic benefits of Bt cotton, the same controversy prevails: Advocates claim sustainable socio-economic benefits and associated social development [33,36], while opponents claim Bt cotton to be responsible for farmer debt [50], thereby contributing to India's notoriety for farmers' suicides [37,51], a linkage which has been criticized as reductionist and invalid [52]. However, comparisons are mainly drawn between Bt and non-Bt cotton under conventional management in high-input farming systems. Organic cotton production systems - holding a minor percentage of the cotton growing area in India - are often neglected, and little information exists on the productivity and profitability of organic farming in India [53]. However, organic cotton production is slowly gaining momentum in the global cotton market [27]. GM cultivars are not compatible with the guidelines of organic agriculture [54]. Therefore, organic cotton producers have to refrain from Bt cotton hybrids. In addition, organic producers and processors have to take all possible measures to avoid contamination with Bt cotton in order not to lose organic certification.

While organic farming systems have attracted considerable interest of the scientific community [16,17,21,26], biodynamic farming systems are less common and little investigated. The biodynamic agricultural movement started in the early 1920s in Europe [55] and developed the international certification organization and label DEMETER. In India, the biodynamic movement started in the early 1990s (www.biodynamics.in). Preparations made from manure, minerals and herbs are used in very small quantities to activate and harmonize soil processes, to strengthen plant health and to stimulate processes of organic matter decomposition. Most biodynamic farms encompass ecological, social and economic sustainability and many of them work in cooperatives. One of the first initiatives in India was bioRe India Ltd. in Madhya Pradesh state (formerly called Maikaal cotton project), where several thousand farmers (2007–2010 between 4'700 and 8'800) produce organic cotton mainly for the European market (www.bioreindia.com). Although the farmers in this cooperative are trained in biodynamic farming, and follow the taught practices to a certain extent, the system is not certified as biodynamic. Nonetheless the products are declared as organic. The farming systems comparison trial presented here was set up in 2007 in Madhya Pradesh state, central India, and is embedded at the training and education center of bioRe Association (www.bioreassociation.org/bioresearch.html). The main aim of the trial is to assess the agronomic, economic and ecological performance of cotton-based farming systems under organic, biodynamic and conventional management including non-Bt and Bt cotton. In this paper, we present the yield and gross margin of cotton, soybean and wheat of the four different farming systems within the first four years after inception of the trial (considered as conversion period in this paper).

Materials and Methods

1 Site description and socioeconomic context

The trial site is located in the plains of the Narmada river belt in the Nimar Valley, Khargone district, Madhya Pradesh state, India (22°8′30.3″N, 75°4′49.0″E), at an altitude of 250 meters above sea level. The climate is subtropical (semi-arid), with an average annual precipitation of 800 mm, which occurs in a single peak monsoon season usually lasting from mid-June to September. Temperatures range from 15°C to 49°C with a yearly average of 25°C, and are highest in May/June and lowest in December/January. Climatic data from 2007–2010 obtained near the trial are shown in Figure 1. The trial is located on a fertile Vertisol soil characterized by an average clay content of 600 g kg^{-1} soil, pH (H_2O) of 8.7, organic C content of 5.0 g kg^{-1} soil, and available P content (Olsen) of 7.0 mg kg^{-1} soil at the start of the trial. Vertisols have shrink-swell characteristics; they cover about 73 million ha of the subtropical (semi-arid) regions of India and are the predominant soil type in Madhya Pradesh [56].

Agriculture is the main livelihood activity in the project area. Farm sizes range from less than 1 ha to more than 10 ha, and soil fertility as well as access to irrigation water vary greatly throughout the region. The major crops in the region are cotton, soybean and wheat. Since 2002, Bt cotton has become very popular and is currently grown on more than 90% of the total area under cotton cultivation in Madhya Pradesh [57,58]. About 50% of India's organic cotton is produced in Madhya Pradesh [59]. The year consists of three seasons with distinctly different climatic characteristics: The Kharif (monsoon) season is characterized by the monsoon and lasts from June to October. Crops which require humid and warm condition are grown, for example cotton, or soybean. The Rabi (winter) season is characterized by lower temperatures and less rainfall; it lasts from November to March. Crops which require cool temperatures for vegetative growth are grown, for example wheat or chick pea. Finally, the Zaid (summer) season is characterized by hot temperatures and an extensive dry spell; it lasts from March to June. Only farmers with access to irrigation facilities or near river banks grow crops such as melons, gourds or cucumbers in this season. Longer duration crops such as cotton are cultivated during both Kharif and Rabi seasons.

2 Trial description

The farming systems comparison trial was established in 2007, and is expected to run for a period of 20 years. Before trial setup, the site was under conventional management by a local farmer. The homogeneity of the terrain was assessed before the implementation of the different farming systems with a test crop of unfertilized wheat (HA (0)) grown from December 2006 to

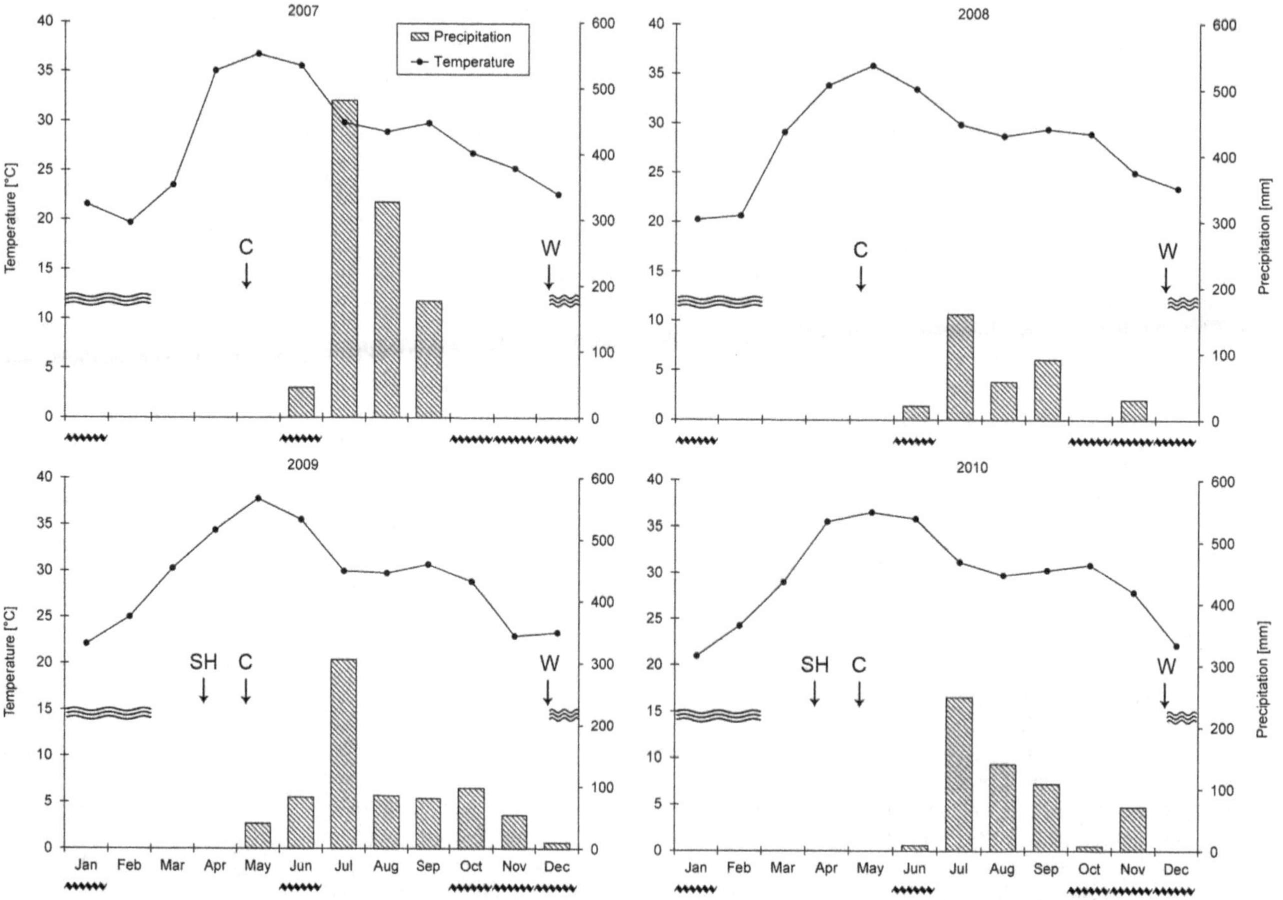

Figure 1. Temperature and precipitation recorded near the trial, Madhya Pradesh, India, 2007–2010, and irrigation practices in the farming systems comparison trial. Vertical arrows (↓) indicate flood irrigation prior to sowing of cotton (C), wheat (W) and sunn hemp (SH). Sunn hemp (green manure) was only grown in 2009 and 2010 on BIODYN and BIOORG plots before cotton sowing. Single closed undulating lines indicate period of drip and flood irrigation in cotton, multiple open undulating lines indicate period of flood irrigation in wheat (wheat received four to five flood irrigations).

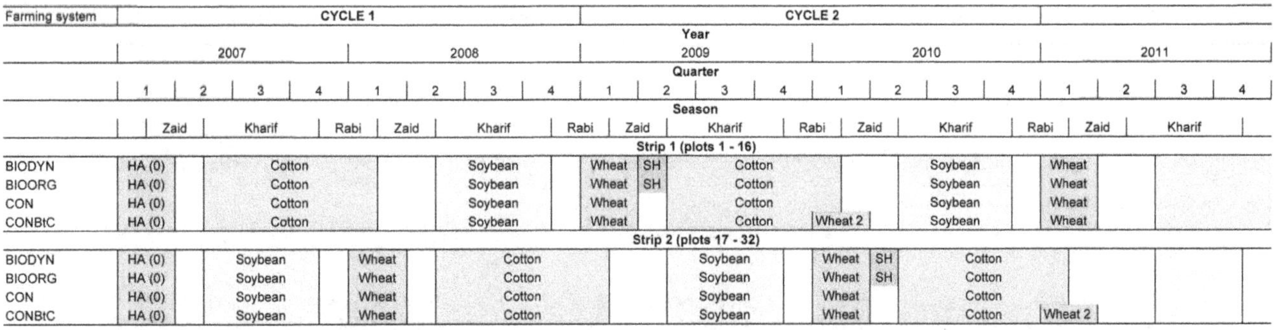
Farming system
CYCLE 1
CYCLE 2
Year
2007
2008
2009
2010
2011
Quarter
Season
Zaid
Kharif
Rabi
Strip 1 (plots 1 - 16)
Strip 2 (plots 17 - 32)
BIODYN
BIOORG
CON
CONBtC
HA (0)
Cotton
Soybean
Wheat
SH
Wheat 2

Figure 2. Sequence of crops in different farming systems of the farming systems comparison trial 2007–2010. Seasons: Zaid (summer): March to June, Kharif (monsoon): June to October, Rabi (winter): November to March. HA (0) indicates the homogeneity assessment performed with unfertilized wheat before the implementation of the different farming systems. In 2009 and 2010 Bt cotton was uprooted 2 months earlier to grow a second wheat crop (wheat 2) to reflect common practice of local Bt cotton farmers.

March 2007 (Figure 2). The test crop was harvested using a 5×5 m grid. Data of wheat grain yield, organic C and pH of the soil were used for allocation of strips, blocks and plots (Figure S1).

The trial comprises two organic farming systems (biodynamic (BIODYN), organic (BIOORG)) and two conventional farming systems (conventional (CON), conventional including Bt cotton (CONBtC)). Details of the farming systems are shown in Tables 1 and S1. Organic and biodynamic farming were carried out according to the standards defined by the International Federation of Organic Agriculture Movements (IFOAM) [60] and DEMETER-International [61], respectively. Conventional farming systems followed the recommendations of the Indian Council of Agricultural Research (ICAR) [62] with a slight adjustment to represent local conventional farming systems: farmyard manure (FYM) was applied to account for the integrated nutrient management of local conventional farmers. BIODYN represented the predominant local organic practices, as farmers associated to bioRe India Ltd. (see above) are provided with the respective inputs and trained in biodynamic farming as practiced in the field trial. BIOORG represented general organic practices as practiced in various regions of India where organic cotton is grown (mainly Madhya Pradesh, Maharashtra and Gujarat [59]). CON represented the local conventional practices in Madhya Pradesh before the introduction of Bt cotton in 2002, and CONBtC represented the current local conventional practices.

The four farming systems mainly differed in the following aspects: Genetic material (cotton only), type and amounts of fertilizer inputs, green manures, plant protection, the use of biodynamic preparations (Table 1, Table S1), and crop sequence (Figure 2). Farming systems are extremely complex, whereby individual management practices are closely linked and interdependent. For instance, it is well known that chemical plant protection is in most cases only economically feasible under conditions of optimal fertilization. That means that we mirror to a certain extent the complexity of a system rather than analyzing effects of single factors, and we intended to mimic common regional practices for the respective farming systems with respect to all management practices, as specified above. This approach is quite common in farming systems research and reflects effects of the system as a whole [20,21], but does not allow to trace potential differences to individual practices. As a basis for the design of the organic and conventional farming systems served a farm survey of Eyhorn *et al.* [9] in the same region.

The two-year crop rotation consisted of cotton (*Gossypium hirsutum* L.), soybean (*Glycine max* (L.) Merr.) and wheat (*Triticum aestivum* L.) (Figure 2). While in organic farming systems green gram (*Vigna radiata*) was grown between cotton rows in all four years and sunn hemp (*Crotalaria juncea*) was used as a preceding green manure crop for cotton in 2009 and 2010 (crop cycle 2), none of these practices were followed in conventional farming systems. Both green manure crops were cut at flowering and incorporated to the soil. In order to compare the CON and CONBtC farming systems as a whole (rather than the effect of the Bt gene), both the fertilizer dose and crop sequence was adapted (Figure 2).

Fertilizer inputs relied mainly on synthetic products in conventional farming systems (depending on crop between 68 and 96% of total nitrogen (N_{total}) applied (Table S1)), whereas organic farming systems received nutrients from organic sources only (Table 1). Organic fertilizers were compost, castor cake, and FYM. Compost was prepared using crop residues, weeds, FYM, and slurry from biogas plants (fed with fresh FYM) as raw materials. FYM was also applied in both conventional farming systems. The relatively high levels of organic fertilizer inputs (Table 1) reflect practices of local smallholder farmers who usually apply some 18.5 t ha^{-1} fresh matter of compost to cotton. On average, compost and FYM contained 0.8-0.6-1.5% and 0.8-0.6-1.6% of N_{total}-P_2O_5-K_2O, respectively, whereas castor cake contained 3.3-0.9-0.9% of N_{total}-P_2O_5-K_2O. Compost and FYM were broadcasted on the field after land preparation and subsequently incorporated to the soil by bullock-drawn harrows in all farming systems; However, in the organic farming systems in cotton, only 50% of the compost was applied as basal fertilizer input, and the remaining 50% were applied in two equal split applications as top dressings, at square formation and flowering, respectively. Castor cake was applied plant to plant. Nutrient inputs by N-fixing green manure crops were not considered, but will be assessed in future studies (Table S1). Synthetic fertilizers applied in both conventional farming systems were Diammonium phosphate (DAP), Muriate of Potash (MOP), Single Super Phosphate (SSP) and Urea. MOP, SSP and Urea/DAP were applied as basal fertilizer input at sowing time, except in cotton where only 50% of Urea/DAP was applied as basal fertilizer input, and the remaining 50% as a single top dressing at flowering. Across all crops and years, input of N_{total} was 65 kg ha^{-1} in organic farming systems (BIODYN, BIOORG), 105 kg ha^{-1} in CON and 113 kg ha^{-1} in CONBtC (Table S1). The lower inputs of N_{total} in organic compared to conventional farming systems represent local organic practice. The difference in inputs of N_{total} between CON and CONBtC arises from adhering to

Table 1. Management of the different farming systems compared in a two-year rotation in central India (2007–2010).

Practices	Organic farming systems[1]		Conventional farming systems[2]	
	BIODYN (biodynamic)	BIOORG (Organic)	CON (conventional)	CONBtC (conventional including Bt cotton)
Genetic material (difference in cotton only)				
	Non-Bt cotton	Non-Bt cotton	Non-Bt cotton	Bt cotton
Fertilizer input				
Type and level (for nutrient inputs see Table S1)	aerobically composted crop residues, weeds, farmyard manure (FYM), and slurry; 19.5-7.7-12.0 t ha^{-1} to cotton-soybean-wheat	aerobically composted crop residues, weeds, farmyard manure (FYM), and slurry; 19.5-7.7-12.0 t ha^{-1} to cotton-soybean-wheat	mineral fertilizers (MOP, SSP, Urea, DAP (wheat only))	mineral fertilizers (MOP, SSP, Urea, DAP (wheat only))
	stacked FYM; 2.8-1.6-2.2 t ha^{-1} to cotton-soybean-wheat	stacked FYM; 2.8-1.6-2.2 t ha^{-1} to cotton-soybean-wheat	stacked FYM; 8.1-3.9-1.6 t ha^{-1} to cotton-soybean-wheat	stacked FYM; 8.1-3.9-1.6 t ha^{-1} to cotton-soybean-wheat
	castor cake; 0.1 t ha^{-1} to cotton (2007 & 2008 only)	castor cake; 0.1 t ha^{-1} to cotton (2007 & 2008 only)		
Green manure				
Type and timing of green manure	broadcasted sunn hemp (*Crotalaria juncea*) before cotton in 2009 and 2010 only	broadcasted sunn hemp (*Crotalaria juncea*) before cotton in 2009 and 2010 only	None	None
	hand sown green gram (*Vigna radiata*, 9'070 plants ha^{-1}) between cotton rows in all years	hand sown green gram (*Vigna radiata*, 9'070 plants ha^{-1}) between cotton rows in all years	None	None
Plant protection				
Weed control	bullock-drawn blade or tine harrows	bullock-drawn blade or tine harrows	bullock-drawn blade or tine harrows	bullock-drawn blade or tine harrows
	Hand weeding in cotton	Hand weeding in cotton	Hand weeding in cotton	Hand weeding in cotton
	None	None	Herbicide (2009 and 2010 in soybean and wheat only)	Herbicide (2009 and 2010 in soybean and wheat only)
Insect control and average number of applications per crop rotation (detailed product list, see Table S1)	organic (natural) pesticides 12.5	organic (natural) pesticides 12.25	synthetic pesticides 11.5	synthetic pesticides 11.0
Disease control	None	None	None	None
Special treatments	biodynamic preparations[3]	None	None	None

[1]in the text, BIODYN and BIOORG are referred to consistently as organic farming systems, [2] in the text, CON and CONBtC are referred to consistently as conventional farming systems, average dry matter content of organic fertilizers: 70%, DAP: Diammonium phosphate, MOP: muriate of potash, SSP: single super phosphate, [3]biodynamic preparations entailed cow dung (BD-500) and silica powder (BD-501) both stored for six months, and a mixture of cow dung, chicken egg shell powder, basalt rock powder, and plant materials (yarrow, chamomile, stinging nettle, oak bark, dandelion, valerian) stored for 6 months in an open pit (cow pat pit = CPP).

recommendations by ICAR [62] who advocate systems with Bt cotton to be managed more intensively than systems with non-Bt cotton.

Pest management - including seed treatment - was done with organic (natural) pesticides in organic farming systems, while in conventional farming systems synthetic pesticides were used (Table 1). The type and number of pesticide applications in CON and CONBtC was the same to reflect local farmers' practices [63]. This practice was also confirmed in the survey of Kathage and Qaim [36] comparing conventional Bt and non-Bt cotton in the period 2006–2008 conducted in the four states Maharashtra, Karnataka, Andhra Pradesh, and Tamil Nadu.

The BIODYN system received small amounts of biodynamic preparations (Table 1) consisting of organic ingredients (cow manure, medicinal plants), and mineral compounds (quartz, basalt) which are intended to activate the soil and increase plant health [64]. No significant amounts of nutrients were added by these applications. For further details of biodynamic practices see Carpenter-Boggs *et al.* [64].

With cotton, soybean and wheat the trial represents a cash crop-based farming system in a two-year crop rotation, which is typical for the Nimar Valley in the plains of the Narmada river belt, were the trial is located. Cotton was grown from May to February, except in 2009 and 2010 (crop cycle 2) in CONBtC; in these two years Bt cotton was uprooted two months earlier than in the other three farming systems in order to grow an additional wheat crop (wheat 2) in the Rabi (winter) season (Figure 2). This was done to account for local practices; local conventional farmers noticed that Bt cotton matures earlier than non-Bt cotton, and produces the majority of the yield during the first three months of the harvesting period. Therefore, they started between 2007 and 2010 to grow another wheat crop before the start of the Zaid (summer) season, a practice which was also confirmed by Brookes & Barfoot [65]. Soybean was grown from July to October and followed by wheat from December to March. The land was prepared with bullock-drawn ploughs, harrows and levelers. Cotton was sown by hand at a rate of 0.91 plants m^{-2} (9'070 plants ha^{-1}). Soybean and wheat were sown with bullock-drawn seed drills. The inter row and intra row spacing were 30 cm and 4 cm, respectively for both soybean and wheat. In 2007, heavy monsoon rains led to severe waterlogging in the plots which stunted soybean growth and necessitated re-sowing the whole trial. Cultivars were selected according to local practice and availability. In cotton, these were Maruti 9632 (2007), Ankur 651 (2008), Ankur AKKA (2009) and JK Durga (2010) in all farming systems, except in CONBtC where isogenic Bt lines of the same hybrids were used. Non-GM soybean, variety JS-335, and non-GM wheat, variety LOK-1 were cultivated in all farming systems and years. The whole trial was irrigated and all plots received similar amounts of irrigation water; prior to sowing, flood irrigation was carried out on sunn hemp (green manure), cotton and wheat plots (Figure 1). After the monsoon, cotton received additional drip irrigation and two to three flood irrigations to ensure continuous water supply throughout the cropping season. Sunn hemp and wheat received three to four and four to five flood irrigations, respectively. Soybean was grown purely rainfed during the Kharif (monsoon) season. Weeding was done mechanically at 20 (cotton) and 45 (soybean, wheat) days after sowing, using bullock-drawn blade or tine harrows in all farming systems. In cotton, additional hand weeding was carried out. No hand weeding was carried out in soybean and wheat. No synthetic herbicides were applied in conventional farming systems except in soybean and wheat in 2009 and 2010, which reflects the situation of most smallholder cotton farmers in India [66]. Cotton was harvested by several manual hand pickings. Soybean and wheat were harvested manually with sickles, and bound to bundles which were removed from the field and subsequently threshed with a threshing machine.

In order to obtain data from each crop during each year, the layout was doubled with shifted crop rotation in two strips, resulting in a total of 32 plots, and 16 plots per strip (Figure S1). Each farming system was replicated four times in a randomized block design in each of the two strips. Plots are sized 16 m×16 m (= gross plot) and time measurements of activities were recorded for gross plots. The outermost 2 m of each plot served as a border, and yield data were only obtained in the inner sampling plot sized 12 m×12 m (= net plot) in order to avoid border effects. The distance between two plots within a strip and between the two strips is 6 m and 2 m, respectively. Data was obtained from 2007 to 2010. Data from 2007–2008 belongs to the complete crop rotation of the 1st crop cycle (cycle 1), and data from 2009–2010 to the 2nd crop cycle (cycle 2).

As Bt cotton was commercially released in India in 2002, no official approval of the study was required. The land needed for the farming systems comparison trial was purchased and belongs to bioRe Association. No protected species were sampled.

3 Data consolidation and economic calculations

Calculations of gross margins required consolidation of production costs. We only considered variable (operational) production costs in our study, excluding interest rates for credits. We included input costs, labor costs for field activities (including e.g. compost preparation), and costs associated with the purchase of inputs from the local market. Time measurements on gross plots and farmers' fields were complemented with data obtained in expert meetings with experienced farmers and local extension officers. Variable production costs for cotton (Table S2), soybean (Table S3), and wheat (Table S4) were cross-checked with the values reported by the Ministry of Agriculture, Government of India [67]. Gross margins were obtained by subtracting the variable production costs from the gross return (= yield * price per unit). Prices (products, inputs, labor) corresponded to local market conditions and were adapted each year (Table 2, Table S2). A premium price for organic cotton was considered in 2010 only (after three years conversion period according to IFOAM standards).

4 Statistical analysis

Data exploration revealed four outliers which were removed from the dataset. The reason was heavy monsoon rains and subsequent water-logging in four plots in 2009 (Plots 11 and 27 (both BIOORG), and plots 12 and 28 (both BIODYN), Figure S1).

Yield and gross margin data of each crop, and of the complete crop rotation (cotton+wheat 2+soybean+wheat) were analyzed separately with linear mixed effect models using the function lme from the package nlme [68] of the statistical software R version 2.15.2 [69]. We checked our data for model assumptions graphically (normal Q-Q of fixed and random effects, Tukey-Anscombe and Jitter plots) and no violation was encountered. We used a model with *System*, *Cycle*, the interaction of *System*×*Cycle* and *Strip* as fixed effects, and *Year* (n = 4), *Block* (n = 4) and *Pair* (n = 16) as random intercepts.

The fixed effect *Cycle* was included in the model to account for repeated measures on the same plot (e.g. cotton on plot 1 in 2007 and in 2009) and allows a partial separation of *Cycle* and *Year* effects due to the shifted crop rotation in the two strips as proposed by Loughin [70] for long-term field trials. *Cycle* effects give an indication how the situation changes across the timeframe of the trial. However, as we only have two levels of *Cycle* (thus Df = 1 for *Cycle* in the ANOVA) at this stage of the trial, we have little statistical power to detect *Cycle* effects. The same applies to the fixed effect *Strip*. We nevertheless included *Cycle* and *Strip* into our model to separate the *System* effect from possible *Cycle* and *Strip* effects. To account for similar conditions of neighboring plots (e.g. Plots 1 and 17, 2 and 18, etc., Figure S1) we included the random intercept *Pair* with 16 levels.

For yield and gross margin data of the complete crop rotation, the random intercept *Year* was removed from the model, as data from two years were compiled. Significant *System*×*Cycle* interactions suggested that the main effects of *System* and *Cycle* have to be interpreted with caution; As the effects of the different systems were not consistent across cycles, we split the datasets and performed post-hoc multiple comparisons for the fixed effect *System* separately for each cycle (method: Tukey, superscript letters after cycle-wise values in Tables 3 and 4). In the case of gross margin data of soybean, no significant *System*×*Cycle* interaction was encountered. Therefore, we performed post-hoc multiple comparisons on the whole dataset of cycle 1 and cycle 2 together (superscript letters after average values in Table 4). We defined a difference to be significant if $P < 0.05$ ($\alpha = 0.05$).

Results and Discussion

1 Yield

Cotton yields (seed cotton, picked bolls containing seed and fiber) were, averaged across the four years, 14% lower in organic (BIOORG, BIODYN) compared to conventional farming systems (CON, CONBtC). This is in the same range as the findings of a study conducted in Kyrgyzstan [27]. The *System*×*Cycle* interaction had a significant effect ($P<0.001$) on cotton yields (Table 3). The difference in yield was very pronounced in cycle 1 (2007–2008, +42% yield increase in conventional farming systems), while yields were similar among all farming systems in cycle 2 (2009–2010)

Table 2. Domestic market prices of cotton, soybean and wheat, premium prices on organic cotton and prices per working hour 2007–2010 in Khargone district, Madhya Pradesh, India.

Year	Commodity				
	Cotton [INR kg^{-1}]	Cotton premium price [INR kg^{-1}]	Soybean [INR kg^{-1}]	Wheat [INR kg^{-1}]	Labor [INR h^{-1}]
2007	23.3	4.7 (n.c.)	15.5	10.4	7.5
2008	26.8	3.3 (n.c.)	20.0	11.0	9.0
2009	31.5	3.3 (n.c.)	22.5	12.0	11.3
2010	49.0	4.0 (c.)	22.5	12.0	12.5

n.c.: not considered in economic calculations (conversion = first three years, according to IFOAM standards), c.: considered in economic calculations; No premium exists for organic soybean and wheat due to local market structures; Exchange rate INR: USD = 50:1 (source: http://eands.dacnet.nic.in/AWIS.htm, stand October 2012).

(Figure 3). CONBtC consistently showed higher yields than the three other farming systems, except in 2010. This is in line with the findings of several international meta-studies, which also reported generally higher yields and increased profitability in Bt cotton compared to non-Bt cotton production [34,65,71]. However, cotton yield increases through the use of Bt seeds may vary greatly among regions (from zero in Australia, up to 30% in Argentina) due to e.g. varieties used in Bt and non-Bt production, and effectiveness of chemical plant protection in non-Bt production [65]. Glover [72] also points out that the performance and impacts of Bt crops have been highly variable, socio-economically differentiated and contingent on a range of agronomic, socio-economic and institutional factors, thus underlining that the contextual interpretation of results is of paramount importance. The cotton yields per hectare of CONBtC in cycle 1 in our study were in the same range reported by Konduru *et al.* [73]. The severe decline in yield observed for CONBtC in cycle 2 when compared to cycle 1 (Figure 3) can be partly explained by the fact that Bt cotton plants were uprooted two months earlier than plants in other farming systems in cycle 2 (see chapter 2.2). However, this does not explain the decline in yield observed from 2007 to 2010 in the CON farming system, in which cotton plants were not uprooted. In cycle 1 the cotton yields in CONBtC were 16% higher than in CON (Table 3) which could be due to both the effect of the Bt gene products on pests (as isogenic hybrids were used) as well as the higher input of fertilizer (166 kg N ha^{-1} *vs.* 146 kg N ha^{-1}) recommended for Bt cotton. The difference in yield between Bt and non-Bt cotton in our study was much smaller than the differences in yield reported by others for India [33,36,38–40,74], indicating that the chemical plant protection applied to the CON system in our experiment was relatively effective. In contrast to the conventional systems, both of the organic farming systems showed rather stable cotton yield throughout the entire experimental period (Figure 3). As cycle effects and the *System×Cycle* interaction are confounded by year effects, we have to consider that in 2009 and 2010 the cotton yield was generally lower than in 2007 and 2008, as was confirmed by statistical yield data of the state Madhya Pradesh [67]. Apparently, the conventional farming systems could not realize their yield potential due to the less advantageous growing conditions in cycle 2 (rainfall and water logging in the harvest period October – December, Figure 1). The organic farming systems however, were not affected by the disadvantageous conditions in cycle 2 (Figure 3). An additional nitrogen fixing green manure pre-crop, planted before cotton in the organic systems in cycle 2 (Figure 2), may have contributed to the observed stability in yield in these systems through the consistent provision of nitrogen to the plants. Cotton yields of future crop cycles will thus determine whether productivity in conventional systems will reach their initial high level as well as determine whether yields of organic farming systems will start to increase. A yield depression is usually observed during the conversion to organic farming in India [75]. However, in our trial no such trend was oberserved between cycle 1 and cycle 2.

Non-GM soybean and wheat varieties were cultivated in both CON and CONBtC systems. In 2007, average soybean yields across all farming systems were 45% lower compared to the other three years (Figure 3), as the whole trial had to be re-sown due to severe water logging. Soybean yields were, averaged across the four years, 7% lower in organic compared to conventional farming systems. The *System×Cycle* interaction had a significant effect ($P<0.05$) on soybean yields (Table 3). No significant difference in yield could be identified between farming systems in cycle 1. However, in cycle 2 CON and CONBtC showed significantly higher yields than BIOORG (Table 3). This is likely due to higher pest incidences and thus lower yields in organic systems in 2009. BIODYN produced similar soybean yields as both conventional systems throughout the experimental period ($P>0.05$). The 1% and 11% lower yields in organic farming systems in cycle 1 and 2, respectively, are considerably lower than the 18% lower yields reported for organic soybean in the Karnataka region [76]. These results indicate similar productivity of conventional and organic soybean production systems under subtropical (semi-arid) conditions and suggest that in similar settings no further yield gains can be achieved through the provision of synthetic inputs compared to organic management practices. The smaller difference in yield between conventional and organic soybean - when compared to cotton and wheat (see below) - could be explained by considering the plant type. Soybean is the only legume crop in the crop rotation, possessing the ability to fix atmospheric N, thereby avoiding potential nitrogen shortage for optimal plant growth. These results confirm the findings of Seufert *et al.* [17] whose meta-analysis showed a lower yield gap between conventional and organic legume crops when compared to non-legume crops, and indicate that cotton and wheat yields in organic farming systems in our trial may be restricted by nitrogen limitation in the soil.

Wheat yields were, averaged across the four years, 15% lower in organic compared to conventional farming systems, which is similar to the 20% yield gap reported for Uttarakhand [76]. The *System×Cycle* interaction had a significant effect ($P<0.001$) on wheat grain yield. Similar to cotton, there was a significant yield gap between conventional and organic farming systems in cycle 1 (+37% yield increase in conventional farming systems), but not in cycle 2 due to both slightly lower yields in the conventional systems and slightly higher yields in the organic systems compared to cycle

Table 3. Mean yields [kg ha^{-1}] of cotton, soybean and wheat, and total productivity per cycle and across four years (2007–2010) in the farming systems compared in central India.

Farming system	Crop								Total productivity of crop rotation	
	Seed cotton	SEM	Wheat 2 grains	SEM	Soybean grains	SEM	Wheat grains	SEM	Seed cotton + Wheat 2 grains + Soybean grains + Wheat grains	SEM
Cycle 1 (2007–2008)										
BIODYN	2'047 [c]	68	-	-	1'399 [a]	158	2'997 [c]	153	6'443 [b]	104
BIOORG	2'072 [c]	49	-	-	1'536 [a]	192	2'831 [c]	121	6'440 [b]	187
CON	2'700 [b]	141	-	-	1'483 [a]	155	4'262 [a]	221	8'444 [a]	146
CONBtC	3'133 [a]	176	-	-	1'473 [a]	195	3'730 [b]	272	8'336 [a]	254
Cycle 2 (2009–2010)										
BIODYN	1'894 [a]	108	-	-	1'807 [ab]	87	3'338 [a]	207	7'039 [b]	268
BIOORG	1'942 [a]	103	-	-	1'739 [b]	117	3'303 [a]	191	6'984 [b]	239
CON	1'614 [a]	43	-	-	1'993 [a]	108	3'273 [a]	175	6'880 [b]	119
CONBtC*	1'834 [(a)]	179	1'573	169	1'997 [a]	161	3'481 [a]	182	8'885 [a]	390
Average (2007–2010)										
BIODYN	1'971	64	-	-	1'603	104	3'167	132	6'741	270
BIOORG	2'007	56	-	-	1'638	114	3'067	125	6'712	257
CON	2'157	157	-	-	1'738	113	3'767	187	7'662	455
CONBtC	2'484	207	(787)	-	1'735	140	3'605	161	8'610	376
ANOVAs of linear mixed effect models										
Source of variation	*P* value	Df	-	-	*P* value	Df	*P* value	Df	*P* value	Df
System (S)	<0.001	3	-	-	0.102	3	<0.001	3	<0.001	3
Cycle (C)	<0.001	1	-	-	0.066	1	0.686	1	0.912	1
Strip	0.141	1	-	-	0.472	1	0.960	1	0.002	1
S×C	<0.001	3	-	-	0.039	3	<0.001	3	<0.001	3

SEM: standard error of the mean, BIODYN: biodynamic, BIOORG: organic, CON: conventional, CONBtC: conventional with Bt cotton, different superscript letters indicate significant difference between farming systems within one Cycle (Tukey test, $P<0.05$), * in 2009 and 2010 Bt cotton was uprooted 2 months earlier to grow a second wheat crop (wheat 2) to reflect common practice of local Bt cotton farmers (for the sequence of crops in different farming systems see Figure 2), *P* value and degrees of freedom (Df) of fixed effects in linear mixed effect models, random factors in the model: Year (n = 4), Block (n = 4), Pair (n = 16), for total productivity random factor Year was excluded as data from two years were compiled.

Table 4. Mean gross margins [INR ha^{-1}] of cotton, soybean and wheat, and total gross margin per cycle and across four years (2007–2010) in the farming systems compared in central India.

Farming system	Crop								Total gross margin of crop rotation	
	Seed cotton	SEM	Wheat 2 grains	SEM	Soybean grains	SEM	Wheat grains	SEM	Seed cotton + Wheat 2 grains + Soybean grains + Wheat grains	SEM
Cycle 1 (2007–2008)										
BIODYN	38'243 [c]	2'226	-	-	19'211	4'210	26'044 [c]	1'584	83'498 [b]	8'021
BIOORG	38'676 [c]	1'203	-	-	21'830	4'858	24'420 [c]	1'291	84'926 [b]	8'268
CON	51'792 [b]	2'779	-	-	18'401	4'093	37'099 [a]	2'221	107'292 [a]	4'546
CONBtC	60'811 [a]	4'851	-	-	18'147	4'719	31'361 [b]	2'832	110'319 [a]	8'308
Cycle 2 (2009–2010)										
BIODYN	62'786 [a]	7'217	-	-	32'176	1'683	30'764 [a]	2'374	125'726 [a]	6'721
BIOORG	64'490 [a]	6'714	-	-	30'812	2'278	30'443 [a]	2'181	125'745 [a]	9'354
CON	42'962 [b]	4'653	-	-	28'949	3'724	24'773 [b]	2'028	96'633 [b]	6'701
CONBtC*	43'810 [(b)]	2'995	4'837	2'190	29'399	4'644	27'037 [b]	2'117	105'082 [b]	5'232
Average (2007–2010)										
BIODYN	50'514	4'918	-	-	25'694 [ab]	2'758	28'404	1'507	104'612	7'461
BIOORG	51'583	4'789	-	-	26'321 [a]	2'865	27'432	1'450	105'335	7'008
CON	47'377	2'852	-	-	23'675 [b]	2'780	30'936	2'155	101'988	3'089
CONBtC	52'310	3'165	(2'418)	-	23'773 [b]	3'311	29'199	1'797	107'701	3'849
ANOVAs of linear mixed effect models										
Source of variation	*P* value	Df	-	-	*P* value	Df	*P* value	Df	*P* value	Df
System (S)	0.115	3	-	-	0.006	3	0.022	3	0.298	3
Cycle (C)	0.046	1	-	-	0.158	1	0.606	1	<0.001	1
Strip	0.001	1	-	-	0.469	1	0.805	1	<0.001	1
S×C	<0.001	3	-	-	0.150	3	<0.001	3	<0.001	3

SEM: standard error of the mean, BIODYN: biodynamic, BIOORG: organic, CON: conventional, CONBtC: conventional with Bt cotton, different superscript letters indicate significant difference between farming systems within one Cycle (Tukey test, $P<0.05$), * in 2009 and 2010 Bt cotton was uprooted 2 months earlier to grow a second wheat crop (wheat 2) to reflect common practice of local Bt cotton farmers (for the sequence of crops in different farming systems see Figure 2), *P* value and degrees of freedom (Df) of fixed effects in linear mixed effect models, random factors in the model: Year (n = 4), Block (n = 4), Pair (n = 16), for total gross margin random factor Year was excluded as data from two years were compiled.

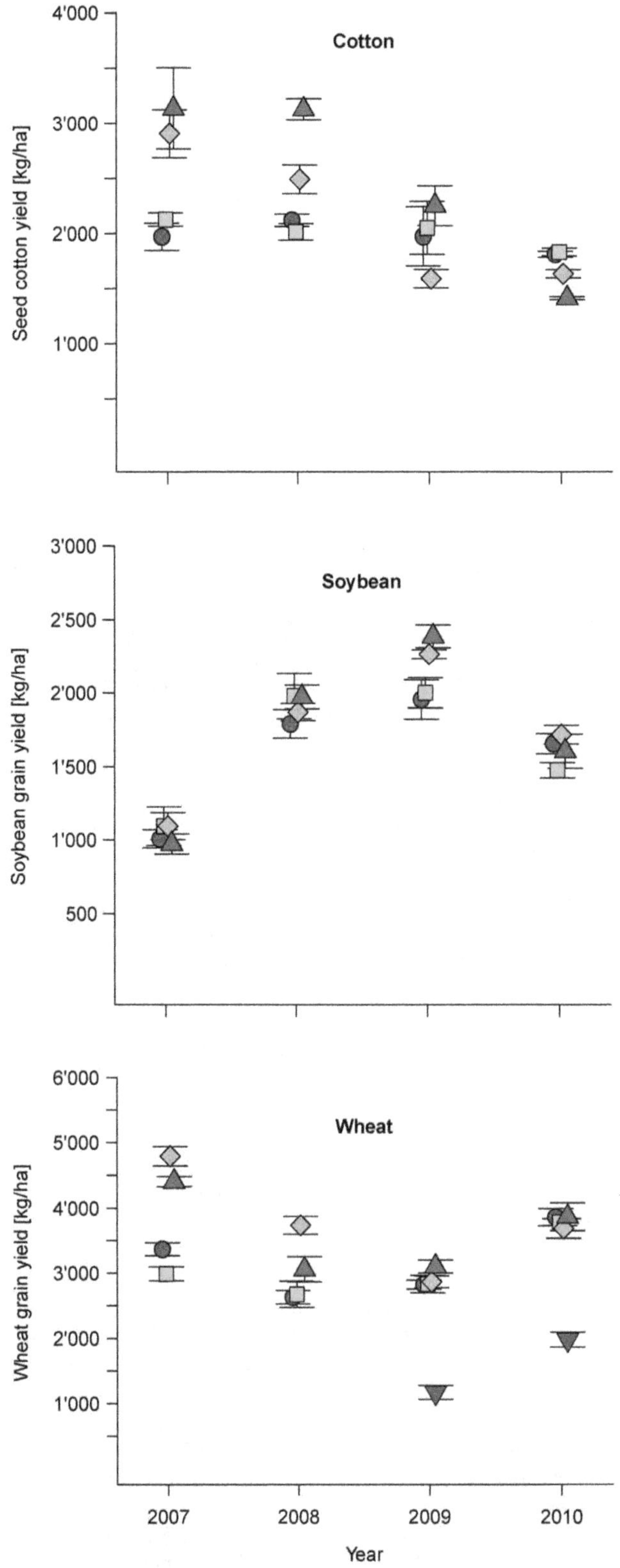

Figure 3. Yield (mean ± standard error) 2007-2010 in cotton, soybean and wheat. Farming systems: (●) biodynamic (BIODYN), (■) organic (BIOORG), (♦) conventional (CON), (▲) conventional with Bt cotton (CONBtC), (▼) wheat after Bt cotton (wheat 2); In 2009 and 2010 Bt cotton was uprooted 2 months earlier to grow a second wheat crop (wheat 2) to reflect common practice of local Bt cotton farmers. Non-GM soybean and wheat varieties were cultivated in the CON and CON-BtC plots throughout the trial. Note the different scales on y-axes in the different panels of the graph.

1 (Table 3, Figure 3). For both soybean and wheat no yield differences were detected between CON and CONBtC farming systems, except for significantly higher wheat yields in CON in cycle 1 (Table 3).

Regarding the total productivity per crop rotation in terms of summed-up dry matter yields of cotton, soybean and wheat (including wheat 2 in CONBtC in cycle 2), a significant effect ($P<0.001$) of the *System×Cycle* interaction was found (Table 3). Both of the conventional farming systems were significantly more productive (+30%, Table 3) than the organic farming systems in cycle 1. However, in cycle 2 only CONBtC showed significantly higher productivity (+28%, Table 3) when compared to the other three farming systems, due to the additional wheat crop (wheat 2). Differences in yield between BIODYN and BIOORG were minor and not statistically significant for all crops and total productivity of the whole crop rotation (Table 3). Unexpectedly, there was a significant *Strip* effect for total productivity across the whole crop rotation. This may be explained by the fact that different crops were cultivated on the two strips in a given year (Figure 2). The compilation of whole crop rotations (compiling years) thus led to the combination of the observed variability for each crop across the four years, and subsequently to the *Strip* effect becoming significant (Table 3).

In general, the first four years of the farming systems comparison trial in India revealed that there was a significant yield gap in cycle 1 (2007–2008) for cotton (−29%) and wheat (−27%), in organic compared to conventional farming systems, whereas in cycle 2 yields of the three crops were similar in all farming systems due to low yields in the conventional systems (Table 3). Because there was no clear trend of yield development for cotton and wheat during the four year period in any of the systems, observed results rather reflect growth conditions in respective years than long-term yield trends of cotton and wheat. However, the marginal yield gap between the BIOORG system and the conventional systems, and the par soybean yields of BIODYN and the conventional systems show that leguminous crops are a promising option for conversion to organic systems under the given conditions. The yield development across the whole crop rotation needs to be verified during future crop cycles.

2 Economic analysis

The production costs (i.e. labor and input costs) in our trial (Tables S2, S3 and S4) were in a similar range as reported by the Ministry of Agriculture, Government of India [67]. The variable production costs of conventional (CON, CONBtC) compared to organic (BIOORG, BIODYN) farming systems were on average 38%, 66%, and 49% higher in cotton, soybean and wheat (Table S2, S3 and S4). This is in agreement with findings for cotton in Gujarat, but contradicts findings for wheat in Punjab and Uttar Pradesh [53]. The main reason for the differences observed in our study were the higher input costs (fertilizer, pesticides) in the conventional farming systems, which is in accordance with the findings of a study conducted in Kyrgyzstan [27]. Labor costs were similar among all farming systems, as organic and conventional farming systems did not differ greatly with regard to time requirements of activities (Table S2, S3 and S4). For instance, weeding was done manually in all systems and no herbicides were applied in the conventional farming systems except for soybean and wheat in cycle 2, reflecting the common practice of most smallholder cotton farmers in India [66]. This practice, however, might change in the near future, as labor costs in Indian agriculture are on the rise [77]. The variable production costs of the two organic farming systems were similar for all crops. This was also true for the two conventional farming systems, except for cotton, where the variable production costs of CONBtC were 17% higher compared to CON due to both the higher seed price of Bt cotton (Table S2) [34,36,47] and production cost of the additional wheat crop in cycle 2. The prices we present here for Bt seed material are in the same range as reported by Singh *et al.* [78].

Cotton was the most important cash crop and accounted for 48% of the total gross return in the crop rotation, irrespective of the system. The *System*×*Cycle* interaction had a significant effect ($P<0.001$) on cotton gross margins (Table 4). Due to much higher yields, conventional cotton led to 32% higher gross margins compared to organic cotton in cycle 1 (2007–2008), which is in accordance with several international meta-studies [34,65,71]. However, the opposite was true in cycle 2 (2009–2010), where we observed 32% lower gross margins in conventional cotton, supporting the findings of Bachmann [27]. The significant *Strip* effect for cotton gross margin can be explained by the highly variable cotton prices across the four years (Table 2).

For soybean, the *System*×*Cycle* interaction was not significant (Table 4) which allowed for an analysis of gross margin data across both cycles (see 2.4). Considerably higher gross margins were obtained in organic systems (+10%) compared to conventional systems between 2007 and 2010. The difference was statically significant for BIOORG (+11%, $P<0.05$) and almost significant for BIODYN (+8%, $P<0.1$). These results indicate that the slightly lower productivity of organic soybean was balanced out by lower production costs rendering soybean production considerably more profitable in organic systems when compared to conventional farming systems.

For wheat gross margins, the *System*×*Cycle* interaction was found to be significant ($P<0.001$, Table 4). Under organic farming, wheat obtained significantly lower gross margins in cycle 1 (−26%), but significantly higher gross margins (+18%) in cycle 2 (Table 4). The earlier removal of Bt cotton from the field in order to grow another wheat crop in CONBtC, before the start of the Zaid (summer) season in cycle 2, only provided minor economic benefits compared to CON (Table 4). This was mainly due to low yields of wheat 2 (<50% compared to regular wheat crop, Figure 3) and lower market prices for wheat compared to cotton (Table 2). Thus, the additional wheat crop could not compensate for the missed cotton yield of the last picking period with respect to economic profitability, a result contradicting total yield performance across all crops.

A highly significant ($P<0.001$) *System*×*Cycle* interaction was found for the total gross margin per crop rotation. In cycle 1, favorable weather conditions allowed for the realization of the anticipated yield potential in conventional farming systems, and thus led to both higher cotton and wheat yields (Figure 3), and concomitantly significantly higher gross margins (+29%, Table 4). However, In cycle 2 the gross margins of the organic farming systems were significantly higher (+25%) (Table 4, Figure 4) due to par yields as measured in the conventional systems (Figure 3), but lower variable production costs (Tables S2 and S4). If the premium price in 2010 would not have been considered, the total gross margin per crop rotation in cycle 2 would not be substantially lower and still be significantly higher in organic farming systems (statistical analysis not shown). This could imply that, in favorable years (e.g. good yield, high price for commodities, etc.), premium prices are not required for achieving comparable economic returns in organic and conventional farming systems. However, the premium is needed in unfavorable years in order to compensate for yield gaps, and to avoid that organic farmers sell their produce to the local conventional market. The significant *Strip* effect for total gross margin can be explained by compiling the individual gross margins of the different crops, thereby transferring the significant *Strip* effect of cotton to the total gross margin (Table 4).

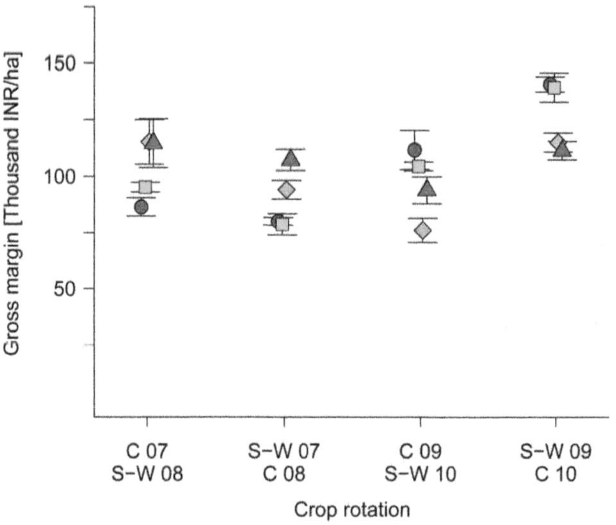

Figure 4. Gross margins (mean ± standard error) of four crop rotations. Farming systems: (●) biodynamic (BIODYN), (■) organic (BIOORG), (♦) conventional (CON), (▲) conventional with Bt cotton (CONBtC) (includes wheat cultivated after Bt cotton on the same plots in 2009 and 2010); C = cotton, S-W = soybean-wheat; Exchange rate Indian rupee (INR): US Dollar (USD) = 50:1 (stand October 2012), premium price on organic cotton only in 2010.

The results of cycle 2 suggest that under certain conditions, organic farming can be an economically viable and less capital-intensive production system compared to conventional farming systems, which is in accordance with the findings by Ramesh *et al.* [76] and Panneerselvam *et al.* [79]. However, long-term studies are needed in order to substantiate these findings. Moreover, the viability of organic farming systems strongly depends on farmers having access to knowledge, purchased inputs such as organic fertilizers, pesticides and non-GM seeds, and assuming that there is a market demand and well developed certification system. These are vital components for the economic profitability of organic farming systems [27] especially against the backdrop of increasing labor costs in Indian agriculture [77]. The costs for organic certification are substantial in case individual farmers have to undergo this process, and premium prices may also have to cover these costs. Up to now, certification costs are usually covered by the cotton organization that is purchasing seed cotton from smallholders (here: bioRe India Ltd.). This includes extensive testing of seeds and seed cotton for GM contamination, as well as the implementation of Tracenet, an internet based electronic service offered by the Agricultural and Processed Food Products Export Development Authority (APEDA) for facilitating certification of organic export products from India which comply with the standards of the (National) Programme for Organic Production (NPOP/NOP). This is a big challenge of certified organic cotton compared to fair trade cotton [15], and further organic cotton initiatives rely on cost-efficient and trustful certification and education programs as well.

3 Transferability of field trial results

So far little is known about the comparative performance of cotton-based farming systems under organic and conventional management. To our knowledge, this is the first publication comparing the agronomic and economic performance of biodynamic, organic, conventional and conventional with Bt cotton-based farming systems. The few studies published to date compared either organic *vs.* conventional [9,28,80] or conventional *vs.* conventional with Bt cotton production systems [36].

By including two organic (BIOORG and BIODYN) and two conventional (CON and CONBtC) farming systems in our trial, we were able to cover a wide range of current cotton-based farming systems in central India (see 2.2). Forming close links to local partners and having practitioners in the steering committee of our systems comparison trial guaranteed that the various agronomic activities that were involved represented local farmers' practice. Due to the cooperative initiative of bioRe, cotton farmers are trained in compost preparations, and organic inputs are purchased collectively and distributed among the farmers. Farmers associated with bioRe may not face the various problems commonly observed during conversion to organic farming [75] to a similar extent as do farmers without affiliation to similar institutions. This is likely due to the experienced and well-functioning extension service of bioRe. Drip irrigation is strongly promoted and subsidized by the Indian government and is not specific to our experiment. However, a direct extrapolation of our results to the reality of smallholder farmers is not possible due to the fact that farmers are confronted with several obstacles not considered in our study; These are for example market access, access to inputs and know-how and in particular costs associated with the organic certification process (see 3.2). One also has to consider that yield estimates from optimally managed trial plots are usually higher than the average yield of smallholder farmers. This is due to the fact, that such optimal crop management, as it was applied in this farming systems comparison trial, might not always be possible under the real-world smallholder conditions. This is especially true as the trial was conducted on a fertile Vertisol soil. Based on a survey of more than 1'000 smallholders in Madhya Pradesh, the average yield levels in the respective time period (2007–2009) were 1'416 kg ha^{-1}, 1'285 kg ha^{-1}, and 2'426 kg ha^{-1} for seed cotton, soybean, and wheat [67], as compared to 2'585 kg ha^{-1}, 1'761 kg ha^{-1}, and 3'658 kg ha^{-1} found in our trial. In a survey performed between 2006 and 2008 among 700 smallholder cotton farmers in India, average yields of 1'743 and 2'048 kg ha^{-1} seed cotton were reported for conventional (without Bt) and Bt cotton, respectively [36], as compared to 2'700 kg ha^{-1} and 3'133 kg ha^{-1} in our trial in the same time period (2007–2008). Thus, our yields might be generally overestimated, but within the range of other field trials in India [78].

The following examples show that comparative findings on yield and economics between organic and conventional cotton are highly contextual. Eyhorn *et al.* [9] surveyed more than 50 conventional (without Bt cotton) and 30 organic cotton farmers in the Nimar Valley, Madhya Pradesh, India, during the period of 2003–2004. Their findings support our results of cycle 2 (years three and four, 2009–2010): yields of cotton and other cash crops were on par with conventional farmers, but with economic benefits for the organic farmers due to lower production costs. These findings also underline the practical relevance of our results for cotton production in the in the smallholder context in Madhya Pradesh. Likewise, Venugopalan *et al.* [81] reported similar or slightly higher cotton yields in an organic compared to a non-organic system under low input and semi-arid conditions in the Yavatmal district, Maharashtra, India (observation phase 2001–2005). In a long-term trial under rainfed conditions in Nagpur (Maharashtra, India), Menon [82] reported a yield gap of 25% of organic cotton compared to the modern method of cultivation (= conventional without Bt cotton) within the first six years after conversion (1994–2000). Thereafter (2002–2004), the organic farming systems outyielded the conventional systems by up to 227 kg seed cotton ha^{-1} [28].

However, our findings from India are in contrast to the results from a survey on cotton farms in Northern San Joaquin Valley, California [80]. There, averaged over a six-year observation period, yields of organic cotton were 19 and 34% lower ($P<0.05$) than those of cotton under conventional and integrated pest management (reduced insecticide input). It has to be taken into account that for two out of the six years assessed in their study, different varieties were compared under conventional and organic management. Production costs of organic cotton were 25 and 60% higher than those of cotton under conventional and integrated pest management, respectively. This was mainly due to the higher labor costs for manual weeding. In our trial, there was less difference in weed control, as manual weeding is still the common practice of most smallholder cotton farmers in India [66]. This example underlines the contextual nature of the findings concerning the agronomic and economic performance between organic and conventional cotton, which was also pointed out by Seufert *et al.* [17] and de Ponti *et al.* [16] for other crops than cotton.

Building on unique panel data on Indian cotton farming of smallholder farmers between 2002 and 2008, Kathage and Qaim [36] showed that the use of conventional Bt cotton led to a 24% yield increase and a 50% gain in cotton profit compared to smallholders growing conventional non-Bt cotton. In contrast to the systematic farm survey by Kathage and Qaim [36], our study represents a pairwise comparison of cotton-based farming systems under identical pedo-climatic conditions. While the study by Kathage and Qaim [36] can better depict the actual situation on real farms for a given region, our results can better represent the potential outcomes that are achievable under optimal conditions with respect to inputs and knowledge access. In our study, there were comparatively little differences between the CONBtC and the CON farming systems regarding cotton yield (+16 and +14% in cycle 1 and 2, respectively) and cotton gross margin (+17 and +2% in cycle 1 and 2, respectively). Furthermore, it needs to be taken into account that with the introduction of Bt cotton to India in 2002, the provincial governments began to subsidize Bt cotton considerably, especially between the years of 2002 and 2008. This led to the rapid spread of Bt cotton and the breakdown of the non-Bt cotton seed chain. The relatively weak performance of the non-Bt cotton in the farm survey by Kathage and Qaim [36] could partly be explained by the poor quality of non-Bt cotton seeds, as propagation of non-Bt cotton was abandoned and led to limited availability of non-Bt cotton seeds from old stocks of probably poor quality and mainly older cultivars [59,83].

In contrast to Kathage and Qaim [36], our study also includes other cash crops such as soybean and wheat as part of cotton-based farming systems. These are essential components for enabling long-term cotton cropping and for securing livelihoods of smallholders, as they enable the distribution of risks. According to our findings, the investigated organic farming systems also showed a significant yield gap compared to the conventional farming systems in wheat in cycle 1, as well as for total productivity per crop rotation (including additional wheat crop after Bt cotton (wheat 2)) in cycle 2. Furthermore, a smaller yet significant yield gap was observed for soybean in cycle 2 for the BIOORG system (but not for the BIODYN system) (Table 3). Nevertheless, as in our study organic farming systems were less capital-intensive than conventional ones for all crops, they may be of particular interest to smallholder farmers who often do not have the financial means to purchase inputs and would thus need to seek loans. If this can be verified in on-farm trials, organic farmers might be less exposed to financial risks associated with fluctuating market prices of synthetic fertilizers and crop protection products [27,79]. Additional on-farm investigations have been started in order to classify regional farms into several farm typologies with corresponding levels of available production factors. This should allow for the assessment of the perspectives of each farm type regarding conversion to organic farming systems. If organic farming is to be adopted more widely, more inter- and transdisciplinary research giving focus to the problems and benefits of organic management practices needs to be undertaken [84]. Furthermore, large efforts have to be made to gather and disseminate knowledge on production techniques. Intensifying research on organic farming systems to a similar extent as was the case for research on GM crops [85] may help to provide additional relevant information to policy makers, advisors and farmers about comparative advantages and limitations of different cotton production systems.

Conclusions

With this publication we respond to the urgent need for farming systems comparison trials in the tropics and subtropics [17,26]. Here we show results from the conversion period (first four years after inception of the trial) of cotton-based farming systems representative for Vertisol soils in Madhya Pradesh, central India. Due to the short-term nature of our results and the observed *System×Cycle* interactions (no clear trend of system performance over time) for yield and gross margin data of cotton and wheat, definitive conclusions about the comparative agronomic and economic performance of the investigated farming systems cannot be drawn. However, our results show that organic soybean productivity can be similarly high as in conventional systems at lower input levels, which can make organic soybean production - as part of cotton-based crop rotations - more profitable. Future research will bring further insights on agronomic and economic performance of the different farming systems after the conversion period, thus providing indications on the long-term sustainability across the whole crop rotation. Furthermore, the effects of the farming systems on biodiversity, soil fertility, other ecological co-benefits such as climate change mitigation by means of C sequestration, and product quality need to be elucidated.

Supporting Information

Figure S1 Experimental design of the farming systems comparison trial in Madhya Pradesh, India. Farming systems: biodynamic (BIODYN), organic (BIOORG), conventional (CON), conventional with Bt cotton (CONBtC), CONBtC includes wheat cultivated after Bt cotton on the same plots in 2009 and 2010, open squares belong to Strip 1, closed squares belong to Strip 2, distance between two plots within a strip = 6 m, distance between the two strips = 2 m.

Table S1 Fertilizer and plant protection practices in the farming systems compared in central India (2007–2010). BIODYN: biodynamic, BIOORG: organic, CON: conventional, CONBtC: conventional with Bt cotton, Ntotal: total nitrogen, OF: organic fertilizers (compost, FYM and castor cake), Ntotal includes only fertilizer derived N, nutrient inputs by green manures were not considered, DAP: Diammonium phosphate, MOP: muriate of potash, SSP: single super phosphate, 1Beavicide®: organic pesticide containing Beauveria bassiana,2GOC: slurry made from rotten garlic, onion and chili with water, 3NeemAzal®: insecticide made from neem kernels, 4Top Ten: slurry made from leaves of ten wild plants and water, 5Verelac: organic pesticide containing Verticillium lecanii.

Table S2 Detailed list of variable production costs in cotton of the farming systems compared in central India (2007–2010). [1] in the text, BIODYN and BIOORG are referred to consistently as organic farming systems. [2] in the text, CON and CONBtC are referred to consistently as conventional farming systems. [3] figures include time for preparation of organic fertilizers to account for their market value. [4] figures represent subsidized prices for mineral fertilizers set by the Government of India. [5] longer time required for soil cultivation in CON and CONBtC due to soil compaction. [6] figure includes application of biodynamic preparations. [7] figures include uprooting cotton and removing the straw from the field. [8] figures include time required to purchase inputs (organic/synthetic) from the market and to produce organic (natural) pesticides and biodynamic preparations.

Table S3 Detailed list of variable production costs in soybean of the farming systems compared in central India (2007–2010). [1] in the text, BIODYN and BIOORG are referred to consistently as organic farming systems. [2] in the text, CON and CONBtC are referred to consistently as conventional farming systems. [3] figures include time for preparation of organic fertilizers to account for their market value. [4] figures represent subsidized prices for mineral fertilizers set by the Government of India. [5] longer time required for soil cultivation in CON and CONBtC due to soil compaction. [6] figure includes application of biodynamic preparations. [7] figures include removing soybean bundles from the field and threshing. [8] figures include time required to purchase inputs (organic/synthetic) from the market and to produce organic (natural) pesticides and biodynamic preparations.

Table S4 Detailed list of variable production costs in wheat of the farming systems compared in central India (2007–2010). [1] in the text, BIODYN and BIOORG are referred to consistently as organic farming systems. [2] in the text, CON and CONBtC are referred to consistently as conventional farming systems. [3] figures include time for preparation of organic fertilizers to account for their market value. [4] figures represent subsidized prices for mineral fertilizers set by the Government of India. [5] longer time required for soil cultivation in CON and CONBtC due to soil compaction. [6] figure includes application of biodynamic preparations. [7] figures include removing wheat bundles from the field and threshing. [8] figures include time required to purchase inputs (organic/synthetic) from the market and to produce organic (natural) pesticides and biodynamic preparations.

Acknowledgments

Special thanks go to Andreas Gattinger (Research Institute of Organic Agriculture, FiBL) for helping to prepare the manuscript and for his many valuable inputs. We thank Kulasekaran Ramesh (Indian Institute of Soil Science, IISS), Padruot Fried (Swiss Federal Agricultural Research Station Agroscope, ART), Monika Schneider, Franco Weibel, Andreas Fliessbach (Research Institute of Organic Agriculture, FiBL), Georg Cadisch (University of Hohenheim), and Philipp Weckenbrock (die Agronauten) for fruitful discussions. The field and desktop work of the whole bioRe Association team is also gratefully acknowledged. We thank Christopher Hay, Ursula Bausenwein and Tal Hertig for the language editing of the manuscript. We acknowledge the inputs by Fränzi Korner and Bettina Almasi regarding statistical analysis and data interpretation. Finally, we sincerely thank the anonymous reviewers for their very constructive and helpful comments and suggestions.

Author Contributions

Conceived and designed the experiments: DF CZ PM. Performed the experiments: DF RV CZ. Analyzed the data: CA MM. Wrote the paper: CA DF MM PM CZ RV.

References

1. Trewavas A (2002) Malthus foiled again and again. Nature 418: 668–670.
2. Badgley C, Moghtader J, Quintero E, Zakem E, Chappell MJ, et al. (2007) Organic agriculture and the global food supply. Renewable Agriculture and Food Systems 22: 86–108.
3. Singh RB (2000) Environmental consequences of agricultural development: a case study from the Green Revolution state of Haryana, India. Agriculture Ecosystems & Environment 82: 97–103.
4. Pimentel D (1996) Green revolution agriculture and chemical hazards. Science of the Total Environment 188: 86–98.
5. IAASTD (2009) International assessment of agricultural knowledge, science and technology for development (IAASTD): Executive summary of the synthesis report. Washington, DC: Island Press. 606 p.
6. Kilcher L (2007) How organic agriculture contributes to sustainable development. In: Willer H, Yussefi M, editors. The world of organic agriculture - Statistics and emerging trends 2007. Rheinbreitbach: Medienhaus Plump. pp. 82–91.
7. Altieri MA (1999) The ecological role of biodiversity in agroecosystems. Agriculture Ecosystems & Environment 74: 19–31.
8. Valkila J (2009) Fair Trade organic coffee production in Nicaragua - Sustainable development or a poverty trap? Ecological Economics 68: 3018–3025.
9. Eyhorn F, Ramakrishnan M, Mäder P (2007) The viability of cotton-based organic farming systems in India. International Journal of Agricultural Sustainability 5: 25–38.
10. Lyngbaek AE, Muschler RG, Sinclair FL (2001) Productivity and profitability of multistrata organic versus conventional coffee farms in Costa Rica. Agroforestry Systems 53: 205–213.
11. Mendez VE, Bacon CM, Olson M, Petchers S, Herrador D, et al. (2010) Effects of Fair Trade and organic certifications on small-scale coffee farmer households in Central America and Mexico. Renewable Agriculture and Food Systems 25: 236–251.
12. Borlaug NE (2000) Ending world hunger. The promise of biotechnology and the threat of antiscience zealotry. Plant Physiology 124: 487–490.
13. Trewavas AJ (2001) The population/biodiversity paradox. Agricultural efficiency to save wilderness. Plant Physiology 125: 174–179.
14. Nelson L, Giles J, MacIlwain C, Gewin V (2004) Organic FAQs. Nature 428: 796–798.
15. Makita R (2012) Fair Trade and organic initiatives confronted with Bt cotton in Andhra Pradesh, India: A paradox. Geoforum 43: 1232–1241.
16. de Ponti T, Rijk B, van Ittersum MK (2012) The crop yield gap between organic and conventional agriculture. Agricultural Systems 108: 1–9.
17. Seufert V, Ramankutty N, Foley JA (2012) Comparing the yields of organic and conventional agriculture. Nature 485: 229–U113.
18. Reganold JP (2012) Agriculture Comparing apples with oranges. Nature 485: 176–176.
19. Watson CA, Walker RL, Stockdale EA (2008) Research in organic production systems - past, present and future. Journal of Agricultural Science 146: 1–19.
20. Reganold JP, Glover JD, Andrews PK, Hinman HR (2001) Sustainability of three apple production systems. Nature 410: 926–930.
21. Mäder P, Fliessbach A, Dubois D, Gunst L, Fried P, et al. (2002) Soil fertility and biodiversity in organic farming. Science 296: 1694–1697.
22. Hepperly P, Douds Jr. D, Seidel R (2006) The Rodale faming systems trial 1981 to 2005: longterm analysis of organic and conventional maize and soybean cropping systems. In: Raupp J, Pekrun C, Oltmanns M, Köpke U, editors. Long-term field experiments in organic farming. Bonn: International Society of Organic Agriculture Resarch (ISOFAR). pp. 15–32.
23. Fliessbach A, Oberholzer H-R, Gunst L, Maeder P (2007) Soil organic matter and biological soil quality indicators after 21 years of organic and conventional farming. Agriculture Ecosystems & Environment 118: 273–284.
24. Teasdale JR, Coffman CB, Mangum RW (2007) Potential long-term benefits of no-tillage and organic cropping systems for grain production and soil improvement. Agronomy Journal 99: 1297–1305.
25. Birkhofer K, Bezemer TM, Bloem J, Bonkowski M, Christensen S, et al. (2008) Long-term organic farming fosters below and aboveground biota: Implications for soil quality, biological control and productivity. Soil Biology & Biochemistry 40: 2297–2308.
26. Gattinger A, Muller A, Haeni M, Skinner C, Fliessbach A, et al. (2012) Enhanced top soil carbon stocks under organic farming. Proceedings of the National Academy of Sciences of the United States of America 109: 18226–18231.
27. Bachmann F (2012) Potential and limitations of organic and fair trade cotton for improving livelihoods of smallholders: evidence from Central Asia. Renewable Agriculture and Food Systems 27: 138–147.
28. Blaise D (2006) Yield, boll distribution and fibre quality of hybrid cotton (Gossypium hirsutum L.) as influenced by organic and modern methods of cultivation. Journal of Agronomy and Crop Science 192: 248–256.
29. Rasul G, Thapa GB (2004) Sustainability of ecological and conventional agricultural systems in Bangladesh: an assessment based on environmental, economic and social perspectives. Agricultural Systems 79: 327–351.
30. FAO (2013) FAOSTAT database on agriculture. Available: http://faostatfaoorg. Accessed 11 October 2013.
31. Herring RJ (2008) Whose numbers count? Probing discrepant evidence on transgenic cotton in the Warangal district of India. International Journal of Multiple Research Approaches 2: 145–159.
32. James C (2012) Global status of commercialized biotech/GM crops: 2012. ISAAA Brief 44. ISAAA: Ithaca, NY. ISBN: 978-1-892456-53-2. Available: http://www.isaaa.org/resources/publications/briefs/44/executivesummary/pdf/Brief%2044%20-%20Executive%20Summary%20-%20English.pdf. Accessed 11 October 2013.
33. Qaim M KS (2013) Genetically Modified Crops and Food Security. PLoS ONE 8(6): e64879 doi:101371/journalpone0064879.
34. Finger R, El Benni N, Kaphengst T, Evans C, Herbert S, et al. (2011) A Meta Analysis on Farm-level Costs and Benefits of GM Crops. Sustainability 3: 743–762.
35. Krishna VV, Qaim M (2012) Bt cotton and sustainability of pesticide reductions in India. Agricultural Systems 107: 47–55.
36. Kathage J, Qaim M (2012) Economic impacts and impact dynamics of Bt (Bacillus thuringiensis) cotton in India. Proceedings of the National Academy of Sciences of the United States of America 109: 11652–11656.
37. Stone GD (2011) Field versus Farm in Warangal: Bt Cotton, Higher Yields, and Larger Questions. World Development 39: 387–398.
38. Crost B, Shankar B, Bennett R, Morse S (2007) Bias from farmer self-selection in genetically modified crop productivity estimates: Evidence from Indian data. Journal of Agricultural Economics 58: 24–36.
39. Bennett R, Kambhampati U, Morse S, Ismael Y (2006) Farm-level economic performance of genetically modified cotton in Maharashtra, India. Review of Agricultural Economics 28: 59–71.
40. Qaim M, Subramanian A, Sadashivappa P (2010) Socioeconomic impacts of Bt (Bacillus thuringiensis) cotton. In: Zehr UB, editor. Biotechnology in Agriculture and Forestry 65: Cotton. pp. 221–240.
41. ICAR (2002) All India coordinated cotton improvement project: Annual report 2001-02. Coimbatore: Indian Council for Agricultural Research. 97 p.
42. Singh NB, Barik A, Gautam HC (2009) Revolution in Indian Cotton. Directorate of Cotton Development, Ministry of Agriculture, Department of Agriculture & Cooperation, Government of India, Mumbai & National Center of Integrated Pest Management, ICAR, Pusa Campus, New Delhi. Available: www.ncipm.org.in/NCIPMPDFs/Revolution_in_Indian_Cotton.pdf. Accessed 11 October 2013.
43. Gruère GP, Sun Y (2012) Measuring the contribution of Bt cotton adoption to India's cotton yields leap. IFPRI Discussion Paper 01170. Available: http://www.ifpri.org/sites/default/files/publications/ifpridp01170.pdf. Accessed 11 October 2013.
44. Kranthi K (2011) 10 years of Bt in India - Three-part feature story on the history of Bt cotton in India.Cordova: Cotton Media Group. Available: http://www.cotton247.com/article/27520/part-ii-10-years-of-bt-in-india. Accessed 11 October 2013.
45. Aris A, Leblanc S (2011) Maternal and fetal exposure to pesticides associated to genetically modified foods in Eastern Townships of Quebec, Canada. Reproductive Toxicology 31: 528–533.
46. Marvier M, McCreedy C, Regetz J, Kareiva P (2007) A Meta-Analysis of Effects of Bt Cotton and Maize on Nontarget Invertebrates. Science 316: 1475–1477.
47. Azadi H, Ho P (2010) Genetically modified and organic crops in developing countries: A review of options for food security. Biotechnology Advances 28: 160–168.
48. Qaim M (2009) The economics of genetically modified crops. Annual Review of Resource Economics: 665–693.
49. Jacobsen S-E, Sørensen M, Pedersen S, Weiner J (2013) Feeding the world: genetically modified crops versus agricultural biodiversity. Agronomy for Sustainable Development: 1–12.
50. Radhakrishnan S (2012) 10 years of Bt cotton: False hype and failed promises. Coalition for a GM-Free India. Available: http://indiagminfo.org/wp-content/uploads/2012/03/Bt-Cotton-False-Hype-and-Failed-Promises-Final.pdf. Accessed 11 October 2013.
51. Herring RJ, Rao NC (2012) On the 'failure of Bt cotton' - Analysing a decade of experience. Economic & Political Weekly XLVII: 45–54.
52. Gruere G, Sengupta D (2011) Bt Cotton and Farmer Suicides in India: An Evidence-based Assessment. Journal of Development Studies 47: 316–337.
53. Charyulu K, Biswas S (2010) Economics and efficiency of organic farming vis-à-vis conventional farming in India. Ahmedabad: Indian Institute of Management. Available: http://www.iimahd.ernet.in/publications/data/2010-04-03Charyulu.pdf. Accessed 11 October 2013.

54. IFOAM (2012) The IFOAM norms for organic production and processing: Version 2012. Bonn: Die Deutsche Bibliothek. 134 p.
55. Koepf HH, Petersson BD, Schaumann W (1976) Biodynamic Agriculture: An Introduction. Anthroposophic Press, Hudson, New York. 430 p.
56. Kanwar JS (1988) Farming systems in swell-shrink soils under rainfed conditions in soils of semi-arid tropics. In: Hirekerur LR, Pal DK, Sehgal JL, Deshpande CSB, editors. Transactions of International Workshop on Swell-Shrink Soils. National Bureau of Soil Survey and Land Use Planning, Nagpur. pp.179–193.
57. Ministry of Agriculture GOI (2011a) State-wise estimates of area and production of cotton released. Bt cotton constitutes about 90% of total area under cotton cultivation. Press release of Department of Agriculture & Cooperation, Government of India, Mumbai. Release ID: 73448. Available: http://pib.nic.in/newsite/erelease.aspx?relid = 73448. Accessed 11 October 2013.
58. Choudhary B, Gaur K (2010) Bt Cotton in India: A Country Profile. ISAAA Series of Biotech Crop Profiles. ISAAA: Ithaca, NY. ISBN: 978-1-892456-46-X. Available: http://www.isaaa.org/resources/publications/biotech_crop_profiles/bt_cotton_in_india-a_country_profile/download/Bt_Cotton_in_India-A_Country_Profile.pdf. Accessed 11 October 2013.
59. Nagarajan P (2012) Fiber production report for India 2010-11. Textile exchange. Available: http://farmhub.textileexchange.org/upload/library/Farm%20and%20fiber%20report/Regional%20Reports%20-%20India%20-%20English%202010-11-FINAL.pdf. Accessed 11 October 2013.
60. IFOAM (2006) The IFOAM norms for organic production and processing: Version 2005. Bonn: Die Deutsche Bibliothek. 136 p.
61. Demeter International e.V. (2012) Production standards for the use of Demeter, biodynamic® and related trademarks. Demeter International production standards: The standards committee. 45 p. Available: http://www.demeter.net/certification/standards/production. Accessed 11 October 2013.
62. ICAR (2009) Handbook of Agriculture. New Delhi: Indian Council of Agricultural Research. 1617 p.
63. Beej SA, Hamara BA (2012) A decade of Bt cotton in Madhya Pradesh: A report. India Environment Portal, Centre for Science and Environment (CSE), National Knowledge Commission (NKC), Government of India. Available: http://www.indiaenvironmentportal.org.in/files/file/MP-DECADE-OF-BT-COTTON-2012.pdf. Accessed 11 October 2013.
64. Carpenter-Boggs L, Kennedy AC, Reganold JP (2000) Organic and biodynamic management: Effects on soil biology. Soil Science Society of America Journal 64: 1651–1659.
65. Brookes G, Barfoot P (2011) The income and production effects of biotech crops globally 1996-2009. International Journal of Biotechnology 12: 1–49.
66. Majumdar EG (2012) CICR technical bulletin - Mechanisation of cotton production in India. Nagpur: Central Institute for Cotton Research (CICR). Available: http://www.cicr.org.in/pdf/mechnaisation_cotton.pdf. Accessed 11 October 2013.
67. Ministry of Agriculture GOI (2011b) Cost of cultivation/production & related data. Directorate of Economics and Statistics, Department of Agriculture & Cooperation, Government of India, Mumbai. Available: http://eands.dacnet.nic.in/Cost_of_Cultivation.htm. Accessed 11 October 2013.
68. Pinheiro J, Bates D, DebRoy S, Sarkar D and the R Development Core Team (2013) nlme: Linear and Nonlinear Mixed Effects Models.R package version 3.1-110. Available: http://cran.r-project.org/web/packages/nlme/nlme.pdf. Accessed 11 October 2013.
69. R Core Team (2012) A language and environment for statistical computing. R Foundation for Statistical Computing, Vienna, Austria. ISBN 3-900051-07-0. Available: http://www.R-project.org/. Accessed 11 October 2013.
70. Loughin TM (2006) Improved experimental design and analysis for long-term experiments. Crop Science 46: 2492–2502.
71. Carpenter JE (2010) Peer-reviewed surveys indicate positive impact of commercialized GM crops. Nature Biotechnology 28: 319–321.
72. Glover D (2010) Is Bt Cotton a Pro-Poor Technology? A Review and Critique of the Empirical Record. Journal of Agrarian Change 10: 482–509.
73. Konduru S, Yamazaki F, Paggi M (2012) A Study of Indian Government Policy on Production and Processing of Cotton and Its Implications. Journal of Agricultural Science and Technology B 2: 1016–1028.
74. Sadashivappa P, Qaim M (2009) Bt cotton in India: Development of benefits and the role of government seed price interventions. AgBioForum 12: pp. 172–183. Available: http://agbioforum.org/v12n2/v12n2a03-sadashivappa.pdf. Accessed 11 October 2013.
75. Panneerselvam P, Halberg N, Vaarst M, Hermansen JE (2012) Indian farmers' experience with and perceptions of organic farming. Renewable Agriculture and Food Systems 27: 157–169.
76. Ramesh P, Panwar NR, Singh AB, Ramana S, Yadav SK, et al. (2010) Status of organic farming in India. Current Science 98: 1190–1194.
77. Ministry of Agriculture GOI (2011b) Agricultural wages in India. Directorate of Economics and Statistics, Department of Agriculture & Cooperation, Government of India, Mumbai. Available: http://eands.dacnet.nic.in/Cost_of_Cultivation.htm. Accessed 11 October 2013.
78. Singh RJ, Ahlawat IPS, Kumar K (2013) Productivity and profitability of the transgenic cotton-wheat production system through peanut intercropping and FYM addition. Experimental Agriculture 49: 321–335.
79. Panneerselvam P, Hermansen JE, Halberg N (2011) Food Security of Small Holding Farmers: Comparing Organic and Conventional Systems in India. Journal of Sustainable Agriculture 35: 48–68.
80. Swezey SL, Goldman P, Bryer J, Nieto D (2007) Six-year comparison between organic, IPM and conventional cotton production systems in the Northern San Joaquin Valley, California. Renewable Agriculture and Food Systems 22: 30–40.
81. Venugopalan MV, Rajendran TP, Chandran P, Goswami SN, Challa O, et al. (2010) Comparative evaluation of organic and non-organic cotton (Gossypium hirsutum) production systems. Indian Journal of Agricultural Sciences 80: 287–292.
82. Menon M (2003) Organic cotton re-inventing the wheel. Hyderabad: Booksline, SRAS Publications. Available: http://www.ddsindia.com/www/PDF/Organiccotton_Cover_Text.pdf. Accessed 11 October 2013.
83. Nemes N (2010) Seed security among organic cotton farmers in South India. University of Hohenheim: Department of Rural Communication and Extension. Available: http://www.organiccotton.org/oc/Library/library_detail.php?ID = 305. Accessed 11 October 2013.
84. Forster D, Adamtey N, Messmer MM, Pfiffner L, Baker B, et al. (2012) Organic Agriculture—Driving Innovations in Crop Research. In: Bhullar G, Bhullar N, editors. Agricultural Sustainability - Progress and Prospects in Crop Research: Elsevier. pp 21–46.
85. Vanloqueren G, Baret PV (2009) How agricultural research systems shape a technological regime that develops genetic engineering but locks out agroecological innovations. Res Policy 38: 971–983.

10

Climate Change and Apple Farming in Indian Himalayas: A Study of Local Perceptions and Responses

Basavaraj Basannagari, Chandra Prakash Kala*

Ecosystem and Environment Management, Indian Institute of Forest Management, Bhopal, Madhya Pradesh, India

Abstract

Apple farming is an important activity and profession of farmer communities in the Himalayan states of India. At present, the traditional apple farming is under stress due to changes in climate. The present study was undertaken in an Indian Himalayan state, Himachal Pradesh, with the major aim of studying perceptions of farmers on the effects of climate change on apple farming along the altitudinal gradient. Through questionnaire survey, the perceptions of farmers were recorded at low hills (<2500 m), mid-hills (2500–3000 m), and upper hills (>3000 m). At all elevation range the majority of farmers reported that there was increase in atmospheric temperature, and hence at low hills 72% farmers believed that this increase in temperature was responsible for decline in fruit size and so that the quality. Thirty five percent farmers at high hills and 30% at mid hills perceived frost as a major cause for damaging apple farming whereas at low hills 24% farmers perceived hailstorm as the major deterrent for apple farming. The majority of farmers, along the altitude (92% at high hills, 79% at mid hills and 83% at low hills), reported decrease in snowfall. The majority of farmers at low altitude and mid altitude reported decline in apple farming whereas 71% farmers at high hill areas refused decline in apple farming. About 73–83% farmers admitted delay in apple's harvesting period. At mid hills apple scab and at low hills pest attack on apple crops are considered as the indicators of climate change. The change in land use practices was attributed to climate change and in many areas the land under apple farming was replaced for production of coarse grains, seasonal vegetables and other horticulture species. Scientific investigation claiming changes in Indian Himalayan climate corroborates perceptions of farmers, as examined during the present study.

Editor: Matteo Convertino, University of Florida, United States of America

Funding: The authors have declared that no competing interests exist.

Competing Interests: No.

* E-mail: cpkala@yahoo.co.uk

Introduction

The mountain ecosystem is one of the most vulnerable ecosystems to the climate change and so that the mountain communities, especially those mainly depend on animal husbandry, marginal agriculture and horticulture products. The Himalayan mountain ecosystem, at present, is facing the challenges created due to increasing aridity, warmer winter season, variability in precipitation, and unexpected frosts and storms [1], [2], which largely affect the entire range of biodiversity, including agriculture and horticulture crops [1], [3]. Though, the Himalaya harbours rich biodiversity and is one of the most vulnerable mountain ecosystems to climate change [4], [5], there is paucity of systematic analysis of climate change and its impacts on the Himalayan ecosystems, biodiversity and local people's livelihoods [6].

Farmers of Indian Himalayan region grow many fruit crops, including pomes (apple and pear) and stone fruits (peach, plum, apricot and cherry) in considerable quantity [7]; however apple has the preference over all other horticultural crops [8]. Worldwide, there are over 7,500 known cultivars of apples [9]. China dominates the world in production of apple, followed by the United States, and India ranks seventh with average yield of about 7.24 tonnes per hectare [9]. All three north-west Himalayan states of India - Himachal Pradesh, Jammu-Kashmir, and Uttarakhand are the major apple producing states of India. In these states, the apples are grown at altitude ranging from 1200 m to 3500 m above mean sea level [10].

In Himachal Pradesh, the area under apple has increased from 400 ha in 1950–51 to 3,025 ha in 1960–61 and further 99,564 ha in 2009–10 [11], [12]. Though the production of apple in this state has steadily increased by bringing more areas into apple farming, the productivity has declined [13]. The present study, therefore, aims to understand the causes of reducing apple farming in the state despite high preference of local people to continue apple farming. Apple being highly sensitive to adversities of climate [14–16], the centre of attention of this study is to access the perceptions of farmers on effects of climate change on the apple farming along the altitudinal gradient.

Methods

Study Area

The present study was carried out in the Kinnaur district of Himachal Pradesh, which is located in the northern India. Himachal Pradesh is bordered by Jammu-Kashmir on north, Punjab on west and south-west, Haryana on south, Uttarakhand on south-east and China on the east [12]. The state is second largest producer of apples in the country. About 66% of area in the state is under forest cover, which is rich in biodiversity including medicinal and aromatic plants. The main occupation of

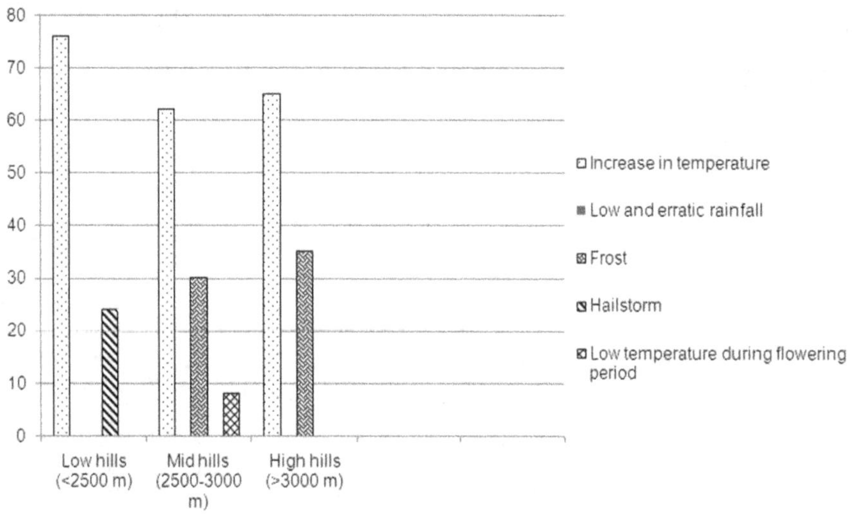

Figure 1. Perceptions of respondents on the various factors responsible for damage of apple farming along the altitudinal gradient in the Kinnaur district of Himachal Pradesh.

the people of this state is agriculture, horticulture and allied sector. The topographical variations and altitudinal differences provide congenial environment for cultivation of temperate to sub-tropical fruits.

Intensive investigations at village level were carried out in the Kinnaur district of Himachal Pradesh. Kinnaur district is situated on both sides of river Satluj from 31°-05′-50″ to 32°-05′-15″ north latitude and between 77°-45′ to 79°-00′-35″ east longitudes. Of the 12 districts of Himachal Pradesh, Kinnuar being third largest district spans over 6,520 km^2. Total human population of this district is 84,298 and the population density is estimated as 15 per km^2 [17]. It is situated between 2,100 m to 3,600 m above mean sea level, and it is popularly known as the 'apple bowl' of the state. Kinnaur district is composed of 3 administrative blocks such as Nichar, Kalpa and Pooh.

Field Surveys

The field surveys were undertaken in 2012. The approval of ethical committee was not required as the present study does not deal with the clinical trials on humans and animals. The study rather deals with the agro-biodiversity, which falls in the jurisdiction of Biological Diversity Act 2002. The regulations under 'Assess to Biological Diversity', Chapter 2, Section 3.0 and Section 4.0, authorize authors, being the Indian citizens, to carry out research and publish research papers with respect to agro-biodiversity in India. The present study complies with all relevant regulations.

Random survey was carried out in all three development blocks of Kinnuar district - Kalpa, Pooh and Nichar along the altitudinal gradient (e.g., upper hills, mid-hills, and low hills). Two villages at each altitudinal range (e.g., upper hills >3000 m; mid-hills

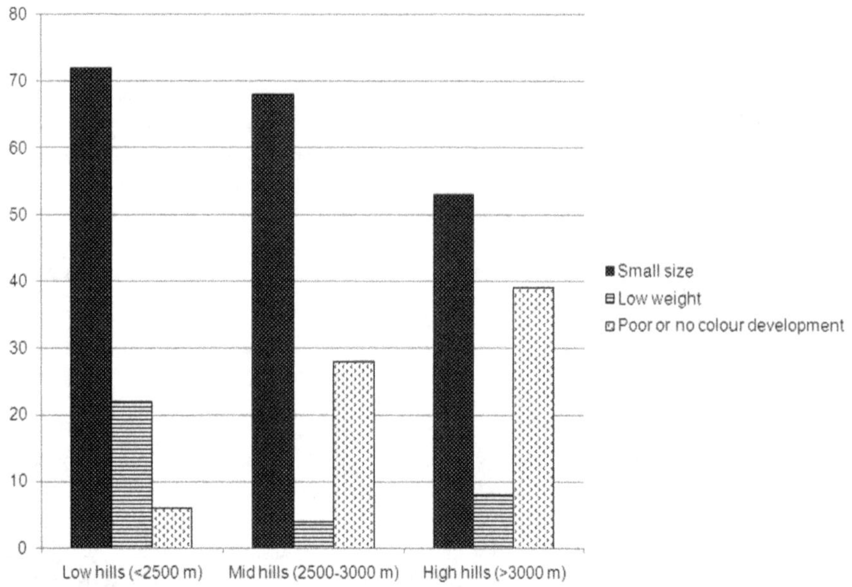

Figure 2. Perceptions of respondents along the altitudinal gradient in the Kinnaur district of Himachal Pradesh on the impact of fruit quality due to climate change.

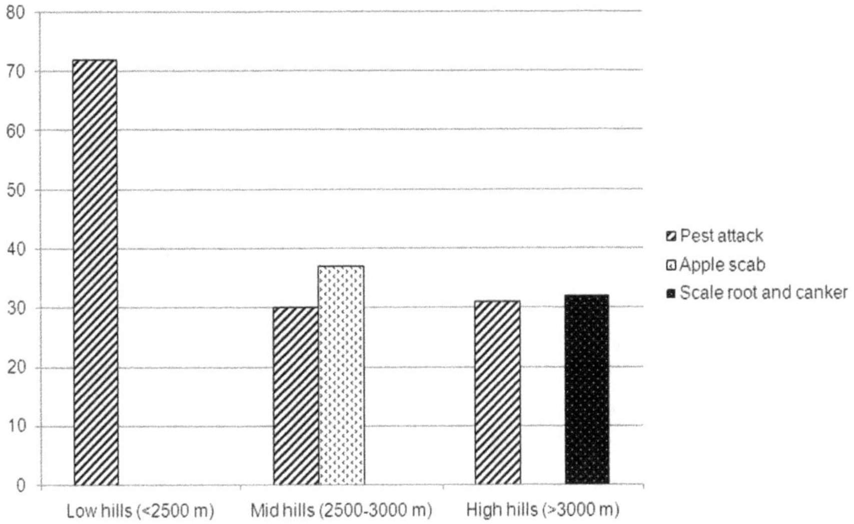

Figure 3. Indicators of climate change impacts on the apple crop as per the perceptions of respondents along the altitudinal gradient in the Kinnaur district of Himachal Pradesh.

between 2500–3000 m; low hills <2500 m) were selected. Villages Pooh and Chango were sampled at the high hills (>3000 m), Kalpa and Sangla at the mid hills (2500 m–3000 m) and Nichar and Urni were selected for sampling at the low hills (<2500 m). A total of 300 respondents inhabited in 6 villages were interviewed during the present investigations along the altitudinal gradient in Kinnaur district.

The data were obtained with the help of structured questionnaire survey. Before conducting the survey, we informed and discussed with the participants about the nature of the research. The subject's participation was voluntary. Fifty households in each village were randomly sampled and information was gathered on various parameters, including perceptions of the farmers on the effects of temperature and precipitation on apple farming. Besides, Participatory Rural Appraisal was also used for collection of data through interviews/discussions with individual and focus groups within the farmer's community. Farmers were also interviewed on the trends in snowfall, extreme events, and changes in seasonality, cropping system and pest attack.

Secondary Information

An extensive literature survey was carried out on the parameters related to research study. Different Departments (e.g., State Forest Department, State Horticulture Department, State Universities and other Government Departments) were approached for collection of present and past information related to apple farming in the state of Himachal Pradesh.

Results

The horticulture has emerged as the main profession of inhabitants in the study area, followed by agriculture and animal

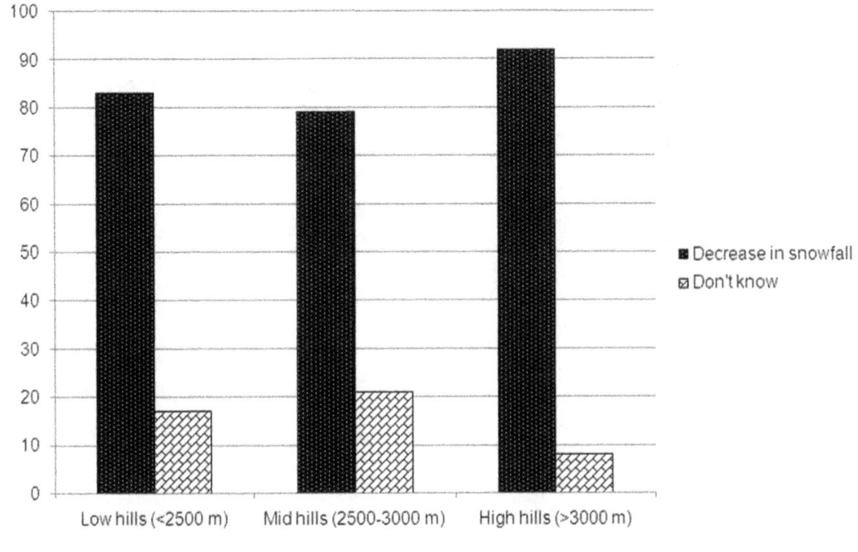

Figure 4. Perceptions of respondents along the altitudinal gradient in the Kinnaur district of Himachal Pradesh on the snowfall pattern.

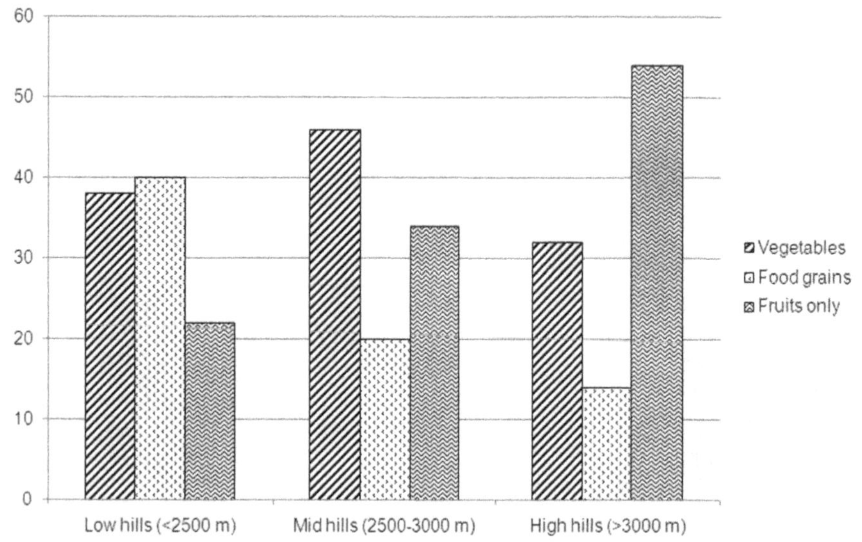

Figure 5. Conversion of apple farming into other land-use practices including cereals and vegetables along the altitudinal gradient in the Kinnaur district of Himachal Pradesh.

husbandry. Of the total people interviewed, 73% admitted horticulture as a primary occupation, followed by agriculture (23%). The majority of farmers (78%) at low altitude (<2500 m) and 72% at mid altitude (2500–3000 m) reported in decline of apple farming whereas majority of farmers (71%) of high hill areas (>3000 m) admitted that there was no decline in the apple farming.

Along the altitudinal gradient the majority of respondents reported that there was increase in atmospheric temperature. About 24% of farmers at low hills perceived hailstorm as the major deterrent for apple farming whereas 35% farmers at high hills and 30% at mid hills perceived frost as a major cause for damaging apple farming (**Figure 1**). The perceptions of growers on the quality of apple fruit production varied along the altitudinal gradient. About 72% farmers at low hills believed that change in climate, especially increasing temperature, was responsible for decline in fruit size and so that the quality. Lacking of appropriate fruit colour, due to climate change, was considered as deterrent factor in maintaining the fruit quality by 39% respondents at high hills (**Figure 2**).

The growers reported many indicators of climate change that impact apple farming along the altitudinal gradient. Infestation of pest and diseases such as apple scab, scale root and canker were some the indicators of climate change that increased the cost of production due to increase in use of pesticides and chemical fertilizers. At low hills majority of respondents reported pest attack on apple crops as one of the indicators of climate change whereas at mid hills apple scab was the most prominent (**Figure 3**). Apple production is known to influence by chilling hours however, the majority of respondent along the altitude (92% respondents at

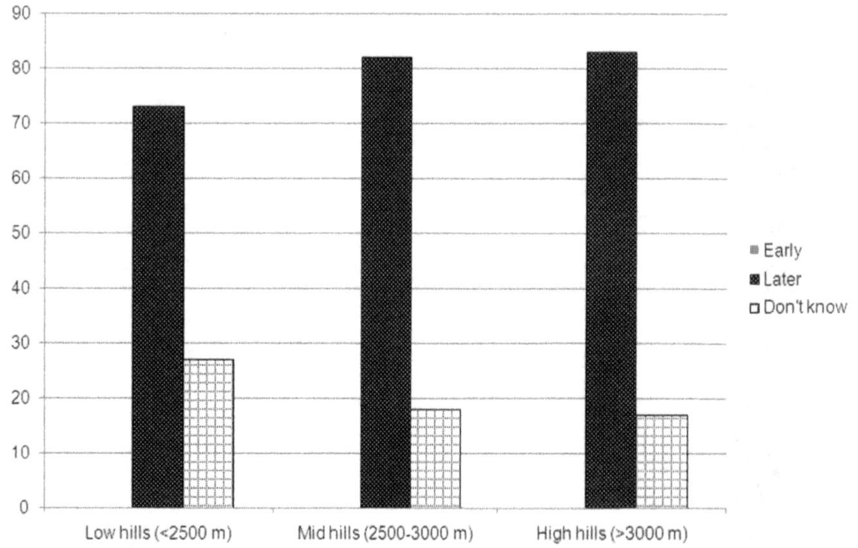

Figure 6. Effects of climate change on the harvesting period of apple along the altitudinal gradient in the Kinnaur district of Himachal Pradesh.

Table 1. Production of apple in Himachal Pradesh; Source: [18], [36].

Year	Production (mt)	Area (ha)
1975-76	200000	35076
1981-82	306798	45335
1989-90	394868	59988
1990-91	342071	62828
1991-92	301730	66767
1992-93	279051	69429
1993-94	294734	72406
1994-95	122782	75469
1995-96	276681	78292
1996-97	288538	80338
1997-98	234253	83056
1998-99	393653	85631
1999-2000	53000	88631
2000-01	376736	90347
2002-03	348263	92820
2003-04	459492	84112
2004-05	527601	86202
2005-06	540356	88560
2006-07	268402	91804
2007-08	592576	94726
2008-09	510161	97438
2009-10	280105	99564

high hills, 79% at mid hills, and 83% at low hills) reported decrease in snowfall. None of the respondents in the study villages admitted that there was increase in snowfall **(Figure 4)**.

The growers indicated that because of climate change the land-use practices were under change. In many areas the land under apple farming was replaced by farming of coarse grains, seasonal vegetables and other horticulture species (**Figure 5**). The majority of respondents reported that there was shift in harvesting period of apple due to climate change, especially of increasing temperature, across the altitudinal gradient (**Figure 6**). Seventy three to eighty three percent respondents admitted that apple harvesting period has been delayed in the study villages.

Discussion

The area brought under apple farming in Himachal Pradesh has increased from 35,076 ha to 99,564 ha during 1975–2010 however, the same time series data show erratic trend in annual apple production, which is low and not in the proportion of land brought under apple farming **(Table 1)**. The erratic trend in apple production may have number of reasons, including farming sites on sloppy and marginal lands, areas prone to adverse weather conditions, rainfed land and small land holdings [18], [19]. Nurseries are being mainly developed by utilizing scion wood of old varieties and seedling rootstock for propagation. The concept of budwood bank for supply of certified quality true-to-type planting material for further propagation is lacking [18]. Apart from this, climate change impacts the apple farming in numerous ways.

Decline in atmospheric moisture and increase in frequent droughts may lead to higher temperature as reported by farmers along the altitudinal gradient during the present course of investigations. The winter temperature and precipitation in the form of snow are very important and sensitive climatic factors for induction of dormancy, bud break and also to ensure proper flowering in apples [20]. Apple requires 1200–1500 hours of chilling depending on type of cultivars. The chilling hours <1000 lead to poor fruit formation [21–23]. Prolonged delay in cold in December and January severely affects the chilling requirements [24]. Apple grows best in the regions where the tree undergoes an uninterrupted winter rest [25].

However, the extreme minimum temperature during the winter causes winter freeze injury in apple fruits, which results poor apple yield [26]. Summer temperature and climate conditions also influence the size and quality of apples as the fruits develop during April to June. The high (>26°C) or low temperature (≤15°C) during flowering phase reduce apple crop [27]. Temperature impacts apples farming throughout the season from immediately after blooming period in the apple orchards to the fruit size at harvest [28]. Delicious group of apple trees, the one which is mostly cultivated in the present study area, under normal conditions require 1234 units of winter chilling [29]. However, some of the low altitude zones under apple cultivation do not fulfil sufficient winter chilling due to rise in temperature. With deficiency in chilling hours, flower buds produce fewer fruit clusters resulting in delay in bloom period.

The increasing pattern of chill unit at 2700 m above mean sea level reports that the area is conducive for apple cultivation and hence there is a shift in apple farming from low hills to middle and high hills [15], [16]. These reports support the farmer's perceptions of the present study area which showed that apple cultivation shifts from low to high elevations with respect to increase in temperature. The harvesting period of apple is also delayed for a week to a fortnight. Apart from temperature, the decline in apple farming at low hills is also attributed to hailstorms, decrease in snowfall and inferior quality of fruit production due to pest attack. In many such areas, the apple farming is being replaced by raising course grains and seasonal vegetables.

Our results with respect to the effects of climate change on apple farming are similar to those inferred for other regions of the world. In Japan, the areas suitable for apple farming (i.e. with an annual average temperature of 7~13°C) are gradually shifting northward besides decline in colour quality of apple due to heat injury [30], [31]. Farmers, here, have also started shifting from apple to peach farming because peach farming is considered less vulnerable to climatic stresses [32]. The apple farming areas have shifted north and or to highlands in the Yangu of Gangwon Province from Daegu of Gyeongbook Province of Korea, as well [31]. The decrease in number of chilling hours in four mountain oases of Oman has made to decline the production of apple, and since such fruit trees barely fulfil their chilling requirements such marginal fruit production is expected to decline further [33]. In the neighbouring country of India, about 90% farmers believe that climate change is the major factor responsible for decline in apple production [34].

A model demonstrating climatically suitable areas for growing apples suggests that as temperature increases the areas suitable for growing apples move from south to north, from coast to midland, from planes to mountains and from urban centre to suburban areas [35]. The Indian Himalaya has warmed by 1.5°C from 1982 to 2006, at an average rate of 0.06°C yr^{-1}, which is considerably higher than the global average [6]. Majority of farmers during the present study believe that there is change in climate, especially increase in atmospheric temperature and decrease in snowfall, which they consider most responsible for decline in fruit size and so that the quality of apples. Scientific investigation claiming that the Indian Himalaya has warmed by 1.5°C from 1982 to 2006 corroborates such beliefs of farmers, as examined during the present study. The increase in temperature throughout the Himalaya makes it likely that our findings with respect to climate change effects on apple farming may represent a trend across the Himalaya. It is assumed that other fruit production regions, where similar sensitive fruit crop species are grown under the climatic conditions similar to the Himalaya, are also susceptible to even slight warming.

Apple is known in Himachal Pradesh as a most significant commercial fruit crop. However, in the study area, the farmers are mainly customized to follow the traditional and age-old practices of cultivation. They are less aware about scientific agro-commercial practices, horticulture schemes and agri-inputs due to lack of communication facilities at high hills. The present changes in climatic conditions such as change in temperature, precipitation, ground frost and hailstorm, and subsequent adversities in terms of proliferation of insect-pest and diseases, loss of soil fertility, water availability, and natural calamities pose serious threats on apple production. Apart from climate change, the apple production may be declined due to continuation of plantations that have crossed their fruit bearing stage. There is a need for re-plantation of apple trees in a systematic manner on a regular basis. Besides, it is important to understand the variations in the patterns of climate change and also to identify management practices and alternatives for farmers in order to cope up the vagaries of changing climate.

Acknowledgments

We thank Director, Indian Institute of Forest Management, Bhopal. The villagers of Kinnaur district are acknowledged for their support and help during fieldwork. Authors thank two anonymous reviewers and the editor for constructive comments on the manuscript.

Author Contributions

Conceived and designed the experiments: CPK. Performed the experiments: BB CPK. Analyzed the data: BB CPK. Wrote the paper: CPK BB. Conducted fieldwork: BB.

References

1. Renton A (2009) Suffering the science: climate change, people and poverty. Oxfam briefing paper number 130. Oxford: Oxfam International.
2. Dash SK, Hunt JCR (2007) Variability of climate change in India. Current Science 93: 782–788.
3. Kala CP (2013) Climate change and challenges of biodiversity conservation. In: Kala CP, Silori CS, editors. Biodiversity, Communities and Climate Change. New Delhi: The Energy and Resources Institute. 259–269.
4. Xu J, Grumbine R, Shrestha A, Eriksson M, Yang X, et al. (2009) The melting Himalayas: cascading effects of climate change on water, biodiversity, and livelihoods. Conservation Biology 23: 520–530.
5. Bawa KS, Koh LP, Lee TM, Liu J, Ramakrishnan P, et al. (2010) China, India, and the Environment. Science 327: 1457–1459.
6. Shrestha UB, Gautam S, Bawa KS (2012) Widespread Climate Change in the Himalayas and Associated Changes in Local Ecosystems. PLOS ONE 7 (5): e36741. doi:10.1371/journal.pone.0036741.
7. Ghosh SP (1999) Deciduous fruit production in India. In: Papademetriou MK, Herath EM, editors. Deciduous fruit production in Asia and the Pacific. Thailand: Regional Office for Asia and the Pacific, Food and Agricultural Organization. 38–56.
8. Kala CP (2007) Local preferences of ethnobotanical species in the Indian Himalaya: Implications for environmental conservation. Current Science 93: 1828–1834.
9. Thamaraikannan M, Palaniappan G, Sengottuval C (2010) Can Indian Apples Beat Imports in Quality and Price? Facts for You. September 2010, 7–11.
10. Deodhar SY, Krissoff B, Landes M (2007) What's keeping the apple away? Addressing price integration issues in India's apple market. Indian Journal of Economics and Business 6: 35–44.
11. Anonymous (2006) Economic Survey of Himachal Pradesh. Shimla: Department of Economics and Statistics, Government of Himachal Pradesh.
12. Anonymous (2012) Economic Survey of Himachal Pradesh 2011–12. Shimla: Economics and Statistics Department. Available: http://himachal.nic.in/economics/pdfs/EconSurveyEng2012_A1b.pdf. Accessed 2013 January 24.
13. Awasthi RP, Verma HS, Sharma RD, Bhardwaj SP, Bhardwaj SV (2001) Causes of low productivity in apple orchards and suggested remedial measures. In: Jindal KK, Gautam DR, editors. Productivity of Temperate Fruits. Solan: Dr. Y.S. Parmar University of Horticulture and Forestry. 1–8.
14. Byrne DH, Bacon TA (1992) Chilling estimation: its importance and estimation. The Texas Horticulturist 18: 8–9.
15. Partap U, Partap T (2002) Warning Signals from the Apple Valleys of the Hindu Kush-Himalayas. In: Shreshtha ABM, editor. Productivity Concerns and Pollination Problems. Kathmandu: Centre for Integrated Mountain Development. 104.
16. Rana RS, Bhagat RM, Kalia V, Lal H (2008) Impact of climate change on shift of apple belt in himachal Pradesh. In: ISPRS Archies XXXVIII-8/W3Workshop Proceedings; Impact of Climate Change on Agriculture. 131–137. Available: http://www.isprs.org/proceedings/XXXVIII/8-W3/b2/10-B10-79_ISRO%20F.pdf. Accessed 2013 February 18.
17. Census of India (2011) Available: www.censusofindia.gov.in.Accessed 2013 February 18.
18. Sharma RP (2006) Current scenario of temperate fruits in Himachal Pradesh. In: Kishore DK, Sharma SK, Pramanick KK, editors. Temperate Horticulture-Current Scenario. New Delhi: New India Publishing Agency. 7–17.
19. Kala CP (2010) Status of an indigenous agro-forestry system in changing climate: A case study of the middle Himalayan region of Tehri Garhwal, India. Journal of Forest Science 56: 373–380.
20. Jindal KK, Chauhan PS, Mankotia MS (2000) Apple productivity in relation to environmental components. In: Jindal KK, Gautam DR, editors. Productivity of Temperate Fruits. Solan: Dr YS Parmar University of Horticulture and Forestry. 12–20.
21. Knight TA (1801) Account of some experiments on the ascent of the sap in trees. Philosophical Transactions of the Royal Society of London 91: 333–353.
22. Samish RM (1954) Dormancy in Woody Plants. Annual Review of Plant Physiology and Plant Molecular Biology 5: 183–204.
23. Saure MC (1985) Dormancy Release in Deciduous Fruit Trees. Horticultural Reviews 7: 239–300.
24. Vedwan N, Robert ER (2001) Climate Change in the Western Himalayas of India: A Study of Local Perception and Response. Climate Research 19: 109–117.
25. Kronenberg HG (1979) Apple growing potentials in Europe: The fulfilment of the cold requirements of apple tree. Netherland Journal of Agriculture Science 27: 131–135.
26. Caprio JM, Quamme HA (1999) Weather conditions association with apple production in the Okanagan valley of British Columbia. Canadian Journal of Plant Science 79: 129–137.
27. Forshey CG (1990) Factors affecting 'Empire' fruit size. Proceedings New York State Horticulture Society 135: 71–74.
28. Beattie BB, Folley RRW (1978) Production variability in apple crops. II. The long term behavior of the English Crops. Scientia Horticulture 7: 325–332.
29. Richardson EA, Seeley SD, Walker DR (1974) A Model for Estimating the Completion of Rest for Redhaven and Elberta Peach Trees. HortScience 9: 331–332.
30. JMAFF (2008) Comprehensive Counterstrategies against Global Warming. Tokyo, Japan: Japanese Ministry of Agriculture, Forestry and Fisheries.
31. Kim C, Lee S, Jeong H, Jang J, Kim Y, Lee C (2010) Impacts of Climate Change on Korean Agriculture and Its Counterstrategies. Seoul, South Korea: Korea Rural Economic Institute, 282 pp.
32. Fujisawa M, Kobayashi K (2013) Shifting from apple to peach farming in Kazuno, norther Japan: perceptions of and responses to climatic and non-climatic impacts. Regional Environmental Change doi: 10.1007/s10113-013-0434-6.
33. Luedeling E, Gebauer J, Buerkert A (2009) Climate change effects on winter chill for tree crops with chilling requirements on the Arabian Peninsula. Climatic Change 96: 219–237. doi: 10.1007/s10584-009-9581-7.
34. Asghar A, Ali SM, Yasmin A (2012) Effect of climate change on apple (*Malus domestica* var. *ambri*) production: A case study in Kotli Satian, Rawalpindi, Pakistan. Pakistan Journal of Botany 44 (6): 1913–1918.
35. Seo HH (2003) Effects of climatic changes influencing fruit tree harvest. Effects of Climatic Changes Influencing the Korean Peninsula. Suwon-si Gyeonggi-do, Republic of Korea: The Rural Development Administration.
36. Anonymous (2013) State Department of Horticulture, Govt. of Himachal Pradesh. Shimla: Himachal Pradesh, India. Available: http://hpagrisnet.gov.in/horticulture/PDF/Horticulture%20At%20A%20-Glance.aspx. Accessed 2013 January 24.

11

Pyrosequencing Reveals the Influence of Organic and Conventional Farming Systems on Bacterial Communities

Ru Li[1], Ehsan Khafipour[2,3]*, Denis O. Krause[2,3], Martin H. Entz[1], Teresa R. de Kievit[4], W. G. Dilantha Fernando[1]*

1 Department of Plant Science, University of Manitoba, Winnipeg, Manitoba Canada, **2** Department of Animal Science, University of Manitoba, Winnipeg, Manitoba, Canada, **3** Department of Medical Microbiology and Infectious Diseases, Winnipeg, Manitoba, Canada, **4** Department of Microbiology, University of Manitoba, Winnipeg, Manitoba, Canada

Abstract

It has been debated how different farming systems influence the composition of soil bacterial communities, which are crucial for maintaining soil health. In this research, we applied high-throughput pyrosequencing of V1 to V3 regions of bacterial 16S rRNA genes to gain further insight into how organic and conventional farming systems and crop rotation influence bulk soil bacterial communities. A 2×2 factorial experiment consisted of two agriculture management systems (organic versus conventional) and two crop rotations (flax-oat-fababean-wheat versus flax-alfalfa-alfalfa-wheat) was conducted at the Glenlea Long-Term Crop Rotation and Management Station, which is Canada's oldest organic-conventional management study field. Results revealed that there is a significant difference in the composition of bacterial genera between organic and conventional management systems but crop rotation was not a discriminator factor. Organic farming was associated with higher relative abundance of *Proteobacteria*, while *Actinobacteria* and *Chloroflexi* were more abundant in conventional farming. The dominant genera including *Blastococcus, Microlunatus, Pseudonocardia, Solirubrobacter, Brevundimonas, Pseudomonas*, and *Stenotrophomonas* exhibited significant variation between the organic and conventional farming systems. The relative abundance of bacterial communities at the phylum and class level was correlated to soil pH rather than other edaphic properties. In addition, it was found that *Proteobacteria* and *Actinobacteria* were more sensitive to pH variation.

Editor: Niyaz Ahmed, University of Hyderabad, India

Funding: The research was funded by Natural Sciences and Engineering Research Council of Canada discovery grants to WGDF and TRdK. The funders had no role in the study design, data collection and analysis, or preparation of the manuscript.

Competing Interests: The authors have declared that no competing interests exist.

* E-mail: Ehsan.Khafipour@ad.umanitoba.ca (EK); Dilantha.Fernando@ad.umanitoba.ca (WGDF)

Introduction

It has long been recognized that maintaining biodiversity of soil microbes is crucial to soil health, which has been defined as soil with the capacity of resilience to stress, sustaining high biological diversity, productivity and high level of internal nutrient cycling, maintaining environmental quality and promoting plant health [1,2]. Bacterial communities are responsible for multifaceted biological functions in soils [3,4,5], and exert an important role in maintaining plant health [6,7,8,9,10]. In turn, soil has a direct impact on the structure and function of soil bacterial communities through perturbations caused by natural or human activities [11,12,13]. It was reported that agricultural soil, perturbed by human activities, has different bacterial diversity, compared to non-disturbed forest and grassland soil [14,15]. However, there is still lack of the detailed information about the bacterial diversity affected by agriculture perturbation.

Over the past decades, conventional agricultural management practices have involved the use of artificial chemical fertilizers and pesticides to increase crop yields. This has led to severe environmental problems such as soil degradation, emission and leaching of fertilizer and pesticide, and the emergence of pesticide resistant species [16,17], resulting in an unsustainable practice [18]. The aim in sustainable management systems is to maintain the biological function of the soil and to promote plant health. Organic farming contributes to these factors using techniques such as crop rotation, green manure, and biological pest control instead of chemical fertilizers and pesticides [19]. Consequently, organic farming systems may have a strong potential for restoring soil health and increase agro-ecosystem resilience to stress [18].

Few studies have evaluated the impact of fertilizer, crop rotation, and crop varieties on microbial community structure when conventional and organic farming systems are compared [20,21,22,23,24]. These studies have found that fertilizers, and crop varieties and rotation could shape the size and structure of soil microbial communities. However, these studies were based on field experiments where the above-mentioned factors varied at the same time between conventional and organic soil management. Therefore, the main discriminator between conventional and organic farming could not be defined. Moreover, the results of these studies were not consistent,

Table 1. Physico-chemical characteristics of Glenlea soil (0–15 cm) on the different treatments (Welsh, 2009; Bell 2012).

Rotation	Management	Total C (g/kg)	Carbonate C (g/kg)	Organic matter (%)	Total N (g/kg)	Olsen P (mg/kg)	pH	C: N Ratio
Grain-Only	Organic	3.0	2.8	6.7	2.6	23.4	7.0	11.3
Grain-Only	Conventional	3.2	2.6	7.2	2.7	21.7	6.6	11.5
Forage-Grain	Organic	3.1	0.4	7.9	2.7	3.6	7.0	11.3
Forage-Grain	Conventional	3.0	1.7	7.9	2.7	14.2	6.6	10.9
SEM		0.06	0.19	0.50	0.008	1.63	0.32	0.19
P-value								
Rotation		NS	<0.0001	0.005	NS	<0.0001	NS	NS
Management		NS	0.0039	NS	NS	0.0035	0.023	NS
Management × Rotation		0.01	0.0001	NS	NS	0.0001	NS	NS

NS = not significant (P>0.05).

which could be due to different analytical methodologies that varied in resolution [14,24,25,26].

Previous studies in our group had investigated the effects of different agriculture management practices (organic versus converntional) and crop rotation systems [(flax-oat-fababean-wheat (Grain-Only rotation) versus flax-alfalfa-alfalfa-wheat (Forage-Grain rotation)] on crop yield, soil edaphic traits, such as nitrate, phosphorus, pH, and organic matter. Results showed that rotation rather than farming managements affect most soil nutrient traits including nitrate-N, Oslen P, and organic matter. Whereas, farming system affect soil pH, with lower pH in conventional farming system than that in organic farming system. However, the effects of farming management and rotation on bacterial communitiy structure of the soil was not evaluated [27,46].

We hypothesized that microbial composition of soil differs between organic and conventional farming sytstems or in different crop rotations. As such, some microorganisms might be present in organic farms while absent or less frequent in conventional farms and vice versa. Similarly, there might be some bacteria that are unique to a specific crop rotation system. Using Roche 454 pyrosequencing methodology, the objective of this study was to identify bacterial populations that are associated with specific farming practice that could potentially influence soil and plant health. This manuscript provides a detailed framework of the soil bacterial composition at the genus level and its possible connection to farming practices.

Materials and Methods

Soil Sampling, Sampling Site and Experimental Design

Soil samples were collected at Glenlea Long-term Crop Rotation and Management Station (GLCRMS) at southern Manitoba, which is Canada's longest running organic-conventional management comparison station commenced in 1992. A detailed description of the location and site management was described previously [27]. In brief, the study site is located 20 km south of Winnipeg, Manitoba, Canada (N 49,39,0/W 97,7,0). The soil type is Rego Black Chernozem and the soil texture is clay (9% sand, 26% silt, and 66% clay) with an organic matter content of 7.7%. The experiment was a randomized complete block design in a split-plot arrangement with three replicates. Two crop rotations, that is, flax-oat-fababean-wheat (Grain-Only rotation) and wheat-alfalfa-alfalfa-flax (Forage-Grain rotation) were used as main plots, and certificated organic and conventional methods served as subplots. The 2×2 combinations of treatments included: Grain-Only Organic (GO), Grain-Only Conventional (GC), Forage-Grain Organic (FO) and Forage-Grain Conventional (FC). All rotation crops appeared in the rotation each year. Both organic and conventional experiments were managed using conventional tillage and plots were tilled with a disc and a field cultivator prior to sowing. Pesticides and chemical fertilizers were applied on the conventional plots but not on the organic plots. Eighteen kg P_2O_5 was banded when wheat was seeded in conventional plots, and 65 kg Nitrogen/ha was

Table 2. Pearson correlation coefficients between soil edaphic factors[1].

Variables	pH	Olsen P	Total N	Total C	Carbonate C	Organic matter	C: N ratio
pH	1.000	−0.380	−0.414	−0.429	−0.527	−0.424	−0.205
Olsen P		1.000	**0.996**	**0.993**	**0.615**	**0.994**	0.256
Total N			1.000	**0.998**	**0.606**	**0.998**	0.258
Total C				1.000	**0.622**	**1.000**	0.297
Carbonate C					1.000	**0.611**	0.135
Organic matter						1.000	0.297
C: N ratio							1.000

[1]Significant correlations between edaphic factors are indicated in bold type when P<0.05.

Table 3. Summary statistics of pyrosequencing 16S rRNA sequences of soil samples.

Rotation	Management	Number of trimmed sequences	Mean (SEM) results for indicated variable[1]						
			OTU[2] (95% distance)	Coverage (%)	Richness[3]		Diversity[4]		
					Chao1	ACE	Shannon	Simpson	Effective number of species
Grain-Only	Organic	30,482	1,917	$78.8^{a,b}$	3,660.0	$4{,}936.4^{a,b}$	7.2	0.0012	562.8
	Conventional	20,473	1,628	84.5^{a}	2,889.7	$3{,}044.2^{b}$	6.8	0.0022	504.4
Forage-Grain	Organic	23,923	2,118	73.2^{b}	4,308.7	$6{,}147.9^{a}$	7.6	0.0005	681.3
	Conventional	31,552	1,860	$80.3^{a,b}$	3,494.4	$4{,}533.6^{a,b}$	7.2	0.0012	592.2
SEM			236.5	2.7	512.1	677.1	0.22	0.01	122.1
P-value									
Rotation			0.35	0.11	0.20	0.08	0.20	0.38	0.43
Management			0.30	0.04	0.13	0.03	0.20	0.41	0.57
Rotation × Management (*P*-value)			0.95	0.83	0.96	0.83	0.97	0.86	0.91

a, b, c Means with different letters are significantly different for management at $P<0.05$.
1 Mean are from statistical models based on 5 to 6 replicate samples.
2 OTU= operational taxonomic units.
3 Based on Chao1 and abundance based coverage estimation (ACE) richness indices.
4 Based on Shannon and Simpson diversity estimators.

Table 4. Phylogenetic composition of bacterial phyla from pyrosequenced 16S rRNA sequences.

Phylum	Rotation (Grain-Only)		Rotation (Forage-Grain)		SEM	*P*-value		
	Management					Rotation	Management	Rotation × Management
	Organic	Conventional	Organic	Conventional				
	Abundant phyla[1]							
Proteobacteria	44.5^{a}	32.2^{a,b}	34.1^{a,b}	27.3^{b}	3.13	0.10	0.05	0.70
Actinobacteria	32.5^{a,b}	39.2^{a,b}	28.4^{b}	43.1^{a}	2.21	0.97	0.0002	0.08
Acidobacteria	8.5	10.3	13.8	12.1	2.35	0.15	0.88	0.44
Gemmatinomadetes	3.5^{B}	3.6^{B}	8.6^{A}	2.9^{B}	1.16	0.15	0.06	0.04
Chloroflexi	3.4^{b}	6.1^{a}	5.2^{a,b}	6.8^{a}	0.98	0.18	0.04	0.42
Bacteroidetes	3.3	2.7	2.7	2.2	0.98	0.54	0.52	0.99
Planctomycetes	1.5	1.7	2.1	1.8	0.35	0.30	0.81	0.49
	Low-abundance phyla[2]							
Firmicutes	0.6	0.9	0.2	1.1	0.34	0.47	0.11	0.32
Fibrobacteres	0.1^{B}	0.2^{B}	0.3^{A}	0.1^{B}	0.04	0.05	0.07	0.01
Nitrospirae	0.2^{b}	0.2^{a,b}	0.3^{a,b}	0.5^{a}	0.07	0.01	0.21	0.35
Verrucomicrobia	0.2^{C}	0.7^{A}	0.8^{A}	0.5^{B}	0.14	0.09	0.42	0.006
OP10	0.2^{B}	0.4^{A,B}	0.5^{A}	0.2^{B}	0.08	0.44	0.36	0.04
TM7	0.1	*	0.1	*	0.13	0.98	0.62	0.86
WS3	*	0.2	0.2	0.1	0.15	0.59	0.85	0.56
Unclassified	1.1	1.3	2.1	1.3	0.52	0.45	0.66	0.40

^{a,b,c}Means for main effects (rotation or management) are significantly different at $P<0.05$.
$^{A,\ B,\ C}$Means for the interaction between rotation and system are significantly different at $P<0.05$.
[1]Percentage of sequences larger than 1.
[2]Percentage of sequences smaller than 1.
*Percentage of sequences below 0.1.

broadcasted in conventional flax plots. One L/ha Buctril M and 0.235 L/ha Horizon were sprayed on conventional wheat plots and 0.2 L/ha Select and 1 L/ha Buctril were sprayed on conventional flax plots. These plots were managed with no external input of manure.

Bulk soil samples were randomly collected from the top level (0–15 cm) throughout wheat and flax plots in June and August 2008. Part of each soil sample was kept at −20°C prior to DNA extraction after sieving (2 mm) to remove roots and stones, while the rest was kept at 4°C for chemical analyses. Samples were analyzed using an elemental analyzer at a commercial soil analysis laboratory (AGVISE, Northwood, ND) for total soil carbon and total soil nitrogen (Vario MAX Carbon-Nitrogen analyzer, Elemetar, Germany). Soil Carbonate carbon was analyzed with a modified pressure technique [28]. Organic matter, soil pH, and Olsen phosphorus (P_{olsen}): sodium bicarbonate-extractable phosphorus was measured as described by Welsh et al. [27].

DNA Extraction

To remove PCR inhibitors, such as humic acids, covalent cations and other easily dissolved organic compounds, from soil samples a pre-lysis washing procedure was introduced before DNA extraction [29]. Soil samples of 0.25 g were mixed with 1.25 ml sodium phosphate (0.1 M, pH 7.5), then incubated in a shaker for 1 hr at room temperature, followed by centrifuging for 10 min at 16000×*g*. Supernatant was discarded. DNA was extracted from pre-washed samples using the PowerSoil DNA Isolate kit, which included a bead-beating step, according to the manufacturer's specifications (Mobio Laboratories, Solana Beach, CA). The DNA purity and quantity were tested by using spectrophotometer (Du 800 Spectrophotometer, BECKMAN COULTER). The average ratio of 260:280 was 1.7. The average DNA yield was 10 ng/µL. The variable regions of V1–V2 of the 16S rRNA genes were successfully amplified using forward primer 27F (AGAGTTT-GATCMTGGCTCAG) and reverse primer 342R (CTGCTGCSYCCCGTAG), indicating that the quality of extracted DNA was sufficient for further PCR application [30]. In order to test the long-term effect of farming practices on the bacterial communities in the soil and reduce the temporal effects of different sampling times, DNA samples of the same treatment collected at different sampling times were pooled before pyrosequencing.

Pyrosequencing

A total of 23 pooled DNA samples were pyrosequenced using the bacterial tag-encoded GS FLX-Titanium amplicon as described by Dowd et al. [31] and Khafipour et al. [32]. In brief, a mixture of Hot Start, HotStar high fidelity Taq polymerases, and Titanium reagents were used to perform a one-step PCR (35 cycles) with primers 28F (GAGTTTGATCMTGGCTCAG) and 519R (GTNTTACNGCGGCKGCTG), which covered the variable regions V1–V3 of the bacterial 16S rRNA genes [31]. The pyrosequencing procedures were carried out at the Research and Testing Laboratory (Lubbock, TX; http://www.Researchandtesting.com).

Table 5. Bacterial taxa showing significant variation under different farming systems generated using pyrosequenced 16S rRNA sequences.

Taxa (family and genus within each phylum or class)	Rotation (Grain-Only)		Rotation (Forage-Grain)		SEM	*P*-value		
	Management					Rotation	Management	Rotation × Management
	Organic	Conventional	Organic	Conventional				
Actinobacteria	32.5^{a,b}	39.2^{a,b}	28.4^{b}	43.1^{a}	2.21	0.97	0.0002	0.08
Geodermatophilaceae; *Blastococcus*	0.8^{b}	1.7^{a}	1.2^{a,b}	1.7^{a}	0.19	0.27	0.002	0.23
Intrasporangiaceae; *Lapillicoccus*	0.3^{b}	0.4^{a,b}	0.2^{b}	0.6^{a}	0.09	0.36	0.01	0.11
Propionibacteriaceae; *Microlunatus*	1.1^{a,b}	2.0^{a}	0.8^{b}	2.1^{a}	0.31	0.78	0.004	0.66
Pseudonocardiaceae; *Pseudonocardia*	0.9^{b}	1.7^{a,b}	1.3^{b}	2.6^{a}	0.24	0.02	0.001	0.31
Solirubrobacteriaceae; *Solirubrobacter*	0.6^{b}	1.7^{a}	0.8^{b}	1.6^{a}	0.29	0.88	0.003	0.48
Rubrobacteriaceae; *Rubrobacter*	0.5^{b}	1.0^{a}	0.3^{b}	1.4^{a}	0.24	0.57	0.005	0.30
Unclassified bacteria	11.9^{a,b}	15.6^{a,b}	9.9^{b}	17.4^{a}	1.81	0.97	0.01	0.33
Proteobacteria	44.5^{a}	32.2^{a,b}	34.1^{a,b}	27.3^{b}	3.13	0.10	0.05	0.70
Alphaproteobacteria	20.5	15.9	18.2	15.0	2.39	0.55	0.15	0.85
Caulobacteraceae; *Brevundimonas*	1.7^{a}	0.1^{b}	0.2^{b}	0.01^{b}	0.11	0.15	0.03	0.78
Xanthobacteraceae; uncultured	0.5^{b}	0.8^{a,b}	0.7^{a,b}	1.1^{a}	0.15	0.10	0.05	0.64
Rhodospirillaceae; *Skermanella*	0.8^{b}	1.5^{a,b}	1.3^{ab}	2.6^{a}	0.30	0.02	0.005	0.36
Gammaproteobacteria	11.4	7.1	7.7	3.6	2.72	0.13	0.08	0.67
Pseudomonadaceae; *Pseudomonas*	4.3	3.9	1.7	0.5	2.02	0.10	0.52	0.44
Xanthomonadaceae; *Stenotrophomonas*	0.7^{a}	0.0^{b}	0.3^{a}	0.0^{b}	0.25	0.67	0.04	0.71
Betaproteobacteria	10.3^{a}	6.3^{b}	8.7^{a,b}	5.1^{b}	2.00	0.42	0.04	0.93
Burkholderiaceae; *Burkholderia*	0.6	0.0	0.1	0.0	0.19	0.29	0.14	0.99
Deltaproteobacteria	1.0^{b}	2.7^{a,b}	3.4^{a}	3.5^{a}	0.05	0.009	0.10	0.13
Chloroflexi	3.4^{b}	6.1^{a}	5.2^{a,b}	6.8^{a}	0.98	0.18	0.04	0.42
Chloroflexaceae; *Roseiflexus*	0.6	1.5	1.1	2.3	0.28	0.02	0.002	0.52

^{a,b,c}Means for main effects (rotation or management) are significantly different at $P<0.05$.
$^{A, B, C}$Means for the interaction between rotation and system are significantly different at $P<0.05$.
[1]Percentage of sequences larger than 1.
[2]Percentage of sequences smaller than 1.
*Percentage of sequences below 0.1.

Bioinformatics of Pyrosequencing Data

Sequence editing, categorical transformation/ classification. Pyrosequencing data were edited, categorically transformed and classified as described by Khafipour et al. [32]. Briefly, all low quality sequences, tags, non-bacterial ribosomal sequences, and chimeras were removed from the database. In total, 123, 316 sequences were generated in this step. Then, the mothur software package [33] was utilized to perform the second round of sequence quality control and assignments of operational taxonomic units (OTU). All sequences shorter than 200 bp, or sequences having one or more ambiguous base, or containing a homopolymer length equal or greater than 8bp were removed from the dataset. The minimum, median and maximum lengths of sequences were 200, 471 and 647 bp, respectively. The unique sequences were then identified and aligned against a database of high quality 16S rRNA bacterial sequences derived from Silva (version 106) [34]. Through screening, filtering, and pre-clustering processes, columns containing a gap were removed in all sequences to reduce noise from pyrosequencing data. The remaining 987 columns (with an actual sequence length varying from 203 to 342 bp) and 39,283 sequences were used to build a distance matrix with a distance threshold of 0.1. Using the furthest neighbor algorithm with a cutoff of 95% similarity, these sequences were clustered to OTU. Representative sequences from each OTU were taxonomically classified with a confidence level of 60% using RDP Bayesian approach [35].

Alfa diversity analysis. An OTU-based approach was performed to calculate the richness, diversity and coverage at OTU cutoff of 0.05, which characterizes the biodiversity of the bacterial population in the soil samples at the genus level. Richness indices, Chao1 and abundance based coverage estimation (ACE), were calculated to estimate the number of species or OTU that were present in the sampling assemblage. The diversity within each individual sample, which is made up of richness and species abundance, was estimated using Simpson and non-parametric Shannon diversity indices. Good's non-parametric coverage estimator was used to estimate the percentage of the total species that were sequenced in each sample. Rarefaction curves for treatment groups were created in mothur [33], based on a re-sampling without replacement approach.

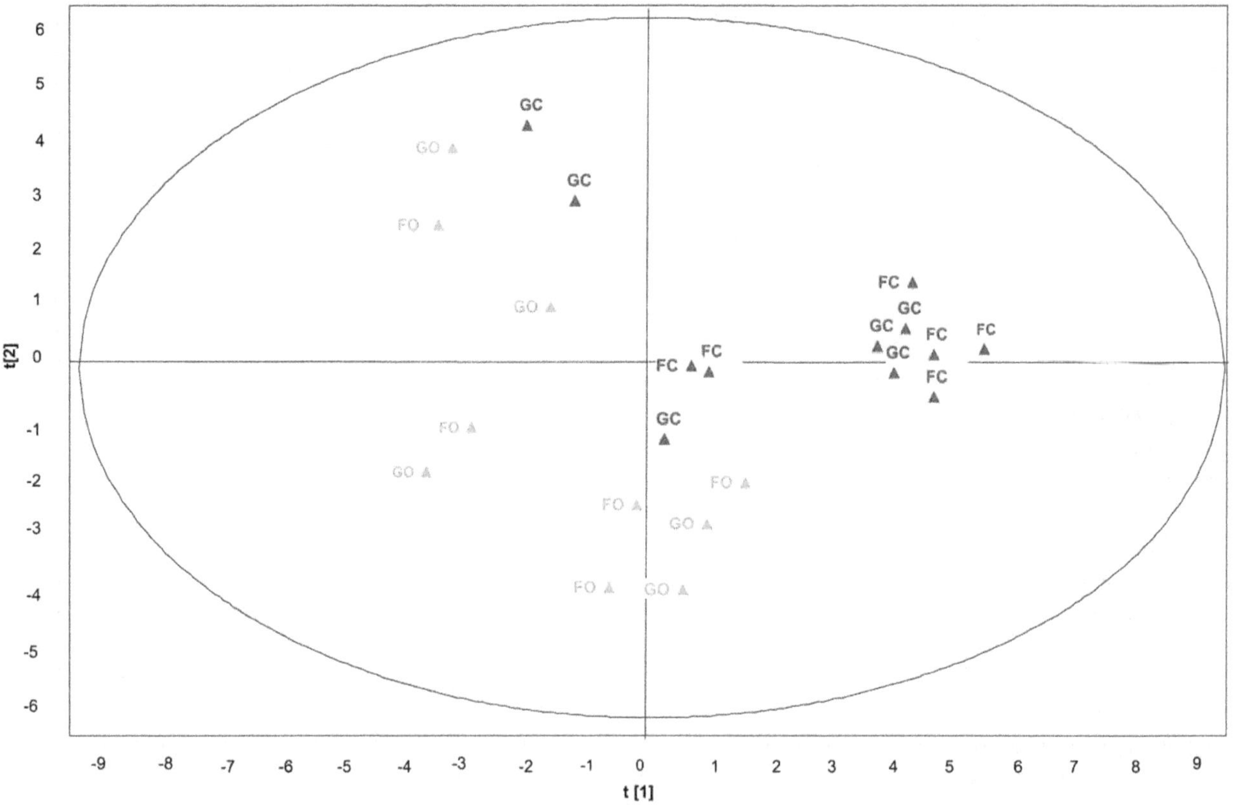

Figure 1. Partial least square discriminant score plot of soil bacteria under organic and conventional treatments. GO: Grain-Only organic; GC: Grain-Only conventional; FO: Forage-Grain organic; FC: Forage-Grain conventional. Model indicated a significant difference in the composition of putative bacterial genera between organic and conventional managements ($R^2X = 0.427$, $R^2Y = 0.882$, $Q^2 = 0.159$). Only genera with VIP>0.35 is included in the model.

Statistical Hypothesis Testing

The UNIVARIATE procedure of SAS [36] was used to test the normality of residuals for Alfa diversity indices. Non-normally distributed data were power transformed using Box-Cox power transformation macro (http://www.datavis.ca/sasmac/boxcox.html) in SAS based on the following models: BoxCox $(y) = (y^{\lambda} - 1)/\lambda$, if $\lambda \neq 0$ OR BoxCox $(y) = \log(y)$, if $\lambda = 0$. The range of λ for each parameter was adjusted by trial and error and the best fitting value of λ was identified using maximum likelihood methods. Normalized data were then used to assess the effect of treatment using MIXED procedure of SAS [36]. The effect of replicates was treated as random in the model.

Percentage data was used to evaluate statistical differences among treatments at the phylum and genus levels. To do so, the count data for each taxon was first transformed to the percentage of that taxon in an individual sample. Then UNIVARIATE procedure of SAS was used to test the normality of residuals for percentage data at each taxonomic level. For non-normally distributed data, Poisson and negative binomial distributions were fitted in the GLIMMIX procedure of SAS [36] to assess the effect of treatment. A log link function was specified for Poisson and negative binomial distributions. The goodness of fit for different distributions was compared using Pearson chi-square/DF (closer to 1 is better). Taxa were categorized as abundant and low-abundant in order to characterize them within each treatment. All taxa above 1% of the population were considered abundant and those below 1% were classified as less-abundant [32]. The differences between treatments were considered significant at $P<0.05$ while trends were observed at $P<0.1$.

Partial Least Square Discriminant Analysis and Redundancy Analysis

Partial least square discriminant analysis (PLS-DA; SIMCA-P+12.0.1, Umetrics, Umea, Sweden) [37] was performed on genus data to identify the effects of crop rotation and management on the bacterial community. The PLS-DA is a particular case of partial least square regression analysis in which Y is a set of binary (0 versus 1) variables describing the categories of a categorical variable on X. In this case, X variables were bacterial genera and binary Y was observations of organic (1) versus conventional (0), or Grain-Only (1) versus Forage-Grain (0) treatments. For this analysis, data were scaled using Unit Variance in SIMCA-P+ [37]. Cross-validation was then performed to determine the number of significant PLS components and a permutation testing was conducted to validate the model. To avoid over parameterization of the model, variable influence on projection value (VIP) was estimated for each genus and genera with VIP<0.35 were removed from the final model [38,39]. R^2X and R^2Y estimates were then used to evaluate the goodness of fit and Q^2 estimate was used to evaluate the predictive value of the model. Scatter- and score-plots were generated only for treatments that were significantly differentiated by the model. The PLS regression coefficients were used to identify genera that were most characteristic of each treatment group. The positive or negative correlations were considered significant when there was no overlap

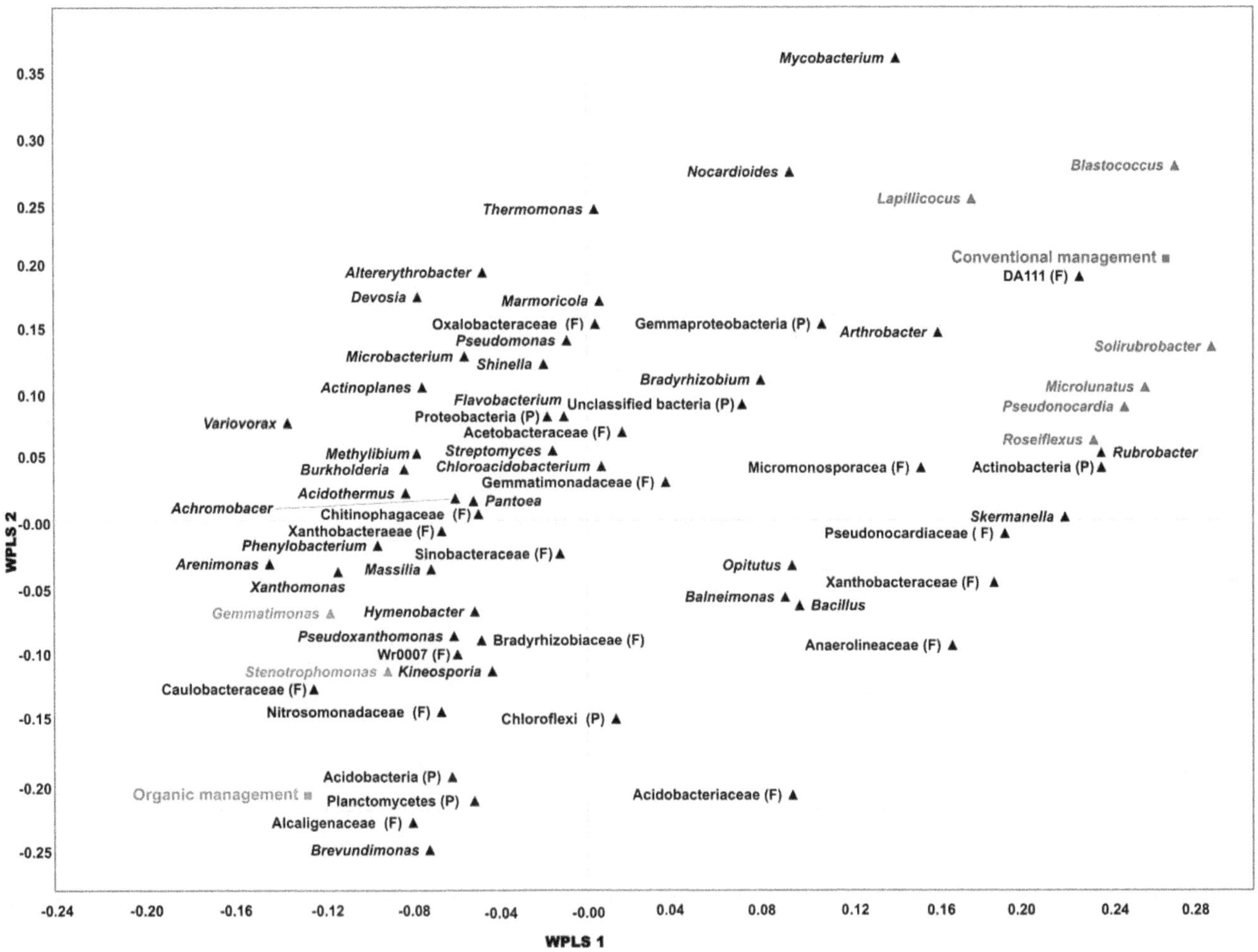

Figure 2. Partial least square discriminant analysis (PLS-DA) loading plot based on the relative abundance of the putative bacterial genera in soil microbiome and their association with organic or conventional treatments. Bacterial genera closer to organic or conventional are highly correlated to either treatment. PLS1 ($R^2X = 0.27$, $R^2Y = 0.525$, $Q^2 = 0.186$) and PLS2 ($R^2X = 0.127$, $R^2Y = 0.218$, $Q^2 = -0.081$). Some sequences could only be affiliated to phylum (P) or family (F) levels.

between the genus 95% confidence interval and the horizontal axis in the PLS regression coefficients graph.

Redundancy analysis (RDA) was carried out using canonical community ordination (CANOCO; Plant Research International BV, Wageningen, The Netherlands) to examine the relationship between abundant phyla and environment variables. Spearman's rank correlations were used to correlate abundant phyla and soil properties using SAS [36].

Results

Cropping Systems and Edaphic Soil Properties

The total soil C, N, and C: N ratio did not vary significantly under different cropping systems, while pH, organic matter, carbonate C and Olsen P were affected to varying degree by cropping systems. Total soil C was 3.0 g/kg, 3.2 g/kg 3.1 g/kg and 3.0 g/kg under GO, GC, FO, and FC farming systems, respectively. Total soil N was 2.7 g/kg in all of four treatments. C: N ratio was 11.3 under GO and FO farming systems, while it was 11.5 and 10.9 under GC and FC farming systems, respectively. In contrast, pH was 7.0 under organic management systems, higher than that of conventional systems with 6.7 ($P = 0.023$). Organic matter was 7.9 under Forage-Grain rotation systems, significantly higher than that of Grain-only rotation systems with 6.7 and 7.2 ($P = 0.005$). Both rotation and management affected carbonate C and Olsen P (Table 1). Carbonate C and Olsen P were much lower under Forage-grain rotation, compared to Grain-only system. Significant correlations existed between total N, C, carbonate C, organic matter and Olsen P, as well as between organic matter, carbonate C, total C and total N (Table 2).

Bacterial α-diversity

Bacterial diversity and richness in individual samples under different treatments were calculated (Table 3). Statistical differences in richness and diversity were only observed for coverage and ACE at the management level. Percentage of coverage for conventional treatment was higher than that of organic treatment ($P = 0.04$). The GC had the highest percentage of coverage (84.5%), followed by FC (80.3%), GO (78.8%), and FO (73.2%). The ACE richness was highest for FO (6,147.9), and lowest for GC (3,044.2). The rarefaction curve (Figure S1) generated with mothur demonstrated that observed numbers of OTU of FO and GO groups were higher than that of FC and GC groups, with FO having the highest number of observed OTU.

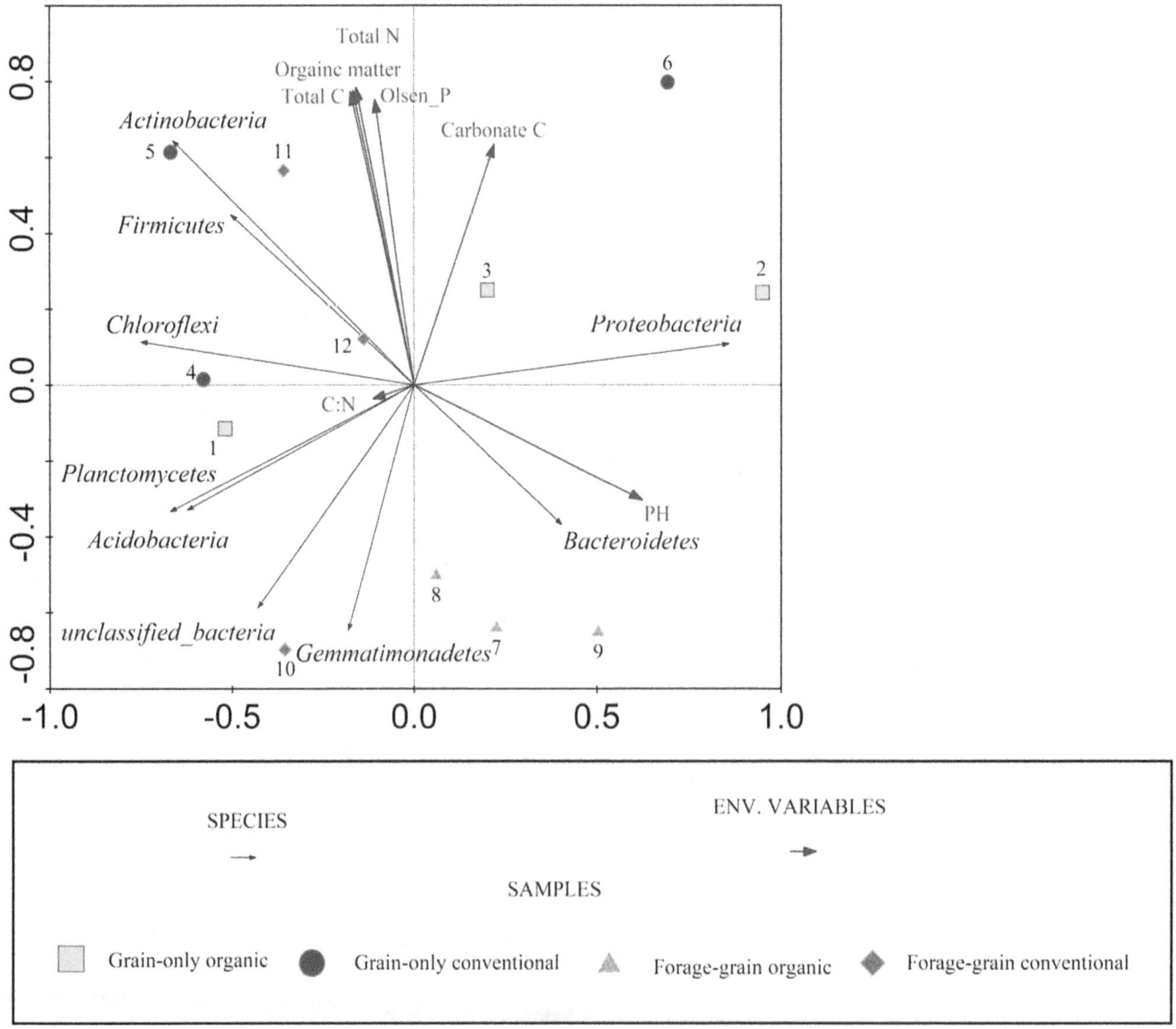

Figure 3. Redundancy analysis ordination plots of abundant phyla for individual sample.

Bacterial Community Composition

A total of 14 bacterial phyla were found in all the samples, of which seven were abundant (>1%) (Table 4). Ninety six percent of soil bacterial sequences belonged to these abundant phyla including: *Proteobacteria*, *Actinobacteria*, *Acidobacteria*, *Gemmatinomadetes*, *Chloroflexi*, *Bacteroidetes* and *Planctomycetes*. *Firmicutes*, *Fibrobacteres*, *Nitrospirae*, *Verrucomicrobia*, *P10*, *TM7* and *WS3* were in low abundance. The phylum distribution fluctuated under different farming disturbances. *Proteobacteria* accounted for 44.5% of total bacterial communities under the GO system, while it was only 27.3% under FC. In contrast, *Actinobacteria* made up 43.1% of total bacterial communities under the FC system, but were present in lower percentage (32.5%) under GO. Phylum *Chloroflexi* was also significantly influenced by management, with the highest percentage (6.8%) found in FC compared to the lowest (3.5%) in GO. A significant interaction between rotation and system was observed for *Gemmatinomadetes*, *Fibrobacteres*, *Verrucomicrobia*, and *P10*. Percentage of *Nitrospirae* was higher under Forage-Grain farming system. Other phyla did not show significant differences under different treatments. Unassigned bacterial sequences at the phylum level were approximately 1% of the total. In total, eight out of 14 phyla showed significant differences under different farming systems.

The relative abundance of different genera showing significant difference under different treatments was listed in Table 5. In phylum *Actinobacteria*, several putative genera including *Blastococcus*, *Lapillicoccus*, *Microlunatus*, *Pseudonocardia*, *Solirubrobacter*, and *Rubrobacter* showed significant differences among the treatments. The relative abundance of these genera was highest in the FC farming system, followed by GC, FO, and GO. The percentage of different class and genera belonging to the phylum *Proteobacteria* was higher under organic farming system compared to the conventional farming conditions with the exception of *Skermanella*, which was 2.6% and 1.5% under FC and GC farming systems, respectively. Classes *Alphaproteobacteria*, *Betaproteobacteria* and *Gammaproteobacteria* were higher in the Grain-Only organic farming system, although the difference was not statistically significant. Class *Deltaproteobacteria* showed the opposite pattern, being highest in Forage-Grain conventional systems, and lowest in Grain-Only organic system. *Pseudomonas* was the predominant genus in *Gammaproteobacteria* with

Table 6. Spearman's rank correlations between abundant phyla with soil properties[1].

Abundant phyla	Correlation						
	pH	Olsen P	Total N	Total C	Carbonate C	Organic matter	C: N ratio
Proteobacteria	**0.61**	0.07	−0.04	−0.03	−0.06	−0.03	0.05
Alphaproteobacteria	0.46	−0.47	−0.09	−0.12	−0.17	−0.11	−0.17
Betaproteobacteria	**0.62**	−0.18	−0.24	−0.38	−0.36	−0.20	−0.36
Gammaproteobacteria	0.25	−0.004	−0.05	−0.15	−0.13	0.04	−0.13
Deltaproteobacteria	−0.47	0.18	−0.21	−0.15	−0.13	−0.04	−0.13
Actinobacteria	**−0.65**	0.44	0.51	0.50	0.41	0.50	−0.17
Bacteroidetes	0.33	−0.37	−0.37	−0.36	−0.53	−0.36	0.38
Chloroflexi	−0.53	0.33	0.40	0.39	0.40	0.39	0.09
Firmicutes	−0.21	0.47	0.53	0.52	0.33	0.52	0.35
Gemmatimonadetes	−0.08	−0.46	−0.45	−0.45	−0.30	−0.45	−0.03
Planctomycetes	−0.18	−0.28	−0.21	−0.24	−0.31	−0.24	−0.16
Acidobacteria	−0.06	−0.23	−0.16	−0.20	−0.35	−0.20	−0.15
Nitrospirae	−0.004	−0.24	−0.19	−0.27	−0.40	−0.27	−0.23
Unclassified	−0.39	−0.47	−0.41	−0.41	−0.19	−0.41	−0.14

[1]Significant correlations between edaphic factors are indicated in bold type when $P<0.05$.

4.3% in GO, 3.9% in GC, 1.7% in FO, and 0.5% in FC. Within Phylum *Chloroflexi*, genus *Roseiflexus* was significantly influenced by interaction of rotation and management. Other genera in the phyla of *Actinobacteria* and *Proteobacteria* did not show statistical variation among the treatments (Table S1 and Table S2). There was no significant fluctuation under different farming systems in other genera within *Acidobacteria*, *Bacteroidetes*, *Firmicutes*, and *Planctomycetes* (Table S3).

The PLS-DA analysis showed that there is a significant difference in the composition of bacterial genera between organic and conventional managements ($R^2X=0.427$, $R^2Y=0.882$, $Q^2=0.159$) (Figure 1). However, crop rotation (Forage-Grain versus Grain-Only) was not a discriminator factor. Genera that were the most characteristic of each management system were identified using a scatter-plot (Figure 2). Among the putative bacterial genera included in the model, *Blastococcus* spp. *Microlunatus* spp. *Pseudonocardia* spp. and *Solirubrobacter* spp. were significantly correlated with conventional treatment, whereas *Gemmatimonas* spp. and *Stenotrophomonas* spp. were highly associated with organic management (Figure 2 and Figure S2).

Effect of Soil Edaphic Properties on Abundant Phyla

Canonical correspondence analysis tested the effect of soil edaphic properties on samples and bacterial populations by using an unconstrained analysis (RDA) (Figure 3). pH explained 24% of the variance ($P=0.06$), CaCO3 C explained 19% ($P=0.02$), and the C: N ratio accounted for less than 5% of the variance ($P=0.52$). Other soil edaphic variables were highly correlated with each other and were not able to explain variance separately. We also used Spearman's rank order correlation to evaluate relationships between abundant phyla and soil edaphic properties (Table 6). It was found that the relative abundance of *Proteobacteria* phylum and *Betaproteobacteria* class was positively correlated with soil pH, while the abundance of *Actinobacteria* was negatively correlated with soil pH. Other phyla did not show significant correlations with soil edaphic properties.

Discussion

In this study, crop rotation and management strategies did not alter total C, total N and C:N ratio but significantly affected organic matter, soil pH, carbonate C and Olsen P (Table 1). This indicates that farming systems gradually but not dramatically change soil edaphic properties [40]. We observed that organic management led to a neutral soil pH compared to conventional practices (7.0 versus 6.6; Table 1). This might be due to the application of synthetic fertilizer that could acidify the soil in the conventional systems [41]. Other reports indicated that soil pH could be influenced by other soil traits such as C:N ratio [42], vegetation, or soil type [43]. However, in our study, the field trials were run under identical condition, and the soil pH was not significantly correlated with other soil edaphic characteristics (Table 2). Therefore, the farming system was the sole factor to change the soil pH. As we expected, soil organic matter was higher under Forage-Grain rotations (Table 1) [42,44]. However, organic farming system did not increase organic matter in the soil surface, compared to conventional farming system. In contrast, other studies have shown that organic matter was higher in the top 0.3 m of soil under organic management [21,45]. This discrepancy could be due to different crops contributing to different amount of biomass and no additional manure added to our trials [46].

We used high-resolution power of 454-pyrosequencing to obtain insight into the effects of farming management styles (organic, conventional) and crop rotations (Grain-Only, Forage-Grain) on the diversity, richness and composition of soil bacterial communities. In total, pyrosequencing identified 14 phyla and 178 putative genera of bacteria in different soil samples. We found that organic and conventional farming management had major influence on soil bacterial communities while the effects of crop rotation were of smaller magnitude. We were also able to identify putative genera that were correlated with either organic or conventional farming management.

It has been reported that organic farming systems enhance microbial diversity in soil compared to the conventional systems

[21]. Although not statistically significant, we found a similar trend in this study (Table 3). Previous research indicated that soil pH might be the primary factor influencing richness and diversity of bacterial communities [47] with the highest richness and diversity found to be near the neutral pH. Lauber et al. [48] proposed that bacterial diversity had a strong negative relationship with soil pH when it was lower than 6.5. In this study, soil pH ranged from 6.6 to 7.0, and was significantly higher for organic compared to conventional system. This indirectly indicates that organic management might favor higher bacterial diversity. And it is important to notice that standard diversity parameters only based on OTU without taxonomic identity of the different groups is not sensitive enough to detect the influence of agriculture management on the soil bacterial community, because changes in some taxonomic groups might be compensated by changes in others [49]. Lending support to this hypothesis, we detected a significant shift in soil bacterial communities due to farming systems when sequences were taxonomically ranked (Figure 1).

When sequences were affiliated to taxonomic level, bacterial populations fluctuated under different farming systems. An interesting observation in this study was the greater percentage of phylum *Proteobacteria*, including classes *Alphaproteobacteria*, *Betaproteobacteria* and *Gammaproteobacteria* in organic farming management (39.3%) compared to the conventional system (29.7%). To interpret these findings in an ecological context and to explain why some bacterial phyla are more abundant in soil than others, some researchers have used the concept of copiotrophic versus oligotrophic bacteria [50,51]. Copiotrophic bacteria (fast growing) flourish in soils with large amounts of available nutrients, while oligotrophic groups (slow growing) predominate in soil having low nutrient availability. It has been proposed that oligotrophic bacteria are more associated with organically than conventionally farmed soils due to low availability of organic carbon and nitrogen [2,40]. Among *Proteobacteria*, *Betaproteobacteria* are considered as copiotrophic [40,50], and thus, their population is expected to be lower in organic farming. There is no indication if other classes within *Proteobacteria* can be classified into copiotrophic-oligotrophic scheme [50]. In our study we found higher *Betaproteobacteria* in organic farmed soil. As organic farming system did not contribute to higher amount of top bulk soil total C, total N and organic matter compared to conventional system, higher relative abundance of these bacteria could be due to other factors. It was found that *Proteobacteria* and *Betaproteobacteria* was highly correlated with pH in this study ($P<0.05$, Table 6), we assumed that neutral pH could increase the abundance of these bacteria in soil.

We believe that because of enormous phylogenetic and physiological diversity within each bacterial phyla, it is unlikely that an entire phylum demonstrate same ecological characteristics. An example would be *Burkholderia*, a genus in *Betaproteobacteria* that exhibits oligotrophic traits due to their catabolically versatility that enables them to degrade recalcitrant compounds and survive in environments with limited nutrient availability [52]. Thus, the hypothesis that oligotrophic bacteria are more associated with organic farmed soil could simplify the ecological categories of bacterial communities in soil.

In this study, we found a higher population of *Brevundimonas* spp., *Burkholderia* spp., *Pseudomonas* spp., and *Stenotrophomonas* spp. in organic farming systems (Table 5). These genera are ubiquitously in the soil and several of their species have important ecological roles in nutrient cycling and suppression of plant diseases [52,53,54]. For instance, members of *Stenotophomonas, Pseudomonas and Burkholderia* genera can fix nitrogen [52,55]. Higher relative abundance of these genera might help maintaining total N level in organic farming soil without fertilizer supplementation. In addition, many plant growth-promoting bacteria (PGPB) belong to *Burkholeria, Stenotrophomonas* and *Pseudomonas* genera, which were more abundant in organic farming system (Table 5). Interestingly, these genera were abundant in soils planted with alfalfa, wheat, oilseed rape and various weeds [53,54]. Because organic farming systems support more weeds than the conventional farming systems, it might promote these PGPB populations [56]. However, it is important to notice that not all species in these genera are PGPB and there are species, which are pathogenic to humans, animals and plants (i.e. *P. aerugionsa*, *P. syringae*, and *S. maltophilia* K279a*)*. The 16S rRNA marker genes have limitation for identification of bacteria up to the species level, and thus other methodologies with high resolution including metagenomic shotgun sequencing [57] must be applied in order to differentiate PGPB from pathogenic species in the soil bacterial community.

The percentage of *Actinobacteria* and *Chloroflexi* were lower in organic (30.4% and 4.3%, respectively) compared to the conventional system (41.1% and 6.4%, respectively). Our results show that the conventional farming system increases the actinobacterial proportion in the community with no change in their composition, compared to the organic farming system. The PLS-DA loading scatter and coefficient plots (Figure 2 and Figure S2) indicated that conventional farming system supported higher population of several genera within *Actinobacteria*, including *Blastococcus* spp., *Microlunatus* spp., *Pseudonocardia* spp. and *Solirubrobacter* spp. *Actinobacteria* are able to degrade a variety of organic compounds including some herbicides and pesticides [58]. *Pseudonocardia* spp. has been reported to degrade environmental contaminants, particularly aromatic hydrocarbons or compounds that contain aromatic rings [59]. As such, herbicides and pesticides sprayed containing aromatic rings may have favored bacteria, such as *Pseudonocardia* spp., with specific metabolic capabilities that can degrade them. Some *Microlunatus* spp. has high levels of phosphorus accumulating function and phosphate uptake/release activities [60]. Therefore, in a conventional farming system where pesticides and inorganic fertilizers are commonly used to increase the crop yield, high availability of substrate for *Microlunatus* spp. and other actinobacterial species could boost their population.

Actinobacteria also play a major role in organic matter turnover and carbon cycling. They can decompose some recalcitrant carbon sources including cellulose and chitin [15,61]. Organically farmed soils have been reported to be rich in recalcitrant carbon sources [62], and the diversity of *Actinobacteria* would be expected to be higher in those soils than in conventionally farmed soils. However, in our organic farming fields the recalcitrant carbon sources were not higher in the organically surface soil than the conventional one [46]. Therefore, recalcitrant carbon sources could not drive the increasing diversity of *Actinobacteria* in our study.

A number of studies have shown that soil edaphic factors shaped microbial communities [43,63,64,65]. In our study, we found that proportions of abundant phyla were highly affected by soil pH. Our observations were consistent with other studies that demonstrated pH was one of the main drivers of change in soil bacterial communities from continental scale [48] to small landscape [64,65]. At the phylum level, *Proteobacteria* were positively correlated with soil pH, while *Actinobacteria* were negatively correlated and *Acidobacteria* had a very weak correlation with soil pH. Our results are in contrast to some studies that showed *Actinobacteria* significantly increased with higher pH values, and *Acidobacteria* was dependent on soil pH [65,66]. The soil pH value varied significantly from 3 to 8 in other studies, while the soil pH in our experiments only varied from 6.6 to 7 which could be the reason for lack of change in *Acidobacteria* populations. In our study,

Betaproteobacteria and *Alphaproteobacteria* populations increased with higher soil pH, while *Deltaproteobacteria* declined. This result was concomitant with the study by Nacke et al. [65]. Our studies demonstrated that *Proteobacteria* and *Actinobacteria* were more sensitive to pH variation than other bacterial phyla.

Finally, it is important to acknowledge that the choice of target variable regions of 16S rRNA may have affected the outcome of species richness and diversity analyses because the sequence divergence is not distributed evenly along the 16S rRNA gene [67,68]. We deep sequenced the V1–V3 regions of the bacterial 16S rRNA, which covered V2–V3 region, most suitable for distinguishing most bacterial species ranging from the phylum level to the genus level [68,69]. Therefore, even if some bacterial communities might have been missed or overestimated, the overall shifting in the phylogenetic composition of bacterial communities under different treatments have been assessed.

Conclusion

We demonstrated that different farming practices significantly changed the relative abundances of *Proteobacteria* and *Actinobacteria*. Farming management practices (organic versus conventional) rather than crop rotation (Grain-Only versus Forage-Grain) appeared to have a strong impact on shifting the abundance of soil bacterial communities, which could translate to changes in soil quality and productivity. Some bacterial groups, such as *Gemmatinomadetes*, *Fibrobacteres*, *Verrucomicrobia* and *OP10* were influenced by the interaction of crop rotation and management. Most soil properties including C: N ratio, total N, total C, Olsen P, and organic matter, did not play a major role in shaping bacterial communities. However, pH had the strongest effect on the bacterial community structure. Organic farming systems led to a neutral pH, which might be beneficial to *Proteobacteria*. On the other hand, conventional farming systems supported a higher percentage of *Actinobacteria*. Therefore, neither organic farming nor conventional farming can address all the aspects of beneficial soil bacterial communities, which is crucial to soil quality and productivity. Further research is required to investigate the shifts in diversity of beneficial bacterial and fungal pathogens under different farming systems in the long run.

Supporting Information

Figure S1 Rarefaction curves for pooled samples within each treatment at OTU cutoff of 0.05 distance.

Figure S2 Coefficient plot of the bacterial profiles of organic and conventional treatments. Partial least squares discriminant analysis (PLS-DA) coefficient plot based on the relative abundant of the bacterial genera in the microbiome profile of organic and conventional treatments. Genera with significantly positive (>0) or negative (<0) PLS regression coefficients (i.e. no overlap between the 95% confidence interval indicated and the horizontal axis) contribute significantly to the prediction of the organic (green bars) or conventional samples (red bars).

Table S1 Phylogenetic composition of putative bacterial genera in *Actinobacteria* phylum determined using 16S rRNA pyrosequencing

Table S2 Phylogenetic composition of putative bacterial genera in *Proteobacteria* phylum determined using 16S rRNA pyrosequencing

Table S3 Phylogenetic composition of putative bacterial genera in *Acidobacteria*, *Bacteroidetes*, *Chloroflexi*, *Firmicutes*, *Gemmatimonadetes*, and *Planctomycetes* phyla determined using 16S rRNA pyrosequencing

Acknowledgments

Authors would like to thank Bell LW and Welsh C for testing soil edaphic traits.

Author Contributions

Conceived and designed the experiments: RL WGDF MHE TRdK. Performed the experiments: RL. Analyzed the data: RL EK. Contributed reagents/materials/analysis tools: WGDF EK DOK. Wrote the paper: RL EK WGDF.

References

1. Doran JW, Parkin TB (1994) Defining and assessing soil quality; Doran JW, Coleman DC, Bezdicek DF, Stewart BA, editors. Madison, Wisconsin USA: Soil Science Society of America, America Society of Agronomy. pp 3–21.
2. van Bruggen AHC, Semenov AM (2000) In search of biological indicators for soil health and disease suppression. Appl Soil Ecol 15: 13–24.
3. Sprent JI (1979) The biology of nitrogen-fixing organisms. London; New York: McGraw-Hill. 196 p.
4. Dorn E, Hellwig M, Reineke W, Knackmuss HJ (1974) Isolation and characterization of a 3-chlorobenzoate degrading pseudomonad. Arch Microbiol 99: 61–70.
5. Torstensson L (1980) Role of microorganisms in decomposition. In: Hance RJ, editor. Interaction between hercibides and the soil London: Academic Press. 159–177.
6. Ongena M, Duby F, Rossignol F, Fauconnier M-L, Jacques D, et al. (2004) Stimulation of the lipoxygenase pathway is associated with systemic resistance induced in bean by a nonpathogenic *Pseudomonas* strain. Mol Plant Microbe Interact 17: 1009–1018.
7. Zehnder GW, Murphy JF, Sikora EJ, Kloepper JW (2001) Application of rhizobacteria for induced resistance. Eur J Plant Pathol 107: 39–50.
8. Weller DM, Raaijmakers JM, Gardener BBM, Thomashow LS (2002) Microbial populations responsible for specific soil suppressiveness to plant pathogens. Annu Rev Phytopathol 40: 309–348.
9. Suman A, Gaur A, Shrivastava AK, Yadav RL (2005) Improving sugarcane growth and nutrient uptake by inoculating *Gluconacetobacter diazotrophicus*. Plant Growth Regul 47: 155–162.
10. Basak BB, Biswas DR (2010) Co-inoculation of potassium solubilizing and nitrogen fixing bacteria on solubilization of waste mica and their effect on growth promotion and nutrient acquisition by a forage crop. Biol Fertil Soils 46: 641–648.
11. Upchurch R, Chiu CY, Everett K, Dyszynski G, Coleman DC, et al. (2008) Differences in the composition and diversity of bacterial communities from agricultural and forest soils. Soil Bio Biochem 40: 1294–1305.
12. Gattinger A, Ruser R, Schloter M, Munch JC (2002) Microbial community structure varies in different soil zones of a potato field. J Plant Nutr Soil Sci 165: 421–428.
13. Smalla K, Wieland G, Buchner A, Zock A, Parzy J, et al. (2001) Bulk and rhizosphere soil bacterial communities studied by denaturing gradient gel electrophoresis: plant-dependent enrichment and seasonal shifts revealed. Appl Environ Microbiol 67: 4742–4751.
14. Roesch LF, Fulthorpe RR, Riva A, Casella G, Hadwin AK, et al. (2007) Pyrosequencing enumerates and contrasts soil microbial diversity. ISME J 1: 283–290.
15. Acostamartinez V, Dowd S, Sun Y, Allen V (2008) Tag-encoded pyrosequencing analysis of bacterial diversity in a single soil type as affected by management and land use. Soil Biol Biochem 40: 2762–2770.
16. Shafiani S, Malik A (2003) Tolerance of pesticides and antibiotic resistance in bacteria isolated from wastewater- irrigated soil. World J Microb Biot 19.
17. Wasi S, Jeelani G, Ahmad M (2008) Biochemical characterization of a multiple heavy metal, pesticides and phenol resistant Pseudomonas fluorescens strain. Chemosphere 71: 1348–1355.
18. Azadi H, Schoonbeek S, Mahmoudi H, Derudder B, De Maeyer P, et al. (2011) Organic agriculture and sustainable food production system: main potentials. Agric Ecosyst & Environ 144: 92–94.
19. Zhengfei G (2005) Damage control inputs: a comparison of conventional and organic farming systems. Eur Rev Agric Eco 32: 167–189.

20. Gunapala N, Scow KM (1998) Dynamics of soil microbial biomass and activity in conventional and organic farming systems. Soil Biol Biochem 30: 805–816.
21. Mader P, Fliessbach A, Dubois D, Gunst L, Fried P, et al. (2002) Soil fertitity and biodiversity in organic farming. Science 296: 1694–1697.
22. Hartmann M, Fliessbach A, Oberholzer HR, Widmer F (2006) Ranking the magnitude of crop and farming system effects on soil microbial biomass and genetic structure of bacterial communities. FEMS Microbiol Ecol 57: 378–388.
23. Esperschutz J, Gattinger A, Mader P, Schloter M, FlieBbach A (2007) Response of soil microbial biomass and community structures to conventional and organic farming systems under identical crop rotations. FEMS Microbiol Ecol 61: 26–37.
24. Sugiyama A, Vivanco JM, Jayanty SS, Manter DK (2010) Pyrosequencing assessment of soil microbial communities in organic and conventional potato farms. Plant Dis 94: 1329–1335.
25. Wu T, Chellemi DO, Graham JH, Martin KJ, Rosskopf EN (2007) Comparison of soil bacterial communities under diverse agricultural land management and crop production practices. Microbial Ecol 55: 293–310.
26. Joergensen RG, Mäder P, Fließbach A (2010) Long-term effects of organic farming on fungal and bacterial residues in relation to microbial energy metabolism. Biol Fert Soils 46: 303–307.
27. Welsh C, Tenuta M, Flaten DN, Thiessen-Martens JR, Entz MH (2009) High yielding organic crop management decreases plant-available but not recalcitrant soil phosphorus. Agron J 101: 1027.
28. Williams DE (1948) A rapid manometer method for the determination of carbonate in soils. Soil Sci Soc of Am J 13: 127–129.
29. He J, Xu Z, Hughes J (2005) Pre-lysis washing improves DNA extraction from a forest soil. Soil Biol Biochem 37: 2337–2341.
30. Khafipour E, Li S, Plaizier JC, Krause DO (2009) Rumen microbiome composition determined using two nutritional models of subacute ruminal acidosis. Appl Environ Microbiol 75: 7115–7124.
31. Dowd SE, Callaway TR, Wolcott RD, Sun Y, McKeehan T, et al. (2008) Evaluation of the bacterial diversity in the feces of cattle using 16S rDNA bacterial tag-encoded FLX amplicon pyrosequencing (bTEFAP). BMC Microbiol 8: 125.
32. Khafipour E, Berard N, Little A, Tkachuk V, Dowd AS (2012) Microbiome changes during sub-acute ruminal acidosis using high-throughput pyrosequencing and statistical evaluation of sequence data. Appl Environ Microbiol (Accepted with revision).
33. Schloss PD, Westcott SL, Ryabin T, Hall JR, Hartmann M, et al. (2009) Introducing mothur: open-source, platform-independent, community-supported software for describing and comparing microbial communities. Appl Environ Microbiol 75: 7537–7541.
34. Pruesse E, Quast C, Knittel K, Fuchs BM, Ludwig W, et al. (2007) SILVA: a comprehensive online resource for quality checked and aligned ribosomal RNA sequence data compatible with ARB. Nucleic Acids Res 35: 7188–7196.
35. Wang Q, Garrity GM, Tiedje JM, Cole JR (2007) Naive bayesian classifier for rapid assignment of rRNA sequences into the new bacterial taxonomy. Appl Environ Microbiol 73: 5261–5267.
36. SAS (2008) SAS/STAT user guide, release 9.2 Cary, NC.: SAS Institute Inc.
37. SIMCA-P+_12 (2008) SIMCA-P+_12 user guide. Malmo, Sweden: MKS Umetrics AB.
38. Verhulst NO, Qiu YT, Beijleveld H, Maliepaard C, Knights D, et al. (2011) Composition of human skin microbiota affects attractiveness to malaria mosquitoes. PLoS One 6: e28991.
39. Pérez-Enciso M, Tenehaus M (2003) Prediction of clinical outcome with microarray data: a Partial least squares discriminant analysis (PLS-DA) approach. Hum Genet 112: 581–592.
40. Vandiepeningen A, Devos O, Korthals G, Vanbruggen A (2006) Effects of organic versus conventional management on chemical and biological parameters in agricultural soils. Appl Soil Ecol 31: 120–135.
41. Barak P, Jobe A, Krueger A, Peterson A, Laird A (1997) Effects of long-term soil acidification due to nitrogen fertilizer inputs in Wisconsin. Plant Soil 197: 61–69.
42. Kuramae EE, Yergeau E, Wong LC, Pijl AS, Veen JA, et al. (2012) Soil characteristics more strongly influence soil bacterial communities than land-use type. FEMS Microbiol Ecol 79: 12–24.
43. Ziadi N, Sen Tran T (2008) Lime requirement; Carter MR, Gregorich EG, editors. Boca Raton, FL: Taylor & Francis Group. pp 129–134.
44. Su Y (2007) Soil carbon and nitrogen sequestration following the conversion of cropland to alfalfa forage land in northwest China. Soil Till Res 92: 181–189.
45. Pimentel D, Hepperly P, Hanson J, Douds D, Seidel R (2005) Environmental, energetic, and economic comparisons of organic and conventional farming systems. BioScience 55: 573–582.
46. Bell LW, Sparling B, Tenuta M, Entz MH (2012) Soil profile carbon and nutrient stocks under long-term conventional and organic crop and Alfalfa-crop rotations and re-established grassland. Agr Ecosyst Environ 158: 156–163.
47. Ramirez KS, Lauber CL, Knight R, Bradford MA, Fierer N (2010) Consistent effects of nitrogen fertilization on soil bacterial communities in contrasting systems. Ecology 91: 3463–3470.
48. Lauber CL, Hamady M, Knight R, Fierer N (2009) Pyrosequencing-based assessment of soil pH as a predictor of soil bacterial community structure at the continental scale. Appl Environ Microbiol 75: 5111–5120.
49. Hartmann M, Widmer F (2006) Community structure analyses are more sensitive to differences in soil bacterial communities than anonymous diversity indices. Appl Environ Microbiol 72: 7804–7812.
50. Fierer N, Bradford MA, Jackson RB (2007) Toward an ecological classificatin of soil bacterial. Ecology 88: 1354–1364.
51. Meyer O (1994) Functional groups of microorganisms; Schulze E, Mooney H, editors. New York, USA: Springer-verlag. pp 67–96.
52. Suárez-Moreno ZR, Caballero-Mellado J, Coutinho BG, Mendonça-Previato L, James EK, et al. (2011) Common features of environmental and potentially beneficial plant-associated *Burkholderia*. Microb Ecol 63: 249–266.
53. Ryan RP, Monchy S, Cardinale M, Taghavi S, Crossman L, et al. (2009) The versatility and adaptation of bacteria from the genus *Stenotrophomonas*. Nat Rev Microbiol 7: 514–525.
54. Haas D, Défago G (2005) Biological control of soil-borne pathogens by fluorescent pseudomonads. Nat Rev Microbiol 3: 307–319.
55. Park M, Kim C, Yang J, Lee H, Shin W, et al. (2005) Isolation and characterization of diazotrophic growth promoting bacteria from rhizosphere of agricultural crops of Korea. Microbiol Res 160: 127–133.
56. Entz M, Penner K, Vessey J, Zelmer C, Thiessen-Martens J (2004) Mycorrhizal colonization of flax under long-term organic and conventional management. Can J Plant Sci 84.
57. Segata N, Waldron L, Ballarini A, Narasimhan V, Huttenhower C. et al. (2012) Metagenomic microbial community profiling using unique clade-specific marker genes. Nat Methods 9: 811–814.
58. De Schrijver A, De Mot R (1999) Degradation of pesticides by Actinomycetes. Crit Rev Microbiol 25: 85–119.
59. Lee SB (2004) *Pseudonocardia chloroethenivorans* sp. nov., a chloroethene-degrading actinomycete. Int J Syst Evol Micr 54: 131–139.
60. Akar A, Akkaya EU, Yesiladali SK, Çelikyilmaz G, Çokgor EU, et al. (2005) Accumulation of polyhydroxyalkanoates by *Microlunatus phosphovorus* under various growth conditions. J Ind Microbiol Biot 33: 215–220.
61. Jenkins S, Waite I, Blackburn A, Husband R, Rushton S, et al. (2010) Actinobacterial community dynamics in long term managed grasslands; Brisbane, Australia.
62. Fließbach A, Oberholzer HR, Gunst L, Mäder P (2007) Soil organic matter and biological soil quality indicators after 21 years of organic and conventional farming. Agr, Ecosyst Environ 118: 273–284.
63. Lauber CL, Strickland MS, Bradford MA, Fierer N (2008) The influence of soil properties on the structure of bacterial and fungal communities across land-use types. Soil Biol Biochem 40: 2407–2415.
64. Singh BK, Dawson LA, Macdonald CA, Buckland SM (2009) Impact of biotic and abiotic interaction on soil microbial communities and functions: A field study. Appl Soil Ecol 41: 239–248.
65. Nacke H, Thürmer A, Wollherr A, Will C, Daniel R, et al. (2011) Pyrosequencing-based assessment of bacterial community structure along different management types in German forest and grassland soils. PLoS One 6: 1–12.
66. Jones RT, Robeson MS, Lauber CL, Hamady M, Knight R, et al. (2009) A comprehensive survey of soil acidobacterial diversity using pyrosequencing and clone library analyses. ISME J 3: 442–453.
67. Liu Z, DeSantis TZ, Andersen GL, Knight R (2008) Accurate taxonomy assignments from 16S rRNA sequences produced by highly parallel pyrosequencers. Nucleic Acids Res 36: e120–e120.
68. Kim M, Morrison M, Yu Z (2011) Evaluation of different partial 16S rRNA gene sequence regions for phylogenetic analysis of microbiomes. J Microbiol Meth 84: 81–87.
69. Chakravorty S, Helb D, Burday M, Connell N, Alland D (2007) A detailed analysis of 16S ribosomal RNA gene segments for the diagnosis of pathogenic bacteria. J Microbiol Meth 69: 330–339.

12

Spatial Variation in Carbon and Nitrogen in Cultivated Soils in Henan Province, China: Potential Effect on Crop Yield

Xuelin Zhang[1]*, Qun Wang[1], Frank S. Gilliam[2], Yilun Wang[1], Feina Cha[3], Chaohai Li[1]

1 The Incubation Base of the National Key Laboratory for Physiological Ecology and Genetic Improvement of Food Crops in Henan Province, Zhengzhou, China; Agronomy College of Henan Agricultural University, Zhengzhou, China, **2** Department of Biological Sciences, Marshall University, Huntington, West Virginia, United States of America, **3** Meteorological Bureau of Zhengzhou, Zhengzhou, China

Abstract

Improved management of soil carbon (C) and nitrogen (N) storage in agro-ecosystems represents an important strategy for ensuring food security and sustainable agricultural development in China. Accurate estimates of the distribution of soil C and N stores and their relationship to crop yield are crucial to developing appropriate cropland management policies. The current study examined the spatial variation of soil organic C (SOC), total soil N (TSN), and associated variables in the surface layer (0–40 cm) of soils from intensive agricultural systems in 19 counties within Henan Province, China, and compared these patterns with crop yield. Mean soil C and N concentrations were 14.9 g kg^{-1} and 1.37 g kg^{-1}, respectively, whereas soil C and N stores were 4.1 kg m^{-2} and 0.4 kg m^{-2}, respectively. Total crop production of each county was significantly, positively related to SOC, TSN, soil C and N store, and soil C and N stock. Soil C and N were positively correlated with soil bulk density but negatively correlated with soil porosity. These results indicate that variations in soil C could regulate crop yield in intensive agricultural systems, and that spatial patterns of C and N levels in soils may be regulated by both climatic factors and agro-ecosystem management. When developing suitable management programs, the importance of soil C and N stores and their effects on crop yield should be considered.

Editor: Dafeng Hui, Tennessee State University, United States of America

Funding: This study was supported by grants from Henan Science and Technology Department of China under the Key Research Project (30200051). The funder had no role in study design, data collection and analysis, decision to publish, or preparation of the manuscript.

Competing Interests: The authors have declared that no competing interests exist.

* Email: xuelinzhang1998@163.com

Introduction

Safeguarding food security and ensuring sustainable development are two fundamental goals of intensive agriculture in China [1,2]. Increasing soil C and N sequestration while reducing C and N emissions from agricultural fields are important aspects of sustainable farming and these goals can be achieved through improvement in soil quality [1,3]. This requires a better understanding of the functional relationship between crop yield and soil organic C and N stores.

Indeed, variations in soil C and N stores may closely regulate crop yield, although published data on the relationship between these parameters are inconsistent. Some studies have reported a positive correlation between soil C and N and crop yield [4,5], whereas other studies have found no significant relationship between these parameters [6,7]. Lal (2006) reported that the relationship between soil organic C and crop yield may vary between patterns that are sigmoidal, linear, or exponential [8]. Clearly, the existence of such variability warrants further investigation.

Soil C and N stores in crop lands, especially in the topsoil layer, are potentially greatly affected by human activity; thus, understanding the spatial pattern of soil C and N stores on a regional scale is crucial to developing a management strategy for improving soil fertility [1,2]. Spatial variation in soil C and N stores in agro-ecosystems has been widely reported [9,10,11], including from the northern [12,13], eastern [14], and southern [15,16] regions of China. Since these reports from China were based on two national surveys from 1960 and 1983, such data may have limited use in helping to develop management strategies based on current practices [17]. Therefore, in order to better understand the spatial patterns and their relationship to crop yield, it is necessary to update regional soil organic C and N information with contemporary measurements, especially for intensively-used crop land.

Henan Province is the second largest area of crop production in China (China National Bureau of Statistics). To produce an adequate supply of food for the domestic population, unsustainable production methods have often been used in this province. Historically, intensive production based on an annual wheat-maize system has been used to achieve high crop yield. This practice, however, has resulted in badly degraded agricultural soils, causing erosion and a loss of good soil structure. More than 600 kg N ha^{-1} annually has been applied in this production area, resulting in an increase in soil acidity [18]. Based on the determination that crop yields in China will need to increase from 50 billion in 2010 to 65 billion kg in 2020, the provincial crop lands in Henan Province will continue to play an important role in food production. Such goals create the challenge of improving soil quality, enhancing soil fertility, and mitigating C and N loss, while achieving food security and practicing sustainable agriculture. A better understanding of the spatial variability of soil organic C and N, and their relationship to crop yield, should help to develop management practices that are designed to meet this challenge [1,19].

The objective of the present study was to characterize the spatial distribution of C and N stores in intensively cultivated counties within the Henan Province of China and to determine the relationship between crop yield and soil organic C and N.

Materials and Methods

Statement: We have field permits for sampling soil in each of the field sites within each county of Henan Province, China. All of the sampling sites are privately owned, and there was no potential impact on any endangered or protected species among these sampling sites.

Study site

The study was carried out in 19 counties within Henan Province, located in central China (Figure 1). Map data were obtained from the National Geomatics Center of China (http://ngcc.sbsm.gov.cn/) using ArcGIS software. As of 2009, the human population of Henan was about 9.9×10^7 persons. The Province is

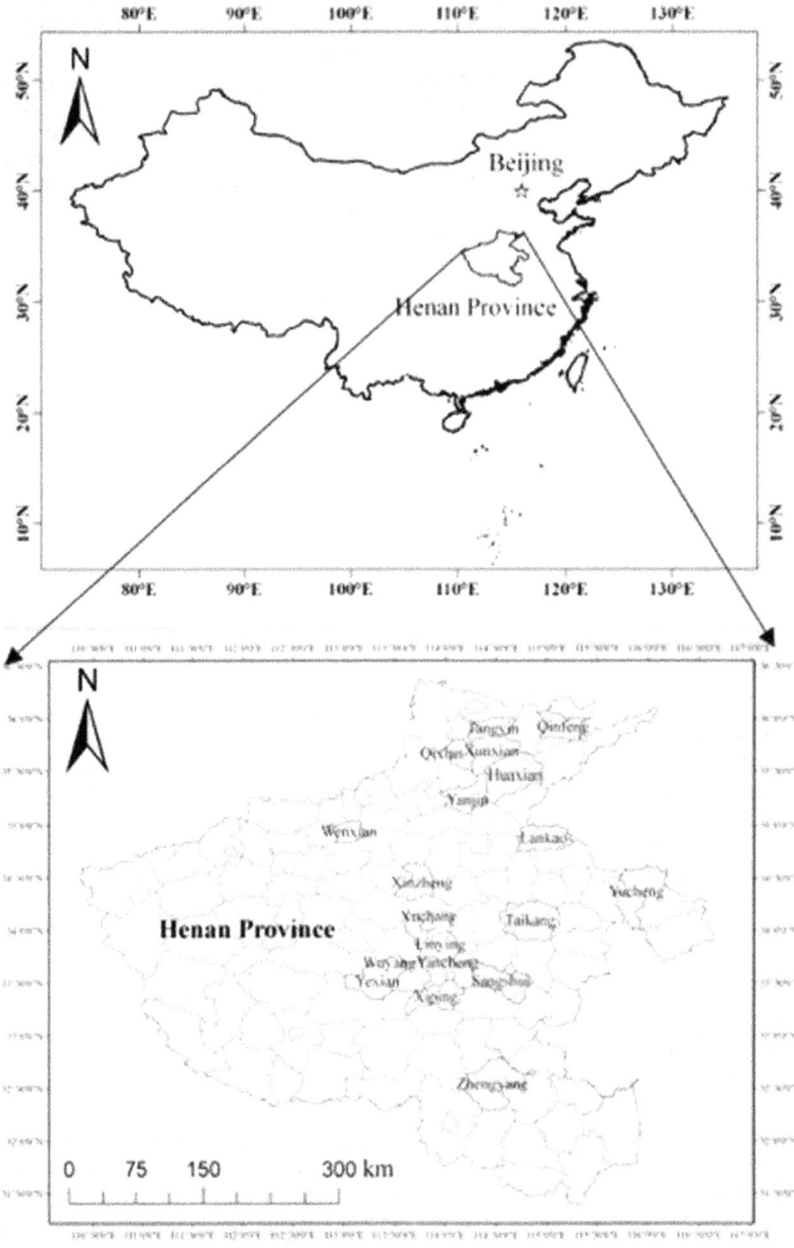

Figure 1. Map of China (top) showing location of Henan Province and counties (bottom) within Henan Province used in this study.

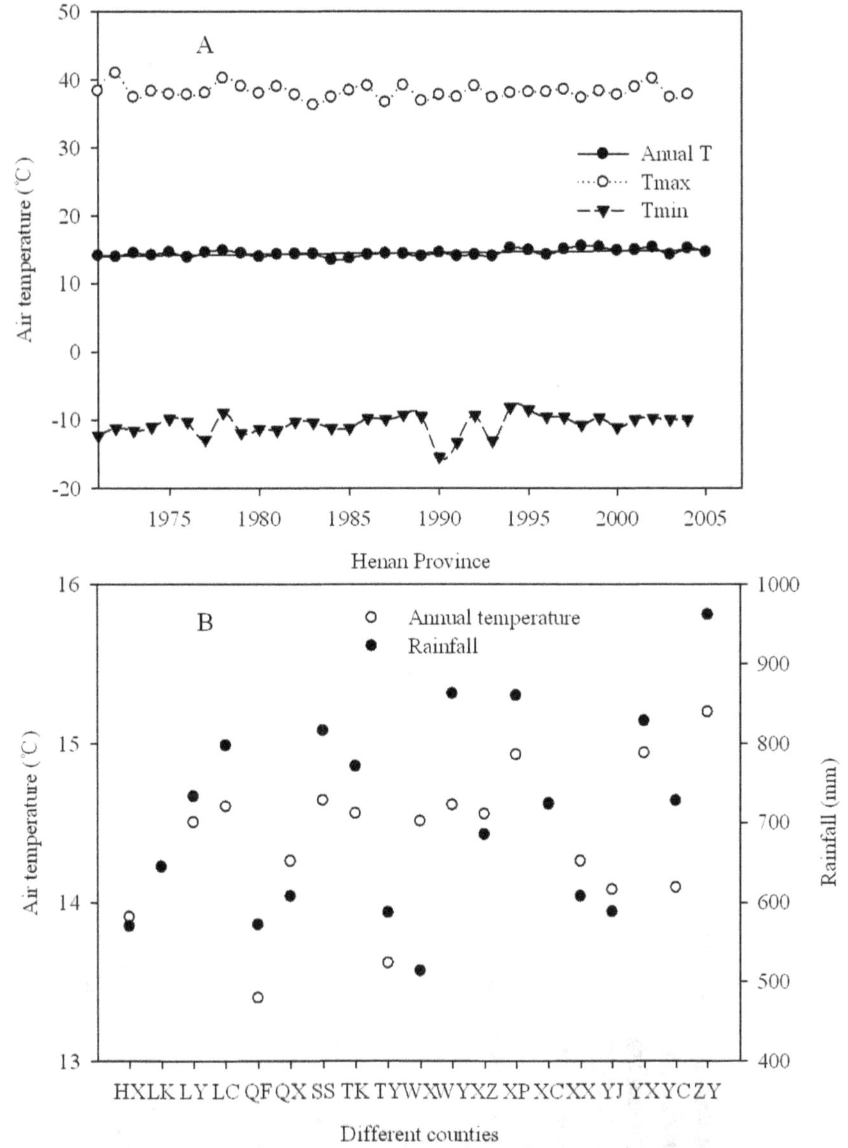

Figure 2. Average annual, maximum, and minimum temperature from 1971 to 2004 in Henan Province, China (A), and (B) average temperature and rainfall from 1975 to 2006 in different counties within Henan Province, China (B). See Table 1 for key to county name abbreviations. All these counties were arranged in English alphabetical order.

approximately 167,000 km^2 in land area, lying within the monsoonal temperate zone. It has a cultivated land area of 79, 260 km^2 for the production of wheat and maize. There are three dominant soil types in Henan Province: Yellow-cinnamon soil (Eutric Cambisols in FAO taxonomy), Sajiang black soil (Eutric Vertisols/Gleyic Cambisol), and Fluvo-aquic soil (Fluvisols in FAO taxonomy) [20]. Mean annual precipitation ranges from 400 to 1000 mm among the counties of the study, with ~70% of it occurring from July to September; mean annual temperature ranges from 13.6 to 15°C (Figure 2). Cultivated agricultural fields are the predominant land use, representing 60% of the total land area in Henan Province. A double cropping system of winter wheat (early October-early June) and maize (mid-June–later September) is the most common planting system used in this region.

Collection of crop yield and soil sampling and analysis

Data on total crop production (including wheat, maize and millet) and wheat yield from 1978–2009 (Figure. 3A) were obtained from the Henan Statistical Yearbook 2010 (13–17) (http://www.ha.stats.gov.cn/hntj/index.htm). Annual yield data for winter wheat and total crop production in 2009 were also obtained from Henan Statistical Yearbook 2010 (29-7) and the Agricultural Bureau of each of the 19 counties in which soil sampling took place (Figure. 3B). These counties, along with basic climatic information, are listed in Table 1. Climatic data of each county were obtained from Meteorological Bureau of Zhengzhou. All counties will be referred to by the two-letter codes presented in Table 1.

The 19 counties were selected as representative of the main agro-ecosystems of Henan Province. Soil samples were collected during June 1–15, 2009 following the wheat harvest but prior to the sowing of maize. Six representative, replicate field plots,

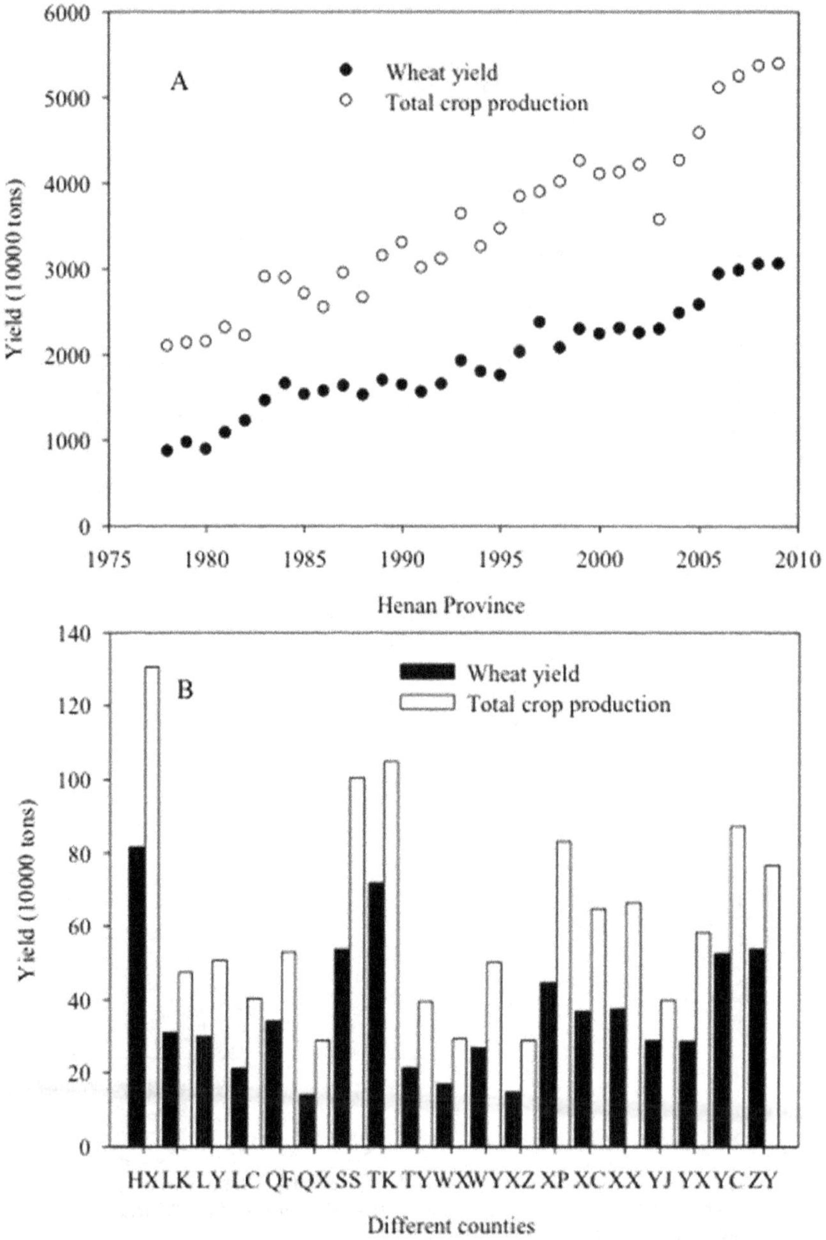

Figure 3. Wheat yield and total crop production (including wheat, maize, millet,) in Henan Province from 1978–2009 (A), and wheat yield and total crop production in different counties within Henan Province in 2009 (B). See Table 1 for key to county name abbreviations.

located at least 6 km apart, were selected within each county based on four criteria: (1) the field plots had been continuously cultivated for at least 30 yr with a native variety, (2) the cropland area was located within 5 km of native vegetation with a similar landscape, soil type and texture, and a relatively flat terrain, and (3) all of the sampling sites are privately owned, and (4) there was no potential impact on any endangered or protected species in the sampling site. Geographic coordinates of each sampling site was recorded by handed GPS of Magellan eXplorist 210(USA), and all of these data were attached in the supporting information.

Sample areas of ~1300 m^2 were established in each plot, with sixteen sampling points taken at random in each of two layers (0–20 cm and 20–40 cm) using a 70 mm - diameter auger. All of the soil samples taken at each layer within a sample plot were mixed together and treated as one sample to represent the value of the plot, yielding 114 soil samples at each layer.

Residual plant material was removed from the soil samples after the samples were air-dried at room temperature. The soil samples were then ground to pass a 2 mm sieve, and a portion of the ground sample was subsequently ground again in a porcelain mortar in order to pass through a 0.15–mm sieve. Organic C and total N measurements were obtained from the twice-ground soil samples. Soil organic C (SOC) was measured using a modified Mebius method. Briefly, 0.1 g soil samples were digested for 5 min

Table 1. Basic geographic coordinates for each county, along with climate data for 19 counties within Henan Province, China.

County	Latitude	Longitude	Sea level (m)	Average Temp (°C)	Rainfall (mm)	Sunshine (h)
Huaxian (HX)	35°44′	114°28′	68	13.9	570.0	2060.9
Lankao (LK)	34°55′	114°46′	70	14.2	644.5	2183.2
Linying (LY)	33°55′	113°55′	63	14.5	732.9	2141.3
Luoheyancheng (LC)	33°35′	114°02′	65	14.6	797.2	2273.0
Qingfeng (QF)	35°53′	115°06′	51	13.4	571.9	2209.1
Qixian (QX)	35°35′	114°12′	72	14.3	607.5	2133.8
Shangshui (SS)	33°39′	114°34′	52	14.6	815.8	1902.0
Taikang (TK)	34°05′	114°50′	53	14.6	770.9	1998.4
Tangyin (TY)	36°03′	114°19′	103	13.6	587.1	2159.3
Wenxian (WX)	35°01′	113°03′	109	14.5	513.2	2302.2
Wuyang (WY)	33°36′	113°32′	77	14.6	862.3	2060.4
Xinzheng (XZ)	34°30′	113°39′	159	14.6	684.6	2058.7
Xiping (XP)	33°29′	113°59′	65	14.9	859.8	2084.7
Xuchang (XC)	34°04′	113°52′	72	14.6	722.7	1959.8
Xunxian (XX)	35°40′	114°32′	59	14.3	607.5	2133.8
Yanjin (YJ)	35°13′	114°11′	69	14.1	588.0	2287.8
Yexiang (YX)	33°38′	113°21′	88	14.9	827.8	1972.4
Yucheng (YC)	34°25′	115°52′	46	14.1	727.3	2244.6
Zhengyang (ZY)	32°37′	114°24′	70	15.2	961.8	2004.4

Note: Counties are arranged in English alphabetical order.

with 5 mL of 1N $K_2Cr_2O_7$ and 10 mL of concentrated H_2SO_4 at 150°C, followed by titration of the digests with standardized $FeSO_4$. Total soil N (TSN) was measured using a modified Kjeldahl wet digestion procedure and a Tector Kjeltec System 1026 distilling unit. Soil available N was analyzed using a micro-diffusion technique after alkaline hydrolysis (1.8 mol L^{-1} NaOH). The Olsen method was used to determine available soil phosphorus (P), and available soil potassium (K) was measured in 1 mol L^{-1} NH_4OAc extracts by flame photometry (Table 2).

Three sampling points were used to determine soil bulk density in each plot. Samples were collected separately from four layers within a depth of 0–40 cm in each sampling point. Soil bulk density was measured using 100-cm^3 soil cores obtained from the four layers. Soil porosity was calculated from soil bulk density and specific gravity, with any stone material removed and not considered in bulk density calculations.

Calculation of soil organic C and N stores and SOC and TSN

Total soil organic C store (TSOCS) and total soil N stores (TSNS) at 0–40 cm depth were calculated as follows:

$$\mathrm{TSOC}(\mathrm{g.m^{-2}}) = $$
$$\text{Soil organic C}(\mathrm{g.kg^{-1}}) \times \text{soil bulk density}(\mathrm{g.cm^{-3}}) \times \text{sampling depth}(\mathrm{cm})$$
$$\mathrm{TSN}(\mathrm{g.m^{-2}}) = $$
$$\text{Soil total N}(\mathrm{g.kg^{-1}}) \times \text{soil bulk density}(\mathrm{g.cm^{-3}}) \times \text{sampling depth}(\mathrm{cm})$$

Given the cultivated area, the total cultivated topsoil (0–40 cm) C and N stocks of each county were estimated by the equation:

$$\mathrm{CS} = \sum area_i \times \mathrm{TSOC}$$
$$\mathrm{NS} = \sum area_i \times \mathrm{TSN}$$

where *area* is the given total cultivated area of each county, and CS and NS are C and N stocks, respectively. SOC and TSN were means of six sampling sites of each county.

Statistics

Analysis of variance was used to assess the significance of location (county) on soil C and N concentration and storage; means were compared using Duncan's multi-range test at $\alpha = 0.05$. Linear regression was used to determine the relationships between C and N stock versus wheat and total crop production. Principle components analysis was used to assess patterns of similarity/dissimilarity among counties with respect to several environmental variables [21]. All statistical analyses were performed using SPSS 10.0 (Chicago IL, USA).

Results

Wheat yields increased more than 250% from 1978 to 2009 while total annual crop production in Henan Province increased from 21 to 54 million tons over the same time period (Figure 3A). Wheat yield varied from 143 to 729 thousand tons among the different counties in 2009 (Figure 3B).

The absolute value of SOC concentration in the top 40 cm of soil varied from 8.13 to 27.89 g kg^{-1} among the 19 counties in 2009 (Table 2) while TSN concentration varied from 0.84 to 2.2 g kg^{-1}. Soil C/N varied from 6.4 to 20 (Table 2). Soil organic C stores (TSOCS) in the 0–40 cm soil layer varied from 2,322 g m^{-2} to 8,038 g m^{-2}, whereas total N stores (TSNS) varied from 221 to

Table 2. Spatial variation in soil (0–40 cm depth) properties, soil organic C (SOC), total soil N (TSN) concentration (g kg^{-1}), and C/N in the 0–40 cm soil layer in 19 counties within Henan province, China.

	Alkaline-extractable N (mg kg^{-1})	Olsen-extractable P (mg kg^{-1})	NH_4OAc-extractable K (mg kg^{-1})	Bulk density (g cm^{-3})	Soil porosity (%)	SOC (g kg^{-1})	TSN (g kg^{-1})	C/N
HX	48.9±3.2abc	1.8±0.7a	80.1±9.2abc	1.44±0.03de	38.3±1.4abcd	12.4±0.9abc	1.4±0.05abcd	8.8±0.7abc
LK	56.5±2.4abcd	7.6±1.9ab	71.9±11.9abc	1.42±0.02bcde	40.7±1.0bcdef	11.2±0.7ab	1.4±0.09abcd	7.9±0.5a
LY	49.9±3.1abc	4.2±0.7a	145.4±24.1ef	1.36±0.02abc	41.7±1.4cdefg	15.5±1.0bcd	1.1±0.07a	14.2±0.2efg
LC	49.0±1.6abc	11. 5±2.1abc	103.6±8.9abcde	1.39±0.02bcd	38.7±1.6abcd	14.6±1.4abcd	1.4±0.14bcd	10.5±1.1abcd
QF	47.5±2.8abc	10.9±4.9abc	71.7±6.3abc	1.39±0.01bcd	41.8±0.4cdefg	11.8±0.6ab	1.4±0.08abcd	8.7±0.8abc
QX	51.9±3.7abc	6.3±2.2ab	82.1±10.8abc	1.44±0.01de	38.7±0.3abcd	21.1±1.8f	1.5±0.21cd	16.2±3.1g
SS	45.1±1.9abc	11.7±3.7abc	169.3±33.9f	1.35±0.02abc	37.9±1.3abcd	14.5±0.9abcd	1.3±0.08abc	11.4±0.8bcde
TK	59.3±4.5cd	17.7±8.2bcd	140.6±23.9def	1.35±0.02ab	41.5±1.5cdef	13.4±1.0abcd	1.1±0.05ab	11.9±0.7cde
TY	59.2±2.4cd	6.4±2.6ab	110.9±12.8bcde	1.45±0.03de	38.6±1.3abcd	15.0±0.4abcd	1.7±0.09de	8.9±0.5abc
WX	56.8±3.5abcd	11.0±1.9abc	82.1±7.7abc	1.3±0.03a	43.6±1.5fg	17.1±1.9de	1.5±0.11bcd	11.5±0.6bcde
WY	47.9±1.1abc	10.3±2.9ab	84.2±10.5abc	1.38±0.02bcd	36.8±1.2ab	14.9±1.6abcd	1.6±0.09cd	9.5±0.6abc
XZ	72.9±7.4e	7.8±2.1ab	95.8±18.9abcd	1.47±0.02e	38.3±0.7abcd	16.1±1.6cd	1.1±0.12ab	14.5±1.1efg
XP	72.2±4.2e	27.3±5.8d	117.5±17.7cde	1.43±0.01cde	37.8±1.2abc	19.9±1.9ef	1.3±0.07abc	15.4±1.3fg
XC	43.2±3.4a	17.2±5.8bcd	66.2±9.6ab	1.42±0.03bcde	42.9±0.9efg	16.6±2.3cde	1.3±0.18abc	12.7±0.7def
XX	53.1±4.2abcd	5.1±0.5ab	89.9±3.0abc	1.41±0.02bcde	35.3±1.0a	14.8±0.6abcd	1.9±0.11e	7.9±0.3a
YJ	49.5±2.9abc	12.8±4abc	77.0±9.6abc	1.39±0.02bcd	38.2±1.0abcd	10.8±0.6a	1.3±0.07abc	8.2±0.4ab
YX	44.5±2.1ab	22.9±5.9cd	89.5±11.4abc	1.3±0.04a	45.5±1.9g	15.4±0.9bcd	1.2±0.07ab	13.6±1.1defg
YC	66.4±11.3de	3.5±1.1a	59.9±8.3a	1.44±0.02de	39±1.3abcde	14.8±0.9abcd	1.1±0.08ab	13.1±0.2defg
ZY	58.7±2.4bcd	6.3±1.4ab	73.8±5.7abc	1.35±0.02ab	41.9±0.9defg	12.7±0.3abc	1.5±0.06bcd	8.8±0.5abc

Different letters indicate significant differences ($p = 0.05$) among the 19 counties. Counties are arranged in English alphabetical order.

Table 3. Total C (TSOCS) and N (TSNS) stores in the surface soil layer (0–40 cm) of soils in 19 counties in Henan Province, China.

	C store (g m^{-2})	N store (g m^{-2})
HX	3541±261.8 abcd	410.1±18.9 cde
LK	3118.9±189.3 ab	399.7±23.6 bcde
LY	4106.7±294.1 abcd	290.4±22.2 a
LC	4023.8±372 abcd	398.6±35.6 bcde
QF	3229.7±140 abc	381.8±22.3 abcd
QX	5977.9±524.3 e	429.1±63.1 de
SS	3881.8±219 abcd	348.3±24.0 abcd
TK	3605.1±328.1 abcd	303.4±19.2 ab
TY	4300.4±98.5 bcd	494.3±24.9 ef
WX	4396.2±451.9 cd	379.4±23.4 abcd
WY	4081.3±455.2 abcd	429.3±25.3 de
XZ	4528.2±516.5 d	320.3±40.1 abc
XP	5709.2±582.9 e	369.8±21.9 abcd
XC	4614.5±609.9 d	366.3±47.7 abcd
XX	4323.8±224.7 bcd	558.2±35.1 f
YJ	3072.5±178.3 a	378.7±19.2 abcd
YX	4190.7±241.3 abcd	315.4±17.4 abc
YC	3926.2±304.8 abcd	299.3±26.3 ab
ZY	3413.4±104.3 abcd	398.1±21.7 bcde

Counties are arranged in English alphabetical order.

659 g m^{-2}. The highest value was in XX County and the lowest in LY County in N reserves (Table 3).

Linear regression analysis indicated that total crop production was significantly and positively correlated with SOC and TSN (Figure 4A), soil C and N store (Figure 4B), and soil C and N stocks (Figure 4C). Soil bulk density was significantly and positively correlated with soil N concentration ($r = 0.25$, $p = 0.008$, n = 114), soil C ($r = 0.21$, $p = 0.03$, n = 114) and N store ($r = 0.43$, $p = 0.001$, n = 114). While soil porosity was significantly and negatively correlated with soil N concentration ($r = -0.19$, $p = 0.05$, n = 114), soil C ($r = -0.25$, $p = 0.007$, n = 114) and N store ($r = -0.32$, $p = 0.001$, n = 114).

Principle components analysis revealed that Axis 1, which explained 98% of the variation in all data (eigenvalue = 0.98), was highly correlated with soil C, whereas Axis 2, explaining 1% of the variation (eigenvalue = 0.09), was highly correlated with soil N. Thus, counties such as QX and XP located highly positive on Axis 1 with high levels of soil C, but other counties, such as LK, YJ, and QF, occupied positions toward the negative end of Axis 1 with low soil C (Figure 5).

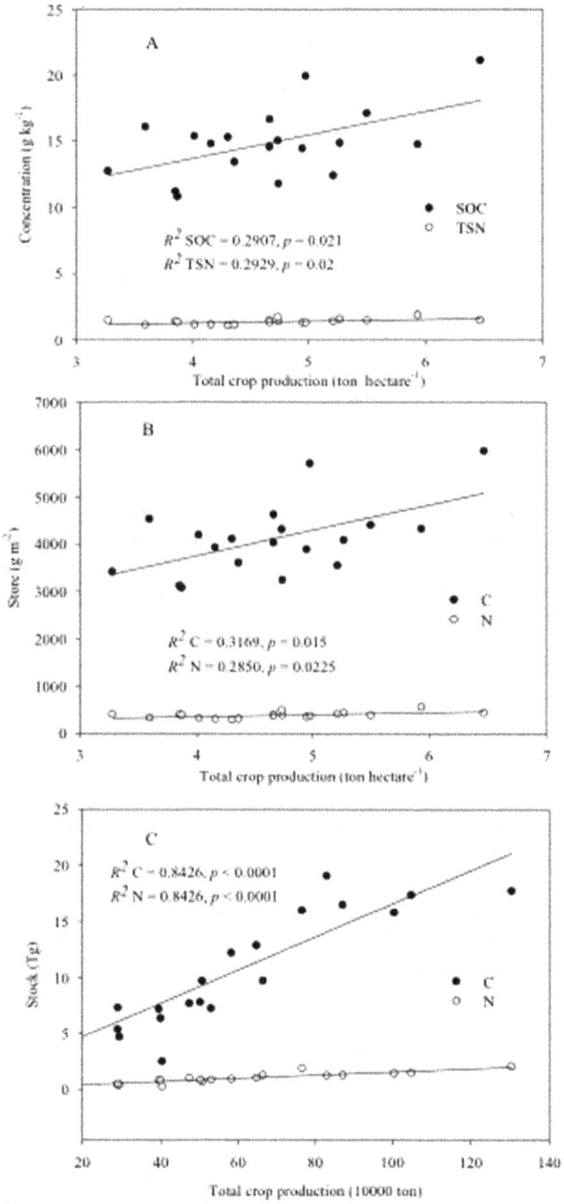

Figure 4. Linear regression analysis of total crop production in each county (ton ha^{-1}) with SOC and TSN (A) and with soil C and N store (0–40 cm) (B), and the total crop production of each county (10000 ton) with their soil C and N stock (C) (n = 19).

Discussion

Potential influences on crop yield

It is notable that 14 environmental (e.g., mean annual temperature and precipitation –Table 1) and soil variables (including extractable nutrients-Table 2) examined in our analysis of the data from the 19 counties in Henan Province were correlated with either wheat or total crop yield (data not shown), and total crop production were significantly, positively related to SOC and TSN, soil C and N store, and soil C and N stock (Figure 4). Part of this is likely related to the highly integrated nature of the measures of C and N stocks, i.e., their calculations combine soil concentrations of C and N, soil bulk density, sampling depth, and area of cultivation. However, all of these have been shown to directly influence crop performance. For example, increases in soil C have been shown to increase crop yield in other studies. Lal (2004, 2006) reported increases in yield from 20 to 70 kg ha^{-1} and 10 to 300 kg ha^{-1} for wheat and maize, respectively, following increases of 1 MT of C in agricultural soils in Africa [1,8]. Similarly, loss of soil C has been shown to decrease yield in agricultural soils of Canada and the U.S. [4,5].

Soil C-mediated increases of crop yields also may arise from improvements in soil structure and available water-holding

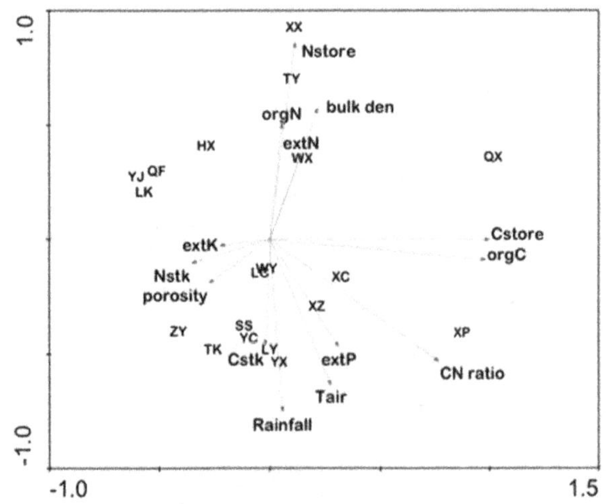

Figure 5. Principle components analysis of environmental and soil variables for agricultural soils in 19 counties within Henan Province. Length of arrows is directly proportional to their importance in explaining spatial patterns in the counties. Direction of the arrows indicates increasing values. Thus, the x-axis is primarily a gradient in soil C, whereas the y-axis is primarily a gradient in soil N and rainfall and secondarily a gradient in soil N. See Table 1 for key to county name abbreviations.

capacity. Enhanced soil structure, via increased soil C, generally arises from several processes, including increasing stability of soil aggregates [22,23,24]. As a result of the increased stability of the aggregates, soils become less prone to crusting, compaction, and erosion [25,26,28]. Emerson (1995) demonstrated that an increase of 1 g of soil organic matter (~50% of which is C) can increase available soil moisture by up to 10 g [27], which is enough to maintain crop growth between periods of rainfall of 5 to 10 days [8].

Spatial variation in cultivated soils

In this study, soil organic C concentration averaged 14.9 g kg^{-1} and total N averaged 1.4 g kg^{-1} in the 0–40 cm layer across all sites, while soil C and N stores averaged 4.1 kg C $m^{-2}$and 0.38 kg N m^{-2}, respectively. These values are comparable to published values from other regions of China, including 9–15 g C kg^{-1} and 1.2–1.8 g N kg^{-1} in northern China [12,29], and 16.1 g C $kg^{-1}$and 1.04 g N kg^{-1} in eastern and southern China [14,16,30]. Liu et al. (2011) reported soil C stores of 4.57 kg C m^{-2} in the Loess Plateau region in northwestern China [13].

Principal components analysis separated the 19 counties primarily along a gradient in soil C, with counties LK, YJ, QF, ZY, HX, and TK (mean soil C = 12.1 g C kg^{-1}) toward the lower end and XP and QX (mean soil C = 20.5 g C kg^{-1}) toward the upper end of Axis 1, which accounted for nearly 80% of the variation in soil and environmental data (Figure 5). Spatial variation in soil organic C in agricultural systems can be influenced by several factors, including microclimate, soil type, topography, and especially human activity [31].

Spatial variation in soil N was essentially orthogonal to that of soil C. This was surprising since typically, the two are highly correlated in terrestrial ecosystems [32]. As a result, the secondary gradient (i.e., Axis 2) was one of soil N, with counties TK, YC, SS, LY, YX, and XP (mean soil N = 1.15 g N kg^{-1}) located toward the lower end of Axis 2 (accounting for <10% of variation) and XX and TY (mean soil N = 1.81 g N kg^{-1}) located toward the upper end of Axis 2 (Figure 5). Although C and N are often correlated through their organic forms in plant detritus, spatial variation of N in soils of agro-ecosystems can also be greatly influenced by the extensive use of N fertilizers.

Management methods used in crop production systems, including tillage practices and fertilizer use, can affect soil C and N on broad spatial scales, including that of an entire Province [33]. Over the course of repeated seasons of crop growth in Henan Province, agricultural fields are repeatedly subjected to soil tillage, planting, fertilization, irrigation, and harvest, all of which potentially influence soil C and N stores [30,34]. In contrast, Zhang et al. (2012) reported that raised-bed planting, a viable alternative to conventional tillage, can significantly enhance the yield of summer maize while simultaneously improving soil structure, as well as the structure and function of microbial communities essential to the quality of agricultural soils [22].

Results presented in the current study underscore the complexity of factors that can impact agricultural soils and their ability to produce crops to meet the ever-increasing demand in China resulting from population growth. Some of the spatial pattern exhibited in ordination space (Figure 5) is clearly related to regional factors, such as microclimate. For example, WY and LC are adjacent to each other in Henan Province (Figure 1) and are also closely clustered in ordination space, indicating that they are very similar with respect to environmental and soil characteristics. XP and SS, however, are also adjacent counties; yet occur distant from each other in ordination space, indicating great dissimilarity in environmental and soil factors. Agronomists should take into account the large spatial variability in important components of the soils in Henan Province, especially in the variation of soil C and N, when considering appropriate agronomic management practices.

Author Contributions

Conceived and designed the experiments: XLZ QW CHL. Performed the experiments: XLZ QW YLW. Analyzed the data: XLZ QW FSG. Contributed reagents/materials/analysis tools: XLZ QW YLW FNC. Contributed to the writing of the manuscript: XLZ FSG CHL.

References

1. Lal R (2004) Soil carbon sequestration impacts on global climate change and food security. Science 304: 1623–1627.
2. Liu DW, Wang ZM, Zhang B, Song KS, Li XY, et al. (2006) Spatial distribution of soil organic carbon and analysis of related factors in croplands of the black soil region, Northeast China. Agriculture, Ecosystems and Environment 113: 73–81.
3. Smith WN, Desjardins RL, Pattey E (2000) The net flux of carbon from agricultural soils in Canada 1970–2010. Global Change Biology 6: 557–568.
4. Bauer A, Black AL (1994) Quantification of the effect of soil organic matter content on soil productivity. Soil Science Society of America Journal 58: 185–193.
5. Larney FJ, Janzen HH, Olson BM, Lindwall CW (2000) Soil quality and productivity response to simulated erosion and restorative amendments. Canadian Journal of Soil Science 80: 515–522.
6. Hairiah K, Van Noordwijk M, Cadisch G (2000) Crop yield, C and N balance of the three types of cropping systems on an Ultisol in northern Lampung. Netherland Journal of Agricultural Science 48: 3–17.
7. Duxbury JM (2001) Long-term yield trends in the rice-wheat cropping system: results from experiments in Northwest India. Journal of Crop Production 3: 27–52.
8. Lal R (2006) Enhancing crop yields in the developing countries through restoration of the soil organic carbon pool in agricultural lands. Land Degradation and Development 17: 197–209.
9. Batjes NH (2002) Carbon and nitrogen stocks in the soils of Central and Eastern Europe. Soil Use and Management 18: 324–329.

10. Maia SMF, Ogle SM, Cerri CC, Cerri CEP (2010) Changes in soil organic carbon storage under different agricultural management systems in the Southwest Amazon Region of Brazil. Soil and Tillage Research 106: 177–184.
11. Piao SL, Fang JY, Ciais P, Peylin P, Huang Y, et al. (2009) The carbon balance of terrestrial ecosystems in China. Nature 458, doi:10.1038/nature 07944.
12. Wang ZM, Zhang B, Song KS, Liu DW, Ren CY (2010) Spatial variability of soil organic carbon under maize monoculture in the Song-Nen plain, Northeast China. Pedosphere 20: 80–89.
13. Liu ZP, Shao MA, Wang YQ (2011) Effect of environmental factors on regional soil organic carbon stocks across the Loess Plateau region, China. Agriculture, Ecosystems and Environment 142: 184–194.
14. Liao QL, Zhang XH, Li ZP, Pan GX, Smith P, et al. (2009) Increase in soil organic carbon stock over the last two decades in China's Jiangsu Province. Global Change Biology 15: 861–875.
15. Zhang HB, Luo YM, Wong MH, Zhao QG, Zhang GL (2007) Soil organic carbon storage and changes with reduction in agricultural activities in Hong Kong. Geoderma 139: 412–419.
16. Feng S, Tan S, Zhang A, Zhang Q, Pan G, et al. (2011) Effect of household land management on cropland topsoil organic carbon storage at plot scale in a red earth soil area of South China. Journal of Agricultural Science 149: 557–566.
17. Harper RJ, Gilkes RJ (1995) Some factors affecting the distribution of carbon in soils of a dry land agricultural system in southwestern Australia. In: Lal R, Kimble JM, Follett RF, Stewart BA (editors). Assessment Methods for Soil Carbon. CRC Press. Boca Raton, FL, USA. PP.577–591.
18. Guo JH, Liu XJ, Zhang Y, Shen JL, Han WX, et al. (2010) Significant Acidification in Major Chinese Croplands. Science 327: 1008–1010.
19. Pan GX, Li LQ, Wu LS, Zhang XH (2003) Storage and sequestration potential of topsoil organic carbon in China's paddy soils. Global Change Biology 10: 79–92.
20. Wu HB, Guo ZT, Gao Q, Peng CH (2009) Distribution of soil inorganic carbon storage and its changes due to agricultural land use activity in China. Agriculture, Ecosystems and Environment 129: 413–421.
21. Gilliam FS, Saunders NE (2003) Making more sense of the order: A review of Canoco for Windows 4.5, PC-ORD version 4 and SYN-TAX 2000. Journal of Vegetation Science 14: 297–304.
22. Zhang XL, Ma L, Gilliam FS, Wang Q, Liu T, et al. (2012) Effects of raised-bed planting for enhanced summer maize yield on soil microbial functional groups and enzyme activity in Henan Province, China. Field Crops Research 130: 28–37.
23. Feller C, Beare MH (1997) Physical control of soil organic matter dynamics in tropics. Geoderma 79: 69–116.
24. Haynes RJ, Naidu R (1998) Influence of lime, fertilizer and manure applications on soil organic matter content and soil physical conditions: a review. Nutrient Cycling in Agroecosystems 51: 123–137.
25. Diaz-Zorita M, Grosso GA (2000) Effect of soil texture, organic carbon and water retention on the compatibility of soils from the Argentinean Pampas. Soil and Tillage Research 54: 121–126.
26. Schertz DL, Moldenhauer WC, Livingston SJ, Weeisies GA, Hintz AE (1989) Effect of past soil erosion on crop productivity in Indiana. Journal of Soil and Water Conservation 44: 604–608.
27. Emerson WW (1995) Water-retention, organic-carbon and soil texture. Australian Journal of Soil Research 33: 241–251.
28. Powlson DS, Hirsch PR, Brookes PC (2001) The role of soil micro-organisms in soil organic matter conservation in the tropics. Nutritional Cycling in Agroecosystems 61: 41–51.
29. Du ZL, Ren TS, Hu CS (2010) Tillage and residue removal effects on soil carbon and nitrogen storage in the North China Plain. Soil Science Society of American Journal 74: 196–202.
30. Pan GX, Li LQ, Zhang Q, Wang XK, Sun XB, et al. (2005) Organic carbon stock in topsoil of Jiangsu Province, China, and the recent trend of carbon sequestration. Journal of Environmental Sciences 17: 1–7.
31. Post WM, Pastor J, Zinke PJ, Stangenberger AG (1985) Global patterns of soil nitrogen storage. Nature 317: 613–616.
32. Gilliam FS, Dick DA, Kerr ML, Adams MB (2004) Effects of silvicultural practices on soil carbon and nitrogen in a nitrogen saturated Central Appalachian (USA) hardwood forest ecosystem. Environmental Management 33: S108–S119.
33. Pan GX, Zhao QG (2005) Study on evolution of organic carbon stock in agricultural soils of China: facing the challenge of global change and food security. Advances in Earth Science 20: 384–393 (in Chinese).
34. Dersch G, Böhm K (2001) Effects of agronomic practices on the soil carbon storage potential in arable farming in Austria. Nutrient Cycling in Agroecosystems 60: 49–55.

13

Roles of Extension Officers to Promote Social Capital in Japanese Agricultural Communities

Kosuke Takemura[1]*, Yukiko Uchida[2], Sakiko Yoshikawa[2]

1 Graduate School of Management, Kyoto University, Kyoto, Japan, **2** Kokoro Research Center, Kyoto University, Kyoto, Japan

Abstract

Social capital has been found to be correlated with community welfare, but it is not easy to build and maintain it. The purpose of the current study is to investigate the role of professional coordinators of social relationships to create and maintain social capital in a community. We focused on extension officers in Japanese agricultural communities, who help farmers in both technical and social matters. A large nation-wide survey of extension officers as well as two supplementary surveys were conducted. We found that (1) social capital-related activities (e.g., assistance for building organizations among farmers) were particularly effective for solving problems; (2) social capital (trust relationships) among community residents increased their life quality; (3) social capital in local communities was correlated with extension officers' own communication skills and harmonious relationships among their colleagues. In sum, social capital in local communities is maintained by coordinators with professional social skills.

Editor: Kimmo Eriksson, Mälardalen University, Sweden

Funding: This work was supported by grants from the Kokoro Research Center at Kyoto University, Japan (Project Number: 08-1-01, 10-1-07); and the Japan Agricultural Development and Extension Personnel Association. The funders had no role in study design, data collection and analysis, decision to publish, or preparation of the manuscript.

Competing Interests: The authors have declared that no competing interests exist.

* E-mail: boz.takemura@gmail.com

Introduction

As humans live in societies, they are more or less social and interwoven in the society [1]. Given that humans significantly rely on social relationships to survive [2], mutual cooperation within community is essential. However, a state of mutual cooperation is not necessarily easy to establish. Tragedy of the commons, which was initially proposed by Hardin [3], is a typical example that highlights the conflicting nature of individual self-interests versus public goods. It has been suggested that one of the key factors to deal with this difficulty of social life and establish mutual cooperation efficiently is social capital [4]. The purpose of the current research is to investigate consequences (e.g., social capital brings improvement of the living condition in a community) as well as antecedents (e.g., personal social skills or positive social interactions with others) of social capital, by focusing on roles of professional social coordinators in agricultural communities.

Social capital

The term social capital captures the idea that social bonds and norms are important for people and communities [4–5]. Social capital can be broadly defined as the benefits of investing in social relationships [6], similar to financial capital and human capital (investing in individual capacities such as education). An extensive literature shows that human welfare depends heavily on social capital and also that social capital varies widely among human social environments. A recent study by Gutiérrez, Hilborn, and Defeo [7] have found, for instance, that social capital was a contributing factor in preventing the tragedy of the commons and leaded to the success of community-based co-management fisheries. Also, Sampson, Raudenbush, and Earls [8] showed that neighborhood-level collective efficacy, comprised of social capital and informal social control, reduced violent acts such as homicide. Social capital has also been suggested to increase regional incomes, life satisfaction, and life expectancy (see [4,9,10], for reviews).

Why does social capital or social cohesion improve the quality of life in the community? For example, a trust relationship, one feature of social capital [4,9], enhances cooperation, and so reduces transaction costs between people. Instead of having to invest in monitoring others, individuals are able to trust them to act as expected, thus saving money and time [4]. Accordingly, it becomes possible for them to invest money and time in other things such as health (see also [11–12], for reviews on how social relationships promote physical health). Also, mutual cooperation that social capital promotes is essential to build and/or maintain public goods such as irrigation systems in agricultural communities and sustainable resources [7,13].

Although it is becoming well-known and accepted that social capital is important for people and communities, it is not easy to build and maintain it. Imagine someone has altruistic intention and is willing to have reciprocal relationships with others. Even in this instance, it is not always easy to promote reciprocal relationships because his or her intention is not transparent to others. He or she needs to prove his or her good will so that others feel safe to step into the relationship. Norms, which are another important component of social capital that help establish mutual cooperation [9], generally take time to be established. Sampson et al. [8] have found that residential instability is negatively associated with neighborhood-level collective efficacy, which, as mentioned above, comprised mutual control against norm violation. Thus, in neighborhoods where residents change frequently, collective efficacy tends to be low and consequently violence tends to be more prevalent. Moreover, even if mutual cooperation is established, maintaining it is not easy. It is generally observed in public goods game experiments that cooperation decreases as the game proceeds [14–15]. Trust relationships, which promote cooperation, are also very sensitive and easy to be broken. Once installing a monitoring and sanctioning system against free-riders, group members' trust ironically becomes even lower than initial levels after removing the sanctioning system [16].

Role of the intermediary

The current paper examines roles of individuals playing as an *intermediary* or *coordinator* of social relationships, which has been suggested to create and maintain social capital. For example, it has been suggested that the existence of an intermediary promotes the building of trust relationships [17]. Suppose that one person needs to decide whether he or she trusts another person whom he or she has just met. In this situation, trust tends to be promoted if there is someone whom both the trustor and trustee have already established relationships with. The first reason is that uncertainty for the trustor about the trustee's personality is reduced through confidence in the intermediary as a good judge of people. Second, the existence of the third person changes incentive structures for the trustee. If the trustee cheats the trustor, the relationship between the trustor and the intermediary may be damaged. Thus, cheating the trustor will bring negative consequences not only to the trustor, but also to the intermediary. As a result, the relationship between the trustee and the intermediary will also be potentially damaged by the trustee's ill-intended behavior against the trustor. Knowing such potential consequences, the trustee has no (or at least smaller) incentive to cheat the trustor. Given this incentive structure, the trustor now feels secure to trust this person (there is *assurance* in Yamagishi's [18] terminology). After all, the existence of an intermediary promotes the building of trust between two parties who have no established relationship. These mechanisms pointed out by Coleman [17] are examples of how intermediaries serve in constructing social capital.

In line with Coleman's argument, Harada et al. [19] found that public health nurses, professionals that support community residents' health, played the role of intermediary in social relationships. They suggested that public health nurses helped to strengthen relationships within local communities (i.e., bonding social capital; [9]), and contributed to establishing relationships between community residents and the government as well as other types of professionals such as nursery school teachers (i.e., bridging social capital; [9]). Though an untrained third person, such as the mutual acquaintance in Coleman's [17] argument, may help social ties to be formed, the findings of Harada et al. [19] suggests that the existence of professionals who have skills for construction and maintenance of social capital is an important key for community welfare. However, Harada et al.'s [19] study, which employed a semi-structured interview method, was limited because of sample size (only 20 public health nurses). Further investigations about roles of professional intermediaries are undoubtedly needed. To this end, the current study examined roles played by a different professional coordinator of social relationships: extension officers working in an agricultural community.

Extension officers in Japanese agricultural communities

According to Zakaria and Nagata [20], since the end of World War II, Japan started to embark on a concerted effort to revitalize its agriculture sector in order to boost production to meet the escalating demand for food. The central and prefectural governments worked closely to enhance the training of farmers to promote their technical and managerial skills and to ensure sustainability, and this was carried out through the activities and programs by the agricultural extension services.

The Japanese extension system for agriculture, which started in 1948, was meant to help farmers acquire useful, appropriate, and practical knowledge in the domain of agriculture. This system was adapted from the American extension system but modified when applying it in Japan, so that the extension works could well function in Japanese culture for local needs and requirements [21].

As of 2010, there were approximately 7,000 officers in Japan, each of them belonging to a prefectural government. To become an extension officer, one has to pass a national exam. The job of extension officers is to help farmers in person to acquire technical and managerial knowledge and other skills in the domain of agriculture [22]. They do this job in collaboration with the Ministry of Agriculture, Forestry and Fisheries, research institutions, agricultural universities, and the prefectural government.

The Japanese Ministry of Agriculture, Forestry and Fisheries officially states that extension officers are expected to serve two functions: 1) "specialist" function, and 2) "coordinator" function [23–24]. Specialist function means "extension activities to provide farmers with advanced techniques and related knowledge (including managerial knowledge and skills), according as appropriate to local environments." On the other hand, coordinator function means to "help local farmers and related parties share future goals, clarify tasks they need to address, develop an approach to the tasks, and conduct it, under the cooperation with leading farmers as well as relevant organizations within and around local communities." Thus, extension officers are supposed to help farmers not only with skills and knowledge directly related to agriculture, but also help farmers build and maintain bonding-type social capital (social capitals within local communities) as well as bridging-type social capital (connections with related parties outside of communities).

In fact, Zakaria and Nagata [20] pointed out that extension officers in Japan "as intermediaries and catalysts, are the key links between farmers and the relevant agencies in terms of providing personalized and need-based information for decision making by all parties concerned" (in abstract [20]). Also, "the agricultural extension organizations naturally provide the place or the Japanese concept of 'ba' [25], which means 'a shared space that serves as a foundation for knowledge creation' for the promotion of active interactions, consultations and exchanges between extension and farmers" (p. 34 [20]). Also, Fukuda [26,27] points out that one of the roles extension officers play in Japanese agricultural communities is to facilitate communications and interactions among farmers as well as between farmers and related agencies in order to help innovate techniques and spread knowledge. These arguments suggest the significance of social ties within and around agricultural communities, and roles played by extension officers to construct and maintain them.

Importance of social capital in agricultural communities

Prior research has suggested that residents of agricultural communities are particularly interdependent and rely on social relationships compared to communities engaged in other types of economic activities such as herding [28]. This implies that social capital is highly important for agricultural communities. Indeed, there are several studies showing the association between social capital and collective actions as well as welfare of agricultural communities. Fukushima et al. [29], for example, found that trust toward other residents in the same local community was positively related with participation in collective management of local resources such as irrigation systems and commons (see also [30], for similar findings in Thailand). Also, trust relationships within agricultural communities have been found to be associated with residents' self-rated health [31–32] as well as settlement in the communities [33]. It was also shown that damages by animals, such as monkeys, which brought serious harms to agricultural communities, could be efficiently mitigated by routinely employing community level cooperation to scare away the animals [34]. More generally, Gyawali, Fraser, Bukenya, and Banerjee [35] found that increased human well-being (composite of income, education, and employment) was associated with social capital, through a survey conducted in the west-central Black Belt region of Alabama, which contains vast amounts of forest resources and fertile agricultural land. As a whole, in line with the social capital literature, social capital such as trust relationships have been found to relate to human welfare in agricultural communities.

There are at least two issues awaiting empirical study. First, though it has been suggested that social capital is associated with human welfare in agricultural communities, much of the prior research did not prove that the observed associations are causal. The current study addressed this by analyzing panel data collected through multiple surveys targeting extension officers in agricultural communities (Analyses 1 and 2). Second, we explored what kind of characteristics and situations the extension officers should possess for building social capital efficiently, by analyzing large survey data of extension officers all over Japan (Analysis 3).

Overview

We conducted three survey studies, and performed several analyses by combining the datasets when necessary. Data 1 constituted our major dataset, which was created by a nation-wide survey for more than 4,000 extension officers all over Japan, conducted in 2010. Data 2 and 3 were studies that included responses of the extension officers in two different areas (Data 2: Kinki area in 2009 and Data 3: Aichi prefecture in 2011). In Data 2 and Data 3, we could identify those who were also included in Data 1 in 2010 and thus we can examine that data as a panel data for a time-series analysis.

We conducted analyses mainly based on Data 1, but used combined datasets as well, depending on the purpose of the analysis. Analyses 1 and 2 addressed the effect of social capital in communities, such as trust relationships among community residents. In Analysis 1 with Data 1, we investigated the effect of extension service activities related to establishing social capital for solving farmers' problems. Though previous studies on extension officers in Japan [20] and the literature on the importance of social capital suggest the significant roles of the officers' coordinator function, this is not necessarily a widely shared notion even among the officers themselves [36], and thus needs to be examined empirically. Analysis 1 compared the performance of extension activities related to social capital and the other types of activities. In Analysis 2, we examined whether trust relationships among community residents (one of the important components of social capital) would increase life quality or not through a time-series analysis based on combined datasets of Data 1, 2, and 3.

In Analysis 3, we investigated what kinds of extension officers were more likely to contribute to trust relationships among community residents. We analyzed Data 1 to examine effects of personal traits of extension officers such as communication skills and the social relationships they had.

We first describe the three datasets we used. Then, we report the results of our Analyses 1 to 3.

Method

Ethics statement

The survey for Data 1 was approved by the Japan Agricultural Development and Extension Personnel Association. The survey for Data 2 was approved by the Kinki Regional Agricultural Administration Office. The survey for Data 3 was approved by the Aichi Agricultural Development and Extension Personnel Association. All participants gave consent by completing the survey. Data files are not available due to the consent agreement with the organization of extension officers. Upon request, however, we could provide the detailed information about the data.

Data 1

Respondents. With the cooperation of the Japan Agricultural Development and Extension Personnel Association, we called all the extension officers in Japan ($N = 7{,}241$) for the study, and 4,355 extension officers participated in the study (response ratio was 60.0%). Respondents from all but two prefectures completed the online survey (for Saitama and Wakayama prefectures, the surveys were sent and returned through the mail because computers at their workplaces could not access the survey website due to access limitations). Data collection was conducted between September and October 2010. Table 1 provides information on gender, age, and years of working experience as extension officers of respondents in Data 1.

Measures. Extension activities they had conducted. The respondents were asked to recall one of their recent experiences in which they were faced with a difficulty in the case they were charged with. Then they were instructed to check all the extension activities they had conducted in that agricultural situation from the list of 11 types of activities, which were derived from a previous study [37]: 1) Assistance to foster the sustainable workforce, 2) Assistance to establish the desirable area of productions, 3) Assistance to conduct eco-friendly agriculture, 4) Assistance regarding food safety, 5) Assistance for the development of agricultural communities, 6) Introduction of agricultural techniques, 7) Assistance for sales promotion, 8) Collaboration and coordination with relevant organizations, 9) Assistance for building organizations and collaboration among farmers, 10) Providing a vision for the future, and 11) Identifying specific problems the community has.

After indicating all the conducted activities, they completed the following four items, all of which were considered to reflect their assessment of their activities; the first two items asked how satisfied they were, and how satisfied they thought the community residents were with their activities as a whole in that situation. For both items, they indicated the level of satisfaction using the same scale ranging from 0 to 100. The other two questions were about positive feedbacks from the community residents. The first item asked how often the community residents showed their gratitude to the respondents (from 0 = Never to 3 = Very often). The second one asked how pleased the community residents were about the respondent's activities in total (from -3 = Not pleased at all to 3 = Fairly pleased). To give equal weights to the four items, we rescaled the items so that each ranged from 0 to 1. The average of the rescaled items forms our measure of performance ($\alpha = .87$).

Perceived state of the community. Another part of the survey was about the state of the community where the respondents were working at that time. The respondents were instructed to indicate life quality of the community residents (2 items, "The community residents are satisfied with their circumstances of life," "The living conditions of the community residents were all right"; $\alpha = .78$). The second set of this part was about trust

Table 1. Summary of respondent characteristics.

		Data 1	Data 2	Data 3[a]
Study period		**September-October, 2010**	**July-August, 2009**	**October, 2011**
Population		**Extension officer in Japan**	**Extension officer in Kinki area (6 prefectures)**	**Extension officer in Aichi prefecture**
Sample size		4,355	319	101
Response rate		60%	52%	54%
Gender	Female	23%	18%	68%
	Male	60%	63%	26%
	No response	17%	19%	6%
Age	20s	6%	3%	6%
	30s	18%	18%	16%
	40s	37%	35%	61%
	50s	24%	24%	16%
	60s	1%	2%	0%
	No response	13%	18%	0%
Years of experience working for the current job	3 or less	11%	6%	16%
	Between 4–10	16%	15%	16%
	Between 10–15	15%	17%	6%
	15 or longer	38%	44%	58%
	No response	20%	18%	3%

[a]Data 3 was planned to be merged with Data 1. To reduce the burden of respondents, we did not ask respondents in Data 3 on their gender, age, or years of working experience. For Data 3, we report information on these variables based on 31 respondents who were identified in Data 1.

relationships within community (7 items, e.g., "I think the community residents trust each other," "The interpersonal relationships among the community residents are generally smooth"; $\alpha = .82$). Response options were provided on 7-point scales (1 = Strongly disagree to 7 = Strongly agree).

The respondent's skills and social relationships. We administered a modified version of Tsutsui's [38] collaboration activity scale, which was designed to measure a behavioral tendency for collaborations and communications with other related parties such as agricultural cooperatives and local government. Sample items included "I hear about services and actual situations of other related organizations (including resident organizations) from themselves," "I know what kind of professionals are in other related organizations (including resident organizations)," "I ask for cooperation of other related organizations (including resident organizations)," ($\alpha = .84$). Response options were provided on 4-point scales (1 = Not at all to 4 = Very much). We call this the collaboration index hereafter.

We also asked about their self-evaluation of communication skills ("What do you feel about your own current communication skills as an extension officer?") and self-evaluation of their knowledge and technical skills ("What do you feel about the current level of your knowledge and techniques which are directly related to extension activities?"), which related to the major functions of extension officers ("coordinator" function and "specialist" function, respectively). For both items, response options were provided on 7-point scales (-3 = Far from enough to 3 = Good enough). They also worked on a 10-item scale of extraversion developed by Goldberg [39] including sample items such as "I am the life of the party" and "I start conversations" (1 = Strongly disagree to 7 = Strongly agree; $\alpha = .91$). Extraversion was measured in this study as one of the control variables of personality traits that might promote their coordinator function.

Two questions were administered to assess social relationships the respondents had. The first one was about their relationships with the community they were working for (hereafter referred to as "tie with community"), ranged from -3 (I am independent and separate from the community) to 3 (I am connected with the community). The second one was about relationship harmony at their workplace (i.e., extension center) (-3 = Very bad to 3 = Very good).

Data 2

Respondents. With the cooperation of the Research Group of Specialists in Extension Activities of Kinki, we called all the extension officers in six prefectures of the Kinki area ($N = 616$), and 319 extension officers participated in this study (response ratio was 51.8%). The URL of the survey website was announced and the Excel form of the questionnaire was also sent to them via email just in case they could not work on the survey online. Data collection was conducted between July and August, 2009, approximately one year before Data 1.

For both Data 1 and 2, respondents were asked to create a unique identification code for themselves. Out of the 319 respondents of Data 2, 61 people were identified in Data 1 as well, and among them, 29 people had worked for the same community between the two data collections, and completed all the relevant scales.

Measures. To measure the perceived trust relationships among the community residents as well as the perceived life quality of the community residents, the identical measures to those of Data 1 were used for Data 2 (αs $= .83$ and .76 for trust relationships and life quality of the community residents, respectively). The respondents in Data 2 also indicated how long they had been working for the same community. This information helped us to select the respondents who worked for the same community in Data 1 and 2.

Data 3

Respondents. With the cooperation of the Aichi Agricultural Development and Extension Personnel Association, we called all the extension officers in Aichi prefecture ($N = 188$) for the study, and 101 of them participated in the study (response ratio was 53.7%). The URL of the survey website was announced to all the extension officers in Aichi. They worked on the survey during October 2011, approximately one year after Data 1.

Respondents of Data 3 were asked to create a unique identification code based on the same rule from Data 1 and Data 2. Out of the 101 respondents of Data 3, 47 people were found to be identical in Data 1 as well, and 31 of them had worked for the same community between the two data collections, completed all the relevant scales, and did not learn about the results of Data 1.

Measures. Just like Data 2, the same measures for the perceived trust relationships among the community residents and the perceived life quality of the community residents were used for Data 3 (αs $= .77$ and $.73$, respectively). Also the respondents in Data 3 indicated how long they had been working for the same community so that we could select the respondents who worked for the same community in Data 1 and 3.

Results

Analysis 1: Types of extension activities and their effects

In Analysis 1, we analyzed Data 1 to see what kind of extension activities were efficient to solve problems that farmers faced. We predict that activities related to social capital (e.g., activities promoting coordination among farmers) are effective to solve the problems among farming communities.

As aforementioned, respondents of Data 1 were asked to recall one of their recent experiences in which they had faced a difficulty, and indicate all the extension activities they conducted in the situation. The most frequently conducted activity was collaboration and coordination with relevant organizations (63%), followed by introduction of agricultural techniques (61%; see Table 2 for the other extension activities).

What kinds of activities have a positive effect in solving problems? We first performed an exploratory factor analysis (principal factor solution) with varimax rotation on the 11 activity items to see convergence among the activities. The scree test suggested two factors, which together accounted for 38.4% of the variance. Factor 1 accounted for 24.7% of the variance, and Factor 2 accounted for 13.6% of the variance. As shown in Table 3, the six activity items that were related to social capital (e.g., "assistance for building organizations and collaboration among farmers") had high factor loadings on Factor 1 (hereafter called social capital-related activity). On the other hand, the other five activity items that were related to agricultural skills and business management (e.g., "assistance regarding food safety", "introduction of agricultural techniques") had high loadings on Factor 2 (hereafter called agricultural business management activity). We constructed the social capital-related activity indicator and the agricultural business management activity indicator by taking the mean of the respective items.

We then examined correlations of these two types of activity indicators with the performance score. The analysis revealed that social capital-related activity was positively related with the performance score, $r = .39$, $p < .001$. In addition, agricultural business management activity was also positively correlated with the performance score, $r = .28$, $p < .001$. As expected, however, the effect of social capital-related activity was greater than the other $t(3899) = 6.62$, $p < .001$. This suggests that extension activities which enhance social capital are especially effective and lead to good performance to solve problems in communities.

Analysis 2: Social capital (trust relationships) and life quality of community residents

To examine how social capital promotes the quality of life in agricultural communities, Analysis 2 focused on the effect of perceived trust relationships among the community residents (the indicator of social capital) on perceived life quality of the community residents.

First, we examined simple correlation between these two variables (Table 4). As predicted, they were found to be positively correlated with each other consistently across three datasets. This suggests that trust relationships among the community residents have a positive effect on their quality of life.

However, the analyses above do not indicate the causal association. To examine the causal relationships showing if

Table 2. Extension activities and implementation rate in difficult situations respondents experienced.

Extension activities	Implementation rate
Assistance to foster the sustainable workforce	50%
Assistance to establish the desirable area of productions	44%
Assistance to conduct eco-friendly agriculture	24%
Assistance regarding food safety	24%
Assistance for the development of agricultural communities	31%
Introduction of agricultural techniques	61%
Assistance for sales promotion	26%
Collaboration and coordination with relevant organizations	63%
Assistance for building organizations and collaboration among farmers	44%
Providing a vision for the future	36%
Identifying specific problems the community has	38%

Table 3. Rotated factor matrix of the extension activity items.

Item	Factor loading: Social capital-related (Factor 1)	Factor loading: Agricultural business management (Factor 2)
Providing a vision for the future	**.57**	.09
Assistance for building organizations and collaboration among farmers	**.54**	.06
Identifying specific problems the community has	**.51**	.17
Assistance for the development of agricultural communities	**.49**	.06
Collaboration and coordination with relevant organizations	**.41**	.17
Assistance to foster the sustainable workforce	**.37**	.09
Assistance regarding food safety	.05	**.65**
Assistance to conduct eco-friendly agriculture	.05	**.64**
Assistance to establish the desirable area of productions	.27	**.35**
Introduction of agricultural techniques	.08	**.34**
Assistance for sales promotion	.24	**.32**

Factor loadings .30 or greater are shown in bold.

perceived trust relationships really had an effect on perceived life quality, we conducted a time-series analysis by combining Data 1, 2, and 3. As mentioned above, 29 respondents in Kinki area completed the relevant scales both in Data 1 and 2. Similarly, 31 respondents in Data 3 (Aichi prefecture) were also found in Data 1, and completed the relevant scales. For both Kinki area and Aichi prefecture, the first data collection was conducted approximately one year before the second data collection. For Kinki area, responses at Time 1 were from Data 2, and responses at Time 2 came from Data 1. For Aichi prefecture, Data 1 and 3 were used for Time 1 and 2, respectively. Finally, we combined these two datasets about the two areas into single datasets to see if perceived trust relationships at Time 1 has significant effect on perceived life quality at Time 2 even after controlling for the effect of perceived life quality at Time 1.

We regressed perceived life quality at Time 2 on perceived trust relationships at Time 1, perceived life quality at Time 1, a dummy-coded variable of area (0 = Kinki, 1 = Aichi), and the interaction effect terms between perceived trust relationships at Time 1 and area, as well as perceived life quality at Time 1 and area (see Table 5; adjusted R^2 = .08, $p = .085$). As expected, perceived trust relationships at Time 1 had marginally significant positive effect on perceived life quality at Time 2. Standardized regression coefficient of perceived trust relationships at Time 1 suggests that an increase of one standard deviation in this variable led to an increase of 0.34 of one standard deviation in perceived life quality measured at approximately one year later. Area did not moderate the effect of perceived trust relationships. The other effects were also found to be non-significant. In sum, the results suggest the positive effect of trust relationships on life quality of community residents.

Analysis 3: Correlates of social capital (trust relationships) of communities

As shown in Analysis 2, trust relationships among community residents, one aspect of social capital, has a positive effect on the residents' life quality. The next question is, "what enhances trust relationships among community residents?"

By analyzing Data 1, we examined correlations of perceived trust relationships among community residents with respondent's collaboration index, extraversion, communication skills, knowledge and technical skills, tie with community, and interpersonal relationships at the workplace. As shown in Table 6, collaboration index, extraversion, and communication skills were positively correlated with perceived trust relationships. Though knowledge and technical skills also had a positive association with perceived trust relationships, the effect size was quite small.

Tie with community and interpersonal relationships at the workplace also had positive correlations with perceived trust relationships. Furthermore, these variables, which were about social relationships surrounding respondents, had positive correlations with perceived trust relationships among community residents even after controlling for the respondent's own internal traits on social relationships such as the collaboration index, extraversion, and communication skills (tie with community: r_ps = .31, .32, .30, ps <.001; interpersonal relationships at the workplace: r_ps = .22, .22, .21, ps <.001, by controlling for the collaboration index, extraversion, communication skills, respectively). These results indicate that the effects of social relationships surrounding extension officers are not spurious correlations caused by the extension officer's personal traits.

Discussion

The concept of social capital including trust relationships and social networks has served to bring researchers' attention to the significance of social bonds for human welfare. Prior research has actually demonstrated the associations of social capital with several domains of human life, such as financial incomes, life expectancy, life satisfaction, decreased violence, maintenance of public goods, and so on [4,9,10].

The current study was conducted to investigate consequences and antecedents of social capital in Japanese agricultural communities by focusing on roles of professional extension officers. Extension officers are involved in many kinds of activities to help farmers, such as introducing agricultural techniques, providing managerial knowledge, and building and maintaining trust

Table 4. Perceived trust relationships and perceived life quality of community residents (means, standard deviations, and Pearson's *r* coefficients).

	N	Perceived trust relationships among community residents		Perceived life quality of community residents		Correlation	
		M	*(SD)*	*M*	*(SD)*	*r*	*p*
Data 1	3268	5.27	(0.75)	3.83	(1.26)	.14	.000
Data 2	163	5.10	(0.65)	3.83	(1.16)	.19	.016
Data 3	97	5.36	(0.68)	4.40	(1.19)	.31	.002

Note. For this analysis, we included not only respondents who had data both in Data 1 and Data 2 (or Data 3) but also respondents who completed the relevant scales only in Data 2 or 3.

relationships and collaboration inside and around agricultural communities.

Our analyses, based on data collection including a nation-wide survey of extension officers, showed that extension activities related to social capital are particularly important. Analysis 1 revealed that to solve problems that farmers are faced with extension activities for enhancing social capital had greater effects compared to other activities such as the introduction of agricultural techniques. This finding suggests social capital plays essential roles for life in agricultural communities.

In line with the results of Analysis 1, trust relationships (one important aspect of social capital) and life quality of community residents were found to have positive association across three survey data (Analysis 2). Moreover, by analyzing panel data, we validated the causal relationship between them: Trust relationships among community residents promote their life quality.

Furthermore, to study antecedents of social capital, we explored which factors or skills of extension officers were associated with trust relationships among community residents. Analysis 3 revealed that extension officers' collaboration with related parties and communication skills were positively correlated with trust relationships within communities. In addition, interpersonal relationships at extension officers' workplace (i.e., extension centers) were positively connected with trust relationships in the local communities where the extension officers worked. This suggests a "chain effect" of social capital, meaning positive relationships in one place (extension officers' workplace) also facilitate positive relationships in another place (an agricultural community) presumably through extension officers' activities. Taken together, the current research demonstrates the importance of extension officers' work in promoting social capital in agricultural communities.

Extension activities related to agricultural techniques must not be viewed as unimportant. In fact, the demand for "specialist" function is still high. Fukuda's [27] research that collected farmers' opinions found that one of the highest-priority needs of farmers is extension of innovative techniques. Yet, the current study suggests that the other function of extension officers that have not received broad attention—"coordinator" function—has to do with a very important resource of agricultural communities, namely, social capital.

Limitations and future directions

It is important to emphasize that our data was collected through self-report and perceptions about states of communities by extension officers. This means that the current paper relies only on the service providers' point of view, rather than the service recipients (i.e., farmers). However, it is also important to note that relying only on the farmers' point of view is not sufficient either to investigate roles of extension activities. Some extension officers we interviewed emphasized that, to motivate farmers, it is sometimes important to hide the roles of their activities from farmers. Thus, the farmers may not be aware about the functions of extension activities. It is therefore of importance to investigate associations between extension activities and communities' welfare from *both* sides. In addition to this, it would be an important future work to include objective measures, such as objective indices of farmers' health, economic success, and the actual number of cooperative interactions among farmers within communities. It is suggested that relying only on the same type of measurement (e.g., self-report likert scale) from the same source may exaggerate observed correlations due to the common method variance bias [40]. Some of our findings, however, cannot be explained solely by this bias. We found the predicted associations even when we had covariates

Table 5. Effects of trust relationships among community residents (Time 1), life quality of community residents (Time 1), area, and their interactions on life quality of community residents (Time 2), $N = 60$.

	b	*(SE)*	β	*t*	*p*
Perceived trust relationships (Time 1)	0.65	(0.36)	.34	1.82	.075
Perceived life quality (Time 1)	0.29	(0.20)	.29	1.49	.142
Area (0 = Kinki, 1 = Aichi)	−0.19	(0.32)	−.08	−0.61	.543
Perceived trust relationships (Time 1) x Area	−0.61	(0.49)	−.23	−1.25	.217
Perceived life quality (Time 1) * Area	0.07	(0.26)	.05	0.28	.778

that should share the same bias (see Analyses 2 and 3). Yet, it is undoubtedly desirable to have objective measures as well and to examine the robustness of the findings.

Another limitation of the current study is the small sample size of the panel data for Analysis 2. We needed to include only respondents who had worked for the same community for both Time 1 and Time 2 in the analysis. We scarcely had a fair amount of respondents since extension officers' working terms for one community are generally short (in Data 1, we asked the respondents how long they had been working for the same community; the length of the mean time was 2.18 years, median was 1.50 years, and mode was 0.50 years). Future studies collecting panel data through those who stay in the same community for a longer period of time (e.g., farmers) are needed. It is also important to point out that our finding from the time-series analysis (Analysis 2) is not conclusive about the causality. By controlling for the effect of life quality at Time 1, we could show that the opposite causality (i.e., life quality promotes trust relationships among community residents) cannot fully explained the observed association. However, it is still possible that a third variable explains the association between trust relationships and life quality. For example, it may be possible that existence of strong leadership in a community promotes both trust relationships among residents and their life quality. Future studies that examine effects of such potential third variables are needed.

Additionally, it is also important to investigate potential negative effects of social capital on welfare of agricultural communities. For example, it has been suggested that excessive levels of bonding-type social capital (social capitals within a group) may promote distrust toward outsiders and inhibit the group's economic growth [41]. Future studies need to investigate what kinds of extension activities promote (or inhibit) social networks crossing a boundary of local communities.

There is another important question that future research needs to address: How can we train good coordinators? Though it is suggested by the current study that good coordinators (e.g., extension officers who have high communication skills) can help communities enhance trust relationships and collaborations, knowledge on how to foster such good coordinators is requisite to keep communities benefiting from them in the future. Thus, we need to know, for example, how to obtain ability for collaboration, how to acquire good communication skills, and how to recruit those who are (or have potential to be) good coordinators. Also, though the current study targeted extension officers and social capital in Japanese agricultural communities, presumably other types of communities, organizations, and groups face similar problems and thus coordinators may play crucial roles. How to achieve efficient problem solving in groups is one of the questions social psychological research has extensively addressed. From the findings of the current study, skilled coordinators are expected to play significant roles in groups and organizations that need cooperation and collaboration among members, such as medical institutions and educational institutions. Future research is needed to find ways to build systems that can sustainably provide coordinators who support building connections between people.

Table 6. Collaboration index, extraversion, communication skills, knowledge and technical skills, tie with community, and interpersonal relationships at the workplace (means, standard deviations, and Pearson's *r* coefficients with perceived trust relationships among community residents).

	M	*(SD)*	**Correlation with perceived trust relationships among community residents**	
			r	*p*
Respondent's personal traits				
Collaboration index	2.63	(0.34)	.17	.000
Extraversion	3.87	(1.08)	.14	.000
Communication skills	0.28	(1.41)	.17	.000
Knowledge and technical skills	−0.50	(1.58)	.08	.000
Social relationships surrounding respondent				
Tie with community	0.63	(1.15)	.34	.000
Interpersonal relationships at the workplace	1.03	(1.29)	.24	.000

Note. Scales ranged from 1 to 4 for Collaboration index, from 1 to 7 for Extraversion, and from −3 to 3 for the other scales.

Acknowledgments

The authors would like to thank the Japan Agricultural Development and Extension Personnel Association, the Kinki Regional Agricultural Administration Office, the Research Group of Specialists in Extension Activities of Kinki, and the Aichi Agricultural Development and Extension Personnel Association, for their help with data collection. We also would like to thank researchers at the Kokoro Research Center for their helpful comments on the projects.

Author Contributions

Conceived and designed the experiments: KT YU SY. Performed the experiments: KT YU. Analyzed the data: KT. Wrote the paper: KT YU SY.

References

1. Aronson E (1972) The social animal. New York: Viking Adult. 338 p.
2. Baumeister RF, Leary ML (1995) The need to belong: Desire for interpersonal attachments as a fundamental human motivation. Psychol Bull 17: 497–529.
3. Hardin G (1968) The tragedy of the commons. Science 162: 1243–1248.
4. Pretty J (2003) Social capital and the collective management of resources. Science 302: 1912–1914.
5. Coleman JS (1988) Social capital in the creation of human capital. Am J Sociol 94: S95–S120.
6. Wilson DS, O'Brien DT, Sesma A (2009) Human prosociality from an evolutionary perspective: Variation and correlations at a city-wide scale. Evol Hum Behav 30: 190–200.
7. Gutiérrez NL, Hilborn R, Defeo O (2011) Leadership, social capital and incentives promote successful fisheries. Nature 470: 386–389.
8. Sampson RJ, Raudenbush SW, Earls F (1997) Neighborhoods and violent crime: A multilevel study of collective efficacy. Science 277: 918–924.
9. Putnam RD (2000) Bowling alone: The collapse and revival of American community. New York: Simon and Schuster. 541 p.
10. Wilkinson RG (1999) Health, hierarchy, and social anxiety. Ann N Y Acad Sci 896: 48–63.
11. House JS, Landis KR, Umberson D (1988) Social relationships and health. Science 241: 540–545.
12. Uchino BN, Cacioppo JT, Kiecolt-Glaser JK (1996) The relationship between social support and physiological processes: A review with emphasis on underlying mechanisms and implications for health. Psychol Bull 119: 488–531.
13. Pretty J, Ward H (2001) Social capital and the environment. World Dev 29: 209–227.
14. Fehr E, Gächter S (2002) Altruistic punishment in humans. Nature 415: 137–140.
15. Puurtinen M, Mappes T (2009) Between-group competition and human cooperation. Proc. R. Soc. B 276: 355–360.
16. Mulder LB, van Dijk E, De Cremer D, Wilke HAM (2006) Undermining trust and cooperation: The paradox of sanctioning systems in social dilemmas. J Exp Soc Psychol 42: 147–162.
17. Coleman JS (1990) Systems of trust and their dynamic properties. In: Coleman JS, editor. Foundations of social theory. Cambridge, MA: Harvard University Press. pp. 175–196.
18. Yamagishi T (2011) Trust: The evolutionary game of mind and society. New York: Springer. 153 p.
19. Harada H, Konishi M, Teraoka S, Ura M (2011) Professional skills in the process of establishing human relations within a support framework: Focusing on building community systems supported by public health nurses. Jpn J Exp Soc Psychol 50: 168–181 (in Japanese with English Summary).
20. Zakaria S, Nagata H (2010) Knowledge creation and flow in agriculture: The experience and role of the Japanese extension advisors. Libr Manag 31: 27–35.
21. Japan Agricultural Development and Extension Association (1992) Susumeyo jiko kenshu - shokuba kenshu [Let's do self-development training and on-job training]. The Extension Information Center, Japan Agricultural Development and Extension Association (in Japanese, the title translated by the current authors). 212 p.
22. Ministry of Agriculture, Forestry and Fisheries, Japan (n.d.) What is extension service? Available: http://www.maff.go.jp/j/seisan/gizyutu/hukyu/h_about/index.html (in Japanese). Accessed 2012 August 18.
23. Ministry of Agriculture, Forestry and Fisheries, Japan (2010) Guideline for management of agricultural extension services: Ministerial Notification No. 590 of MAFF. Available: http://www.maff.go.jp/kinki/seisan/keieishien/fukyu/pdf/h22_shishin.pdf (in Japanese). Accessed 2012 August 18.
24. Ministry of Agriculture, Forestry and Fisheries,Japan (2012) Guideline for management of agricultural extension services: Ministerial Notification No. 848 of MAFF. Available: http://www.maff.go.jp/j/seisan/gizyutu/hukyu/h_tuti/pdf/h24_shishin.pdf (in Japanese). Accessed 2012 August 18.
25. Nonaka I, Konno N (1998) The concept of "ba": Building a foundation for knowledge creation. Calif Manage Rev 40: 40–54.
26. Fukuda K (2003) A study on the functions of agricultural extension: Focusing on recent studies of agricultural extension theory overseas and in Japan. J Rural Community Stud 97: 82–90 (in Japanese with English Summary).
27. Fukuda K (2007) The direction of extension activities in formulating rural agriculture from the viewpoint of the needs of farmers: Based on the case of vegetable production areas in Yamagata prefecture and Chiba prefecture. J Rural Community Stud 105: 25–40 (in Japanese with English Summary).
28. Üskül AK, Kitayama S, Nisbett RE (2008) Ecocultural basis of cognition: Farmers and fishermen are more holistic than herders. Proc Natl Acad Sci U S A 105: 8552–8556.
29. Fukushima S, Yoshikawa G, Saizen I, Kobayashi S (2012) Analysis on rerated (sic) factors of participation in regional resource management in rural areas in northern Kyoto prefecture: A comparison of the two factors: Bonding and bridging social capital. J Rural Plan Assoc 31: 84–93 (in Japanese with English Summary).
30. Matsushita K, Asano K (2007) The effect of social capital on the efficiency of irrigation water management. Proc Agric Econ Soc Jpn 2007 482–489. (in Japanese).
31. Fukushima S, Yoshikawa G, Ichida Y, Saizen I, Kobayashi S (2009) Comparison of the influence between the generalized trust and the trust in members of one's community on self-rated health. Pap Environ Inf Sci 23: 269–274 (in Japanese).
32. Fukushima S, Yoshikawa G, Saizen I (2012) Determining the relationship between the frequency of contact with friends and self-rated health based on the proximity of residence: Aiming at the integration of individual and area level social capital studies. Environ Inf Sci 40: 31–39 (in Japanese with English Summary).
33. Yamaguchi S, Nakatsuka M, Hoshino S (2007) Study on region characteristic and settlement in rural area: A case of Sasayama City, Hyogo prefecture. J Rural Plan Assoc 26: 287–292 (in Japanese with English Summary).
34. Yamabata N (2010) Mitigate effect on damage to food crops achieved by collaboration of a whole village for chase-off of monkeys. J Rural Plan Assoc 28: 273–278 (in Japanese with English Summary).
35. Gyawali BR, Fraser R, Bukenya J, Banerjee SB (2010) Spatial relationship between human well-being and community capital in the Black Belt region of Alabama. J Agric Ext Rural Dev 2: 167–178.
36. Research Group of Specialists in Extension Activities of Kinki (2009) "Wakate fukyu shidoin no ikusei shuho" ni kansuru chosa kenkyu [Research on training methods for junior extension officers]. Research Group of Specialists in Extension Activities of Kinki (in Japanese, the title translated by the current authors). 40 p.
37. Uchida Y, Takemura K, Yoshikawa S (2011) The coordination roles of extension officers within Japanese agricultural communities. Sociotechnica 8: 194–203 (in Japanese with English Summary).
38. Tsutsui T (2005) Chiiki hoken sa-bisu no tantou shokuin ni okeru renkei hyouka shihyou kaihatsu ni kansuru toukei teki kenkyu [Statistical research to develop collaboration index of community health services officials]. Research Report of Health Labour Sciences Research Grant (the Integrated Research Project for Health Science) (in Japanese, the title translated by the current authors). 117 p.
39. Goldberg LR (1992) The development of markers for the Big-Five factor structure. J Pers Soc Psychol 4: 26–46.
40. Podsakoff PM, Organ DW (1986) Self-reports in organizational research: Problems and prospects. J Manage 12: 531–544.
41. Svendsen GLH, Svendsen GT (2004) The creation and destruction of social capital. Cheltenham, UK: Edward Elgar Publishing. 207 p.

14

Representation of Ecosystem Services by Terrestrial Protected Areas: Chile as a Case Study

América P. Durán[1]*, Stefano Casalegno[1], Pablo A. Marquet[2,3,4,5], Kevin J. Gaston[1]

1 Environment and Sustainability Institute, University of Exeter, Penryn, Cornwall, United Kingdom, **2** Departamento de Ecología, Facultad de Ciencias Biológicas, Pontificia Universidad Católica de Chile, Santiago, Chile, **3** Instituto de Ecología y Biodiversidad (IEB), Santiago, Chile, **4** Santa Fe Institute, Santa Fe, New Mexico, United States of America, **5** Laboratorio Internacional en Cambio Global (LINCGlobal), Facultad de Ciencias Biológicas, Pontificia Universidad Católica de Chile, Santiago, Chile

Abstract

Protected areas are increasingly considered to play a key role in the global maintenance of ecosystem processes and the ecosystem services they provide. It is thus vital to assess the extent to which existing protected area systems represent those services. Here, for the first time, we document the effectiveness of the current Chilean protected area system and its planned extensions in representing both ecosystem services (plant productivity, carbon storage and agricultural production) and biodiversity. Additionally, we evaluate the effectiveness of protected areas based on their respective management objectives. Our results show that existing protected areas in Chile do not contain an unusually high proportion of carbon storage (14.9%), agricultural production (0.2%) or biodiversity (11.8%), and also represent a low level of plant productivity (Normalized Difference Vegetation Index of 0.38). Proposed additional priority sites enhance the representation of ecosystem services and biodiversity, but not sufficiently to attain levels of representation higher than would be expected for their area of coverage. Moreover, when the species groups were assessed separately, amphibians was the only one well represented. Suggested priority sites for biodiversity conservation, without formal protection yet, was the only protected area category that over-represents carbon storage, agricultural production and biodiversity. The low representation of ecosystem services and species' distribution ranges by the current protected area system is because these protected areas are heavily biased toward southern Chile, and contain large extents of ice and bare rock. The designation and management of proposed priority sites needs to be addressed in order to increase the representation of ecosystem services within the Chilean protected area system.

Editor: Eric Gordon Lamb, University of Saskatchewan, Canada

Funding: This work has been funded by APD's studentship, which is a Chilean studentship under the Becas Chile program of the Comisión Nacional de Investigación Científica y Tecnológica, Gobierno de Chile (CONICYT). PAM acknowledges support from grants ICM P05-002 and PFB-23. The funders had no role in study design, data collection and analysis, decision to publish, or preparation of the manuscript.

Competing Interests: The authors have declared that no competing interests exist.

* E-mail: paz.duran.moya@gmail.com

Introduction

Ecosystem services, the benefits that humans derive from ecosystems, are vital for sustaining human well-being [1–3]. However, these services are also increasingly threatened by human activities [1]. It is thus critical to evaluate to what extent current conservation strategies capture ecosystem services, and therefore might ensure their provision in the future [4,5]. Due to their vast terrestrial coverage and historical success in conserving natural ecosystems, protected areas are increasingly considered to play a key role in the maintenance of the ecosystem processes that promote ecosystem service provision [1,6,7]. However, most existing protected areas have not been designated, established or managed to meet this specific objective, and might reasonably be expected in some instances to be inappropriate for doing so (e.g. agricultural and timber production). Indeed, whilst the representation of biodiversity within protected areas has been widely assessed [8–13], only a few studies have evaluated to what extent these are capturing ecosystem services [14–16]. Moreover, those studies that have been conducted have tended to focus on the representation of a single ecosystem service [16], or have been carried out at a rather coarse spatial resolution [15]. Assessments considering multiple services at a finer resolution are limited to developed countries (i.e. highly human-dominated regions) [17]. A broader range of studies are required to help understand the nature of the gaps in ecosystem service conservation and where they occur, and thus to aid systematic planning to designate and establish future protected areas to redress these gaps.

The provision of key ecosystem services can present trade-offs (e.g. carbon storage vs agricultural production) making their conservation within the same areas challenging [18,19]. Ecosystem services that involve active management practices can influence the potential for "disservices", often harming biodiversity and reducing the production of other services. For example, agriculture is a highly valuable provisioning service [1], providing food, forage, fibre, bioenergy and pharmaceuticals, but due to the often intensive form of associated land management, it is commonly considered a negative pressure on biological conservation [20,21]. How well protected area systems represent ecosystem services will depend, therefore, on what services are considered valuable to include or exclude, and this threshold is generally determined by the socioeconomic conditions and climatic region in which the protected area system is located. For instance, in a highly human-dominated region, where a higher proportion of land has been converted and species assemblages may have long been shaped by human activities, agriculture might be promoted, or at least tolerated, as an ecosystem service within protected areas [17]. In contrast, in a less developed region with relatively pristine ecosystems, this activity might be excluded from protected areas [22].

There is a particular paucity of data for appropriate evaluation of protected area effectiveness in capturing ecosystem services in poor and developing countries [23]. These are often also countries that are particularly rich in natural resources (renewable and non-renewable) and whose economies depend on their extraction, which makes the establishment of protected areas, strict management objectives and the assessment of their performance particularly challenging. Chile provides one such example [24,25]. Its economy depends strongly on extractive activities such as wood pulp production, agricultural production and mining, and the establishment and performance of strict environmental management strategies has been poor [24,26]. Indeed, the Chilean National System of Protected Areas (SNASPE) is known to be inefficient in providing adequate coverage of the country's biodiversity [27–30] and is underfunded, receiving only 0.03% of the national budget [CONAF 2005, unpublished data]. In response, the Chilean Ministry of Environment has made an urgent call to assess and improve the current protected area system [31] and to increase the protection of the country's non-transformed ecosystems. Thus, in collaboration with the Global Environment Facility (GEF) and the United Nations Development Program (UNDP), the Ministry of Environment aims to create an integrated public and private protected area system in order to increase protected area coverage and share responsibilities and costs among the different governmental and private bodies [31]. Private protected areas and priority sites for biodiversity conservation have been identified and suggested to be incorporated into a new integrated protected area system, however the extent to which these locations are valuable for ecosystem service provision is unknown.

This study analyses for the first time to what extent the current and suggested integrated protected area system represent selected ecosystem services of Chile. Specifically, the representation of three ecosystem services - plant productivity, carbon storage, agricultural production - and biodiversity is assessed under three protection scenarios. These scenarios capture the current status and medium-term projections for the protected area system. Given the large extent of pristine forest ecosystem remaining, we would expect that Chilean protected areas tend to represent high levels of net primary production, carbon storage, and biodiversity, but tend to exclude agricultural production. We address three main questions: 1) How are the three ecosystem services and biodiversity distributed across Chile?; 2) To what extent do the three protection scenarios represent the chosen ecosystem services and biodiversity?; and 3) How effective are Chilean protected area categories in representing ecosystem services and biodiversity?

Data and Methods

Protected Area System Coverage

In order to assess the effectiveness of the current Chilean protection system we considered all areas with statutory protection. These are the protected areas belonging to SNASPE, which comprises 33 national parks, 49 national reserves, and 16 natural monuments. We also included nature sanctuaries (n = 31), and lands protected by the Chilean Ministry of National Heritage (n = 18) (Table S1). SNASPE makes up the majority of traditional public protected areas in Chile and is administered by the National Forestry Corporation (CONAF) created by the Chilean government in 1984. Nature sanctuaries include both public and private lands that obtained statutory protection under the Chilean National Environmental Law in 1994 and law No. 17 288 relating to National Monuments in 1970. Those lands administered by the Ministry of National Heritage are public protected areas managed exclusively for conservation and established by decree in 1977. To assess the potential effectiveness of the new suggested sites we considered protected priority sites for biodiversity conservation (PSBC) identified by the National Environmental Commission (CONAMA) in 2011 (n = 68), and private protected areas (n = 295). PSBC identified by the Chilean Ministry of Environment are part of the countrys National Biodiversity Strategy [CONAMA 2003, unpublished data], which aims to improve the representation of biodiversity within the Chilean protected area system. Private protected areas were also defined by the National Environmental Law (Article 35), and these are portions of private land which the owners have voluntarily set aside for conservation objectives. Both PSBC and private protected areas have not received statutory protection yet, but they are suggested as protected areas to be incorporated into the current protected area system and thus create the new *integrated protected area system*. Current protected areas, PSBC and private protected areas datasets were obtained from the Chilean Ministry of Environment in vector format. Current protected areas datasets are freely available at the Chilean Ministry of Environment web site (ide.mma.gob.cl). Considering the entire set used in this study (n = 510), the average size of protected areas was 40,218 ha, varying from a minimum of 0.64 ha to a maximum of 3,677,849 ha. All of the seven protected area groups, except PSBC and private protected areas, are listed under an IUCN category [32]. These management categories differ in the level of human activity allowed, from strict protection where no extractive activity is allowed (IUCN Ia- III) to a more permissive approach where human habitation and sustainable extractive use are accepted (IUCN IV-VI) (see Table S1 for details).

Following a similar approach to that of Pliscoff and Fuentes-Castillo [33], we created three protection scenarios in order to evaluate the effectiveness of the existing protection system and the potential contribution of PSBC and private protected areas to the new integrated protected area system. The three scenarios were (Fig. 1):

- Scenario 1 (current protection system, Fig. 1-A): SNASPE+National Sanctuary+Ministry of Heritage lands.
- Scenario 2 (Fig. 1-B): Scenario 1+ Priority sites for biodiversity conservation (PSBC).
- Scenario 3 (Fig. 1-C): Scenario 2+ Private protected areas.

Distribution of Ecosystem Services and Biodiversity

Carbon storage. We calculated carbon storage by combining an estimate of above and below ground vegetation biomass and a soil organic carbon (SOC) dataset (in kg C). Aboveground vegetation data were obtained following the IPCC GPG Tier-1 method for estimating vegetation carbon stocks using the global default values provided for above ground biomass [34]. Below ground vegetative biomass (root) carbon stock was added using the root-to-shoot ratios for each vegetation type (i.e. shrubland, forest, grassland, steppe) obtained from the same IPCC [34] document, and then total living vegetation biomass was converted to carbon stock using the carbon fraction for each vegetation type. All estimates and conversions were specific to each of the nine ecofloristic zones [35] in Chile, and vegetation type obtained from

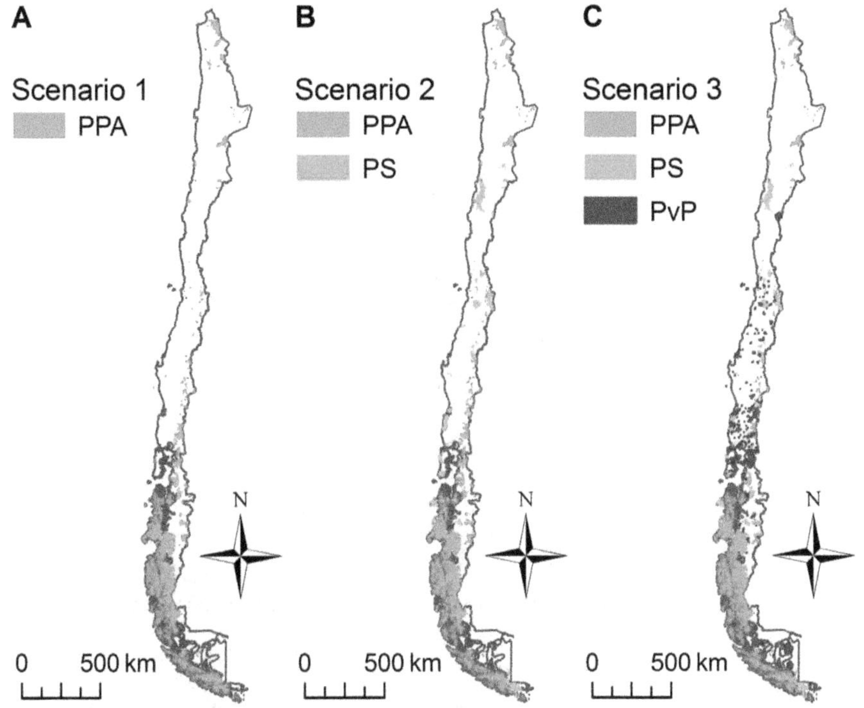

Figure 1. Distribution of three protection scenarios. The scenarios represent alternative conservation approaches. (A) Scenario 1, (B) Scenario 2, (C) Scenario 3. PPA: Public Protected Areas (current PA system in Chile); PS: Priority Sites for Biodiversity; PvP: Private Protected Areas.

the Chilean land use cover at 1.56 km^2 resolution (CONAMA). Thus, a total of 246 carbon zones with unique carbon stock values were compiled based on the IPCC Tier-1 methods.

Soil carbon density data were obtained from the most recent soil carbon database, the Harmonized World Soil Database (HWSD) version 1.1 [36], at 1×1 km resolution.

We used these datasets to create a final carbon storage estimation at 1.25×1.25 km resolution.

Plant productivity. Plant productivity (PP) patterns were determined using the Normalized Difference Vegetation Index (NDVI), which has been widely used for this purpose [16,37–40]. NDVI is a linear estimator of the fraction of photosynthetically active radiation intercepted by vegetation (fAPAR) [41–43], which is the main control of carbon gain (Monteith, 1981) and hence a good estimator of PP. NDVI is derived from the red:near-infrared reflectance ratio [$NDVI = (NIR - RED)/(NIR + RED)$, where NIR and RED are the amount of near-infrared and red light, respectively, reflected by the vegetation and captured by the sensor of the satellite]. The formula is based on the fact that chlorophyll absorbs RED (fAPAR as mentioned above), whereas the mesophyll leaf structure scatters NIR. NDVI values range from -1 to $+1$, where negative values correspond to an absence of vegetation (e.g. water bodies) and values closer to $+1$ correspond to abundant and dense vegetation (e.g. evergreen forest).

Monthly NDVI composites were obtained from the 1 km^2 resolution Global MODIS (TERRA) (Moderate Resolution Imaging Spectroradiometer - LPDAAC, NASA) dataset, available for 2000–2010. For each pixel we calculated the average of the annual NDVI mean for the 10 year period.

Agricultural production. Agricultural production was calculated as the sum of gross production (USA dollar) for 2000. In order to generate a fine resolution layer, a spatial disaggregation process was carried out, in which a coarse resolution dataset is 'disaggregated' in a finer and related resolution dataset. Specifically, the agricultural production layer was calculated as follows (i) We multiplied the harvested area of 32 major crops (i.e. proportion of a grid cell that has been harvested for a specific type of crop) (Table S2) at 10 km × 10 km resolution [44] by crop land cover at 1 km × 1 km resolution (i.e. spatial distribution of agricultural lands) [45]. Thus, through the disaggregation process, we obtained a second 1 km resolution layer showing the area per pixel (i.e. ha) that was harvested for each major crop; (ii) The resultant layers for each major crop were then multiplied by their respective yields (tonnes/ha) [46], obtaining tonnes of crops produced per pixel; and (iii) Finally, tonnes per pixel of each major crop were then multiplied by prices (USD/tonnes) for 2000 (FAOStat, http://faostat3.fao.org/home/index.html), thus obtaining USD of agricultural production per pixel.

Biodiversity. The Chilean biodiversity dataset comprised four taxonomic groups: mammals ($n = 113$), birds ($n = 364$), amphibians ($n = 58$) and vascular plants ($n = 1,061$). Distribution maps for mammals, birds and amphibians that occur in Chile were obtained from the IUCN Global Mammal Assessment, BirdLife International and the Global Amphibian Assessment, respectively. All these are freely available at the IUCN Red List web site [47], and released as polygon vector files. The dataset for plant distributions was obtained from work carried out by the Ministry of the Environment [48]. Plant distributions were generated using the Maximum Entropy Model (MaxEnt), which was based on a dataset comprising georeferenced records from the largest plant collection in Chile (Museum of Concepción) complemented with records derived from available literature. Only species with more than 10 records entered into the analysis. MaxEnt models were developed using the meteorological database for Chile (1961–1990) developed by the Department of Geophysics of the University of Chile [49]. Plant species distributions were modelled using the variables temperature (max., min. and average), precipitation (max., min. and total), altitude, slope and aspect.

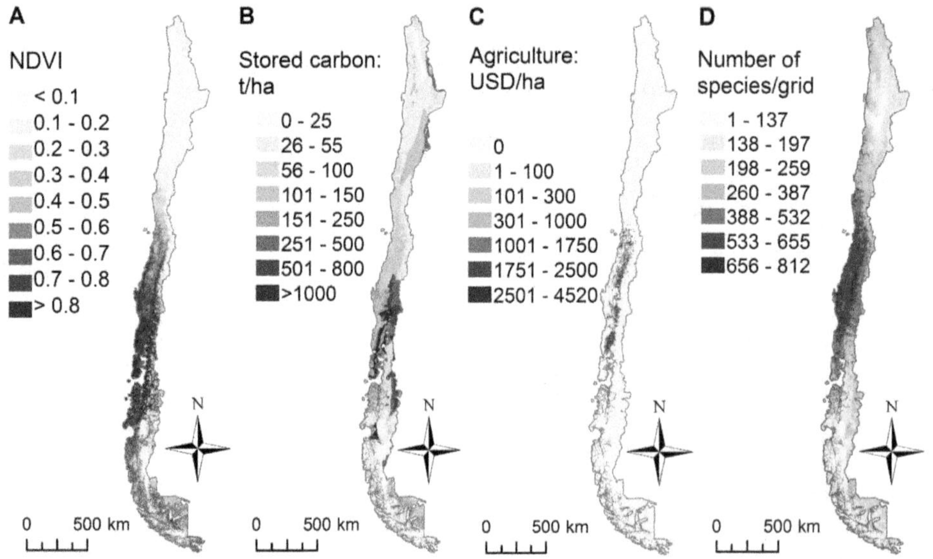

Figure 2. Ecosystem service and biodiversity distribution in Chile. Distribution of (A) net primary production, (B) carbon storage, (C) agricultural production and (D) biodiversity.

The area under the curve (AUC), a criterion used to assess fit in distribution models such as MaxEnt (see [50], was on average 0.978.

Each taxonomic group was analysed separately, using a 1 km × 1 km grid resolution.

Data Analyses

Quantification of ecosystem services and biodiversity within protected areas. A spatial overlap analysis was used to calculate the representation of each ecosystem service and of biodiversity within the Chilean protected area system. The three protection scenario covers were overlapped with each ecosystem service and biodiversity layer, and the spatially coincident coverage extracted. However, as ecosystem service layers and biodiversity were mapped in different units, their representation was calculated in distinct ways as follows:

- The units of carbon storage and agricultural production layers are the total amount of carbon (kg) and USD production, respectively, per pixel. Thus, the representation of these two ecosystem services was calculated as the sum of all those pixels that fell within protected areas.
- The PP captured was estimated from the average of the NDVI values of those pixels that fell within protected areas. As NDVI varies according to vegetation type, we calculated a weighted average in accordance with the proportion of total area of each vegetation type found within the protected area system. Thus, the resulting NDVI average is representative of the extents of different vegetation types within the protected area system. Vegetation types found within protected areas were Forest, Shrubland, Steppe, Wetland, Crop, Peatland and Bare areas (Table S5). For comparison purposes, a weighted NDVI average was also calculated for the entire country (Table S3).
- The representation of biodiversity was calculated as the summed proportion of species' ranges that fell within the protected area coverage. This was calculated per taxon and for all species together.

Assessing effectiveness of protection scenarios and protected area categories. We divided the percentage of each of the measures of ecosystem services and biodiversity contained within each scenario and protected area category by the percentage land area covered by that particular scenario and category [17]. This approach will indicate whether the amount of a given ecosystem service or biodiversity is more or less than would be expected for the protected coverage area. A value greater than one thus indicates that a particular scenario or category contains a disproportionately large amount of a specific ecosystem service or biodiversity group relative to the area that it covers. Our measure of biodiversity within each of the three protection scenario and seven management categories was the summed proportion of the ranges of all species. NDVI is an index and it is thus meaningless to use the same approach, so we calculated a weighted average of NDVI values that fall within each of the seven protected area categories in the same way as indicated above.

Results

The bulk of carbon storage, net primary production and agricultural production were located in the south-central zone of Chile (Fig. 2). Areas with the highest density of stored carbon were located between 36°–41° S, mainly concentrated in the eastern forest (Fig. 2B). Areas with the highest values of NDVI were located between 35°–43° S, particularly in the southern-central coastal range (Fig. 2A). Croplands were grouped in the central valley of Chile between 32°–41° S, and the highest production crops were in the region of Bernardo O'Higgins (32°–34° S) (Fig. 2C). The latitudinal region with the highest species richness was between 31°–40° S (Fig. 2D, Fig.S1).

The proportion of stored carbon varied from a low of 14.9% in Scenario 1, which increased to 19.0% and 19.9% in Scenario 2 and Scenario 3, respectively (Table 1). In the three scenarios carbon was underrepresented as would be expected for their coverage area (i.e. ratio less than 1, Table 1). An NDVI value of 0.38 was represented within the current protected area system (Scenario 1), 0.04 units higher than the whole country average (Table S3). This increased to 0.39 in Scenario 3 when private protected areas were added to the total coverage (Table 1). Only 0.2% of the total agricultural production was captured within the current protected area system (Table 1). However, this representation increased to 2.2% in Scenario 2 and 2.7% with Scenario 3 (Table 1). Again, none of the three representations was as much as would be expected for the area covered by the scenarios (Table 1).

The current protected areas capture 13.9%, 18.2%, 20.7% and 8.9% of mammal, bird, amphibian and plant ranges respectively. Amphibians was the only group well represented, with a ratio slightly higher than one (1.03). All species' representation levels increased substantially in Scenario 2 (i.e. when PSBC sites were included), capturing this time 18.9%, 23.0%, 27.1% and 14.7% of mammal, bird, amphibian and plant species' ranges respectively. In Scenario 2 only amphibian representation ratio was above one (1.07). When Private Protected Areas were included in Scenario 3, representation levels increased by approximately 1% for all species groups, with only that of amphibians' being well represented (Table 1). When species groups were assessed all together, its level of representation was 11.8%, 17.3% and 18.0% in scenario 1, 2 and 3 respectively. In all scenarios biodiversity was underrepresented (Table 1).

Carbon storage was well represented only by PSBC (4.08 times as much as would be expected for the area). The other protected area categories, except private protected areas, had values slightly below 1 (Table 2). All protected area categories together under-represented carbon stock with a value below 1 (Table 2). Private protected areas had the highest NDVI average value (0.54), followed by National Reserves (0.48), Nature Sanctuaries (0.47), and Ministry of Heritage lands (0.45). PSBC, National Parks and Natural Monuments had NDVI values below 0.4, Natural Monuments having the lowest average (Table 2). All categories had an NDVI value (0.39) slightly higher than the national average (0.38). Agricultural production was also well represented only by PSBC (2.09). This time the rest of the categories, including all categories together, had ratios below 1 (Table 2). Finally, biodiversity, the summed proportion of ranges of all species, was under-represented in all categories together, but was over-represented by Natural Monuments and PSBC categories (Table 2). When the species groups were assessed separately, amphibians were best represented by different protected area categories: Ministry of Heritage lands (2.36), National Parks (1.06), Nature Sanctuaries (2.92), and PSBC (5.46). Amphibians were the only group well represented by all categories together (Table S4). Mammals were well represented by Natural Monuments (1.36), Nature Sanctuaries (1.22), and PSBC (4.05). Birds and plants were over-represented only by PSBC (4.16 and 4.69 respectively), this being the most successful category in the representation of biodiversity (Table S4).

Discussion

Previous assessments of the effectiveness of the Chilean protected area system have focused exclusively on biodiversity [7,27,33,51,52]. Here, we document for the first time the effectiveness of the system in capturing both ecosystem services and biodiversity relative to its area of coverage (Table 1). We found that existing protected areas in Chile do not contain an unusually high proportion of the total national carbon storage (14.9%), agricultural production (0.2%) or species' ranges (11.8%). Also, PP representation (0.38) was low with regard to the maximum value range (−1 to +1) and with respect to the national forest cover PP (0.63, Table S3). This was, however, slightly higher than the national average (0.34). When the levels of representation were assessed relative to the percentage of land area covered by existing protected areas, we found that amphibians was the only conservation feature overrepresented. The underrepresentation by existing protected areas seems to result from the strong spatial bias of current protected areas toward southern Chile (Fig. 1-A), which raises three key points regarding the resulting representation of ecosystem services and because of their relatively small geographic ranges.

Table 1. Provision of ecosystem services and biodiversity under three protection scenarios.

	Scenario 1		Scenario 2		Scenario 3	
	% of total	ratio	% of total	ratio	% of total	ratio
PP[a]		0.38		0.38		0.39
Carbon	14.9	0.73	19.0	0.75	19.9	0.76
Agriculture	0.2	0.01	2.2	0.9	2.7	0.1
Biodiversity	11.8	0.59	17.3	0.68	18.0	0.69
Mammals	13.9	0.69	18.9	0.74	19.7	0.75
Birds	18.2	0.91	23.0	0.91	24.0	0.92
Amphibians	20.7	**1.03**	27.1	**1.07**	27.8	**1.06**
Plants	8.9	0.44	14.7	0.58	15.4	0.59

A ratio of >1 (in bold) indicates that an ecosystem service is over-represented compared with what would be expected for the area; values <1 indicate under-representation. The percentage of the total ecosystem services and biodiversity (summed proportion of ranges) in each of the three scenarios is given. Scenario 1: current protection system; Scenario 2: scenario 1+ suggested priority sites for biodiversity conservation; Scenario 3: scenario 2+ suggested private protected areas.

[a]Weighted average of NDVI pixels within protected areas (see methods).

Table 2. Provision of ecosystem services and biodiversity under seven protected area categories.

Protected area categories	Carbon storage	PP*	Agriculture	Biodiversity
Natural Monument	0.88	0.17	0.06	**1.13**
National Parks	0.75	0.33	0.006	0.55
National Reserve	0.97	0.48	0.02	0.59
Nature Sanctuary	0.75	0.47	0.12	0.88
Ministry of Heritage lands	0.84	0.45	0.0003	0.67
PSBC	**4.08**	0.39	**2.09**	**4.51**
Private PAs	0.37	0.54	0.22	0.21
All PA categories	0.76	0.39	0.10	0.69

A ratio of >1 (in bold) indicates that an ecosystem service is over-represented compared with what would be expected for the area; values <1 indicate under-representation. The percentage of the total amount of biodiversity (summed proportion of ranges) and other ecosystem services in Chile is given for each protected area category. PSBC: Priority sites for biodiversity conservation; PAs: protected areas; 'All PA categories' refers to the area covered by all seven categories.

*Weighted average of NDVI values that fall within each of the seven protected area categories (see methods for details).

First, as forest coverage, as well as protected areas, is concentrated in southern Chile, we would have expected a higher representation of carbon storage (Table 1). The c. 15% of carbon storage represented reflects, therefore, that southern protected areas are mainly protecting lands devoid of vegetation, such as ice and rock (Table S5). This is also reflected in the level of PP represented within protected areas, which despite being slightly higher than the national average, is closer to zero than to one, indicating a predominance of poorly vegetated lands within the Chilean protected area system.

Second, the underrepresented crop production found within the existing protected areas suggests that these are displacing or avoiding areas of agricultural production, which could reasonably be argued as reflecting their effectiveness. Chilean protected areas conserve a significant proportion of untransformed landscape, facing the challenge of displacing human activities beyond their boundaries. What is not clear however is whether the low agricultural activity within protected areas is due to the management strategy of conserving these lands intact, or because the spatial bias of protected areas towards southern regions renders them unsuitable for agriculture.

Third, while the largest coverage by protected areas is concentrated in the Austral Chilean zone (44°–56° S), our results show that the highest species richness areas are located in the central (28°–36° S) and south-central (36°–43° S) zones of Chile (Fig. S1, see [53]), which is reflected in the underrepresentation of biodiversity within the current protected area system. In fact, central and south-central zones include a hotspot of global biodiversity [54], which is characterized by a large number of endemic plants and vertebrate species [29,51]. However, amphibians is the only group overrepresented, likely because a relatively high proportion of their distribution ranges covers southern areas.

Adding PSBC and private protected areas to the current protected area system (i.e. Scenarios 2 and 3) enhances the representation of ecosystem services and total biodiversity (Table 1). This increase, however, was not sufficient to attain a representation higher than would be expected for their respective areas of coverage (Table 1). Interestingly, PSBC increase the representation of both carbon storage and crop production, suggesting that current croplands are located in rich organic carbon soil areas, an important proportion of the total calculated carbon storage (see methods). Given that PSBC represent multiple ecosystem services and biodiversity, a multi-goal management strategy will be required in order to optimize the supply of carbon storage and biodiversity as much as agriculture. Thus, conservation planning exercises that include both biodiversity and ecosystem services [55] may be required to improve the Chilean PA network.

When protected areas were evaluated based on their management objective categories, our results showed that no existing protected area category with statutory protection represents the level of ecosystem services and biodiversity one would have expected based on their coverage, except the Natural Monument category that over-represented biodiversity (Table 2). This over-representation is likely related to the small coverage of the Natural Monument category, the smallest of all categories (Tables S6). PSBC was the only category with no statutory protection that over-represented carbon storage (4.08 times as much as would be expected for their area), agricultural production (2.09 times) and biodiversity (4.51 times) (Table 2). Despite the under-representation of existing protected area categories together, our results show that National Parks, the strictest protection category (IUCN, Ia), represent a carbon ratio close to one (0.75), and a very low representation value for agriculture (0.006), which is also reflected in the proportion of land use cover within this category (Table S6). This is consistent with the strict and single land use management aim of this category, which is apparently mainly promoting carbon storage. By contrast, Nature Sanctuary sites, the more permissive category (IUCN, VI), represent exactly the same ratio of carbon storage as National Parks, but also 20 times more crop production, indicating the multi-use landscape nature of this category (Table S6). Only Natural Monument and PSBC categories over-represent biodiversity, indicating that these protected area categories are well placed with regard to species' range distributions (Table 2), however around 20% of amphibian are gap species, not yet represented in protected areas [29]. When species groups were assessed separately, amphibians was the only one overrepresented by all protected area categories (1.06), although the bird representation ratio was very close to one (0.92) (Table S4).

The existing Chilean protected area network does not perform well in representing all biodiversity groups together, but achieves a good representation of amphibians. Also, its provision of ecosystem services is poor. It is highly likely that this gap would need to be addressed principally by the expansion of the coverage of the protected area system. In this regard we suggest two measures. First, a re-evaluation of the already suggested new sites for the integrated protected area system, as these do not significantly increase ecosystem service representation. Second, a systematic assessment plan of current conservation management objectives and strategies, in order to enhance ecosystem service supply by existing protected areas.

Supporting Information

Figure S1 Distribution maps of species richness for four taxonomic groups at 1 km^2 grid resolution. a) Amphibians, b) Mammals, c) Birds and d) Plants.

Table S1 Protected area categories used in this study, and their associated management strategies defined under the International Union for Conservation of Nature (IUCN) regulatory framework.

Table S2 Summary of values used in calculating agricultural production (FAO, 2000).

Table S3 Average of NDVI values and coverage characteristics of different vegetation types in Chile.

Table S4 Biodiversity representation by species group in the five management categories and the suggested sites for the new integrated protection system (PSBC and Private protected areas). A ratio of >1 indicates that a particular group is over-represented relative to what would be expected for its area; values <1 indicate under-representation. 'All management strategies' refers to the area covered by all the seven categories. PA: Protected Area; PSBC: Priority sites for biodiversity conservation.

Table S5 Land use cover within the current Chilean protected areas system (Scenario 1).

Table S6 Land cover within each of the five management categories and the suggested sites for the new integrated protection system (PSBC and Private protected areas). PA: Protected Area; PSBC: Priority sites for biodiversity conservation.

Acknowledgments

The LP DAAC MODIS Data were obtained through the online Data Pool at the NASA Land Processes Distributed Active Archive Center (LP DAAC), USGS/Earth Resources Observation and Science (EROS) Center, Sioux Falls, South Dakota (https://lpdaac.usgs.gov/get_data).

Author Contributions

Conceived and designed the experiments: APD SC. Analyzed the data: APD SC. Wrote the paper: APD PAM KJG.

References

1. MA (2005) Ecosystems and human well-being: synthesis. Washington DC: Island Press. 137 p.
2. Daily GC (1997) Natures services: Societal dependence on natural ecosystems. Washington DC: Island Press. 412 p.
3. De Groot RS, Wilson MA, Boumans RMJ (2002) A typology for the classification, description and valuation of ecosystem functions, goods and services. Ecological Economics 41: 393–408.
4. Daily GC, Matson PA (2008) Ecosystem services: from theory to implementation. Proceedings of the National Academy of Sciences of the United States of America 105: 9455–9456.
5. Pimm SL (2001) Can we defy nature's end? Science 294: 788–788.
6. Perrings C, Naeem S, Ahrestani F, Bunker DE, Burkill P, et al. (2010) Ecosystem services for 2020. Science 330: 323–324.
7. Turner RK, Daily GC (2008) The ecosystem services framework and natural capital conservation. Environmental and Resource Economics 39: 25–35.
8. Brooks TM, Bakarr MI, Boucher T, Da Fonseca GAB, Hilton-Taylor C, et al. (2004) Coverage provided by the global protected-area system: is it enough? Bioscience 54: 1081–1091.
9. Bruner AG, Gullison RE, Rice RE, Da Fonseca GAB (2001) Effectiveness of parks in protecting tropical biodiversity. Science 291: 125–128.
10. Cantú-Salazar L, Gaston KJ (2010) Very large protected areas and their contribution to terrestrial biological conservation. Bioscience 60: 808–818.
11. Chape S, Harrison J, Spalding M, Lysenko I (2005) Measuring the extent and effectiveness of protected areas as an indicator for meeting global biodiversity targets. Philosophical Transactions of the Royal Society B-Biological Sciences 360: 443–455.
12. Klorvuttimontara S, McClean CJ, Hill JK (2011) Evaluating the effectiveness of protected areas for conserving tropical forest butterflies of Thailand. Biological Conservation 144: 2534–2540.
13. Rodrigues ASL, Andelman SJ, Bakarr MI, Boitani L, Brooks TM, et al. (2004) Effectiveness of the global protected area network in representing species diversity. Nature 428: 640–643.
14. Eigenbrod F, Anderson BJ, Armsworth PR, Heinemeyer A, Gillings S, et al. (2010) Representation of ecosystem services by tiered conservation strategies. Conservation Letters 3: 184–191.
15. Naidoo R, Balmford A, Costanza R, Fisher B, Green RE, et al. (2008) Global mapping of ecosystem services and conservation priorities. Proceedings of the National Academy of Sciences of the United States of America 105: 9495–9500.
16. Tang Z, Fang J, Sun J, Gaston KJ (2011) Effectiveness of protected areas in maintaining plant production. Plos One 6.
17. Eigenbrod F, Anderson BJ, Armsworth PR, Heinemeyer A, Jackson SF, et al. (2009) Ecosystem service benefits of contrasting conservation strategies in a human-dominated region. Proceedings of the Royal Society B-Biological Sciences 276: 2903–2911.
18. Anderson BJ, Armsworth PR, Eigenbrod F, Thomas CD, Gillings S, et al. (2009) Spatial covariance between biodiversity and other ecosystem service priorities. Journal of Applied Ecology 46: 888–896.
19. Eigenbrod F, Armsworth PR, Anderson BJ, Heinemeyer A, Gillings S, et al. (2010) The impact of proxy-based methods on mapping the distribution of ecosystem services. Journal of Applied Ecology 47: 377–385.
20. Mascia MB, Pailler S (2011) Protected area downgrading, downsizing, and degazettement (PADDD) and its conservation implications. Conservation Letters 4: 9–20.
21. Power AG (2010) Ecosystem services and agriculture: tradeoffs and synergies. Philosophical Transactions of the Royal Society B-Biological Sciences 365: 2959–2971.
22. Soares-Filho B, Moutinho P, Nepstad D, Anderson A, Rodrigues H, et al. (2010) Role of Brazilian Amazon protected areas in climate change mitigation. Proceedings of the National Academy of Sciences of the United States of America 107: 10821–10826.
23. Tallis H, Kareiva P, Marvier M, Chang A (2008) An ecosystem services framework to support both practical conservation and economic development. Proceedings of the National Academy of Sciences of the United States of America 105: 9457–9464.
24. Asmüessen MV, Simonetti JA (2007) Can a developing country like Chile invest in biodiversity conservation? Environmental Conservation 34: 183–185.
25. Pauchard A, Villarroel P (2002) Protected areas in Chile: history, current status, and challenges. Natural Areas Journal 22: 318–330.
26. Armesto JJ, Manuschevich D, Mora A, Smith-Ramirez C, Rozzi R, et al. (2010) From the Holocene to the Anthropocene: a historical framework for land cover change in southwestern South America in the past 15,000 years. Land Use Policy 27: 148–160.
27. Armesto JJ, Rozzi R, Smith-Ramirez C, Arroyo MTK (1998) Conservation targets in South American temperate forests. Science 282: 1271–1272.
28. Squeo FA, Estevez RA, Stoll A, Gaymer CF, Letelier L, et al. (2012) Towards the creation of an integrated system of protected areas in Chile: achievements and challenges. Plant Ecology and Diversity 5: 233–243.
29. Tognelli MF, de Arellano PIR, Marquet PA (2008) How well do the existing and proposed reserve networks represent vertebrate species in Chile? Diversity and Distributions 14: 148–158.
30. Tognelli MF, Fernandez M, Marquet PA (2009) Assessing the performance of the existing and proposed network of marine protected areas to conserve marine biodiversity in Chile. Biological Conservation 142: 3147–3153.
31. CONAMA-PNUD (2006) Creación de un Sistema Nacional Integral de Áreas Protegidas. Available: http://www.pnud.cl/proyectos/fichas/areas-protegidas.asp. Accessed 2011 Dec. Santiago, Chile.
32. IUCN (1994) Guidelines for protected area management categories. Cambridge: IUCN. 106 p.
33. Pliscoff P, Fuentes-Castillo T (2011) Representativeness of terrestrial ecosystems in Chile's protected area system. Environmental Conservation 38: 303–311.
34. Aalde H, Gonzalez P, Gytarsky M, Krug T, Kurz WA, et al. (2006) Forest land. In: IPCC Guidelines for national greenhouse gas inventories: Agriculture, forestry and other land use: IPCC. pp. Volume 4.
35. FAO (2000) Global ecological zoning for the global forest resources assessment 2000.
36. FAO/IIASA/ISRIC/ISSCAS/JRC (2012) Harmonized World Soil Database (version 1.2). Rome, Italy and IIASA, Laxenburg, Austria.
37. Paruelo JM, Epstein HE, Lauenroth WK, Burke IC (1997) ANPP estimates from NDVI for the central grassland region of the United States. Ecology 78: 953–958.
38. Pettorelli N, Chauvenet ALM, Duffy JP, Cornforth WA, Meillere A, et al. (2012) Tracking the effect of climate change on ecosystem functioning using protected areas: Africa as a case study. Ecological Indicators 20: 269–276.
39. Running SW, Thornton PE, Ramakrishna N, Glassy JM (2000) Global Terrestrial Gross and Net Primary Productivity from the Earth Observing System. In: Sala OE, Jackson R, Mooney H, Howarth R, editors. Methods in Ecosystem Science. New York: Springer-Verlag. 44–57.
40. Turner W, Spector S, Gardiner N, Fladeland M, Sterling E, et al. (2003) Remote sensing for biodiversity science and conservation. Trends in Ecology and Evolution 18: 306–314.
41. Garbulsky MF, Paruelo JM (2004) Remote sensing of protected areas to derive baseline vegetation functioning characteristics. Journal of Vegetation Science 15: 711–720.
42. Roldan M, Carminati A, Biganzoli F, Paruelo JM (2010) Are private refuges effective for conserving ecosystem properties? Ecologia Austral 20: 185–199.
43. Di Bellat CM, Paruelo JM, Becerra JE, Bacour C, Baret F (2004) Effect of senescent leaves on NDVI-based estimates of fAPAR: experimental and modelling evidences. International Journal of Remote Sensing 25: 5415–5427.
44. Monfreda C, Ramankutty N, Foley JA (2008) Farming the planet: 2. Geographic distribution of crop areas, yields, physiological types, and net primary production in the year 2000. Global Biogeochemical Cycles 22: GB1022.
45. European Commission JRC (2003) Global land cover 2000 database. http://gem.jrc.ec.europa.eu/products/glc2000/glc2000.php. Accessed 2011 Dec.
46. Ramankutty N, Evan AT, Monfreda C, Foley JA (2008) Farming the planet: 1. Geographic distribution of global agricultural lands in the year 2000. Global Biogeochemical Cycles 22: GB1003.
47. IUCN (2012) IUCN Red List of Threatened Species (Version 2012.1). Available: http://www.iucnredlist.org. Accessed 2012 Feb.
48. Marquet PA AM, Labra F, Abades S, Cavieres L, et al. (2011) Estudio de vulnerabilidad de la biodiversidad terrestre en la eco-región mediterránea, a nivel de ecosistemas y especies, y medidas de adaptación frente a escenarios de cambio climático. Ministerio de Medio Ambiente. Santiago, Chile.

49. DFG-CONAMA (2006) Estudio de la variabilidad climática en Chile para el siglo XXI. Comisión Nacional del Medio Ambiente. Santiago, Chile.
50. Elith J, Graham CH, Anderson RP, Dudik M, Ferrier S, et al. (2006) Novel methods improve prediction of species' distributions from occurrence data. Ecography 29: 129–151.
51. Arroyo MTK, Cavieres L (1997) The Mediterranean typeclimate flora of central Chile. What do we know and how can we assure its protection? Noticiero de Biologia 5: 48–56.
52. Luebert F, Becerra P (1998) Representatividad vegetacional del Sistema Nacional de Áreas Silvestres Protegidas del Estado (SNASPE) en Chile.
53. Samaniego H, Marquet PA (2009) Mammal and butterfly species richness in Chile: taxonomic covariation and history. Revista Chilena De Historia Natural 82: 135–151.
54. Myers N, Mittermeier RA, Mittermeier CG, da Fonseca GAB, Kent J (2000) Biodiversity hotspots for conservation priorities. Nature 403: 853–858.
55. Chan KMA, Shaw MR, Cameron DR, Underwood EC, Daily GC (2006) Conservation planning for ecosystem services. Plos Biology 4: 2138–2152.

Do We Produce Enough Fruits and Vegetables to Meet Global Health Need?

Karen R. Siegel[1,2]*, Mohammed K. Ali[2], Adithi Srinivasiah[3], Rachel A. Nugent[4], K. M. Venkat Narayan[1,2]

1 Nutrition and Health Sciences, Laney Graduate School, Emory University, Atlanta, Georgia, United States of America, 2 Hubert Department of Global Health, Emory University, Atlanta, Georgia, United States of America, 3 Emory College, Emory University, Atlanta, Georgia, United States of America, 4 Department of Global Health, University of Washington, Seattle, Washington, United States of America

Abstract

Background: Low fruit and vegetable (FV) intake is a leading risk factor for chronic disease globally, but much of the world's population does not consume the recommended servings of FV daily. It remains unknown whether global supply of FV is sufficient to meet current and growing population needs. We sought to determine whether supply of FV is sufficient to meet current and growing population needs, globally and in individual countries.

Methods and Findings: We used global data on agricultural production and population size to compare supply of FV in 2009 with population need, globally and in individual countries. We found that the global supply of FV falls, on average, 22% short of population need according to nutrition recommendations (supply:need ratio: 0.78 [Range: 0.05–2.01]). This ratio varies widely by country income level, with a median supply:need ratio of 0.42 and 1.02 in low-income and high-income countries, respectively. A sensitivity analysis accounting for need-side food wastage showed similar insufficiency, to a slightly greater extent (global supply:need ratio: 0.66, varying from 0.37 [low-income countries] to 0.77 [high-income countries]). Using agricultural production and population projections, we also estimated supply and need for FV for 2025 and 2050. Assuming medium fertility and projected growth in agricultural production, the global supply:need ratio for FV increases slightly to 0.81 by 2025 and to 0.88 by 2050, with similar patterns seen across country income levels. In a sensitivity analysis assuming no change from current levels of FV production, the global supply:need ratio for FV decreases to 0.66 by 2025 and to 0.57 by 2050.

Conclusion: The global nutrition and agricultural communities need to find innovative ways to increase FV production and consumption to meet population health needs, particularly in low-income countries.

Editor: Wagner L. Araujo, Universidade Federal de Vicosa, Brazil

Funding: There are no current funding sources for this study.

Competing Interests: The authors have declared that no competing interests exist.

* Email: krsiege@emory.edu

Introduction

Low fruit and vegetable (FV) intake is a leading risk factor for death and disability globally, estimated to contribute to approximately 16.0 million disability-adjusted life years and 1.7 million deaths worldwide annually [1]. According to a World Health Organization report, current global dietary guidelines recommend that individuals consume at least 5 servings of FV daily [2]. Recent cross-country evidence supports this recommendation, showing a strong dose-response relationship between higher FV consumption and lower all-cause mortality [3] as well as lower risk of major chronic diseases such as cardiovascular disease, diabetes, and certain cancers, which impact every region of the world [4–6].

Much of the world's population, however, does not consume the recommended five servings of FV daily. Data from 52 mainly low- and middle-income countries participating in the 2002–2003 World Health Survey reported that, overall, 77.6% of men and 78.4% of women surveyed consumed less than the recommended five daily servings of FV. The survey also showed that FV consumption patterns vary around the world, but lower-than-recommended reported consumption is common in high, middle, and low-income countries. For example, in a recent report, poor dietary habits, which includes low FV consumption, was *the* leading risk factor in the United States (U.S.), accounting for 26% of all deaths and 14% of all disability [7], and increasing individual FV consumption to up to 600 grams per day (slightly more than 5 servings per day) could reduce the total worldwide burden of disease by 1.8%, and reduce the burden of ischemic heart disease and ischemic stroke by 31% and 19% respectively [2].

Despite a wealth of research on behavioral determinants of FV, it remains unknown whether global production and supply of FV is actually sufficient to meet population needs. We used global population and agriculture databases to compare the global supply of (“supply”) with recommended dietary intake (implied “demand”, hereafter referred to as “need”) globally and in individual countries. Using agricultural production and population projections data, we also project supply and need for FV for 2025 and 2050.

Methods

Data Sources

We used three main data sources for our analysis: (1) Food and Agricultural Organization (FAO) 2009 Food Balance Sheets [8], (2) age-specific FV intake recommendations for individuals [2], and (3) the United Nations (UN) World Population Prospects: The 2012 Revision [9].

The FAO 2009 Food Balance Sheets (the most recent year for which these data were available) report FV (excluding wine) supply by individual country for over 175 countries. These data are calculated by taking into account production, imports and exports, and food losses (through storage, transport, and processing; feed to livestock; or use as seeds and non-dietary purposes). The data reflect "formal" food production, and do not capture FV production from subsistence farming and production, which may not enter formal economies. For the FAO Food Balance Sheets, this estimated national food supply is divided by population size estimates to derive the reported per capita supply of FV (in kg/person/year).

For FV recommendations, we used a World Health Organization (WHO) report on the quantitative comparison of different health risks worldwide [2]. The report cited previously calculated and validated estimates for the average annual weight of the 5 recommended servings of FV per day: 330 grams per day for individuals aged 0–4 years, 480 grams per day for individuals aged 5–14 years, and 600 grams per day for all individuals aged 15 years and older. We converted these data into kilograms.

The UN World Population Prospects: 2012 Revision (the most recent version) provides country-level population estimates, in terms of the total population size as well as the proportion of each country's population by age. Calculations are done yearly using data classified by broad age groups (0–14 years, 15+ years) and for five-year periods (the latest years being 2005 and 2010) using data classified by more specific age groups, including 0–4 years, 5–14 years, and 15 years and older. To align our population estimates with age-specific FV recommendations, we used population estimates from 2010. This data source also provides population projections based on different scenarios for changing fertility levels for the period 2010–2100 for individual countries and globally.

Data Analysis

To calculate "supply" (in kg/year), we multiplied the FAO per-capita estimates by total population estimates for each country from the UN. The equation for supply is:

$$Supply = \frac{FV(kg)}{person} * population$$

To calculate "need" (assuming all individuals are able to meet their daily recommended intake of FV – "perfect need"), we multiplied the UN's age-specific population estimates by recommendations for FV servings per day for the same age-specific groups. Total country-specific population need (in kg/year) was then calculated by summing the recommended FV weights for all three age categories. The equation for "need" is:

$$\begin{aligned} Need = & \left[popn(0-4\ yrs) * \frac{0.33\ kg}{persons(0-4\ yrs)}\right] \\ & + \left[popn(5-14\ yrs) * \frac{0.48\ kg}{persons(5-14\ yrs)}\right] \\ & + \left[popn(15+\ yrs) * \frac{0.60\ kg}{persons(15+yrs)}\right] \end{aligned}$$

Finally, we calculated a supply:need ratio by dividing supply by need, both expressed in kg/year, where a value greater than 1.0 signifies surplus, a value of 1.0 implies balance, and less than 1.0 signifies deficit. Supply, need, and supply:need ratios were calculated for each individual country and globally. We also calculated averages of these supply, need, and supply:need ratio indicators across varying country income levels, defined according to World Bank categories: low-income economies (per capita Gross Domestic Product [GDP] of $1,025 or less), lower-middle-income economies (per capita GDP of $1,026 to $4,035), upper-middle-income economies (per capita GDP of $4,036 to $12,475), and high-income economies (per capita GDP of $12,476 or more).

For the projections for 2025 and 2050, we calculated changes in production ("supply") using agricultural production growth rates to 2030 (1.6% for developing and 0.7% for developed countries) and 2050 (0.9% for developing and 0.3% for developed countries) as estimated by the FAO [10]. Similar to our calculations for current need, we calculated projected need by multiplying age-specific population projections for 2025 and 2050 by recommendations for FV servings per day for the same age-specific groups and summing across all three groups. For this projections analysis, we assumed a medium variant fertility scenario (2–3 children per woman).

All calculations were performed in Excel and data analysis was performed using Statistical Analysis Software (SAS) version 9.3. We used ArcMAP to illustrate the data geographically.

Sensitivity Analyses

To account for need-side food wastage at the household/individual level, we performed a sensitivity analysis to adjust these estimates to account for wastage of 33% in high-income regions/countries and 15% in low- to middle-income regions/countries [11]. For the projections, we also performed a sensitivity analysis in order to account for "best-case" (low fertility, or <2.1 children per woman) and "worst-case" (high fertility, or >5 children per woman) scenarios [9]. In addition to the main projections analysis, we also performed a sensitivity analysis assuming current levels of agricultural production.

Results

Table 1 shows descriptive statistics, overall and by country income level, for all countries for which all data were available (n = 170). Overall, the global supply (not including subsistence production that may not enter formal economies) of available FV falls 22% short of population's need according to nutritional recommendations, and as much as 95% short in some countries (overall supply:need ratio: 0.78 [range: 0.05–2.01]). This ratio varies widely by country income level, with a median supply:need ratio of 0.42 in low-income countries and a median supply:need ratio of 1.02 in high-income countries (Table 1). In a sensitivity analysis in which we accounted for need-side food wastage, similarly insufficient FV supplies were noted, to a slightly greater extent. The global supply:need ratio was 0.66 when need-side

Table 1. Descriptive Statistics of Fruit and Vegetable Supply, Need, and Supply: Need Ratio, Overall and by Country Income Level.

	n	**Supply**	**Need**	**Supply:Need Ratio**
Full Sample, all countries	170	1.15 (0.01–524.25)	1.90 (0.02–282.50)	0.78 (0.05–2.01)
Low Income	34	0.97 (0.05–7.50)	2.36 (0.13–30.18)	0.42 (0.05–0.99)
Lower-middle Income	43	1.01 (0.01–142.51)	1.49 (0.02–241.62)	0.63 (0.19–1.72)
Upper-middle Income	50	1.52 (0.01–524.25)	1.71 (0.02–282.50)	0.87 (0.24–2.01)
High Income	43	1.60 (0.04–71.63)	1.64 (0.05–64.59)	1.02 (0.55–1.86)

Notes: All numbers provided as median (range). Supply and Need are reported in billions of kilograms of fruits and vegetables. Country Income Level defined according to World Bank categories: Low-income economies ($1,025 or less), Lower-middle-income economies ($1,026 to $4,035), Upper-middle-income economies ($4,036 to $12,475), High-income economies ($12,476 or more).

wastage was accounted for, and this varied from 0.37 (low-income countries) to 0.77 (high-income countries) (see Table S1 for results by country and Table S2 for results across country income level).

The supply:need ratio also varies widely by geographical region. The highest ratios of greater than 1.0 (indicating more than sufficient supply to meet the population's needs) are seen in the Mediterranean/North African countries of Montenegro (supply:-need ratio 2.01), Greece (1.86), Turkey (1.78), Egypt (1.72), Libya (1.67), Tunisia (1.52), Italy (1.50), and Portugal (1.48); Middle Eastern countries of Iran (1.78), Israel (1.56), and Lebanon (1.46); Caribbean countries of Bahamas (1.61) and Belize (1.50); Albania (1.59); and China (1.86). The countries with the greatest shortage, where need is far greater than supply, are primarily African countries such as Eritrea (0.05), Chad (0.09), Burkina Faso (0.10), Mozambique (0.12), Ethiopia (0.12).

Table 2 shows projected supply, need and supply:need ratios overall and by country income level, for all countries for which all data was available (n = 169). Assuming medium fertility (2–3 children per woman) and projected agricultural production growth rates, the global supply:need ratio for FV increases slightly to 0.81 by 2025 and to 0.88 by 2050. As with current, the projected supply:need ratio in 2025 and 2050 varies by country income level. The lowest ratio is seen in low-income countries, where it dips to 0.30 in 2050, assuming medium fertility. The projected supply:need ratio is higher in high income countries, where it ranges from an estimated 0.98 to 1.21. In a sensitivity analysis using current levels of FV production (ie, assuming no increase in production), the global supply:need ratio for FV decreases to 0.66 by 2025 and to 0.57 by 2050 assuming medium fertility (2–3 children per woman). As with current, the projected supply:need ratio in 2025 and 2050 varies by country income level, with the lowest ratio of 0.18 in low-income countries by 2050 and the highest ratio of 0.99 in high income countries (see Table S3).

Figure 1 illustrates current and projected supply:need ratios, highlighting the growing gap between supply and need in low-income countries over time.

Discussion

Within the formal agricultural sector, there is an estimated 22% supply gap in meeting current need for FV (34% when considering food wastage at the household/individual level), and this varies from 58% to 13% across low- and upper-middle income countries. High income countries appear to have sufficient supply (supply: need ratio is 1.02). Furthermore, these gaps between high/middle-income and low-income countries will worsen with time. Assuming medium fertility and projected increases in production of FV, the global supply:need ratio for FV increases slightly to 0.81 by 2025 and to 0.88 by 2050, but divergence occurs whereby we estimated a supply gap of 70% and 65% in low-income countries by 2025 and 2050, respectively, while middle- and high-income countries approach a supply:need of 1.0, implying balance of supply and need. Without the projected increase in FV production, however, the global supply:need ratio could decrease to 0.66 by 2025 and to 0.57 by 2050, dipping as low as 0.18 in low-income countries.

There may be several reasons for these findings. Supply-side factors include subsidies and distribution systems for supply, and international trade for addressing imbalances in supply:need ratios across countries and country-income levels [12]. Many countries provide producer-end subsidies for grain crops and meat/dairy, incentivizing farmers to grow these items while dis-incentivizing FV production. In the U.S., the commodity crops receiving the largest amount of agricultural subsidies are grains, livestock, and dairy and under current agricultural policy, farmers are penalized for growing "specialty crops" (FV) if they have received federal farm payments to grow other crops [13,14]. As a result, grains, meat, and dairy are abundant [15], the supply of FV, at least in the US, is insufficient to meet population needs [16]. In low-income countries, where we found FV need to be greatest, the lack of adequate distribution systems may lead to supply-side wastage and disincentives for their production. This is an issue particularly in warm climates like India and Africa, where FV are prone to spoiling before reaching their market destinations [17].

In particular, international trade (and climates ideal for growing FV) could help explain the differences in findings across country-income groups and geographical regions. International trade in FV, which since the 1980s has expanded more rapidly than other agricultural commodities and was 17% of total agricultural trade in 2001, is also an important consideration for increasing supply of FV, particularly in countries where production may be high but supply low due to exports [18]. Climates ideal for growing FV is also a very important supply-side factor when considering FV production. As noted in the results section, there appear to be varying levels of agronomical potentials of countries located in different geographical regions, as highlighted by the large geographical variations in the supply:need ratio, with high ratios seen in many Mediterranean countries. For example, it is known that Mediterranean countries are great producers of fruits for the fresh market due to climatic conditions – drip irrigation combined with dry summers is a perfect scenario for producing high quality crops (although a substantial proportion of this production is exported to other countries).

On the need side of the equation, population size – and relatively large projected increases, particularly in certain low-income countries – helps to explain the large and growing gaps between supply and need in these countries. The projections data show that, assuming an estimated increase in FV production, the

Table 2. Projected Need and Supply:Need Ratios, Overall and by Country Income Level.

		2025			2050		
	n	**Supply**	**Need**	**Supply:Need Ratio**	**Supply**	**Need**	**Supply:Need Ratio**
Full Sample, all countries	169	1.45 (0.02–675.83)			1.79 (0.02–875.25)		
High fertility			2.21 (0.02–310.96)	0.79 (0.04–2.52)		2.74 (0.02–380.34)	0.78 (0.03–3.25)
Medium fertility			2.16 (0.02–302.40)	0.81 (0.04–2.59)		2.48 (0.02–335.52)	0.88 (0.03–3.69)
Low fertility			2.10 (0.02–293.83)	0.84 (0.04–2.67)		2.23 (0.02–293.93)	1.00 (0.03–4.21)
Low Income	34	1.20 (0.07–9.67)			1.55 (0.09–12.52)		
High fertility			3.65 (0.19–37.53)	0.34 (0.04–1.15)		5.89 (0.33–48.38)	0.26 (0.03–1.30)
Medium fertility			3.55 (0.19–36.28)	0.35 (0.04–1.18)		5.28 (0.30–42.11)	0.30 (0.03–1.47)
Low fertility			3.45 (0.18–35.03)	0.36 (0.04–1.22)		4.70 (0.27–36.43)	0.33 (0.03–1.68)
Lower-middle Income	42	1.32 (0.04–183.72)			1.71 (0.06–237.93)		
High fertility			2.35 (0.04–297.58)	0.62 (0.16–1.90)		3.49 (0.05–380.34)	0.58 (0.09–2.34)
Medium fertility			2.28 (0.04–288.77)	0.64 (0.16–1.95)		3.08 (0.05–335.52)	0.66 (0.10–2.65)
Low fertility			2.21 (0.04–279.95)	0.66 (0.17–2.02)		2.70 (0.04–293.93)	0.75 (0.11–3.01)
Upper-middle Income	50	1.96 (0.02–675.83)			2.54 (0.02–875.25)		
High fertility			1.85 (0.02–310.96)	0.94 (0.19–2.52)		1.86 (0.02–327.36)	1.03 (0.12–3.25)
Medium fertility			1.79 (0.02–302.40)	0.97 (0.19–2.59)		1.64 (0.02–290.93)	1.16 (0.14–3.69)
Low fertility			1.74 (0.02–293.83)	1.00 (0.20–2.67)		1.44 (0.02–257.35)	1.33 (0.15–4.21)
High Income	43	1.79 (0.04–80.09)			1.97 (0.04–88.05)		
High fertility			1.91 (0.06–74.60)	1.04 (0.59–2.04)		2.17 (0.07–92.40)	0.98 (0.52–2.15)
Medium fertility			1.86 (0.06–72.69)	1.06 (0.61–2.09)		1.96 (0.07–83.32)	1.08 (0.58–2.38)
Low fertility			1.81 (0.06–70.79)	1.09 (0.63–2.14)		1.76 (0.06–74.67)	1.21 (0.65–2.65)

Notes: All numbers provided as median (range). Need is reported in billions of kilograms of fruits and vegetables. Country Income Level defined according to World Bank categories: Low-income economies ($1,025 or less), Lower-middle-income economies ($1,026 to $4,035), Upper-middle-income economies ($4,036 to $12,475), High-income economies ($12,476 or more). Fertility is defined according to the United Nations World Population Prospects, 2012 Revision: high fertility (more than 5 children per woman), medium fertility (2–3 children per woman), and low fertility (less than 2.1 children per woman).

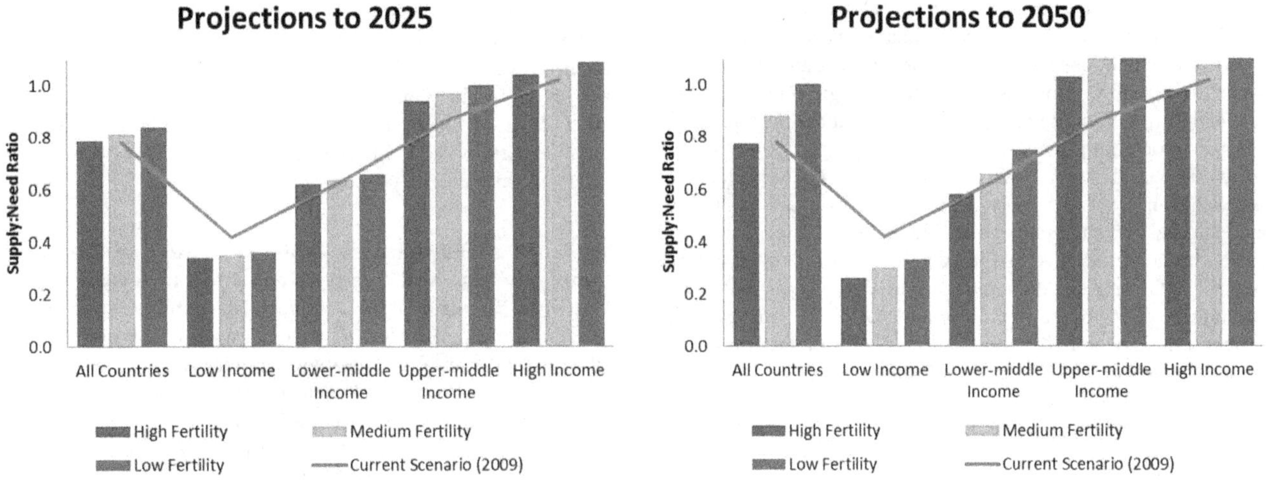

Figure 1. Projected Supply: Need Ratio, 2025 and 2050. Notes: Country Income Level defined according to World Bank categories: Low-income economies ($1,025 or less), Lower-middle-income economies ($1,026 to $4,035), Upper-middle-income economies ($4,036 to $12,475), High-income economies ($12,476 or more). Fertility is defined according to the United Nations World Population Prospects, 2012 Revision: high fertility (5 or more children per woman), medium fertility (2–3 children per woman), and low fertility (less than 2.1 children per woman).

supply:need ratio narrows on a global scale, but that it widens to a considerable extent in low-income countries, primarily as a reflection of higher fertility in these countries and agricultural production growth rates that cannot keep up with population growth. The ability to produce enough FV to meet the needs of large and growing populations, coupled with the supply-side limitations mentioned above, are of particular concern for these countries. In the 18th century, Malthus projected that human population growth would outpace expansion in food production. Since then, with the help of technological advances spurred by the Green Revolution, production and subsequent supply of carbohydrates and grains has increased to meet global population needs. Our projections analysis suggests that high-income countries may be making strides towards increasing production and subsequent supply of FV to meet their population's needs, but that the same cannot be said for the low-income countries, at least within the formal agricultural economy, where the gap in supply not taking subsistence farming into account could widen to 65% by 2050 if not addressed. Of greater concern, if projected increases in agricultural production of FV do not manifest, by 2025 and 2050 high- and low-income countries alike may not able to meet their population's needs for FV.

While ecological data suggests that food availability can influence food consumption patterns and in turn, cardiometabolic health outcomes like diabetes [19,20], to date there has been a relatively limited focus on production and supply of FV. Researchers at America's Farmland Trust investigated supply of FV in the United States (U.S.) alone; they concluded that an estimated 13 million more acres of farmland would be needed to produce a sufficient supply for the U.S. population [21]. Our analysis builds upon these results. The first study to incorporate empirical country-level data and age-specific recommendations for FV consumption to examine global and country-specific FV supply (in the formal sector) as it compares to need, our study highlights inadequate supply of FV as it compares to the population's nutritional needs, from the perspective of preventing chronic diseases, which currently place enormous burdens on countries around the world and are largely preventable through healthy diet and higher FV consumption [2,7].

These findings must be contextualized by limitations to our analysis. First, the data used were macro-level indicators collected at the country-level and may be prone to either over- or under-estimation. The data do not account for how much people actually access FV in various countries nor the quality and diversity of FV consumption, including how these FV are consumed (raw, cooked, or processed FV have different nutrient bioavailability), nor how much *individuals* actually consume. For example, many Mediterranean and Caribbean countries, which were found to have high supply:need ratios, are great citrus producers, but in the latter fruits are processed (for juice) and not sold on the fresh market. Additionally, every fruit and vegetable does not have the same macro- and micro-nutrient content, and even the same fruit or vegetable grown in a different climate or soil may have differing amounts of macro- and micro-nutrients. Additionally, there may also be differences in the quality and validity of the data in high- versus low-income countries. However, the FAO Food Balance Sheets are the most commonly used source of food availability information at the national level, providing standardized estimates of the average amount of food available per person on a daily basis and a useful tool for international comparisons [22]. Second, our analysis is at the country level, and therefore does not take into account urban/rural differences in supply that may result from challenges in distribution (for example, transporting FV from the farm to urban areas. This may be a particular issue in resource-poor settings, where distributional infrastructure may be lacking. Further analyses could investigate these issues, analyzing potential heterogeneity of supply and need within countries and in urban versus rural settings.

A third limitation is that our analysis does not capture local food economies (ie, subsistence farming and food production) in individual countries. That is, it does not take into account the production of FV that may exist outside of the formal agricultural sector (i.e., home gardens), which may vary widely across countries. This may be an additional area of future research. For example, researchers could utilize the powerful technologies of Google Earth to look within countries, at the regional, city, district, or even household level, at the presence or absence of informal community or household gardens. Lastly, our analysis does not incorporate additional economic indicators such as the costs of production or the resulting prices of FV. Our results suggest that insufficient supply exists relative to population needs under current production conditions. We have not taken into account the potential for supply to increase due to technological improvements and supportive government policies. Both those factors could lower FV prices and increase consumption.

Our study adds unique value by underlining the importance of increasing supply of FV and sets the stage for further analyses to delve further into the policy levers for increasing production and supply. In particular, investigating the supply of FV resulting from subsistence farming could augment our analysis. At the same time, continuing efforts to improve demand for FV – for example, through public health education and health promotion programs, proposing taxes on foods of low nutritional value (e.g., soda, high-fat foods) or subsidies on foods of high nutrition value (e.g., FV), improved food labeling, and stricter controls on the marketing of foods [23–27] – is equally important. Without an accompanying increase in supply, however, these efforts may have limited reach. It is hoped that our straightforward analysis, highlighting inadequate formal supply of FV in the context of perfect need (assuming all individuals are able to meet their daily recommended intake of FV), may provide value by offering an understanding of the current and future global disconnect between nutritional recommendations and supply of FV, and guide conversations and future investigations to consider appropriate policy responses. The triumph of grains production over the doom and gloom forecast of Malthus is a major testament to the technological and organizational success of food production and distribution worldwide that has accompanied industrialization and modern development. The current state of affairs presents a challenge to the global nutrition and agricultural communities to increase FV production in the same way, especially in low-income countries. Change is possible.

Supporting Information

Table S1 List of Countries and Their Respective Supply, Need, and Supply:Need Ratios. Notes: All numbers provided as median (range). Need is reported in billions of kilograms of fruits and vegetables. Country Income Level defined according to World Bank categories: Low-income economies ($1,025 or less), Lower-middle-income economies ($1,026 to $4,035), Upper-middle-income economies ($4,036 to $12,475), High-income economies ($12,476 or more). Fertility is defined according to the United Nations World Population Prospects, 2012 Revision: high fertility (more than 5 children per woman), medium fertility (2–3 children per woman), and low fertility (less than 2.1 children per woman).

Table S2 Sensitivity Analysis of Fruit and Vegetable Supply, Need, and Supply:Need Ratio, Overall and by Country Income Level. Notes: All numbers provided as median (range). Supply and Need are reported in billions of kilograms of fruits and vegetables. Country Income Level defined according to World Bank categories: Low-income economies ($1,025 or less), Lower-middle-income economies ($1,026 to $4,035), Upper-middle-income economies ($4,036 to $12,475), High-income economies ($12,476 or more).

Table S3 Sensitivity Analysis of Projected Need and Supply:Need Ratios (Assuming Current Levels of Agricultural Production), Overall and by Country Income Level. Notes: All numbers provided as median (range). Need is reported in billions of kilograms of fruits and vegetables. Country Income Level defined according to World Bank categories: Low-income economies ($1,025 or less), Lower-middle-income economies ($1,026 to $4,035), Upper-middle-income economies ($4,036 to $12,475), High-income economies ($12,476 or more). Fertility is defined according to the United Nations World Population Prospects, 2012 Revision: high fertility (more than 5 children per woman), medium fertility (2–3 children per woman), and low fertility (less than 2.1 children per woman).

Author Contributions

Conceived and designed the experiments: KRS KMN. Performed the experiments: KRS AS. Analyzed the data: KRS MKA KMN RAN. Contributed reagents/materials/analysis tools: KRS AS. Wrote the paper: KRS.

References

1. Lim SS, Vos T, Flaxman AD, Danaei G, Shibuya K, et al. (2012) A comparative risk assessment of burden of disease and injury attributable to 67 risk factors and risk factor clusters in 21 regions, 1990?2010: a systematic analysis for the Global Burden of Disease Study 2010. The Lancet 380: 2224–2260.
2. Lock K, Pomerleau J, Causer L, McKee M (2004) Low Fruit and Vegetable Consumption. In: Ezzati M, Lopez AD, Rodgers A, Murray CJL, editors. Comparative Quantification of Health Risks: Global and Regional Burden of Diseases Attributable to Selected Major Risk Factors. Geneva: World Health Organization.
3. Bellavia A, Larsson SC, Bottai M, Wolk A, Orsini N (2013) Fruit and vegetable consumption and all-cause mortality: a dose-response analysis. Am J Clin Nutr 98: 454–459.
4. FAO/WHO (2003) Diet, nutrition and the prevention of Chronic Diseases. Geneva: Food and Agricultural Organization, World Health Organization.
5. Hung HC, Joshipura KJ, Jiang R, Hu FB, Hunter D, et al. (2004) Fruit and vegetable intake and risk of major chronic disease. J Natl Cancer Inst 96: 1577–1584.
6. Dauchet L, Amouyel P, Hercberg S, Dallongeville J (2006) Fruit and vegetable consumption and risk of coronary heart disease: a meta-analysis of cohort studies. J Nutr 136: 2588–2593.
7. (2013) The state of US health, 1990–2010: burden of diseases, injuries, and risk factors. JAMA 310: 591–608.
8. FAO (2009) Food and Agricultural Organization Food Balance Sheets. FAO.
9. UNDESA (2012) World Population Prospects: The 2012 Revision. United Nations, Department of Economic and Social Affairs: Population Division, Population Estimates and Projections Section.
10. Alexandratos N, Bruinsma J (2012) World Agriculture Towards 2030/2050: The 2012 Revision. Rome: Global Perspective Studies Team, FAO Agricultural Development Economics Division.
11. Joffe M, Robertson A (2001) The potential contribution of increased vegetable and fruit consumption to health gain in the European Union. Public Health Nutr 4: 893–901.
12. Nugent R (2011) Bringing Agriculture to the Table: How Agriculture and Food Can Play a Role in Preventing Chronic Disease. Chicago, IL: The Chicago Council on Global Affairs.
13. Jackson RJ, Minjares R, Naumoff KS, Shrimali BP, Martin LK (2009) Agriculture Policy Is Health Polciy. Journal of Hunger & Environmental Nutrition 4: 393–408.
14. Franck C, Grandi SM, Eisenberg MJ (2013) Agricultural subsidies and the american obesity epidemic. Am J Prev Med 45: 327–333.
15. Pollan M (2003) The (Agri)Cultural Contradictions of Obesity. The New York Times Magazine. New York City: The New York Times.
16. AFT (2010) 13 Milion More Acres.
17. Industry PMsCoTa (2010) Report on Food & Agro Industries Management Policy.
18. Wu Huang S (2004) Global Trade Patterns in Fruits and Vegetables. Washington, DC: Economic Research Service, United States Department of Agriculture.
19. Siegel KR, Echouffo-Tcheugui JB, Ali MK, Mehta NK, Narayan KM, et al. (2012) Societal correlates of diabetes prevalence: An analysis across 94 countries. Diabetes Research and Clinical Practice 96: 76–83.
20. Basu S, Yoffe P, Hills N, Lustig RH (2013) The relationship of sugar to population-level diabetes prevalence: an econometric analysis of repeated cross-sectional data. PLoS One 8: e57873.

21. AFT (2010) American Farmland Trust Says The United States Needs 13 Million More Acres of Fruits and Vegetables to Meet the RDA. Washington, DC: American Farmland Trust.
22. Sekula W, Becker W, Trichopoulou A, Zajkas G (1991) Comparison of dietary data from different sources: some examples. WHO Reg Publ Eur Ser 34: 91–117.
23. Pomerleau J, Lock K, Knai C, McKee M (2005) Interventions designed to increase adult fruit and vegetable intake can be effective: a systematic review of the literature. J Nutr 135: 2486–2495.
24. Wolfenden L, Wyse RJ, Britton BI, Campbell KJ, Hodder RK, et al. (2012) Interventions for increasing fruit and vegetable consumption in children aged 5 years and under. Cochrane Database Syst Rev 11: CD008552.
25. WHO/FAO (2003) Diet, nutrition and the prevention of chronic diseases. Geneva: World Health Organization and Food and Agricultural Organization.
26. Thow AM, Jan S, Leeder S, Swinburn B (2010) The effect of fiscal policy on diet, obesity and chronic disease: a systematic review. Bull World Health Organ 88: 609–614.
27. CDC (2011) Strategies to Prevent Obesity and Other Chronic Diseases: The CDC Guide to Strategies to Increase the Consumption of Fruits and Vegetables. Atlanta: U.S. Department of Health and Human Services.

16

Inorganic Nitrogen Leaching from Organic and Conventional Rice Production on a Newly Claimed Calciustoll in Central Asia

Fanqiao Meng[1]*, Jørgen E. Olesen[2], Xiangping Sun[1], Wenliang Wu[1]

1 College of Resources and Environmental Sciences, China Agricultural University, Beijing, China, **2** Department of Agroecology and Environment, Faculty of Agricultural Sciences, Aarhus University, Tjele, Denmark

Abstract

Characterizing the dynamics of nitrogen (N) leaching from organic and conventional paddy fields is necessary to optimize fertilization and to evaluate the impact of these contrasting farming systems on water bodies. We assessed N leaching in organic versus conventional rice production systems of the Ili River Valley, a representative aquatic ecosystem of Central Asia. The N leaching and overall performance of these systems were measured during 2009, using a randomized block experiment with five treatments. PVC pipes were installed at soil depths of 50 and 180 cm to collect percolation water from flooded organic and conventional paddies, and inorganic N (NH_4-N+NO_3-N) was analyzed. Two high-concentration peaks of NH_4-N were observed in all treatments: one during early tillering and a second during flowering. A third peak at the mid-tillering stage was observed only under conventional fertilization. NO_3-N concentrations were highest at transplant and then declined until harvest. At the 50 cm soil depth, NO_3-N concentration was 21–42% higher than NH_4-N in percolation water from organic paddies, while NH_4-N and NO_3-N concentrations were similar for the conventional and control treatments. At the depth of 180 cm, NH_4-N and NO_3-N were the predominant inorganic N for organic and conventional paddies, respectively. Inorganic N concentrations decreased with soil depth, but this attenuation was more marked in organic than in conventional paddies. Conventional paddies leached a higher percentage of applied N (0.78%) than did organic treatments (0.32–0.60%), but the two farming systems leached a similar amount of inorganic N per unit yield (0.21–0.34 kg N Mg^{-1} rice grains). Conventional production showed higher N utilization efficiency compared to fertilized organic treatments. These results suggest that organic rice production in the Ili River Valley is unlikely to reduce inorganic N leaching, if high crop yields similar to conventional rice production are to be maintained.

Editor: Ben Bond-Lamberty, DOE Pacific Northwest National Laboratory, United States of America

Funding: The study was funded by National Natural Science Foundation of China (No. 30970533) and the National Key Science and Technology Project-Organic Farming Development in the Ili River Valley (No. 2007BAC15B05). The funders had no role in study design, data collection and analysis, decision to publish, or preparation of the manuscript.

Competing Interests: The authors have declared that no competing interests exist.

* E-mail: mengfq@cau.edu.cn

Introduction

Nitrogen (N) leaching is one of the primary pathways of N loss from flooded paddy farmland, representing 30–50% of total N loss from such soils [1]. The amount of leached N may equal up to 15% of the total applied N in rice (*Oryza sativa* L.) production [2,3]. Flooded rice crops typically capture only 20–40% of N applied in the harvested grain [4]. N utilization efficiency can be improved by reducing the various forms of loss, including leaching. Because reactive N is highly soluble, its subsequently high mobility in the environment makes it a major contributor to surface and groundwater contamination. Ju et al. [5] reported that high inputs (550–600 kg N ha^{-1}) associated with current rice production systems in China did not significantly increase crop yields, yet cause a doubling of N losses to the environment, mainly from increased denitrification in waterlogged systems. Accordingly, understanding N loss processes, particularly through leaching and N_2O emission, is necessary to both improve resource use efficiency and to protect the quality of nearby water bodies. In this study, we focused on monitoring N leaching losses at different rice growth stages and soil depths in organic and conventional rice production systems.

Located in the Xinjiang Uygur Autonomous Region of northwestern China, the Ili River Valley has become an important grain production region, owing to its plentiful surface water, well-developed soils, and extensive meadow grasslands. Of the total land area (5.82 million ha), approximately 1 million ha are under cultivation, and the introduction of new irrigation technologies is facilitating the conversion of meadow grasslands to agricultural uses [6]. However, the Ili River is sensitive to pollution, which is of local and international concern as the river flows into Balkhash Lake in Kazakhstan. The river receives pollutants from both surface run-off and drainage from farming activities throughout the valley. In addition to water quality concerns, the terrestrial ecosystem may also be at risk because it contains shallow soils with a coarse texture. As such, organic farming is being promoted for the conservation of soil and water resources to reduce the overall negative environmental impact of agricultural production, as well as to increase farmer's income. However, it is unclear whether organic production can maintain high rice yields, with or without high fertilizer inputs, and what effect shifting from conventional to organic farming may have on N loading to the Ili River.

Many studies have examined N leaching from organic and conventional production [7,8]. Such research has revealed that organic farming may lower N leaching compared to conventional systems, both by reducing N inputs and by including catch crops (cover crops) in the rotation [8,9], although the magnitude of this effect is subject to high uncertainty due to variation in crop yields and input intensities [9,10]. Although rice dominates grain production in many developing countries including China, little research has been conducted in this Central Asia region about N loss in organic versus conventional rice farming, especially regarding N loss through percolation or leaching [2,4,11].

This study compared the inorganic N (NH_4-N and NO_3-N) in percolation/leachate water from organic versus conventional production systems with high yields of rice production, by the application of high-loads of animal manure. We hypothesized that organic rice farming leaches less inorganic N compared with conventional rice in the Ili River Valley. In order to calculate the amount of N leached from paddies, we had to address the technical challenge of quantifying the total volume of leachate from a single-cropping rice paddy. To do so, we designed a filtration system using PVC pipes at two depths (50 and 180 cm) to capture the leachate, and quantified the percolation rates throughout the rice growing season using a water balance based on the difference between incoming (*i.e.*, irrigation and precipitation) and outgoing (evapotranspiration and surface runoff) flows. To better understand the potential environmental effects of the treatments, we evaluated N leaching not only in absolute terms, but also in terms of leachate per unit yield of rice and as a percentage of the total N applied.

Materials and Methods

Experiment site

The field experiment was conducted at Chabuchaer Farm (48°65'N, 80°06'E, 634 m ASL) in the Ili River Valley of the Xinjiang Uygur Autonomous Region, China. The region has a temperate continental climate with a mean annual temperature of 9.5°C and annual precipitation of 260 mm, which falls predominantly in June, July, and August. In 2009 when monitoring was implemented, precipitation mainly happened in April to July, November and December (Fig. 1). The farm is located on an alluvial plain off the south bank of the Ili River. The land use was converted to agriculture by the local government in 2005 as part of a national grain supply project. Because this land was newly reclaimed, we could examine the contrasting effects of conventional and organic farming techniques without the usual conversion period from conventional to organic farming, which can confound the results. Our study was initiated in 2008 and the data presented here were collected in the 2009 season.

The soil was a calciustoll with salt content of 1.4%. Rice had been continuously cultivated on this land since 2005, with the aim of reducing soil salinity. The soil is a sandy loam, with 3.2% clay, 44.7% silt, and 52.1% sand. Other important characteristics of the soil (0–20 cm) are as follows: 14.0 g kg^{-1} soil organic matter, 1.15 g kg^{-1} total N, 11 mg kg^{-1} Olsen phosphorous, 264 mg kg^{-1} available potassium, and a pH of 7.9.

Permission to conduct the experiment and sampling at the site was obtained from the Ili Agricultural Technical Extension Station, and the Farmer Liu Ermao. These field studies did not involve endangered or protected species.

Crop management

Rice (*Oryza sativa* L., cultivar Nonglin-315) was transplanted to the experimental farm on May 29, 2009. The fertilizer used was either a conventional combination of mineral fertilizers with animal manure, as used in the local production system, or only organic fertilizers (composted animal manure [B] with castor (*Ricinuscommunis* L.) bean meal [A]), as used typically in organic rice production. The level of organic fertilization was manipulated to modify the overall N input. Hence, the five treatments were 1) an unfertilized control treatment (CK), 2) low-level organic fertilization (B1A1): composted animal manure at 15,000 kg ha^{-1} and castor bean meal at 2,250 kg ha^{-1}, 3) mid-level organic fertilization (B2A1): composted animal manure at 45,000 kg ha^{-1} and castor bean meal at 2,250 kg ha^{-1}, 4) high-level organic fertilization (B3A1): composted animal manure at 75,000 kg ha^{-1} and castor bean meal at 2,250 kg ha^{-1}, and 5) conventional production system (LS): composted animal manure at 11,250 kg ha^{-1}, urea at 277.5 kg ha^{-1}, diammonium phosphate at 247.5 kg ha^{-1}, and compounded mineral fertilizer at 75 kg ha^{-1} (Table 1). The treatments were arranged in a randomized block design with three replicates.

Composted animal manure was applied on May 9 as the basal fertilizer for all treatments (except CK), 20 days before transplanting. The total amount of castor bean meal was halved and applied in two stages, first as topdressing at the early tillering stage (June 27) and then at the late tillering stage (July 24). Topdressings for LS were applied on June 27 (120 kg ha^{-1} diammonium phosphate with 75 kg ha^{-1} compounded fertilizer), July 6 (127.5 kg ha^{-1} diammonium phosphate with 127.5 kg ha^{-1} urea), and July 15 (urea, 150 kg ha^{-1}). The nutrient contents of the fertilizers used in the experiments are listed in Table 2.

Each paddy block had an area of 8×10 m = 80 m^2. Flood irrigation (about 20 cm of standing water), popular in other Central Asian countries [4], was used continuously from June 1 to September 7, except during the first week after fertilization application. Hand weeding was used in all treatments, along with the addition of herbicides to conventional replicates. The main development stages for the rice were transplant (May 29-Jun 13), tillering and elongation (~Aug 5 for organic paddies / ~Aug 10 for conventional paddies), heading (Aug 5-Sep 1 for organic / Aug 10-Sep 5 for conventional), and ripening (Sep 1-Oct 2 for organic / Sep 5-Oct 13 for conventional).

Measurements of N cycling

Previous studies have shown that inorganic N (NH_4-N+NO_3-N), rather than particulate N, makes up 50–90% of the total N leached from the soils, and for organic fertilizers or fertile soil, this proportion maybe even higher [12,13,14]. Only inorganic N was analyzed in percolation water collected in our experiment. Percolation water was collected from flooded paddies using polyvinyl chloride (PVC) standpipes [14,15,16]. Two PVC pipes (5 cm in diameter, 150 and 250 cm in length) with sealed bottoms were installed roughly in the center of each field plot to collect drainage water from the saturated soil (Fig. 2). Both pipes were perforated 80 times (6-mm internal diameter) within a 10-cm-wide band, 30 cm from the bottom of the pipe. The porous zone of the pipe was wrapped with nylon textile to prevent sand in-filling. As the average soil layer is 2 m deep in this region, we compared the inorganic N leaching in 50 and 180 cm. At the later depth we considered that inorganic N was not usable by the rice. Thus, the pipes were installed at depths of 50 and 180 cm from the surface to the uppermost pore (Fig. 2). The gap between the collection pipe and soil wall in the porous zone was filled with quartz powder to prevent anything except soil solution from entering the pipe. At a

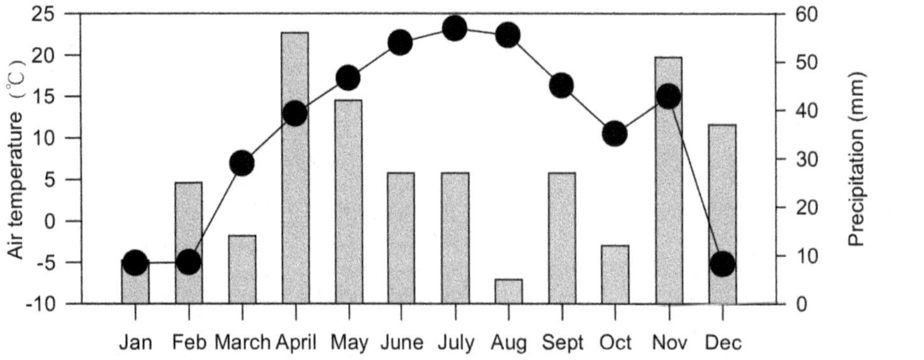

Figure 1. Air temperature and precipitation at the experimental site in Chabuchaer County for 2009.

depth of 20 cm, plastic film was wrapped around the PVC pipe, extending horizontally in a 30-cm radius from the pipe to reduce the preferential flow from the irrigation water.

Sampling was conducted 1 and 3 days after fertilization, and thereafter at intervals of 5 days to 2 weeks, depending on the percolate volume. Water samples were stored in plastic vials and refrigerated until analysis. Percolation and irrigation water were analyzed for inorganic N (NH_4-N+NO_3-N) content using a continuous flow N analyzer (TRAACS 2000). During the irrigation season, average NH_4-N and NO_3-N concentrations in the irrigation water were 0.11 and 0.68 mg L^{-1} respectively; these values were subtracted from the respective inorganic N concentrations in percolation water to calculate net N (i.e., the amount leached).

The volume of drainage water was calculated per unit area of paddy field. We considered the spatial range of percolation into the pipe to be a half-ellipsoid, having a width of 5 cm and a depth of either 50 or 180 cm, with diameters of 27.5 cm (50/2+5/2) or 92.5 cm (180/2+5/2) [15]. The volume of each ellipsoid was calculated as $4/3*\pi*ab^2$, where a and b are the length of the long axis and radius of the ellipsoid, respectively. Each hectare of paddy field included 252,671 or 222,332 half-ellipsoids, at depths of 50 cm and 180 cm, respectively. Thus, total N leached per ha could be calculated by summing these half ellipsoids at each soil depth, as well as across time to calculate total N lost over one growth cycle. In total, percolation water was sampled 15 times from June 1 to September 7, the full growing season of the rice paddies.

The measured volume of percolation water in the paddy field was calibrated using the following equation [17]:

$$P+I=ET+D+R_0+P_{sd}-V_H \quad (1)$$

Where P is the rainfall (mm) during the rice growth season; I is irrigation (mm); ET is evapotranspiration (mm), calculated using a lysimeter installed in the paddy plot; D is drainage through drain pipe (mm), excluding surface runoff and percolation; R_o is surface runoff (mm); P_{sd} is percolation (mm); and V_H is the change of the water table in the paddy field (mm). Briefly, an iron lysimeter (1 m wide×1 m long×1.2 m high), filled with original paddy soil, was installed in the paddy field. The bottom of the lysimeter was perforated to allow percolated water to drip into a storage tank (0.5 m wide×0.5 m long×0.5 m high) welded to the lysimeter for storage of the percolated water. A meter stick was also fixed inside the lysimeter to measure the water table. Evapotranspiration from the paddy field was calculated according to Equation (1). In our study, there was no artificial drainage or surface runoff, so D and R_o were set to 0. All of the terms are expressed in millimeters and then converted into m^3 ha^{-1}. One growth cycle (June 1 to September 7, 2009) was used to bound all calculations. The difference between P_{sd} calculated from Equation (1) and from the measured percolation water volume was less than 10%; hence, here we presented percolated inorganic N as defined by its concentration (NH_4-N and NO_3-N) multiplied by measured percolate volume.

Relative N leaching loss (% of applied N) for a given plot was determined after first subtracting N leached from the nil treatment

Table 1. Experimental treatments in terms of total nutrient inputs for N, P, and K (kg ha^{-1}).

Treatment	Fertilization	Nutrient input (kg ha^{-1})		
		N	P	K
CK	Control with no fertilization	0	0	0
LS	Local conventional system: composted animal manure (27.6% of water content) at 11,250 kg ha^{-1}, urea at 277.5 kg ha^{-1}, diammonium phosphate at 247.5 kg ha^{-1}, and compounded fertilizer at 75 kg ha^{-1}.	297	83	117
B1A1	Composted animal manure (27.6% of water content) at 15,000 kg ha^{-1} plus castor bean meal (22.5% of water content) at 2,250 kg ha^{-1}	193	52	163
B2A1	Composted animal manure (27.6% of water content) at 45,000 kg ha^{-1} plus castor bean meal (22.5% of water content) at 2,250 kg ha^{-1}	498	125	446
B3A1	Composted animal manure (27.6% of water content) at 75,000 kg ha^{-1} plus castor bean meal (22.5% of water content) at 2,250 kg ha^{-1}	802	199	728

Table 2. Nutrient contents of fertilizers used in the experiments (% dry weight).

	N	*P*	*K*
Composted animal manure	1.40	0.34	1.30
Castor bean meal	2.37	0.84	1.10
Urea	46		
Diammonium phosphate	18	20	
Compounded fertilizer	14	6.98	12.45

(CK) and then expressed as a percentage of the total N applied. Similarly, the environmental performance of organic and conventional fertilizer treatments, i.e., N leaching, was also evaluated based on the amount of inorganic N leached during the production of 1 Mg of rice (kg N Mg^{-1} grain). N utilization efficiency (NUE) was estimated using the equation (2):

$$NUE = \left(\begin{array}{l} N\ uptake\ by\ rice\ in\ fertilized\ treatment \\ (kg\ N\ ha^{-1}) - \\ N\ uptake\ by\ rice\ in\ CK\ treatment \\ (kg\ N\ ha^{-1}) \end{array} \right) / N\ input\ from\ the\ fertilizers (kg\ N\ ha^{-1}). \quad (2)$$

Sampling and measurement of fertilizers, soils and plants

Manures were sampled to determine the content of dry matter (DM) and total N before spreading. Plant samples were taken in early October to obtain rice grain with husk and straw, by cutting stems at 2 cm above the ground from two 2 m^2 areas per plot. The plant samples were washed with distilled water, oven-dried at 65°C, and weighed to determine rice biomass. Plant samples were then digested using H_2SO_4-H_2O_2, after which N, P, and K in the digestion solution were determined using the Kjeldahl method, calorimetrically by a Techniconautoanalyzer, and via atomic emission spectrometry (AES), respectively.

Statistical analyses

Cumulative inorganic N leached, net inorganic N leached per Mg of rice produced, and the proportion of net leached inorganic N in total N input were compared among different treatments using the least-significant difference test after a one-way analysis of variance for a randomized block design at the 0.05 significance level. The PROC MIXED procedure in SAS v.9.1 (SAS Institute, Cary, NC, USA) was performed to analyze the effects of fertilization on NH_4-N and NO_3-N concentrations over the sampling period (15 samplings), with the treatment as the fixed effect and sampling time as the random factor. Differences among means were calculated using a Differences of Least Squares Means

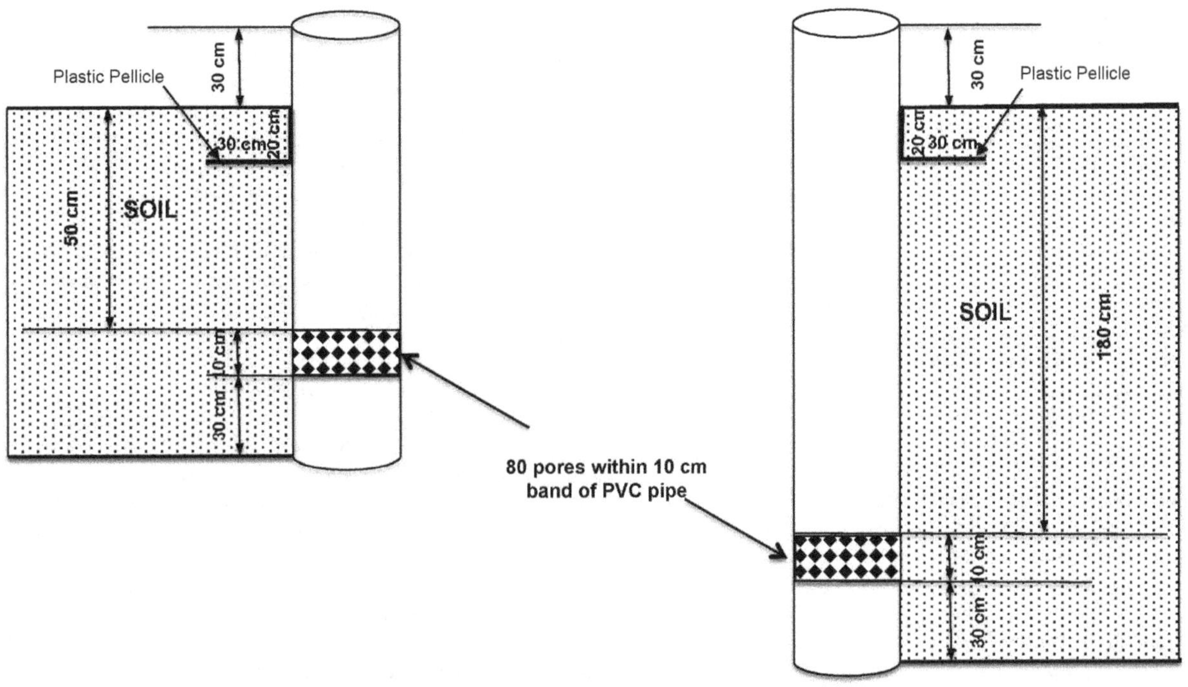

Figure 2. Schematic of standpipes for measuring percolate inorganic N concentrations in the paddy fields. Shaded area indicates the depth below the soil surface. Horizontal lines at 20 cm depth indicate the placement of a plastic film that extended a 30 cm-radium around the pipes.

with the PDIFF option and a Bonferroni adjustment method. The significance level was also at 0.05.

Results

Inorganic N concentrations in percolation water at 50 cm depth

Leaching of NH_4-N and NO_3-N during the rice growing season under different fertilization treatments is shown in Fig. 3. We observed high variance in NH_4-N concentration among three replicates of the same treatment, and this was also the case for NO_3-N (data not shown). Two peak concentrations of NH_4-N were observed in organic treatments, at the beginning of tillering (late June) and at heading (mid-August), and one additional NH_4-N peak at late-tillering (late July) in conventional treatment (LS). By PROC MIXED analysis, we found over the entire growth stage of rice, application of organic fertilizers increased NH_4-N concentration compared to no fertilization (CK, 0.63 mg L^{-1}, DF = 50.9, t = 2.76); this increase was significant for B2A1 (1.37 mg L^{-1}, DF = 50.9, t = 6.04) and B3A1 (1.74 mg L^{-1}, DF = 50.9, t = 7.65) but not for B1A1 (1.14 mg L^{-1}, DF = 50.9, t = 5.04). The concentration of NH_4-N that percolated from LS soils was significantly higher (1.20 mg L^{-1}, DF = 50.9, t = 5.29) than from CK soils, but not different from organic fields (B1A1, B2A1 and B3A1).

Organic and conventional fertilization affected NO_3-N concentrations in percolation water (Fig. 3) differently than they affected NH_4-N. At the 50-cm soil depth, the average NO_3-N concentration was lowest in CK (0.48 mg L^{-1}, DF = 42.2, t = 2.79) and then increased in the order of LS (1.08 mg L^{-1}, DF = 42.2, t = 6.27)< B1A1 (1.96 mg L^{-1}, DF = 42.2, t = 11.33)<B2A1 (2.16 mg L^{-1}, DF = 42.2, t = 12.48)<B3A1 (2.19 mg L^{-1}, DF = 42.2, t = 12.67). Within the percolation waters of both the CK and LS treatments, concentrations of NH_4-N and NO_3-N were similar, whereas in the organic treatments (B1A1, B2A1 and B3A1) average concentrations of NO_3-N were 21–42% higher than NH_4-N. In contrast to the two peak concentrations of NH_4-N, NO_3-N declined from high concentrations at early tillering until the end of the rice harvest

Inorganic N concentrations in percolation water at 180 cm depth

The average NH_4-N concentrations at 180 cm were lowest in the CK treatment (0.41 mg L^{-1}, DF = 48.8, t = 1.75) and then increased as follows: LS (0.88 mg L^{-1}, DF = 48.8, t = 3.75) < B1A1 (1.08 mg L^{-1}, DF = 48.8, t = 4.61) <B2A1 (1.28 mg L^{-1}, DF = 48.8, t = 5.44), and B3A1 (1.57 mg L^{-1}, DF = 48.8, t = 6.67, Fig. 4). These results indicated that NH_4-N leaching increased with the increasing intensity of organic fertilization, although a high temporal variation was also observed over the monitoring period. As was found at 50 cm depth, two NH_4-N peaks were recorded at the beginning of tillering and at the heading stage for the organic treatments, with an additional peak for LS at mid-tilling. Different from the observations at 50 cm, NH_4 concentrations for all treatments increased from early heading, for about 1 month, until the start of paddy filling. NH_4-N concentrations in organic rice leachates at 180 cm were similar to those at 50 cm, except for LS plots, in which values at 180 cm were significantly lower than that at 50 cm depth (0.88 vs. 1.20 mg L^{-1}).

The temporal dynamics of NO_3-N concentrations at 180 cm were similar to those at 50 cm (Fig. 4), *i.e.*, as the quantity of organic fertilizer increased, leached NO_3-N concentrations also increased. However, significantly lower concentrations of NO_3-N were observed for organic treatments than for LS, and this was opposite to 50 cm depth where organic treatments percolated significantly higher concentrations of NO_3-N than LS. Organic treatments had much lower NO_3-N concentrations at 180 cm than at 50 cm (B1A1: 0.59 vs. 1.96 mg L^{-1}, B2A1: 0.85 vs. 2.16 mg L^{-1} and B3A1: 1.17 vs. 2.19 mg L^{-1}) while NO_3-N concentrations in the LS treatment increased from 50 cm to 180 cm (1.08 vs. 1.70 mg L^{-1}).

Total leached inorganic N

The rice fields were flooded during the growing period from transplant (June 1) to harvest (Sep 7). At the 50 cm depth, leached inorganic N varied from 18.5 to 23.1 kg N ha^{-1} for fertilized organic treatments comparing to 12.1 kg N ha^{-1} for LS (Table 3). Of the total amount of leached inorganic N, the fertilized organic treatments showed a slight higher proportion of N as NO_3-N (51.1~58.9%), compared to 64.2% as NH_4-N in LS. Higher organic fertilization rates led to higher amounts of leached NH_4-N, but this was not the case for NO_3-N, i.e., among the three fertilized organic treatments, the quantities of NO_3-N remained similar. For the net total inorganic N leached, B3A1 was significantly higher than other treatments, especially compared to LS.

At 180 cm, far less total inorganic N was percolated than at 50 cm for all treatments. For fertilized organic treatments (B1A1, B2A1 and B3A1), total inorganic N leached at 180 cm was only 9-13% of the amount leached at 50 cm; for the LS treatment, it was 26%. Relatively higher proportions of inorganic N were leached as NH_4-N (>68%) from fertilized organic production systems, whereas 61% of leached inorganic N from LS was NO_3-N. Among all organic treatments, B3A1 leached the highest quantity of inorganic N, significantly higher than other organic treatments, but interestingly, not higher than LS.

For the purpose of calculating N losses, we considered the inorganic N present at the depth lower than 180 cm (Table 4) to be that which could no longer be utilized by crops, and could potentially reach surface or ground water bodies. These data were used to assess the environmental performance of organic versus conventional rice production using two indicators: the amount of inorganic N leached per unit yield of rice (kg N Mg^{-1} grain) and the percentage of inorganic N leached per unit applied N (% loss) at 180 cm depth. Regarding leachate per unit yield, we found that LS and organic treatments were not significantly different, i.e., a similar quantity of inorganic N was leached per production unit yield of rice (0.21~0.34 kg N Mg^{-1} grain). However, in terms of percent N loss, LS leached significantly more inorganic N in total N input (0.78%) than did organic production (0.32~0.60%).

Discussion

Inorganic N concentration in percolation water

In flooded paddy soils, aerobic and anaerobic metabolisms occur in close proximity [18,19,20]. At the 50 cm depth in the sandy loam soil that we studied, in organic treatments, the presence of the oxidized zone close to the reduced soil zone was conductive for the transformation of NH_4-N mineralized from organic manure to NO_3-N. However, the continuous anaerobic conditions (under flood irrigation, until 1 week before harvest) at the 180 cm depth maintained NH_4-N as the main form of inorganic N, more so than NO_3-N. Particularly, in continuously flooded paddy soils with abundant organic substrate and a limited availability of electron acceptors, the reduction of NO_3-N to NH_4-N would be more efficient than the formation of N_2, so NH_4-N concentrations can be an order of magnitude higher than those of NO_3-N [21,22,23,24]. This situation also occurs in soils with low

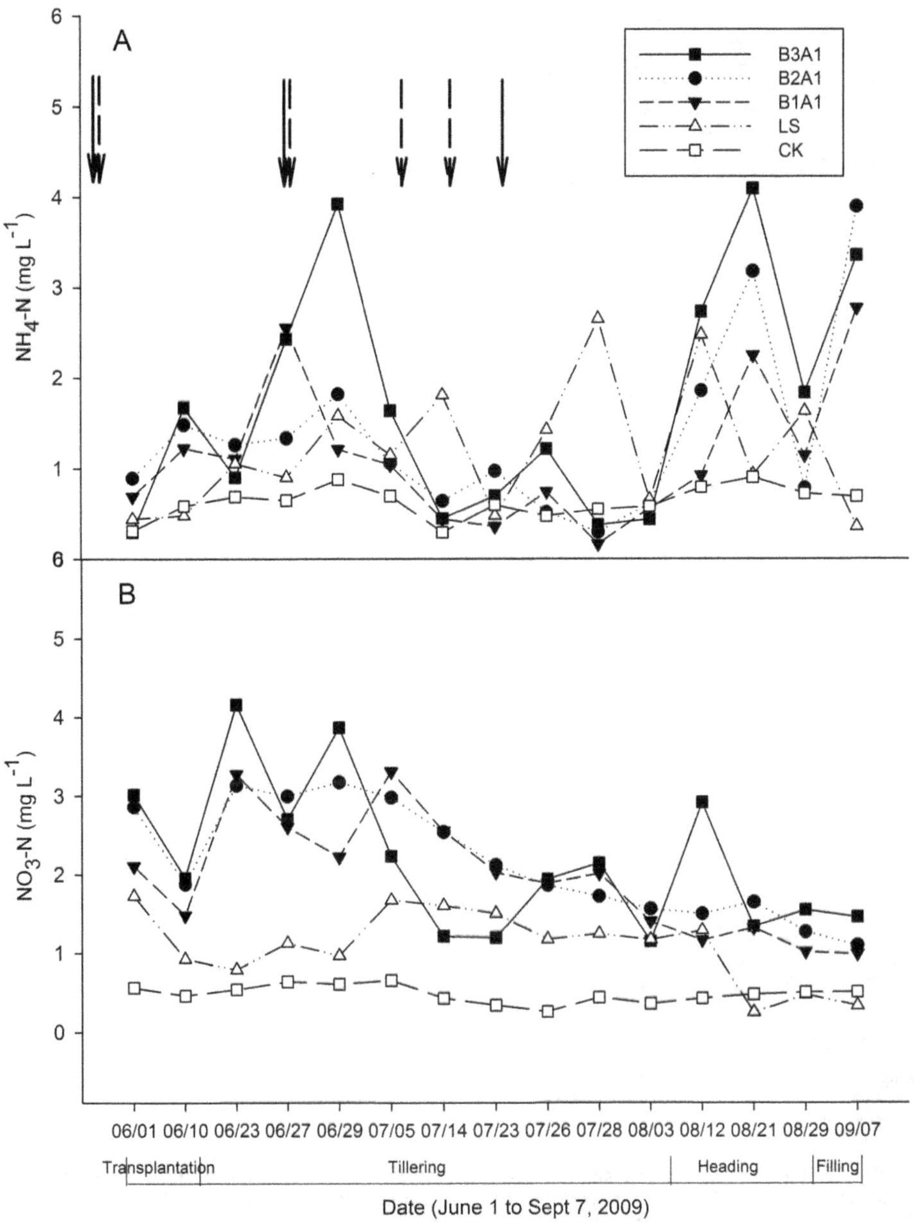

Figure 3. Concentrations of NH_4-N and NO_3-N in leachate at the 50 cm soil depth under different fertilization treatments. A: NH_4-N; B: NO_3-N. The solid and long-dashed arrows indicate the date of fertilizer applications for organic treatments (B1A1, B2A1, and B3A1) and for the conventional treatment (LS), respectively. Bars represent standard deviations ($n = 3$).

CEC and a low content of exchangeable base cations [19,25,26,27]. From the depth of 50 cm to 180 cm, average NH_4-N and NO_3-N concentrations decreased by about 5–10% and 47–70%, respectively, in fertilized organic treatments (B1A1, B2A1 and B3A1). The smaller decrease in NH_4-N concentration, compared to NO_3-N, may resulted from continuous decomposition of organic fertilizer and the limited soil adsorption capacity for NH_4-N [19,25,28]. From 50 cm to 180 cm in LS soils, there was a 27% decrease in NH_4-N, but a 57% increase in NO_3-N, indicating that mineral fertilization may have led to an increased downward movement of NO_3-N, compared to organic fertilizer, and the low dentrification capacity in deeper LS soils would thus have caused NO_3-N concentration to remain high [20]. This substantial loss of inorganic via leaching NH_4-N from continuous flooded rice fields has also been confirmed by other studies, which reported that NH_4-N might account for up to 92% of the total inorganic N in leachate - this large risk of NH_4-N leaching deserves more attention than the risk of NO_3-N loss [23,27].

During the vegetative phase of rice growth, NO_3-N concentrations were higher than those of NH_4-N, whereas the opposite was observed during the reproductive phase (Figs. 3 and 4). Rice requires more NH_4-N than NO_3-N during the vegetative stage, contributing to a suppression of NH_4-N concentrations in leachate [29,30]. As rice plants shifted into their reproductive stage, NH_4-N was continuously mineralized from the organic fertilizers, but the anaerobic conditions prevented nitrification into NO_3-N, both in the upper and lower soil profiles [23,31]. Unlike NH_4-N, the NO_3-N concentration in percolation water declined from transplant

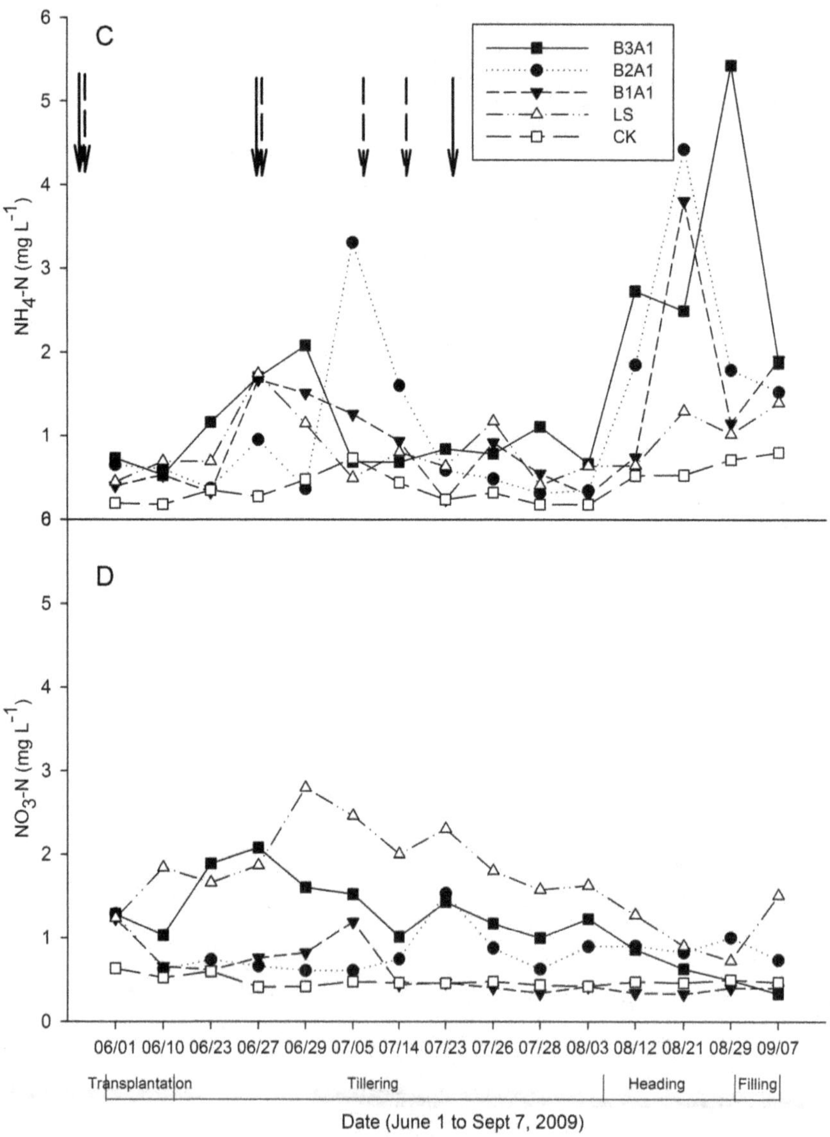

Figure 4. Concentrations of NH_4-N and NO_3-N in leachate at the 180 cm soil depth under different fertilization treatments. C: NH_4-N; D: NO_3-N. The solid and long-dashed arrows indicate the date of fertilizer applications for organic treatments (B1A1, B2A1, and B3A1) and for the conventional treatment (LS), respectively. Bars represent standard deviations ($n = 3$).

until harvest mainly because of continuous uptake by the plants throughout the growing season [32]. The two peaks observed in NH_4-N concentrations were also driven by fertilization and environmental conditions (higher temperature). At the beginning of tillering (~June 29), air temperature was close to its annual maximum (Fig. 1), which may have promoted the mineralization of organic fertilizer. However, the observed increases in NO_3-N concentrations lagged behind those of NH_4-N, because urea or organic fertilizer must first be transformed to amide N, then to ammonium and further nitrified to NO_3-N [28].

Comparison of N utilization and loss from organic and conventional rice production

Organic and mineral fertilizers influenced rice yield differently, through differential effects on yield components such as the number of panicles per hill, number of grains per panicle, and grain weight. In our experiment, topdressing with castor bean meal in combination with increasing manure application resulted in higher grain weight, but mineral fertilizer still produced a larger number of panicles and grains per panicle [33]. A previous study revealed that increased mineral N supply to rice increases the amount of dry matter translocated into the grain [34]. However, if organic and mineral fertilizers are used together, higher numbers of panicles and grains per panicle can be achieved because of improved nutrient balance [35]. As the yield increased, more N was removed from the soil, but not in proportion with the increase in N input, such that N utilization efficiency (NUE) declined. Rice grown under conventional production showed the highest NUE (29.6%), which was significantly higher than that of organic treatments (8.7–17.6%, Table 5). This difference in NUE between conventional and organic production has been demonstrated in many other studies [8,32], which is attributed to slow release of N from organic fertilizers that limits N uptake by crops, and also maintains the continuous loss of N via leaching or other channels [22].

Table 3. Cumulative inorganic N leaching (kg N ha^{-1}) at 50 and 180 cm soil depths for control (CK), organic (B1A1, B2A1, and B3A1) and conventional (LS) treatments during a single rice growth cycle.

Depth (cm)	Treatments	NH_4-N (kg ha^{-1})	NO_3-N (kg ha^{-1})	Total inorganic N leached (kg ha^{-1})	Net inorganic N leached (kg ha^{-1}) [a b]
50	CK	3.4 c	0.0 c	3.4 d	n.a.
	LS	7.8 b	4.3 b	12.1 c	8.7 c
	B1A1	7.6 b	10.9 a	18.5 b	15.2 b
	B2A1	9.6 ab	12.6 a	22.2 a	18.8 ab
	B3A1	11.3 a	11.8 a	23.1a	19.8 a
180	CK	0.4 c	0.0 d	0.4 d	n.a
	LS	1.1 bc	1.7 a	2.7 ab	2.3 ab
	B1A1	1.4 ab	0.2 d	1.6 c	1.2 b
	B2A1	1.7 ab	0.5 c	2.2 bc	1.8 ab
	B3A1	2.0 a	1.0 b	3.0 a	2.5 a

[a]After subtraction of the leakage quantity measured in CK.
[b]Different letters in a column denote a significant difference between treatments at the 0.05 level.

Environmental performance per se, in terms of N loss through leaching is a key concern for maintaining water quality and the integrity of organic farming. When comparing organic and conventional paddy rice production, N leaching losses are a tradeoff with rice yield; for example, if high yields are to be achieved, relatively high amounts of inorganic N must be supplied and may be lost. In our study, the highest yield of organic rice (7550 kg ha^{-1}) was achieved using the highest rate of N application (B3A1, 802 kg N ha^{-1}; Table 4); yield in these plots amounted to 80% of that achieved through conventional treatment (LS), which received inputs of only 297 kg of N ha^{-1}. Furthermore, organic production did not always perform well in terms of inorganic N leaching per unit yield of rice. In fact, if organic fertilization was intensified enough to obtain a "conventional" yield, even more N would be leached per unit yield than currently occurs in conventional production. Organic treatments with relatively higher N input (B2A1 and B3A1) tend to have lower overall rates of inorganic nitrogen loss than the LS treatment, but this came at the cost of a much lower rice yield.

In the current study, the N input levels in organic treatments (except for B1A1) were higher than most conventional rice production in China, where for a single growth cycle N application varies from 200 to 300 kg N ha^{-1} [36]. In the Ili River Valley, rice production had an average N application rate of 234 kg ha^{-1}, equal to the national average. This generates additional doubt that organic production can help to reduce N leaching from paddy production. In our experiments, 2.9–7.9% (8.7–19.8 kg ha^{-1}) of total applied N was lost through leaching at 50 cm, and 0.32–0.78% (1.2–2.5 kg ha^{-1}) was lost at 180 cm, whereas other studies involving organic fertilization reported losses of 0.1 to 15% [13,16,31,37,38]. Based on these findings, we believe that organic rice production in the Ili River Valley would not reduce (or may even increase) N leaching compared to conventional production, especially when organic fertilizer input is high enough to achieve conventional yields.

High reliance on external nutrient supply, especially from non-organic/conventional sources, has become a concern for large-scale organic production in recent years [39,40]. Organic

Table 4. Amount of inorganic N leached at 50 cm and 180 cm soil depths, defined as production of 1 Mg rice grains produced and as a % of N input.

Depth (cm)	Treatments	Rice yield (Mg ha^{-1})	N input (kg ha^{-1})	Net leached inorganic N per Mg of rice grains (kg N Mg^{-1}) [b]	Proportion of net leached inorganic N in total N input (%) [a b]
	CK	3.6 d	n.a.		
	LS	9.4 a	297	0.93 b	2.9 bc
50	B1A1	5.6bc	193	2.69 a	7.9 a
	B2A1	6.9bc	498	2.72 a	3.8 b
	B3A1	7.6 b	802	2.62 a	2.5 c
	CK	3.6	n.a.	n.a.	n.a.
	LS	9.4	297	0.25 a	0.78 a
180	B1A1	5.6	193	0.21 a	0.60 ab
	B2A1	6.9	498	0.26 a	0.36 b
	B3A1	7.6	802	0.34 a	0.32 b

[a]"Net N leached" was calculated by subtracting the amount of N leached in the control treatment (CK) from total treatment values.
[b]Different letters in a column denote a significant difference between treatments at the 0.05 level.

Table 5. Nitrogen budget for rice production.

Treatments	N input from fertilizers (kg ha^{-1})	N removed by rice (kg ha^{-1})	N removed by straw (kg ha^{-1})	Total N removed by rice crop (kg ha^{-1})	N budget (kg ha^{-1})	N utilization efficiency (%) [a]
CK	0	40	13	53	−53	n.a
LS	297	100	41	141	156	29.6 a
B1A1	193	59	28	87	106	17.6 b
B2A1	498	85	29	114	384	12.2 c
B3A1	802	87	36	123	679	8.7 d

[a]Different letters in a column denote a significant difference between treatments at the 0.05 level.

fertilizers are able to meet crop N requirements, but this is achieved through high application rates, which could be 1) mineralized to release a sufficient amount of N for rice growth, and 2) synchronized between N mineralization and crop uptake [10]. However, higher organic N input can cause higher N losses through leaching, as shown in our experiments. In addition, N in paddy soils can also be denitrified to the powerful greenhouse gas N_2O [14], an additional significant environmental impact that was not measured in our study. Ju et al. [5] summarized that rice production in China using 300 kg N ha^{-1} per season contributed denitrification N losses of approximately 36%, which are markedly higher than losses from volatilization (12%) or leaching (0.5%). Here, we assume that denitrification and volatilization losses for mineral fertilizer in our study would be roughly comparable to those reported by Ju et al. [5] and Guan et al. [41]. Besides inorganic N, organic N in the leachate was not measured in our study, but it could eventually enter water bodies and cause of eutrophication. In addition, given that conventional rice has a higher NUE (29.6%), lower N input, and higher crop yield, N leaching from conventional rice production would not pose as high risk of contamination to local water systems as from organic rice production [10].

Conclusions

This study indicates that inorganic N concentrations in leachate decreased as soil depth increased, but the decrease was significantly larger in organic than in conventional paddies. NO_3-N tended to remain in the upper soil profile of organic paddies, whereas in conventional soils, NO_3-N migrated further downward. In organic paddy soils, NH_4-N accounted for a substantial portion of inorganic N in the leachate due to the organic manure decomposition and denitrification process under the continuous flooding conditions. In terms of the N leached per unit yield of rice, organic fertilization did not perform better than conventional fertilization in all cases, particularly for the organic production with higher rice yields. Consistent with this, conventional production showed higher N utilization efficiency (29.6%) compared to organic production (8.7–17.6%). We conclude that converting conventional rice production to organic production in the Ili River Valley of Central Asia will not reduce N leaching into local water systems, especially given the high-load application of organic manure to maintain high rice yields. A longer period study with integrated monitoring of N loss through leaching, volatilization, nitrification and denitrification is necessary to compare the overall performance of organic versus conventional rice production.

Acknowledgments

We thank Luo Xinhu and Liu Fuwang of the Ili Agricultural Technical Station, Xinjiang Uygur Autonomous Region for their support during field work. The authors would also like to thank the two anonymous reviewers for their helpful remarks.

Author Contributions

Conceived and designed the experiments: FqM XpS WlW. Performed the experiments: XpS. Analyzed the data: FqM XpS JEO. Contributed reagents/materials/analysis tools: FqM XpS WlW. Wrote the paper: FqM JEO.

References

1. Ghosh BC, Bhat R (1998) Environmental hazards of nitrogen loading in wetland rice fields. Environmental Pollution 102 S1: 123–126.
2. Khind CS, Meelu OP, Singh Y, Singh B (1991) Leaching losses of urea-N applied to permeable soils under lowland rice. Fertilizer Research 28: 179–184.
3. Zhou S, Sugawara S, Riya S, Sagehashi M, Toyota K, et al. (2011) Effect of infiltration rate on nitrogen dynamics in paddy soil after high-load nitrogen application containing ^{15}N tracer. Ecological Engineering 37: 685–692.
4. Devkota KP, Manschadi A, Lamers JPA, Devkota M, Vlek PLG (2013) Mineral nitrogen dynamics in irrigated rice–wheat system under different irrigation and establishment methods and residue levels in arid drylands of Central Asia. European Journal of Agronomy 47: 65–76.
5. Ju XT, Xing GX, Chen XP, Zhang SL, Zhang LJ, et al. (2009) Reducing environmental risk by improving N management in intensive Chinese agricultural systems. Proceedings of the National Academy of Sciences 106: 3041–3046.
6. Yang L, He GH (2008) The impact of Ili irrigation region construction on environment systems. Water Saving Irrigation 3: 34–35. (In Chinese).
7. Jemison JM, Fox RH (1994) Nitrate leaching from nitrogen-fertilized and manured corn measured with zero-tension pan lysimeters. Journal of Environmental Quality 23: 337–343.
8. Askegaard M, Olesen JE, Rasmussen IA, Kristensen K (2011) Nitrate leaching from organic arable crop rotations is mostly determined by autumn field management. Agriculture, Ecosystems and Environment 142: 149–160.
9. Hansen B, Kristensen ES, Grant R, Høgh-Jensen H, Simmelsgaard SE, et al. (2000) Nitrogen leaching from conventional versus organic farming systems-a systems modelling approach. European Journal of Agronomy 13: 65–82.
10. Kirchmann H, Bergström L (2001) Do organic farming practices reduce nitrate leaching? Communications in Soil Science and Plant Analysis 32: 7–8, 997–1028.
11. Chhabra A, Manjunath KR, Panigrahy S (2010) Non-point source pollution in Indian agriculture: Estimation of nitrogen losses from rice crop using remote sensing and GIS. International Journal of Applied Earth Observation and Geoinformation 12: 190–200.
12. Xing G, Cao Y, Shi S, Sun G, Du L, et al. (2001) N pollution sources and denitrification in waterbodies in Taihu Lake region. Science in China Series B: Chemistry 44: 304–314.
13. Tian Y, Yin B, Yang L, Yin S, Zhu Z (2007) Nitrogen runoff and leaching losses during rice-wheat rotations in Taihu Lake Region, China. Pedosphere 17: 445–456.
14. Li F, Pan G, Tang C, Zhang Q, Yu J (2008) Recharge source and hydrogeochemical evolution of shallow groundwater in a complex alluvial fan

system, southwest of North China Plain. Environmental Geology 55: 1109–1122.
15. Li CF, Cao CG, Wang JP, Zhan M, Yuan WL, et al. (2008) Nitrogen losses from integrated rice-duck and rice-fish ecosystems in southern China. Plant Soil 307: 207–217.
16. Zhu JG, Han Y, Liu G, Zhang YL, Shao XH (2000) Nitrogen in percolation water in paddy fields with a rice/wheat rotation. Nutrient Cycling in Agroecosystems 57: 75–82.
17. Luo LG, Wen DZ (1999) Nutrient balance in rice field ecosystem of northern China. Chinese Journal of Applied Ecology 10: 301–304. (In Chinese).
18. Bauder JW, Montgomery BR (1980) N-source and irrigation effects on nitrate leaching. Agronomy Journal 72: 593–596.
19. Xiong Z, Huang T, Ma Y, Xing G, Zhu Z (2010) Nitrate and ammonium leaching in variable-and permanent-charge paddy soils. Pedosphere 20: 209–216.
20. Keeney DR, Sahrawat KL (1986) Nitrogen transformations in flooded rice soils, Fertilizer Research 9: 15–38
21. George T, Ladha JK, Garrity DP, Buresh RJ (1993) Nitrate dynamics during the aerobic soil phase in lowland rice-based cropping systems. Soil Science Society of America Journal 57: 1526–1532.
22. Ouyang H, Xu YC, Shen QR (2009) Effect of combined use of organic and inorganic nitrogen fertilizer on rice yield and nitrogen use efficiency. Jiangsu Journal of Agricultural Science 25: 106–111. (In Chinese).
23. Wang X, Suo Y, Feng Y, Shohag M, Gao J, et al. (2011) Recovery of ^{15}N-labeled urea and soil nitrogen dynamics as affected by irrigation management and nitrogen application rate in a double rice cropping system. Plant Soil 343: 195–208.
24. Islam MM, Lyamuremye F, Dick RP (1998) Effect of organic residue amendment on mineralization of nitrogen in flooded rice soils under laboratory conditions. Communications in Soil Science and Plant Analysis 29: 7–8, 971–981.
25. Qian C, Cai ZZ (2007) Leaching of nitrogen from subtropical soils as affected by nitrification potential and base cations. Plant Soil 300: 197–205.
26. Weier KL, Doran JW, Power JF, Walters DT (1993) Denitrification and the dinitrogen/nitrous oxide ratio as affected by soil water, available carbon, and nitrate. Soil Science Society of America Journal 57: 66–72
27. Stanford G, Smith SJ (1972) Nitrogen mineralization potentials of soils. Soil Science Society of America Journal 36: 465–472
28. Reddy KR, Patrick WH Jr (1986) Denitrification losses in flooded rice fields. Fertilizer Research 9: 99–116.
29. Reddy KR, Patrick WH, Lindau CW (1989) Nitrification-denitrification at the plant root-sediment interface in wetlands. Limnol. Oceanogr 34: 1004–1013.
30. Uhel C, Roumet C, Salsac L (1989) Inducible nitrate reductase of rice plants as a possible indicator for nitrification in water-logged paddy soils. Plant and soil 116: 197–206.
31. Luo L, Itoh S, Zhang Q, Yang S, Zhang Q, et al. (2010) Leaching behavior of nitrogen in a long-term experiment on rice under different N management systems. Environmental monitoring and assessment 177: 141–150.
32. Cassman KG, Peng S, Olk DC, Ladha JK, Reichardt W, et al. (1998) Opportunities for increased nitrogen-use efficiency from improved resource management in irrigated rice systems. Field Crops Research 56: 7–39.
33. Sun XP, Li GX, Meng FQ, Guo YB, Wu WL, et al. (2011) Nutrients balance and nitrogen pollution risk analysis for organic rice production in Ili reclamation area of Xinjiang. Transactions of the CSAE 27: 158–162. (In Chinese).
34. Gao JS, Qin DZ, Liu GL, Xu MG (2002) The impact of long term organic fertilization on growth and yield of rice. Tillage and Cultivation 2: 31–33. (In Chinese).
35. Li X, Liu Q, Rong XM, Xie GX, Zhang YP, et al. (2010) Effect of organic fertilizers on rice yield and its components. Hunan Agricultural Sciences 5: 64–66. (In Chinese).
36. Xi YG, Qin P, Ding GH, Fan WL, Han CM (2007) The application of RCSODS model to fertilization practice of organic rice and its effect analysis. Shanghai Agricultural Sciences 23: 28–33. (In Chinese).
37. Chowdary VM, Rao NH, Sarma PBS (2004) A coupled soil water and nitrogen balance model for flooded rice fields in India. Agriculture, Ecosystems & Environment 103: 425–441.
38. Ma J, Sun W, Liu X, Chen F (2012) Variation in the stable carbon and nitrogen isotope composition of plants and soil along a precipitation gradient in Northern China. PloS one 7: e51894.
39. Kirchmann H, Tterer KT, Bergström L (2008) Nutrient supply in organic agriculture: plant availability, sources and recycling. In: Kirchmann H, Bergström L (eds) Organic crop production–ambitions and limitations. Springer, Netherlands, 89–116.
40. Rodrigues MA, Pereira A, Cabanas JE, Dias L, Pires J, et al. (2006) Crops use-efficiency of nitrogen from manures permitted in organic farming. European journal of agronomy 25: 328–335.
41. Guan JX, Wang BR, Li DC (2009) Effect of chemical fertilizer applied combined with organic manure on yield of rice and nitrogen using efficiency.Chinese Agricultural Science Bulletin 25: 88–92. (In Chinese).

17

Organic Farming Improves Pollination Success in Strawberries

Georg K. S. Andersson[1,2]*, Maj Rundlöf[2,3], Henrik G. Smith[1,2]

1 Centre for Environmental and Climate Research, Lund University, Lund, Sweden, 2 Department of Biology, Lund University, Lund, Sweden, 3 Department of Ecology, Swedish University of Agricultural Sciences, Uppsala, Sweden

Abstract

Pollination of insect pollinated crops has been found to be correlated to pollinator abundance and diversity. Since organic farming has the potential to mitigate negative effects of agricultural intensification on biodiversity, it may also benefit crop pollination, but direct evidence of this is scant. We evaluated the effect of organic farming on pollination of strawberry plants focusing on (1) if pollination success was higher on organic farms compared to conventional farms, and (2) if there was a time lag from conversion to organic farming until an effect was manifested. We found that pollination success and the proportion of fully pollinated berries were higher on organic compared to conventional farms and this difference was already evident 2–4 years after conversion to organic farming. Our results suggest that conversion to organic farming may rapidly increase pollination success and hence benefit the ecosystem service of crop pollination regarding both yield quantity and quality.

Editor: Dorian Q. Fuller, University College London, United Kingdom

Funding: This work was supported by The Swedish Reseasrch Council FORMAS: http://www.formas.se/default____529.aspx; Grant number in 2009: 1680. The funders had no role in study design, data collection and analysis, decision to publish, or preparation of the manuscript.

Competing Interests: The authors have declared that no competing interests exist.

* E-mail: georg.andersson@biol.lu.se

Introduction

Agricultural intensification, resulting in loss and degradation of natural and semi-natural habitats, threatens biodiversity [1,2,3] and associated ecosystem services [4,5]. Pollination provided by wild pollinators can increase crop production and benefit wild plants [6], though recent declines in pollinator numbers has been suggested as one reason for pollination deficiency in crops ([6,7] but see [8]). Although the major crops are not dependent on pollinators, large proportions of human nutrient supply come from pollinator dependent crops [9]. Hence it is important to understand how pollinators and pollination in the agricultural landscapes can be enhanced.

Organic farming has been proposed to be a means to alleviate the decreasing biodiversity in agricultural landscapes [10,11,12]. It mainly differs from conventional farming by the prohibition of most pesticides and inorganic fertilisers [13], necessitating more elaborate crop-rotations such as the use of nitrogen-fixing plants [14]. Organic farming has been shown to affect biodiversity of several taxonomic groups [15,16,17,18], but effect strength and direction can vary with taxonomic group [19], scale [12] and landscape context [16]. However, current results suggest that management effects benefiting biodiversity may not necessarily translate into improved ecosystem services [20].

Pollinator diversity and abundance often benefit from organic farming [21,22,23], which may lead to improved pollination [17,24,25]. However, how pollination is influenced by farming practice is not thoroughly understood and studies have generated contrasting results [7,26,27,28]. Effects of organic farming on pollination are not necessary simply related to pollinator richness or abundance as pollination rates may be modified by e.g. community composition, visitation frequencies, and foraging behaviour. Therefore it is important to first understand if the effects of organic farming on pollinator diversity and abundance translate into enhanced crop pollination, and then further examine the causes.

Past land-use can affect biodiversity during shorter or longer transition periods [29,30], suggesting that there may be a time-lag between land-use change and resulting changes in biodiversity and abundance [30,31]. The time-lag effect can have two main causes. First, time-lags may arise as a consequence of that organisms respond slowly to changes in habitat extension, distribution and quality, e.g. by slow dispersal, slow reproduction or both [32]. Second, diversity can respond to physical properties of the environment that change slowly themselves under the new regime, e.g. slow build-up of carbon in soil or formation of new patches of habitats. For example, a recent study showed such temporal patterns for transition time to organic farming on butterfly abundance [33]. This was attributed to a possible low carrying capacity of arable land for butterflies and a hence a steady slow increase in abundance over years with habitat improvement. Hence it can be hypothesized that effects of organic farming on pollination may not immediately be evident after conversion to organic farming. To our knowledge, this hypothesis has not been empirically tested.

We studied the effects of transition from conventional to organic farming on the pollination of strawberries, a crop that benefits from biotic pollination[34,35,36] and is visited by a wide range of pollinators, e.g. hoverflies and wild bees [37]. The main questions were *(i)* is the pollination success in strawberries higher on organic compared to conventional farms and *(ii)* is there a time-lag effect in pollination success since transition to organic farming.

Methods

All necessary permits were obtained for the described field studies. The farmers, on whose properties we conducted the study, were informed about, and approved, the studies before they started. The study was conducted on 12 farms in Scania, the southernmost part of Sweden, in 2009. Circular landscapes with a radius of 1 km around farms were digitized using ArcGIS 9.2 and characterized with information from the Integrated Administration and Control System, IACS, a database for agricultural land-use in Sweden administrated by the Swedish Board of Agriculture. We described land-use intensity using an index calculated as the proportion of annual crops and ley of all farmland and forest within the circular landscape, excluding e.g. urban areas and lakes. The index therefore reflects the proportion of the landscape under intensive cultivation. To avoid bias, in terms of organic farming being more common in complex landscapes [16], we selected all farms in landscapes where the proportion of intensively cultivated land was at least 70%.

The twelve farms were either conventionally managed ($n = 4$), recently transformed to organic farming ("new", 2–4 years; $n = 4$) or under organic management for a longer time period ("old", 14–24 years; $n = 4$). The organic farms were certified according to the Swedish certification organisation for organic products, KRAV, which mainly follows the European Council Regulation [13]. To assess pollination we established strawberry (*Fragaria ananassa*) plants as phytometers which we placed in 7.5 litre pots adjacent to field borders of spring wheat or barley to minimize the crop influence on pollinators. We placed the pots halfway down in the ground to reduce water loss and watered them when necessary. This method makes it possible to control for potential confounding factors such as soil-type. To make cross-pollination possible four pots of strawberries were used at each farm and placed just far enough to prevent the flowers to come into physical contact with each other. We used strawberries for two main reasons. First, a strawberry is an aggregated fruit where the pollination success can be assessed on each strawberry. This means that pollination success can be measured independent of total fruit set and thereby all variables apart from pollination success that may influence total fruit set. Strawberry pollination success is also increased by insect pollination [34,35,36]. Second, strawberries are an economically important crop in Sweden. We further address these issues in the discussion.

The phytometers were placed outside between June 2nd and August 8th, to allow visitation of native pollinators in the study landscapes. None of our farms had managed honeybees on them. The flowering period lasted from June 5th to July 3rd and all fully ripe strawberries were continuously collected. Pollination success was estimated by counting the numbers of malformations on each strawberry and the proportion of fully pollinated berries per farm. These malformations are formed when ovaries are not fertilised. The area of the receptacle holding these ovaries does then not swell up which results in a malformed strawberry [36]. The malformations we counted are thus areas of unfertilized achenes, or seeds, on the mature strawberry corresponding to the unpollinated stigmas and unfertilised ovaries, on the flower. On average 18.9 ± 2.1 (SE) strawberries per farm were collected and counted for malformations.

We analysed pollination success (mean number of miss-formation and proportion fully pollinated strawberries) in relation to farming practice (conventional vs. organic) and in relation to age category (new organic vs. old organic) in R ver. 2.13.1 [38]. First, the overall difference between the three categories was tested in an ANOVA. Then we used *a priori* contrasts to test, first the difference between conventional and organic farming, and second the difference between new and old organic farms. To test the proportion of fully pollinated strawberries we used a generalized linear model assuming binomial error followed by same *a priori* contrast tests as above.

Results

The global ANOVA model showed a significant effect of farming practice (conventional, new organic, old organic) on mean number of malformations ($F_{2,8} = 7.5$, $p = 0.015$). The mean number of malformations on strawberries was lower on organic farms (0.63 ± 0.089; mean $\pm$ SE), compared to conventional farms 1.26 ± 0.14 ($t_8 = -3.74$, $p = 0.0057$; Fig. 1). There was also an effect of farming practice on the proportion of fully pollinated berries ($z_{10} = 2.86$, $p = 0.0043$) with a higher proportion of fully pollinated berries on organic farms (0.45 ± 0.081) compared to conventional farms 0.17 ± 0.037 ($z_{10} = 2.77$, $p = 0.0055$; Fig. 2). The number of malformations did not differ between new (0.71 ± 0.08) and old (0.54 ± 0.16) organic farms ($t_8 = -0.96$, $p = 0.37$), nor did the proportion fully pollinated berries, 0.40 and 0.51 on new and old organic farms, respectively ($z = -1.14$, $p = 0.26$).

Discussion

Pollination potential in strawberries was significantly higher on organic compared to conventional farms, shown by the fewer malformations on berries and a higher proportion of fully pollinated berries. However, we found no effect of time since transition to organic farming on either the number of malformations or the proportion of fully pollinated strawberries. This suggests that the increase in pollination success occurs already within a few years after conversion to organic farming. As butterfly and plant species richness has been found to increase rapidly after transition to organic farming [33], our result suggest that pollinator richness may respond rapidly as well.

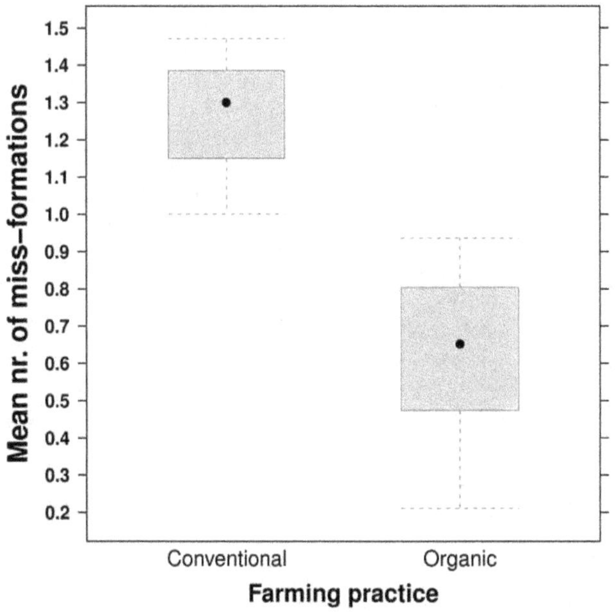

Figure 1. Mean number of malformations. The mean number of malformations on strawberries from plants on organic and conventional farms. Boxes represent 25th and 75th of the sample, dots the median and error-bars the minimum and maximum values respectively.

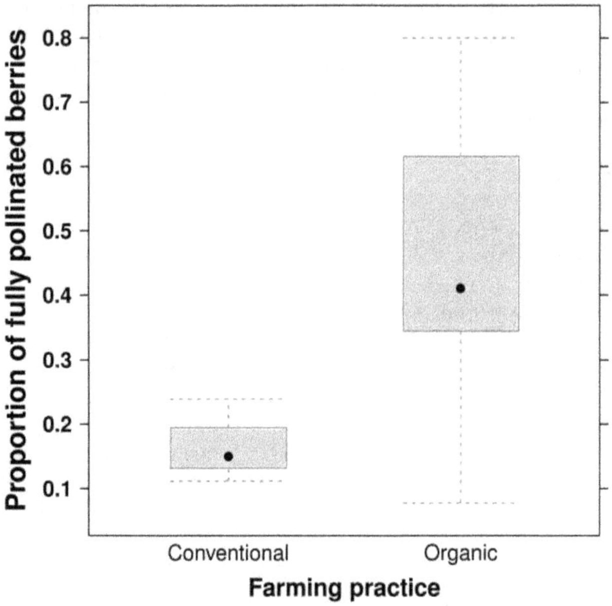

Figure 2. Proportion of fully pollinated strawberries. The proportion of fully pollinated strawberries, i.e. having no malformations, from plants on organic and conventional farms. Boxes represent 25th and 75th of the sample, dots the median and error-bars the minimum and maximum values respectively.

The few studies previously examining the effect of organic farming on seed-set, some of which has been on wild plants [17,39], have found varied results. Turnip rape-seed on organic farms had less pollination deficit than rape-seed on conventional farms [27] whereas seed-set in petunias did not differ between farming practices [26]. Strawberries, as an aggregate fruit, allowed us to assess pollination on each individual berry and consequently compare the actual pollination success between farms. A caveat when estimating pollination success on plants with non-aggregate fruits is that a number of factors other than pollination can influence seed set. Non-aggregated fruits, or simple fruits, can have many seeds but they come from a flower with only one pistil. Thus, as each seed can be affected by predation, be aborted or just not develop for various reasons, it is harder to relate the number off seeds in such types of fruits to the pollination success. Aggregate fruits come from one flower with many pistils where every ovary forms one seed, in strawberries an achene. This has the advantage that, as in the case of the strawberries, each individual strawberry, shows the pollination success. In our study 45% of the strawberries were fully pollinated on organic farms with only 17% being fully pollinated on conventional ones. In one earlier study [7] watermelons were found to receive sufficient pollen deposition to allow production of fully developed fruits on every second organic farm, but not on any conventional farm. However, no significant difference between farming practices in pollination services remained after accounting for isolation from natural habitats [7]. In the present study any confounding landscape factors were explicitly controlled for in the design of the field experiments.

Possible explanations for the higher pollination success may be a higher abundance, visitation frequency or a higher diversity of pollinators, which has been shown to be supported by organic farms [22,23]. Several studies suggest a correlation between pollinator diversity and fruit-set [24,40,41]. However, in strawberries, the influence of farming practice on pollination success may not result from a higher diversity of pollinators *per se* at organic farms, but may be a consequence of community composition of pollinators [7]. Chagnon et al. [42] showed that the quality of strawberries is affected by the composition of pollinator communities, with large- and average-sized apoids pollinating the top and small-sized apoids pollinating the bottom and sides of strawberries. However, single visits by honeybees, large hoverflies and small solitary bees contributed equally to pollination of strawberries [34] indicating that no group alone is most important. Based on our results it remains to be tested if there is a higher number of functionally important pollinator groups at organic farms or if other explanations, such as an increase in pollinator abundance, can account for the increased pollination success.

Our results suggest that increased pollination success mediated by organic farming may increase both crop yield quantity and quality on farms growing strawberries. This is economically important since approximately 11 000 tonnes of strawberries are produced annually in Sweden according to Swedish board of Agriculture statistics, 2011 and 40 million tonnes globally [43]. However, since we use phytometers, our results are not directly related to estimates of differences in crop pollination at a farm scale. First, we did not determine if strawberry plants compensate for low pollination success by e.g. producing more berries. Second, the effects may be scale-dependent hence farms with whole fields of strawberries the pattern in relation to farming practice may be different compared to a limited number of strawberry plants. In order to link our results to production in strawberry fields, simultaneous measurements of pollination in both fields and phytometers will be required.

To further determine the relationship between farming practice and the functions pollinators provide enhances our understanding of measures needed to preserve pollinators. Although the mechanistic cause for the observed differences in pollination success between farming practices remains unknown in our study, effects were present at recently transformed farms. Combined with results on other ecosystem services, our results imply that agri-environment schemes should be evaluated both directly after implementation and over a longer term.

Acknowledgments

We thank the farmers for letting us work on their fields and the field assistants for invaluable support. Thanks also to Johan Ekroos and Klaus Birkhofer for valuable comments and suggestions on the manuscript.

Author Contributions

Conceived and designed the experiments: GA HS MR. Performed the experiments: GA. Analyzed the data: GA HS MR. Contributed reagents/materials/analysis tools: GA HS MR. Wrote the paper: GA HS MR.

References

1. Kleijn D, Kohler F, Baldi A, Batary P, Concepcion ED, et al. (2009) On the relationship between farmland biodiversity and land-use intensity in Europe. Proceedings of the Royal Society B-Biological Sciences 276: 903–909.
2. Krebs JR, Wilson JD, Bradbury RB, Siriwardena GM (1999) The second Silent Spring? Nature 400: 611–612.
3. Potts SG, Biesmeijer JC, Kremen C, Neumann P, Schweiger O, et al. (2010) Global pollinator declines: trends, impacts and drivers. Trends in Ecology & Evolution 25: 345–353.
4. Tilman D, Fargione J, Wolff B, D'Antonio C, Dobson A, et al. (2001) Forecasting agriculturally driven global environmental change. Science 292: 281–284.

5. Geiger F, Bengtsson J, Berendse F, Weisser WW, Emmerson M, et al. (2010) Persistent negative effects of pesticides on biodiversity and biological control potential on European farmland. Basic and Applied Ecology 11: 97–105.
6. Klein AM, Vaissière BE, Cane JH, Steffan-Dewenter I, Cunningham SA, et al. (2007) Importance of pollinators in changing landscapes for world crops. Proceedings of the Royal Society B: Biological Sciences 274: 303–313.
7. Kremen C, Williams NM, Thorp RW (2002) Crop pollination from native bees at risk from agricultural intensification. Proceedings of the National Academy of Sciences 99: 16812–16816.
8. Ghazoul J, Koh LP (2010) Food security not (yet) threatened by declining pollination. Frontiers in Ecology and the Environment 8: 9–10.
9. Eilers EJ, Kremen C, Smith Greenleaf S, Garber AK, Klein A-M (2011) Contribution of Pollinator-Mediated Crops to Nutrients in the Human Food Supply. PLoS One 6: e21363.
10. Hole DG, Perkins AJ, Wilson JD, Alexander IH, Grice PV, et al. (2005) Does organic farming benefit biodiversity? Biological Conservation 122: 113–130.
11. Rundlöf M, Edlund M, Smith HG (2010) Organic farming at local and landscape scales benefits plant diversity. Ecography 33: 514–522.
12. Bengtsson JN, Ahnstrom J, Weibull AC (2005) The effects of organic agriculture on biodiversity and abundance: a meta-analysis. Journal of Applied Ecology 42: 261–269.
13. EC (834/2007) Council Regulation (EC) on organic production and labelling of organic products and repealing. In: Union E, ed. 2092/91. Official Journal of the European Union European Council.
14. Stockdale EA, Lampkin NH, Hovi M, Keatinge R, Lennartsson EKM, et al. (2001) Agronomic and environmental implications of organic farming systems. Advances in Agronomy: Academic Press. pp 261–262.
15. Roschewitz I, Gabriel D, Tscharntke T, Thies C (2005) The effects of landscape complexity on arable weed species diversity in organic and conventional farming. Journal of Applied Ecology 42: 873–882.
16. Rundlöf M, Smith HG (2006) The effect of organic farming on butterfly diversity depends on landscape context. Journal of Applied Ecology 43: 1121–1127.
17. Gabriel D, Tscharntke T (2007) Insect pollinated plants benefit from organic farming. Agriculture, Ecosystems & Environment 118: 43–48.
18. Smith H, Dänhardt J, Lindström Å, Rundlöf M (2010) Consequences of organic farming and landscape heterogeneity for species richness and abundance of farmland birds. Oecologia 162: 1071–1079.
19. Fuller RJ, Norton LR, Feber RE, Johnson PJ, Chamberlain DE, et al. (2005) Benefits of organic farming to biodiversity vary among taxa. Biology Letters 1: 431–434.
20. Diekötter T, Wamser S, Wolters V, Birkhofer K (2010) Landscape and management effects on structure and function of soil arthropod communities in winter wheat. Agriculture, Ecosystems & Environment 137: 108–112.
21. Holzschuh A, Steffan-Dewenter I, Kleijn D, Tscharntke T (2007) Diversity of flower-visiting bees in cereal fields: effects of farming system, landscape composition and regional context. Journal of Applied Ecology 44: 41–49.
22. Rundloef M, Nilsson H, Smith HG (2008) Interacting effects of farming practice and landscape context on bumblebees. Biological Conservation 141: 417–426.
23. Holzschuh A, Steffan-Dewenter I, Tscharntke T (2008) Agricultural landscapes with organic crops support higher pollinator diversity. Oikos 117: 354–361.
24. Hoehn P, Tscharntke T, Tylianakis JM, Steffan-Dewenter I (2008) Functional group diversity of bee pollinators increases crop yield. Proceedings of the Royal Society B: Biological Sciences 275: 2283–2291.
25. Klein A-M, Steffan-Dewenter I, Tscharntke T (2003) Fruit set of highland coffee increases with the diversity of pollinating bees. Proceedings of the Royal Society B: Biological Sciences 270: 955–961.
26. Brittain C, Bommarco R, Vighi M, Settele J, Potts SG (2010) Organic farming in isolated landscapes does not benefit flower-visiting insects and pollination. Biological Conservation 143: 1860–1867.
27. Morandin LA, Winston ML (2005) Wild bee abundance and seed production in conventional, organic, and genetically modified canola. Ecological Applications 15: 871–881.
28. Ekroos J, Hyvönen T, Tiainen J, Tiira M (2010) Responses in plant and carabid communities to farming practises in boreal landscapes. Agriculture, Ecosystems & Environment 135: 288–293.
29. Lindborg R, Eriksson O (2004) Historical landscape connectivity affects present plant species diversity. Ecology 85: 1840–1845.
30. Bissonette JA, Storch I (2007) Temporal Dimensions of Landscape Ecology. First ed. New York: Springer.
31. Andersson GKS, Rundlöf M, Smith HG (2010) Time lags in biodiversity response to farming practices. In: Boatman ND, Green M, Holland J, Marshall J, Renwick A, et al. (2010) Aspects of Applied Biology; 2010; University of Leicester, Oadby, UK Association of Applied Biologists. pp 381–384.
32. With KA (2007) Invoking the Ghost of Landscapes Past to Understand the Landscape Ecology of the present...and the Future. In: Bissonette JA, Storch I, eds. Temporal Dimensions of Landscape Eology First ed. New York: Springer. pp 43–58.
33. Jonason D, Andersson GKS, Öckinger E, Rundlöf M, Smith HG, et al. (2011) Assessing the effect of the time since transition to organic farming on plants and butterflies. Journal of Applied Ecology 48: 543–550.
34. Albano S, Salvado E, Duarte S, Mexia A, Borges PAV (2009) Pollination effectiveness of different strawberry floral visitors in Ribatejo, Portugal: selection of potential pollinators. Part 2. Advances in Horticultural Science 23: 246–253.
35. Lopez-Medina J, Palacio-Villegas A, Vazquez-Ortiz E (2006) Misshaped fruit in strawberry, an agronomic evaluation. Acta Horticulturae 708: 77–78.
36. Free JB (1993) Insect Pollination of Crops. London: Academic Press.
37. Albano S, Salvado E, Duarte S, Mexia A, Borges PAV (2009) Floral visitors, their frequency, activity rate and Index of Visitation Rate in the strawberry fields of Ribatejo, Portugal: selection of potential pollinators. Part 1. Advances in Horticultural Science 23: 238–245.
38. R Development Core Team (2011) R: A Language and Environment for Statistical Computing. 2.14.0 ed. Vienna, Austria: R Foundation for Statistical Computing.
39. Power EF, Stout JC (2011) Organic dairy farming: impacts on insect–flower interaction networks and pollination. Journal of Applied Ecology 48: 561–569.
40. Klein AM (2009) Nearby rainforest promotes coffee pollination by increasing spatio-temporal stability in bee species richness. Forest Ecology and Management 258: 1838–1845.
41. Vergara CH, Badano EI (2009) Pollinator diversity increases fruit production in Mexican coffee plantations: The importance of rustic management systems. Agriculture, Ecosystems & Environment 129: 117–123.
42. Chagnon M, Gingras J, De Oliveira D (1993) Complementary aspects of strawberry pollination by honey and indigenous bees (Hymenoptera). Journal of Economic Entomology 86: 416–420.
43. FAO (2011) Crop production database (2008). Food and Agricultural Organisation of the United Nations.

18

Eco-Label Conveys Reliable Information on Fish Stock Health to Seafood Consumers

Nicolás L. Gutiérrez[1]*, Sarah R. Valencia[2], Trevor A. Branch[3], David J. Agnew[1], Julia K. Baum[4], Patricia L. Bianchi[1], Jorge Cornejo-Donoso[5,6], Christopher Costello[2], Omar Defeo[7], Timothy E. Essington[3], Ray Hilborn[3], Daniel D. Hoggarth[1], Ashley E. Larsen[8], Chris Ninnes[1], Keith Sainsbury[9], Rebecca L. Selden[8], Seeta Sistla[8], Anthony D. M. Smith[10], Amanda Stern-Pirlot[1], Sarah J. Teck[8], James T. Thorson[3], Nicholas E. Williams[11]

1 Marine Stewardship Council, London, United Kingdom, **2** Bren School of Environmental Science and Management, University of California Santa Barbara, Santa Barbara, California, United States of America, **3** School of Aquatic and Fishery Sciences, University of Washington, Seattle, Washington, United States of America, **4** Department of Biology, University of Victoria, Victoria, British Columbia, Canada, **5** Interdepartmental Graduate Program in Marine Science, Marine Science Institute, University of California Santa Barbara, Santa Barbara, California, United States of America, **6** Universidad Austral de Chile, Centro Trapananda, Coyhaique, Chile, **7** UNDECIMAR, Facultad de Ciencias, Montevideo, Uruguay, **8** Department of Ecology, Evolution and Marine Biology, University of California Santa Barbara, Santa Barbara, California, United States of America, **9** University of Tasmania, Tasmanian Aquaculture & Fisheries Inst, Taroona, Tasmania, Australia, **10** Commonwealth Scientific and Industrial Research Organization, Wealth from Oceans Flagship, Hobart, Tasmania, Australia, **11** Department of Anthropology, University of California Santa Barbara, Santa Barbara, California, United States of America

Abstract

Concerns over fishing impacts on marine populations and ecosystems have intensified the need to improve ocean management. One increasingly popular market-based instrument for ecological stewardship is the use of certification and eco-labeling programs to highlight sustainable fisheries with low environmental impacts. The Marine Stewardship Council (MSC) is the most prominent of these programs. Despite widespread discussions about the rigor of the MSC standards, no comprehensive analysis of the performance of MSC-certified fish stocks has yet been conducted. We compared status and abundance trends of 45 certified stocks with those of 179 uncertified stocks, finding that 74% of certified fisheries were above biomass levels that would produce maximum sustainable yield, compared with only 44% of uncertified fisheries. On average, the biomass of certified stocks increased by 46% over the past 10 years, whereas uncertified fisheries increased by just 9%. As part of the MSC process, fisheries initially go through a confidential pre-assessment process. When certified fisheries are compared with those that decline to pursue full certification after pre-assessment, certified stocks had much lower mean exploitation rates (67% of the rate producing maximum sustainable yield vs. 92% for those declining to pursue certification), allowing for more sustainable harvesting and in many cases biomass rebuilding. From a consumer's point of view this means that MSC-certified seafood is 3–5 times less likely to be subject to harmful fishing than uncertified seafood. Thus, MSC-certification accurately identifies healthy fish stocks and conveys reliable information on stock status to seafood consumers.

Editor: Myron Peck, University of Hamburg, Germany

Funding: SRV, JCD, AEL, RLS, SS, SJT, and NEW thank the Henry Luce Foundation and the National Center for Ecological Analysis and Synthesis, which is funded by National Science Foundation (NSF) Grant EF-0553768, the University of California, Santa Barbara, and the State of California. TAB was funded by NSF grant 1041570. The funders had no role in study design, data collection and analysis, decision to publish, or preparation of the manuscript.

Competing Interests: The authors have declared that no competing interests exist.

* E-mail: nicolas.gutierrez@msc.org

Introduction

The global per-capita demand for seafood has reached an all-time high, and is likely to continue to increase [1]. Wild capture seafood harvest has also peaked, and while management measures have led to rebuilding in some fish stocks, one-third of the world's well-studied fisheries are overfished [1–3]. To maintain or enhance wild fish supplies on a sustainable basis management agencies and governments must rebuild fisheries whose stocks are at low biomass and maintain healthy stocks at or above sustainable levels. Fisheries and conservation objectives can be attained by redundancy in management actions, including catch controls, gear modifications, closed areas, and community-based management, depending on local context and specific features [2,4]. One way of influencing these fishery practices is through market-based approaches such as "eco-labeling" which aim to harness consumer preferences to increase market demand, and often prices, for well-managed fisheries and diminish demand for others [5,6]. To further this aim, national and global schemes designed to allow consumers to make informed choices when purchasing seafood have proliferated [5]. These efforts include awareness campaigns such as consumer guides produced by Monterey Bay Aquarium, World Wildlife Fund and Greenpeace, the risk of extinction Red List categories of the IUCN, and certification and eco-labeling programs such as the Marine Stewardship Council (MSC) and Friend of the Sea [7].

Unlike some consumer awareness campaigns, certification programs such as the MSC consider a fishery or fish stock, rather than a species, to be the primary unit of certification. This acknowledges variation in harvest practices among fleets and recognizes those that adopt environmentally sound activities [7]. In theory, eco-labels convey information about these improved fishing practices to consumers, who then make choices about what seafood to buy based on this information. Consumer preference can result in increased prices [8] and indirect non-economic benefits for fishers [9,10] and access to markets looking to exclusively source certified fish products [7,11]. Moreover, leading supermarkets and restaurant chains recognize that consumers increasingly expect retailers to make responsible purchasing decisions as part of their corporate social responsibility, and may require third-party certification in the products they source. These act as incentives for improvement in uncertified fisheries and for continued stewardship in certified fisheries. The conservation value of eco-labels, however, relies on their ability to convey accurate information to consumers about the sustainability of fisheries.

The Marine Stewardship Council is the most prominent global fisheries eco-label program. It arose from a partnership between the World Wildlife Fund and Unilever in 1996, and has operated as an independent non-profit since 1999. There are currently 132 MSC-certified fisheries and 141 more at different stages of the certification process. MSC-certified seafood covers 10% of the annual global harvest of wild capture fisheries and more than 13,000 products, by far the highest representation of eco-labeled seafood in global markets [7,8]. The MSC's rapid growth has stoked debate about what constitutes a sustainable fishery, and the decisions to certify certain fisheries as sustainable have been scrutinized. Recent criticisms have questioned the rigor of the MSC's certification standards for ecosystem impacts of fisheries that damage habitats or result in high levels of bycatch [12–16]. There have been calls to focus certification on small-scale fisheries [12,16–18] based on the perception that they have a lower environmental impact than industrial fisheries [16], despite a paucity of data with which to assess the sustainability of small-scale fisheries. The strongest criticism of the MSC, however, argues that its certification standards fail to accurately identify healthy stocks, with several case studies cited by critics as not being sustainably managed, including Pacific Hake (*Merluccius productus*) and Eastern Bering Sea (EBS) walleye pollock (*Theragra chalcogramma*) [12,16].

The MSC certifies fisheries as sustainable only if they score highly on each of 3 principles [19]: (1) fishing should be conducted in a way that prevents overfishing (depletion of exploited populations beyond biological limits) through the use of target reference points that should maintain the stock at or above the biomass that produces the maximum sustainable yield (MSY), and overexploited stocks must be demonstrably on a path to recovery; (2) fishing operations must maintain the structure, diversity, function, and productivity of associated ecosystems; and (3) the management system must respect national and international regulations. Fisheries applying for MSC certification first undergo a confidential pre-assessment stage to evaluate their potential for meeting the certification standard. Based on this evaluation, fishing industry groups decide whether to undergo a public full assessment by an independent third party. Fisheries meeting the above standards are certified for five years and undergo annual surveillance audits. Those that meet the standard but are weak in certain areas can be certified if they commit to and demonstrate progress toward meeting agreed conditions on improvement. Thus, fisheries must demonstrate continued adherence to, and improvement in, a variety of aspects of sustainability to maintain their certification status.

The term "sustainable" is difficult to define because it encompasses ecological, social, and economic components. At a basic level, however, a renewable resource must be extracted no faster than the level at which it can replace itself for it to be considered sustainable. If certified fisheries are no better at identifying and responding to low biomass levels than uncertified fisheries, then eco-labeling is unlikely to catalyze widespread improvements in fisheries management [14]. Therefore the decline of some certified stocks has cast doubt on the validity of the information conveyed by the MSC label.

Here we assess the performance of fish stocks against Principle 1 (targeted population status), specifically evaluating the status and harvest levels of fish stocks targeted by MSC-certified fisheries, because recent criticisms of MSC have questioned whether these fisheries actually target healthy stocks [12,16]. While all three principles are equally weighted criteria in the MSC's assessment process, Principles 2 and 3 address effects of fishing whose impacts can be difficult to measure directly. Thus assessing the performance of certified vs. uncertified fish stocks is a critical first step in evaluating the effectiveness of the MSC's certification standards.

Methods

Data

We compiled time series of catch data and model estimates of biomass and fishing mortality rates for all stocks where the above information was available (45 certified stocks, Table S1, and 179 uncertified stocks, Table S2). Certified stocks managed under different schemes than single-species MSY (e.g., salmon and invertebrates) or without biomass time series and thus qualitatively assessed under a risk-based framework [19] (e.g., small-scale and data-deficient fisheries) were excluded from our analysis. However, given that some stocks are targeted by multiple certified fisheries, analyzed stocks represented 62% ($n = 82$) of the total certified fisheries ($n = 133$) and 85% of the certified landings (4.5 million tons).

The majority of data was sourced from the RAM Legacy Stock Assessment Database (20), which represents the largest global stock assessment database currently available, but status was updated if newer stock assessments were available (Table S1 and Table S2). For each stock we recorded the biomass (B_{MSY}) and exploitation rate (u_{MSY}) or fishing mortality (F_{MSY}) that results in MSY from published stock assessments. When estimates of one or both reference points were not available, they were estimated by fitting Schaefer surplus production models [2,20] to time series of biomass estimates using a maximum likelihood approach in AD Model Builder [21]. Table S1 includes the method used for each certified stock, and Table S2 lists the method for each uncertified stock. Finally, we collected time series of biomass estimates from stock assessments in order to compare long-term trends in biomass relative to target B_{MSY} for certified and uncertified stocks.

The confidential MSC pre-assessment process screens out many applicants that are unlikely to achieve full certification. Since 1997, 447 fisheries have applied for pre-assessment. Of these fisheries, 55% were not recommended to enter full assessment due either to major management weaknesses (35% of the 447 fisheries) or because they had low biomass, high exploitation rates, or insufficient information to judge stock status (20% of the 447 fisheries) [22]. Of the fisheries not recommended to enter full assessment, 25 stocks for which information was available were included in our analysis (Text S1). Thus our analysis compared certified stocks with both uncertified stocks and the 25 stocks (counted among the 179) that a pre-assessment had suggested would fail the Principle 1 standard. Due to confidentiality, details about the identities of these stocks cannot be released.

Definitions and Analysis

According to international agreements and many national laws, fish stocks should be maintained at or rebuilt to a size that can support MSY [23,24]. This biomass, denoted B_{MSY}, is typically 20–50% of the average population biomass in the absence of fishing [25]. The corresponding annual exploitation rate (u, catch divided by total biomass) that stabilizes the stock around B_{MSY} is u_{MSY}. We follow international convention by using MSY-based reference points, which remain the most widely used method of assessing whether stocks are overfished or not. For example, the UN Convention on the Law of the Sea [23] requires countries to rebuild to MSY levels, and the United Nations Fish Stock Agreement [24] specifies F_{MSY} as a reference point. We acknowledge that there is some debate about whether MSY reference points are an appropriate target for sustainability, and that alternative methods may result in differing estimates of B_{MSY} and F_{MSY}, but these are larger issues than can be addressed in this paper.

For a fishery to be MSC-certified, biomass should be at or fluctuating around B_{MSY}, or if consistently below B_{MSY}, should be under a rebuilding plan (i.e., $F<F_{MSY}$) that will lead to recovery of the stock in the near future [19]. There is also a minimum level of biomass, or limit reference point, below which stocks are considered overfished and certification cannot be obtained irrespective of any rebuilding plan. Limit reference points must be set such that if a stock is maintained above these, there is a very low risk of impaired recruitment. These MSY reference points are currently the most informative benchmark with which to assess status across a global sample of fish stocks [2,20].

We compared the biomass status and exploitation rate in relation to MSY targets of certified, uncertified, and non-recommended fish stocks by plotting B/B_{MSY} vs. u/u_{MSY} or F/F_{MSY} and using kernel density smoothing functions to describe the probability of occurrence in each quadrant. To determine whether B/B_{MSY} and u/u_{MSY} were significantly different between groups we used re-sampling inference (100,000 times without replacement), which allows us to assess how often a difference of the observed magnitude or larger would arise by chance. We also estimated the proportion of stocks in each group that met or exceeded biomass and harvest targets and that were below $0.5B_{MSY}$ (as a proxy for the U.S. legal definition of overfished, or the point below which recruitment can be impaired for certain stocks; [26]) and above a more conservative target of $1.3B_{MSY}$ [27]. The long-term performance of certified and uncertified stocks in relation to B_{MSY} was assessed using time series data from 1970 to the present (available for 165 uncertified stocks and 31 certified stocks). Using an autoregressive model we tested for differences in the conditional mean of B/B_{MSY} between certified and uncertified stocks over time. The model structure was selected through Akaike's Information Criteria (AIC) and model parameters were estimated using Ordinary Least Squares.

Results and Discussion

Statuses of Certified and Uncertified Stocks

Our analysis indicates that the ratio of current biomass to the biomass at MSY ($B_{current}/B_{MSY}$) is significantly different between certified and uncertified fisheries (1.25 vs. 0.87; $P<0.005$), and between certified and non-recommended fisheries (1.25 vs. 0.48; $P<0.005$; Table S3 and Table S4). We found 74% of the certified stocks to be currently above sustainable target biomass levels (i.e., $B_{current}>B_{MSY}$; Fig. 1A) (Text S1), compared with 44% of uncertified stocks, and 16% of non-recommended fisheries (Fig. 1B; Table S3). Given that certification assumes harvesting at MSY levels, which will result in biomass fluctuating around B_{MSY}, we would expect 50% of stocks above B_{MSY} and 50% below B_{MSY}. Thus, our finding that three quarters of stocks targeted by MSC-certified fisheries are above B_{MSY} suggests that managers of these stocks are often aiming to ensure biomass is not kept near B_{MSY} but above B_{MSY}. Additionally, 82% of certified stocks had current exploitation rates that are expected to maintain the stocks at B_{MSY} or allow for rebuilding to B_{MSY} (i.e., $u_{current}<u_{MSY}$) compared with 65% of uncertified stocks and 52% of non-recommended stocks.

Non-recommended Stocks

An analysis of non-recommended stocks revealed that 52% were below $0.5B_{MSY}$ and 48% had exploitation rates higher than u_{MSY} (Fig. 1C). These levels are significantly worse than those for certified stocks ($P<0.005$; Table S3), providing further evidence that the pre-certification process screens out stocks that do not meet internationally recognized standards for stock health. It may seem puzzling that such fisheries would apply for certification in the first place, but this may reflect differing perceptions of what sustainability means, as well as different reasons for pursuing certification. Non-recommended fisheries also show poorer stock status when compared with the global sample of uncertified fisheries. Reasons for such differences may be due to a lack of market incentives to pursue certification for some fisheries targeting healthy stocks [10], or due to weaknesses in other aspects of uncertified fisheries such as management systems or bycatch considerations [22]. Moreover, fisheries with known poor stock status may use pre-assessment as an endorsement to conduct formal stock assessments or as benchmark for amendment of management plans towards MSC certification [22].

Comparing Rates of Poor Performance

Fisheries of most concern are those with biomass below B_{MSY} and continued high harvest rates, which hinder stock rebuilding to sustainable levels ($u>u_{MSY}$; top left sector of Fig. 1A). Four certified stocks (9% of those analyzed) fell into this category, including North Sea saithe (*Pollachius virens*), North Sea sole (*Solea solea*), Atlantic Iberian sardine (*Sardina pilchardus*) and deep-water Cape hake (*Merluccius paradoxus*) (Table S1), compared with 51 (28%) uncertified stocks (Table S3) and 11 (44%) non-recommended stocks. In other words, certified fisheries are 3–5 times less likely to be subject to harmful overfishing than uncertified fisheries. At the time of the original MSC assessment, all three European stocks met certification requirements, as they were above the limit reference points defined by the International Council for the Exploration of the Seas (ICES). However, for the Iberian sardine, the mandatory annual audit revealed poor stock condition (i.e., below the limit reference point where recruitment to the population could be impaired) causing the suspension of the MSC certificate in January 2012, which will remain in place until the stock recovers. The other 3 stocks (North Sea saithe, North Sea sole and deep-water Cape hake) are currently well above their respective limit reference points as defined by the relevant advisory bodies [14,28].

The MSC Standard requires stocks to be above a limit reference point, which itself is above the point at which recruitment is impaired, and requires that if this point is not empirically determined, a biomass level of $0.5B_{MSY}$ could be used as an acceptable proxy. The U.S. also uses $0.5B_{MSY}$ as a default Minimum Stock Size Threshold (MSST), below which stocks are classified as "overfished" [26]. In our analysis, 46 (27%) un-

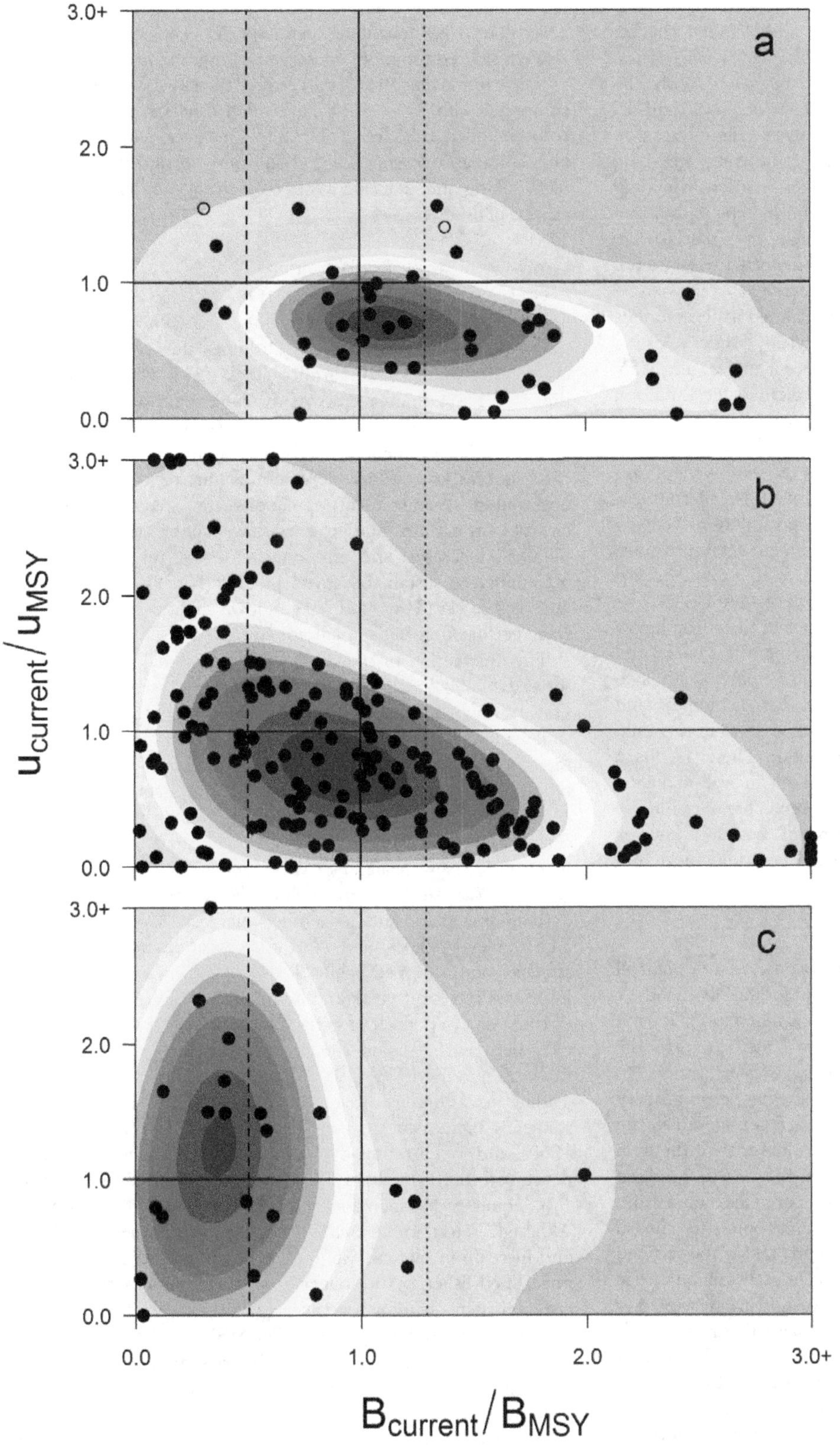

Figure 1. Sustainability of certified and uncertified seafood. Current (i.e., most recent year with available information) biomass and exploitation rate for (**A**) individual certified ($n = 45$); (**B**) all uncertified ($n = 179$); and (**C**) uncertified stocks that went through pre-assessment and were not recommended for certification ($n = 25$). Data are scaled relative to *B*MSY and *u*MSY or *F*MSY (the biomass and exploitation rates or fishing mortality rates that produce maximum sustainable yield). Contour colors show probability of occurrence (red indicates the highest probability and blue the lowest). Vertical and horizontal solid lines represent reference points common to all fisheries (*B/B*MSY = 1 and *u/u*MSY or *F/F*MSY = 1). Dotted lined represents *B* = 1.3*B*MSY and dashed line *B* = 0.5*B*MSY. Footnote: New assessments for some fish stocks were released while this paper was in press, but this figure was not updated to maintain consistency in year of release.

certified stocks had biomass levels below $0.5B_{MSY}$, compared with 4 (9%) certified stocks: North Sea saithe and Atlantic Iberian sardine as previously described, Eastern Baltic cod (*Gadus morhua*) and North Sea haddock (*Melanogrammus aeglefinus*). With the exception of the Atlantic Iberian sardine, which has had its MSC certificate removed, all 3 stocks are above the limit reference points defined by ICES without signs of recruitment overfishing [29]. Moreover, Eastern Baltic cod is under a strict rebuilding plan, which has resulted in an 80% reduction in exploitation rate and a three-fold increase in spawning biomass over the last five years [30] and North Sea haddock has experienced a reduction of the exploitation rate of 55% in the last 5 years [31].

As a target, B_{MSY} is often conditionally defined as the biomass that produces the maximum sustainable yield "under existing environmental conditions" [32]. Thus, even under an MSY control rule where B_{MSY} is the target stock biomass level, natural variation in productivity will result in stock fluctuations, being half of the time below B_{MSY} and half of the time above B_{MSY} [33]. Similarly, harvest rates might exceed u_{MSY} in some years. This natural population variability (typically driven by recruitment fluctuations in marine fishes) precludes the possibility of keeping stocks at constant levels. Successful management therefore must include continuous monitoring and an ability to adjust and enforce harvest levels [33,34]. It is for these reasons that stocks can be certified, and retain their certification even when they drop below B_{MSY}, provided they include proper management feedbacks and precautionary limits. Specifically, fisheries must have a management system in place that will detect decreases in biomass and respond by reducing the exploitation rate to a level that should enable recovery (i.e., harvest control rules). These feedback mechanisms present in certified fisheries result in well-managed fisheries fluctuating around their target reference points. Fisheries must meet these and other criteria in terms of ecosystem impacts and compliance with local and international laws to achieve and maintain certification [19].

Weighting Results by Stock Size

By using individual fish stocks as our unit of analysis we weight all stocks equally. However, this does not account for the fact that their total biomass can differ by several orders of magnitude. Recent annual landings of certified stocks range from 7 metric tons for the North-eastern inshore sea bass (*Dicentrarchus labrax*) fishery to 1.7 million tons for the Pacific skipjack (*Katsuwonus pelamis*) fishery (Table S1). While 179 uncertified fisheries are considered in the present analysis, their combined landings are lower than the total landings of the 45 certified fisheries analyzed (6.8 vs. 8.0 million metric tons; Tables S1 and S2). When we compared the biomass and exploitation rates in relation to MSY reference points for the 10 largest certified and uncertified stocks we found that 8 of the certified stocks, including the largest which represent almost 6 million metric tons of landed seafood, are at or above B_{MSY}, and are harvested at rates that should maintain the stocks above or fluctuating around their reference points (i.e., $u_{current} \leq u_{MSY}$; Table 1). Notably, EBS walleye pollock, which has been the target of many criticisms due to declines in biomass since certification [11,15,17], currently has a biomass level 25% higher than the target level (B_{MSY}) and an exploitation rate that is less than half of u_{MSY} (Table 1). In contrast, most of the 10 largest uncertified stocks have biomass levels substantially lower than B_{MSY} and exploitation rates considerably higher than u_{MSY} (Table 1).

Evaluating Performance against Conservation Targets

The sustainable yield in an ecosystem context depends on both the trophic level of the species [35] and the structure of the ecosystem [36]. As a result, the target biomass associated with an ecologically sustainable yield is unknown for each stock, and may be higher or lower than B_{MSY} depending on the species. To account for this, we also considered the more conservative target biomass of 1.3 B_{MSY} [37]. We found that 49% of the certified stocks were above 1.3 B_{MSY}, compared with 29% of the uncertified fisheries and 4% of non-recommended fisheries (Table S3). However, such conservative target biomass reference points should be evaluated on a case by case basis taking into consideration biological, economic and social aspects of fisheries [38].

Long-term Stock Performance

Given the relatively young age of the MSC, with 40% of stocks and 65% of fisheries certified in the last two years, trends in population biomass for individual fisheries after certification could only be analyzed for 10 certified stocks (23% of those analyzed) that had more than 5 years of available data after certification (Fig. 2; Table S1). For 7 out of 10 stocks, a combination of favorable environmental conditions, improved compliance to catch quotas (total allowable catch) as part of a rebuilding plan, and multiple management regulations (e.g., spatial and temporal closures, minimum landing sizes) have contributed to observed biomass increases and in some cases a rapid recovery in abundance to sustainable levels [30,31] (Fig. 2). The rest of the analyzed stocks (3 out of 10) have shown slight declines in biomass since certification but remain above B_{MSY}.

This fluctuation of stock size over time is evident in the time series data of biomass in relation to B_{MSY} after certification (Fig. 2). However, whether the changes over time seen in the 10 stocks with sufficient post-certification data are the product of good management depends on the initial biomass in relation to the target at the time the fishery was certified. South Georgia Patagonian toothfish (*Dissostichus eleginoides*), for example, declined in biomass since certification, but was still 22% larger than B_{MSY} in 2009. Conversely, New Zealand's eastern and western hoki (*Macruronus novaezelandiae*) stocks were below B_{MSY} and certified under a rebuilding plan, and have since increased 300% in biomass. EBS walleye pollock and Northeast Atlantic mackerel (*Scomber scombrus*) both dropped below B_{MSY} after being certified but have improved in recent years and are now above B_{MSY}. Certification of EBS walleye pollock was criticized [16] because the biomass fell 64% between 2004 and 2009. However, climate regime shifts in this region in the 1970s [39] greatly increased pollock productivity leading to a marked surge in biomass. For example, pollock spawning biomass in the 2000s has averaged 3.3 times that in the 1960s and is currently 5 times higher than when the fishery developed in 1964 [40]. Finally, for those stocks currently below B_{MSY}, biomass has increased since the time of certification (Fig. 2). Although determining a causal connection between certification and increase in biomass would require an evaluation of reference uncertified fisheries and longer time series, the observed patterns are consistent with conservative harvest levels and responsive management systems required under MSC certification standards.

Natural variability, changes in fisheries management systems, and adjustments in the MSC's standards would likely affect the performance of certified fisheries against the standards in a particular year. When we examined four decades (1970-present) of data to characterize long-term trends in biomass relative to B_{MSY} we found that certified stocks on average performed better over the long-term than uncertified fisheries (Fig. 3). Biomass of uncertified fish stocks globally has been below B_{MSY} since the 1970s but shows signs of recovery towards B_{MSY} since 2000, while certified stocks have on average been consistently above B_{MSY} since 1980 (biomass long-term average $= 1.3B_{MSY}$). It is possible that differences early in the time series may reflect differences in

Table 1. Stock status by landings.

Stock name	Species	Large Marine Ecosystem	Landings (MT)	Most recent year with data	$B_{current}/B_{MSY}$	$u_{current}/u_{MSY}$
Certified						
Skipjack tuna	*Katsuwonus pelamis*	Pacific High Seas	1,700,000	2010	2.67	0.34
Herring	*Clupea harengus*	North East Atlantic	1,687,371	2010	1.24	1.05
Bering Sea walleye pollock	*Theragra chalcogramma*	East Bering Sea	813,000	2011	1.25	0.46
North East Atlantic mackerel*	*Scomber scombrus*	Celtic-Biscay Shelf	734,889	2010	1.37	1.27
Barents Sea Atlantic cod	*Gadus morhua*	Barents Sea	523,430	2010	1.02	0.58
Barents Sea saithe	*Pollachius virens*	Barents Sea	520,529	2010	1.08	0.99
Pacific hake	*Merluccius productus*	California Current	216,910	2010	1.75	0.82
Barents Sea haddock	*Melanogrammus aeglefinus*	Barents Sea	200,512	2010	1.20	0.71
North Sea herring	*Clupea harengus*	North Sea	168,443	2010	0.93	0.47
North Sea saithe	*Pollachius virens*	North Sea	161,462	2010	0.37	1.27
Median					***1.22***	***0.77***
Uncertified						
Chilean Jack Mackerel	*Trachurus murphyi*	Humboldt Current	744,495	2010	0.09	3.66
Blue Whiting Northeast Atlantic	*Micromesistius poutassou*	Iceland Shelf	634,978	2010	0.29	1.01
Yellowfin tuna Central Western Pacific	*Thunnus albacares*	Pacific High Seas	413,418	2005	1.29	0.80
Capelin Iceland	*Mallotus villosus*	Iceland Shelf	391,000	2010	0.40	0.01
Yellowfin tuna Indian Ocean	*Thunnus albacares*	Indian Ocean	325,854	2009	1.02	1.15
Capelin Barents Sea	*Mallotus villosus*	Barents Sea	323,000	2010	1.01	0.27
Sandeel North Sea Dogger Bank SA1	*Ammodytes marinus*	North Sea	285,794	2010	1.86	0.28
Yellowfin tuna Eastern Pacific	*Thunnus albacares*	Pacific High Seas	255,923	2010	0.71	1.13
Sardine South Africa	*Sardinops sagax*	Benguela Current	217,138	2006	0.75	0.55
Argentine hake Southern Argentina	*Merluccius hubbsi*	Patagonian Shelf	212,618	2008	0.40	1.49
Median					***0.73***	***0.91***

Biomass status and exploitation rate in relation to MSY reference points for the 10 largest certified and uncertified stocks by landings (metric tons, in 2010). Rows in ***bold italics*** represent median values for certified and uncertified stocks.
*MSC certificate currently suspended.

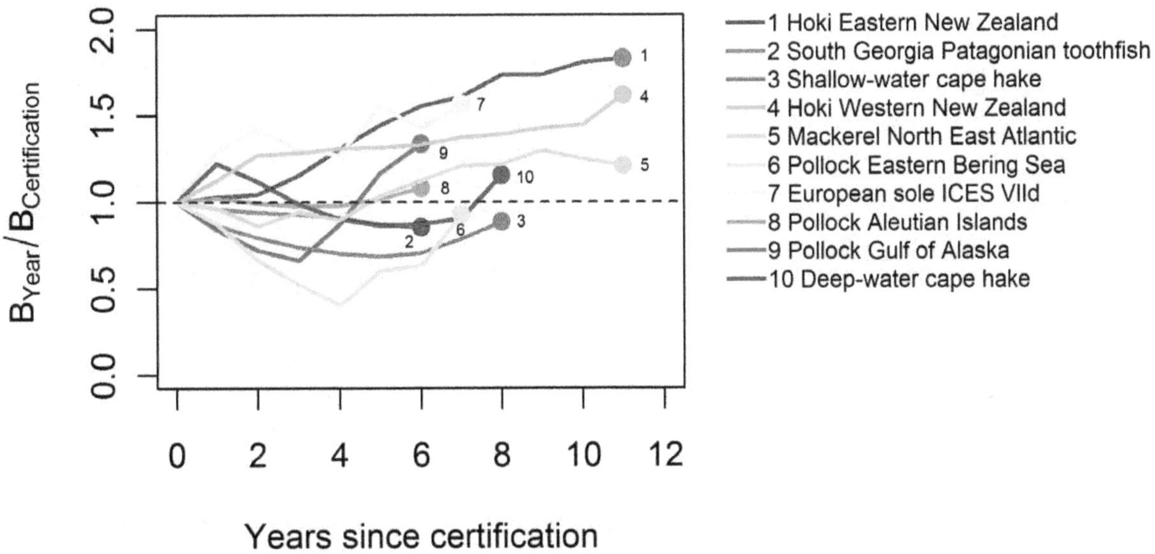

Figure 2. Time trends of current biomass (total or spawning) relative to biomass levels at the time of MSC assessment for stocks with available information and more than 5 years of certification. Colors represent current stock status (green to yellow: $B>BMSY$; orange to red: $B<BMSY$) and dots represent most recent year of available information (as per Table S1).

how and when fisheries developed. However, MSC began certifying fisheries in 1999, while certified and uncertified stocks diverged from each other in the 1980s. This suggests that the stocks certified by MSC were performing well prior to certification. This improved performance has continued over the last 10 years, with certified stocks experiencing an average 45% increase in biomass compared with a 9% increase for uncertified stocks (Fig. 3; Table S5).

Conclusions

Successful single-species management is only one part of fostering sustainable fisheries. There has been increasing support to move away from the single-species paradigm and towards an ecosystem-based approach [41,42] which recognizes that fishing has both direct and indirect effects on marine systems [43]. Conserving ecosystem diversity and structure will play an important role in helping to retain ecosystem function in the face of climate change [44,45]. While the impacts of fishing on ecosystems are difficult to measure, a few studies have attempted to quantify damage done not only by different types of gear but also by the volume of gear in the water [46,47]. Through their certification standards the MSC has the opportunity to recognize fisheries that limit the collateral impacts of fishing on food webs, habitats, and ecosystems structure. However, more research is needed to quantify the performance of certified fisheries in these areas in comparison to uncertified fisheries [22].

Most certified fisheries come from developed countries with strong central governments, sophisticated fisheries management and data-rich situations. The analyzed time series (Fig. 3) suggests that these fisheries were already well managed by their agencies before certification. Given the increase in the number of fisheries seeking MSC certification in the last three years, future analyses will be able to examine the effect of MSC certification on initially less well managed fisheries, particularly small-scale and data-limited fisheries, which are critically important to developing world economies in terms of employment, national food security, and foreign exchange earnings [1]. An open question is whether certification and eco-label programs should raise the bar of sustainability, which may result in decreased market opportunities for small-scale and data-limited fisheries, or attempt to catalyze positive changes in those fisheries with informal and traditional management that are characteristic of many parts of the developing world, and in fisheries with current poor performance. These fisheries must be part of the solution in the pursuit of sustainable fisheries on a global scale.

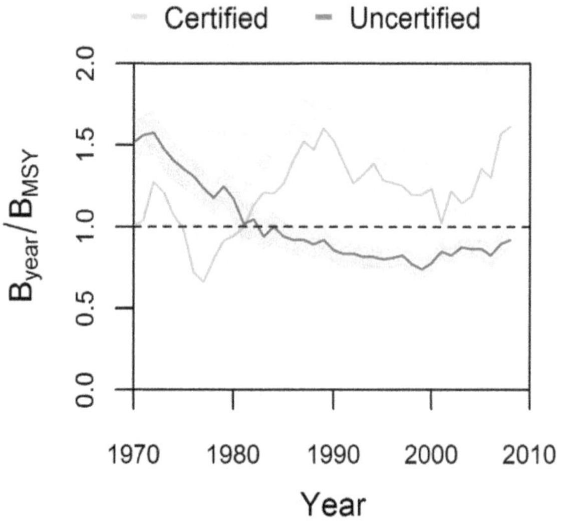

Figure 3. Performance of MSC-certified and uncertified fisheries. Long term trends (1970–2009) of biomass relative to their targets levels (i.e., estimated biomass at which the maximum sustainable yield should be obtained: B_{MSY}; median ±S.E.). B_{MSY} is set to 1 (broken line).

Our study reveals that MSC-certified stocks are on average more likely to meet or exceed MSY-based target reference points, with higher biomass and lower exploitation rates than uncertified stocks. Certified stocks are also more likely to meet more conservative targets than uncertified stocks. Further, for those stocks with lower biomass levels, rebuilding plans are in place to improve stock health. While our time series analysis indicates that the observed difference in performance between certified and uncertified stocks existed prior to certification, the MSC eco-label is a reliable indicator of target stock health to consumers. It is important to note that MSC-certified fisheries not only must pass the sustainability criterion for the target stock but also must minimize ecosystem impact and have robust management systems. As more agencies attempt to implement ecosystem-based management, certified fisheries will need to demonstrate enhanced performance in these other areas to meet evolving definitions of sustainability and maintain the integrity of the MSC eco-label. A key part of MSC certification is the chain of custody to ensure that seafood labeled as MSC-certified indeed comes from the certified fishery and is not mislabeled catch from uncertified fisheries [48,49]. Nevertheless, the current study shows that certification and eco-labeling can effectively recognize healthy stocks and fisheries that are achieving internationally accepted management targets. This is a critical first step in providing a mechanism for consumers to effectively influence change in fishing practices and ensure future ocean health and productivity.

Supporting Information

Table S1 Summary information on certified stocks and their estimated current biomass and exploitation rates relative to MSY reference points ($B_{current}/B_{MSY}$ and $u_{current}/u_{MSY}$ or $F_{current}/F_{MSY}$). Rows in grey indicate certified stocks without available reference points or biomass estimates. These stocks were not used in the analysis (Fig. 1A). "Method used" indicates the method used to estimate B_{MSY} and F_{MSY} or u_{MSY}: 1 = stock assessment model, 2 = surplus production model, and 3 = combination.

Table S2 Summary information on uncertified stocks and their estimated current biomass and exploitation rates relative to MSY reference points (B/B_{MSY} and u/u_{MSY} or F/F_{MSY}). "Method used" indicates the method used to estimate B_{MSY} and u_{MSY} or F_{MSY}: 1 = stock assessment model, 2 = surplus production model, and 3 = combination.

Table S3 Median (±SE) biomass and exploitation rates relative to their targets and differences among certified, uncertified and not-recommended stocks. *denotes statistical significance (*$P<0.05$; **$P<0.005$).

Table S4 Median biomass and exploitation rates relative to their targets and differences among certified, uncertified and not-recommended stocks for those fisheries with both B_{MSY} and F_{MSY} available from stock assessments

Table S5 Results of test for difference in mean B/B_{MSY} over time between certified and uncertified stocks. *denotes significance (*$P<0.05$; **$P<0.005$).

Text S1 This supporting information file includes expanded descriptions of methods used, additional results and model outputs and supporting references.

Acknowledgments

We thank the Certification and Accreditation Bodies that provided confidential information on fisheries pre-assessments. The data reported in this paper are tabulated in the Supporting Information.

Author Contributions

Conceived and designed the experiments: NLG SRV TAB. Analyzed the data: NLG SRV. Wrote the paper: NLG SRV. Discussed the results and contributed to the manuscript: NLG SRV TAB DJA JKB PLB JCD CC OD TEE RH DDH AEL CN KS RLS SS ADMS ASP SJT JTT NEW.

References

1. FAO (2010) *The State of World Fisheries and Aquaculture 2010* (Food and Agriculture Organization of the United Nations, Rome) Available at: http://www.fao.org/docrep/013/i1820e/i1820e.pdf.
2. Worm B, Hilborn R, Baum JK, Branch TA, Collie JS, et al. (2009) Rebuilding global fisheries. Science 325: 578–585.
3. Branch TA, Jensen OP, Ricard D, Ye Y, Hilborn R (2011) Contrasting global trends in marine fishery status obtained from catches and from stock assessments. Conservation Biology 25: 777–786.
4. Gutierrez NL, Hilborn R, Defeo O. (2011) Leadership, social capital and incentives promote successful fisheries. Nature 470: 386–389.
5. Ward T, Phillips B eds. (2008) *Seafood Ecolabelling: Principles and Practice* Ward T, Phillips Beds (Wiley-Blackwell).
6. Hilborn R, Cowan JH (2010) Marine stewardship: high bar for seafood. Nature 467: 531–531.
7. Parkes G, Young JA, Walmsley SF, Abel R, Harman J, et al. (2010) Behind the signs–a global review of fish sustainability information schemes. Reviews in Fishery Sciences 18: 344–356.
8. Roheim CA, Asche F, Santos JI (2011) The elusive price premium for ecolabelled products: evidence from seafood in the UK market. Journal of Agronomic Economy 62: 655–668.
9. Perez-Ramirez M, Phillips B, Lluch-Belda D, Lluch-Cota S (2012) Perspectives for implementing fisheries certification in developing countries. Marine Policy 36: 297–302.
10. Perez-Ramirez M, Ponce-Diaz G, Lluch-Cota S (2012) The role of MSC certification in the empowerment of fishing cooperatives in Mexico: The case of red rock lobster co-managed fishery. Ocean and Coastal Management 63: 24–29.
11. Gulbrandsen L (2009) The emergence and effectiveness of the Marine Stewardship Council. Marine Policy 33: 654–660.
12. Jacquet J, Pauly D (2007) The rise of seafood awareness campaigns in an era of collapsing fisheries. Marine Policy 31: 308–313.
13. Ponte S (2008) in *Seafood labelling: principles and practice*, Ward T, Phillips Beds (Wiley-Blackwell, Oxford, UK), 287–306.
14. Ward TJ (2008) Barriers to biodiversity conservation in marine fishery certification. Fish and Fisheries 9: 169–177.
15. Potts T, Haward M (2007) International trade, eco-labelling, and sustainable fisheries – recent issues, concepts and practices. Environmental Development Sustainability 9: 91–106.
16. Jacquet J, Pauly D, Ainley D, Holt S, Dayton P (2010) Seafood stewardship in crisis. Nature 467: 28–29.
17. Ponte S (2006) Ecolabels and fish trade: Marine Stewardship Council certification and the South African hake industry. TRALAC Working Paper, DIIS, Denmark. 66.
18. Jacquet J, Pauly D (2008) Funding priorities: big barriers to small-scale fisheries. Conservation Biology 22: 832–835.
19. Marine Stewardship Council (2011) *MSC Certification Requirements v1* (London, UK) Available at: http://www.msc.org/documents/scheme-documents/msc-scheme-requirements/msc-certification-requirement-v1.1/view.
20. Ricard D, Minto C, Jensen OP, Baum JK (2011) Examining the knowledge base and status of commercially exploited marine species with the RAM Legacy Stock Assessment Database. Fish and Fisheries Online early: DOI: 10.1111/j.1467-2979.2011.00435.x.
21. Fournier DA, Skaug HJ, Ancheta J, Ianelli J, Magnusson A, et al. (2012) AD Model Builder: using automatic differentiation for statistical inference of highly parameterized complex nonlinear models. Optimization Methods Software 27: 233–249.
22. Martin S, Cambridge T, Grieve C, Nimmo F, Agnew DA (2012) Environmental impacts of the Marine Stewardship Council certification scheme. Reviews Fisheries Sciences 20: 61–69.
23. UNCLOS (1982) United Nations Convention on the Law of the Sea. 1833: 1–186. Available at: http://treaties.un.org/doc/Publication/UNTS/Volume%201833/volume-1833-A-31363-English.pdf.
24. UNFSA (1995) Agreement for the Implementation of the Provisions of the United Nations Convention on the Law of the Sea of 10 December 1982. 1–40. Available at: http://daccess-dds-ny.un.org/doc/UNDOC/GEN/N95/274/67/PDF/N9527467.pdf.
25. Hilborn R (2010) Pretty Good Yield and exploited fishes. Marine Policy 34: 193–196.
26. Rosenberg AA, Swasey JH, Bowman M (2006) Rebuilding US fisheries: progress and problems. Frontiers in the Ecology and Environment 4: 303–308.
27. Froese R, Branch TA, Proelß A, Quaas M, Sainsbury K, et al. (2011). Generic harvest control rules for European fisheries. Fish and Fisheries 12: 340–351.
28. Rademeyer RA (2012) Routine Update of the South African Hake Base Reference Case Assessment. FISHERIES/2012/AUG/SWG-DEM/39. 7.
29. ICES (2011) Report of the ICES Advisory Committee, 2011. ICES Advice, 2011. Books 11 (Copenhagen).
30. Eero M, Köster FW, Vinther M (2012) Why is the Eastern Baltic cod recovering? Marine Policy 36: 235–240.
31. ICES (2011) Report of the ICES Advisory Committee, 2011. ICES Advice, 2011. Books 6 (Copenhagen) Available at: http://www.ices.dk/committe/acom/comwork/report/2011/2011/had-34.pdf.
32. Mangel M, Marinovic B, Pomeroy C, Croll D (2002) Requiem for Ricker: Unpacking MSY. Bulletin of Marine Sciences 70: 763–781.
33. Hilborn R, Stokes K (2010) Defining overfished stocks: have we lost the plot? Fisheries 35: 113–120.
34. Punt AE, Smith ADM (2001) The Gospel of maximum sustainable yield in fisheries management: birth, crucifixion and reincarnation in *Conservation of Exploited Species*, Reynolds JD, Mace GM, Redford KH, Robinson JG eds (Cambridge University Press, Cambridge, UK), 41–66.
35. Walters CJ, Christensen V, Martell SJ (2005) Possible ecosystem impacts of applying MSY policies from single-species assessment. ICES Journal of Marine Sciences 62: 558–568.
36. Smith ADM, Brown CJ, Bulman CM, Fulton EA, Johnson P, et al. (2011) Impacts of fishing low-trophic level species on marine ecosystems. Science 333: 1147–1150.
37. Froese R, Branch TA, Proelß A, Quaas M, Sainsbury K, et al. (2011). Generic harvest control rules for European fisheries. Fish and Fisheries 12: 340–351.
38. Dichmont CM, Pascoe S, Kompas T, Punt AE, Deng R (2012) On implementing maximum economic yield in commercial fisheries. Proceedings of the National Academy of Sciences of the USA 107: 16–21.
39. Mantua NJ, Hare SR, Zhang Y, Wallace JM, Francis RC (1997). A Pacific interdecadal climate oscillation with impacts on salmon production. Bulletin of the American Meteorology Society 78: 1069–1079.
40. Ianelli J, Barbeaux S, Honkalehto T, Kotwicki S, Aydin K, et al. (2010) Assessment of the walleye pollock stock in the Eastern Bering Sea. Available at: http://www.afsc.noaa.gov/refm/docs/2010/EBSpollock.pdf.
41. Hilborn R (2011) Future directions in ecosystem based fisheries management: A personal perspective. Fisheries Research 108: 235–239.
42. Pikitch EK, Santora C, Babcock EA, Bakun A, Bonfil R, et al. (2004) Ecology: ecosystem-based fishery management. Science 305: 346–7.
43. Crowder LB, Hazen E, Avissar N, Bjorkland R, Latanich C, et al. (2008) The impacts of fisheries on marine ecosystems and the transition to ecosystem-based management. Annual Reviews Ecology and Evolution Systems 39: 259–278.
44. Hooper DU, Chapin FS, Ewel JJ, Hector A, Inchausti P, et al. (2005) Effects of biodiversity on ecosystem functioning: a consensus of current knowledge. Ecological Monographs 75: 3–35.
45. Hughes TP, Bellwood DR, Folke C, Steneck RS, Wilson J (2005) New paradigms for supporting the resilience of marine ecosystems. Trends in Ecology and Evolution 20: 380–386.
46. Hixon MA, Tissot BN (2007) Comparison of trawled vs untrawled mud seafloor assemblages of fishes and macroinvertebrates at Coquille Bank, Oregon. Journal of Experimental Marine Biology and Ecology 344: 23–34.
47. Zhou S, Smith ADM, Punt AE, Richardson AJ, Gibbs M, et al. (2010) Ecosystem-based fisheries management requires a change to the selective fishing philosophy. Proceedings of the National Academy of Sciences of the USA 107: 9485–9489.
48. Marko PB, Nance HA, Guynn KD (2011) Genetic detection of mislabeled fish from a certified sustainable fishery. Current Biology 21: 621–622.
49. Martinsohn JT (2011). Tracing fish and fish products from ocean to fork using advanced molecular technologies in *Food Chain Integrity: a holistic approach to food traceability, safety, quality and authenticity* (Cambrige, UK), p. 259.

19

Spatial and Temporal Variation in Fungal Endophyte Communities Isolated from Cultivated Cotton (*Gossypium hirsutum*)

María J. Ek-Ramos[1]*, Wenqing Zhou[1], César U. Valencia[1], Josephine B. Antwi[1], Lauren L. Kalns[1], Gaylon D. Morgan[2], David L. Kerns[1,3¤], Gregory A. Sword[1]

1 Department of Entomology, Texas A & M University, College Station, Texas, United States of America, **2** Department of Soil and Crop Sciences and Texas AgriLife Extension, Texas A & M University, College Station, Texas, United States of America, **3** AgriLife Extension Service, Texas A & M University, Lubbock, Texas, United States of America

Abstract

Studies of fungi in upland cotton (*Gossypium hirsutum*) cultivated in the United States have largely focused on monitoring and controlling plant pathogens. Given increasing interest in asymptomatic fungal endophytes as potential biological control agents, surveys are needed to better characterize their diversity, distribution patterns and possible applications in integrated pest management. We sampled multiple varieties of cotton in Texas, USA and tested for temporal and spatial variation in fungal endophyte diversity and community composition, as well as for differences associated with organic and conventional farming practices. Fungal isolates were identified by morphological and DNA identification methods. We found members of the genera *Alternaria*, *Colletotrichum* and *Phomopsis*, previously isolated as endophytes from other plant species. Other recovered species such as *Drechslerella dactyloides* (formerly *Arthrobotrys dactyloides*) and *Exserohilum rostratum* have not, to our knowledge, been previously reported as endophytes in cotton. We also isolated many latent pathogens, but some species such as *Alternaria tennuissima*, *Epicoccum nigrum*, *Acremonium alternatum*, *Cladosporium cladosporioides*, *Chaetomium globosum* and *Paecilomyces* sp., are known to be antagonists against plant pathogens, insects and nematode pests. We found no differences in endophyte species richness or diversity among different cotton varieties, but did detect differences over time and in different plant tissues. No consistent patterns of community similarity associated with variety, region, farming practice, time of the season or tissue type were observed regardless of the ecological community similarity measurements used. Results indicated that local fungal endophyte communities may be affected by both time of the year and plant tissue, but the specific community composition varies across sites. In addition to providing insights into fungal endophyte community structure, our survey provides candidates for further evaluation as potential management tools against a variety of pests and diseases when present as endophytes in cotton and other plants.

Editor: Zhengguang Zhang, Nanjing Agricultural University, China

Funding: This work was supported by Texas A&M University Department of Entomology start-up funds (to GAS). The funders had no role in study design, data collection and analysis, decision to publish, or preparation of the manuscript.

* E-mail: ekramos@neo.tamu.edu

¤ Current address: Macon Ridge Research Station, LSU AgCenter, Winnsboro, Louisiana, United States of America

Introduction

Fungal endophytes are fungi that internally colonize plant tissues without causing evident damage or disease [1]. Several groups have proposed that fungal endophytes evolved from plant pathogenic fungi that have long latent periods, or have lost their virulence [2–3]. In addition, there are examples of specific environmental conditions triggering pathogenicity of previously asymptomatic endophytes [1] [4]. Alternatively, a number of studies suggest that fungal endophytes can be involved in many beneficial interactions with their hosts, providing protection against a variety of stressors including herbivores, pathogens, heat and drought [1] [5–13]. Studies of beneficial fungal endophytes present in agricultural crops have focused on the analysis of their assemblages *in planta*, physiological interactions with host plants, the production of secondary fungal metabolites, and their potential use in biological control of plant diseases and insects [1] [8] [14–19]. Some studies have shown positive effects of these fungal endophytes on plant growth including higher rates of germination and rooting, and increased tissue biomass and seed production under adverse conditions [20–24].

Fungal endophyte surveys explicitly aimed at isolating candidate beneficial fungal endophytes from cotton (*Gossypium hirsutum*) cultivated in the United States have not been reported to date. However, many studies dating back to the 1920s have been published on the identification of fungi isolated from a variety of cotton tissues, primarily with an emphasis on monitoring fungal diseases [25–29]. Among the species identified in these studies, several have since been found to live as endophytes in healthy plants across a range of different species [30]. It seems likely that at least some of the fungi previously isolated and considered as pathogenic (or at least putatively pathogenic) in cotton may be more appropriately considered as asymptomatic endophytes. Thus, fungal endophyte community studies have the potential to provide an unexplored source of candidate strains for potential beneficial applications.

Recent surveys of fungal endophytes in healthy cotton conducted in Australia and Brazil demonstrated the presence of a diversity of species and broad scale variation in community composition. McGee (2002) [31] surveyed healthy leaves obtained from a cotton-breeding trial in Australia and identified fungal endophytes from ten different genera by morphological identification of their fruiting bodies. Among the genera obtained, an isolate of *Phomopsis* sp. showed promising effects in reducing caterpillar herbivory. Although *Phomopsis* sp. has been reported as a plant pathogen [32], only rarely does it cause disease [26] [31] [33]. Another Australian study by Wang et al. (2007) [34] sought to determine whether fungal endophytes of native *Gossypium* species could be pathogenic to cultivated cotton, *Gossypium hirsutum*. They showed that fungal endophytes were common in four native *Gossypium* species and dominated by six genera including *Phoma*, *Alternaria*, and *Fusarium*. Interestingly, none of these isolates caused disease symptoms when inoculated to cotton under controlled conditions [34]. A more recent study conducted in Brazil compared fungal endophyte communities associated with transgenic and non-transgenic cotton [35]. Their results indicated that the endophytic fungal community was not affected by the expression of Cry1Ac protein from *Bacillus thuringiensis* (*Bt*) in transgenic cotton plants [35]. In total, they isolated 17 genera of endophytes from both *Bt* and non-*Bt* cotton including *Xylaria* sp., *Phoma* sp., *Phomopsis* sp., *Lecanicillium* sp., *Tritirachium* sp., *Pestatiopsis* sp., *Cladosporium* sp., *Fusarium* sp. and *Guignardia* sp. among others [35].

In order to better characterize communities and functional roles of fungal endophytes in cotton and isolate strains with potential beneficial applications, we conducted a survey of endophytes in asymptomatic cotton in Texas, USA. We focused on commercial varieties cultivated at eight sites distributed across two ecologically-distinct growing regions in north and central Texas, each sampled at two different times of the season. We also surveyed cotton grown on organic farms in comparison to conventional farms. We found a total of 69 different endophytic fungal taxa grouped into 44 different genera. We found evidence of differences in community composition depending on temporal variation and plant tissue rather than on location or cultivation practices. Among the fungal endophytes isolated, several are candidates for potential use as beneficial endophytes in the management of plant pathogens, insect pests and nematodes based either on their known effects as endophytes in other plants or their ecological roles outside the plant.

Results

Endophyte Isolation and Identification

The surface sterilization protocol was stringent enough to eliminate epiphytic fungi, as neither fungi nor bacteria grew after making tissue imprints onto the surface of PDA and V8 media control plates. The total number of endophytic isolates obtained from leaves collected in June 2011 was 1259, and from squares that were present at the time at only one locality, Navasota, it was 46. Later in the season in August, the total number of isolates obtained from leaves was 1354, and from squares/bolls it was 802. Endophytic fungi were isolated from plant samples at different frequencies depending on the tissue, variety, location and cultivation practices surveyed (Figures 1 and 2; Table 1). One way ANOVA (SPSS 20.0, IBM North America, New York, USA) of isolation percentages normalized using arcsine transformation [36] indicated that the number of isolates recovered varied significantly by tissue ($F_{249,\ 248} = 8.321$, $P<0.05$), time of the season ($F_{249,\ 248} = 5.142$, $P<0.05$), location ($F_{249,\ 242} = 17.925$, $P<0.05$), cultivation practices ($F_{249,\ 248} = 80.384$, $P<0.05$) and cotton variety ($F_{249,\ 233} = 5.414$, $P<0.05$), but was independent of culture media used ($F_{249,\ 248} = 0.054$, $P = 0.816$). The highest isolation percentages of approximately 100% were obtained from leaves collected in August 2011 in fields under conventional cultivation practices such as Snook, Navasota, Lubbock-RACE and Dawson-RACE. The lowest isolation frequencies (<30%) were obtained from squares/bolls from all organic farms (Muleshoe, Idalou, South Barrier and North Tucker) (Figure 2). Fungal isolates obtained were identified using ITS1 sequence and morphological data (Table 2) resulting in a total of 69 different endophytic fungal taxa (OTUs) grouped into 44 different genera.

Species Richness and Biodiversity of Fungal Endophyte Communities (α-diversity)

Shannon-Wiener biodiversity values (H′) [37–38] compared across varieties within the four conventional farming sites where multiple varieties were sampled were not significantly different for communities isolated from a given tissue type, indicating that the number and relative abundance of recovered endophytic taxa did not vary among cotton varieties (Figure 3A–D). However, within each variety there were clear differences in endophyte diversity between tissue types and times of the season in which they were sampled (Figure 3A–D). With regards to sampling intensity, the taxa accumulation curves indicated that our sampling of 45–50 samples per site was sufficient to isolate rare fungal endophyte taxa regardless of the tissue being sampled (Figure 4). A comparison among varieties within sites practicing organic farming was not possible because only a single variety was grown at each site, but a similar pattern of variation in diversity between samples from different tissues and time of the season was observed (Figure 3E).

Variation in Fungal Endophyte Communities Among Varieties, Tissues and Time of the Season (β-diversity)

Cluster analyses of endophyte community similarity measures [39] presented in two dimensional non-metric multidimensional scale (NMDS) plots [40] revealed that fungal endophyte communities did not vary among different cotton varieties at any of the sampled sites (Figure 5). However, some local endophyte communities appeared to be specific to particular tissues and times of the season (Figure 6A,C,F). Although clusters indicating similar endophyte communities were apparent depending on the tissue type and time of the season sampled at some locations, the effect was not consistent across sites (Figure 6A–L). The use of different ecological similarity measures [41] did not consistently affect the observed patterns of cluster formation (Figure 6A–L). For example, at the Snook site, clusters of similar communities according to tissue type and time of the year were observed using both the Jaccard's index and Euclidean distance measures (Figure 6A,C), whereas at Navasota similar clustering was only observed using the Euclidean distance (Figure 6F). Kruskal's stress values were too high (>0.2) to confidently discern any clustering of communities at the other sites sampled regardless of the distance measure used (Figure 6G–L).

We could not analyze the effect of variety at the four organic farms we surveyed (Muleshoe, Idalou, South Barrier and North Tucker) because there was only one variety cultivated at each site. As a result, the intensity of sampling and corresponding number of

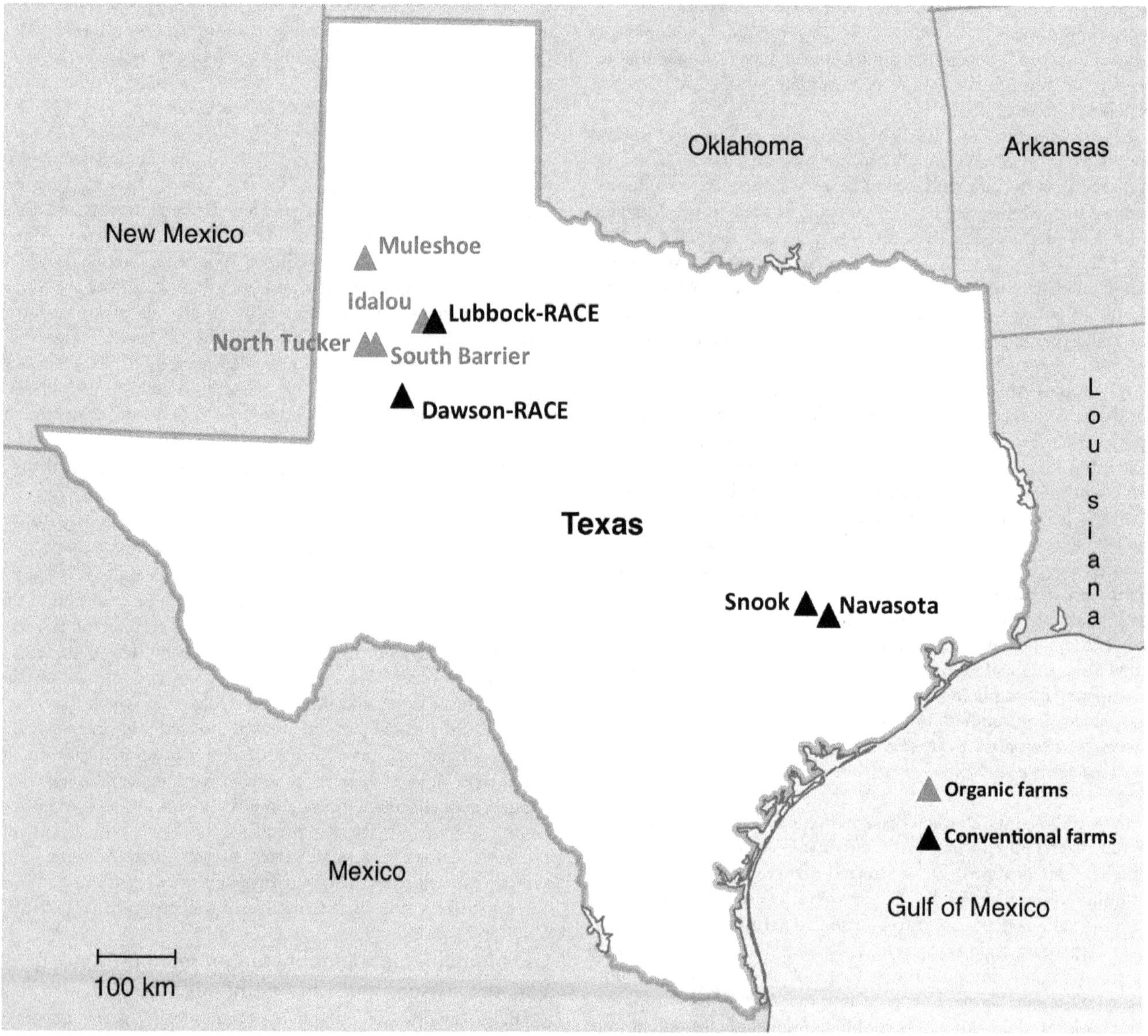

Figure 1. Map of survey area. Geographic location of the eight different cotton farms surveyed during the 2011 growing season in Texas, USA. Green symbols correspond to farms managed following organic practices and black symbols correspond to farms managed using conventional practices.

endophytes isolated was lower at organic sites compared to the conventional cultivated farms where five varieties were sampled, precluding direct comparisons. Therefore, we focused on only identifying the fungal endophyte taxa isolated at the organic farms. Results indicated that there were no unique species of fungal endophytes isolated only from organic farms.

Given that we did not find any effect of cotton variety on fungal endophyte community composition within sites (Figures 3 and 5), we grouped all fungal endophyte taxa obtained at each site and then re-calculated Jaccard's indexes, Bray-Curtis coefficients and Euclidean distances to examine regional variation across sites (Figure 7). We did not observe obvious regional clustering of fungal endophyte communities isolated from the farms located in the Southern Blacklands region of Texas (Snook and Navasota) relative to the communities isolated from farms located to the north in the Southern High Plains region of Texas (Lubbock-RACE and Dawson-RACE) (Figure 7A,D,G). Nor did we observed clustering of fungal endophyte taxa communities isolated from leaves compared to those from squares/bolls (Figure 7C,F,I). The only suggestive pattern of community similarity when considering variation across all sites was in the endophyte communities sampled at different times of the season and compared using Jaccard's index (Figure 6B). However, confidence in this pattern was marginal (Kruskal's stress = 0.208). Increasing the dimensionality of the NMDS analyses increased the confidence in the observed clustering patterns in all the cases, but did not change the interpretation of community similarity patterns as already shown in the two dimension plots (Figure S1).

Discussion

Plants in both natural and agricultural settings can host diverse endophyte communities. All plants surveyed to date contain fungal endophytes, and many isolates have been obtained from a broad

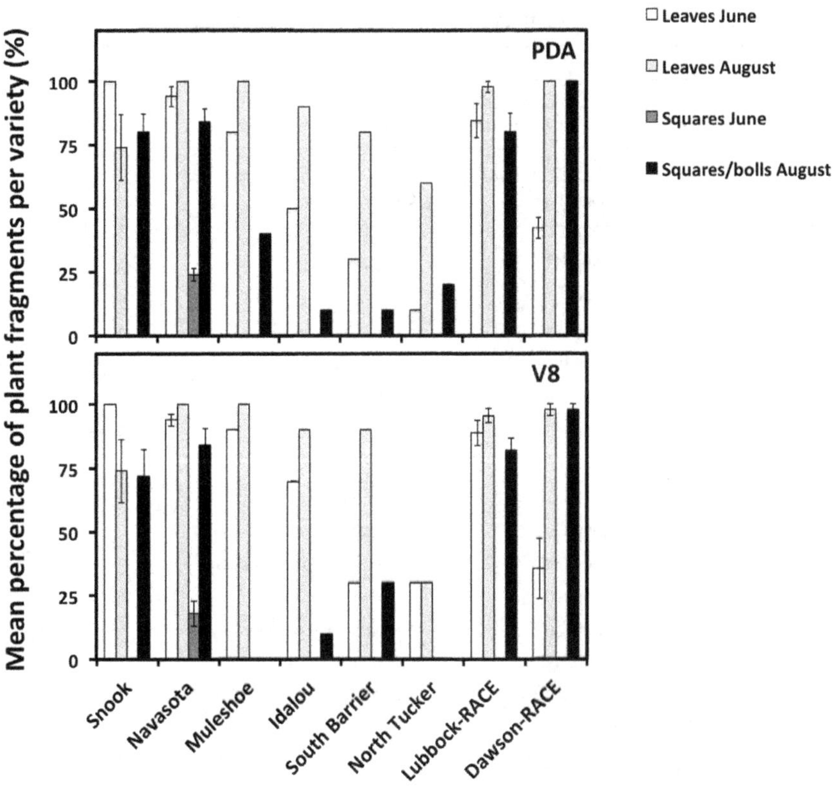

Figure 2. Fungal endophyte isolation efficiency. Fungal endophyte taxa isolation efficiency expressed as the mean % of plant fragments (leaves, squares and bolls as explained in materials and methods) per variety at each site from which at least one endophytic isolate was obtained. Sets of data are also grouped by time of the season and tissue surveyed. (A) Isolates obtained using PDA media. (B) Isolates obtained using V8 media.

range of plants including trees [42–43], palms [44–45], grasses [46], sea grasses [47], crops [30] and lichens [48]. Endophytes have been referred to as a hidden component of fungal diversity [12] [49–50]. The diversity of endophytic fungi within a given plant can be substantial with reports of up to 22 species in a single leaf of the tropical tree *Manilkara bidentata* [1] [51–52] and 51 different operational taxonomic units (OTUs) associated with roots of the arid grassland grass *Bouteloua gracilis* [53]. The aim of this study was to characterize the fungal endophytes present in cotton and to analyze their spatial and temporal patterns of variation (Figure 1; Table 1). There have been many systematic studies investigating *in planta* microbial communities in cotton in the United States dating back to the 1920's (*e.g.*, [25]). However, most have focused on identifying plant pathogens, and studies

Table 1. Location and cotton variety information for the farms surveyed during the 2011 season.

Location	Cultivation	GPS coordinates	Elevation (m)	Varieties sampled	Sampling dates
Snook, TX	Conventional	N30°31.588′,W96° 28.078′	74.98 m	NG4012B2RF ST4498B2F PHY499WRF FM1740B2F DP1044B2F	June 15th and July 24th
Navasota, TX	Conventional	N30°23.835′,W96° 14.376′	61.26 m	PHY375WRF FM1740B2F DP1044B2RF DP0912B2RF 11R115B2R2	June 16th and July 25th
Muleshoe, TX	Organic	N34°16.380′,W102° 22.958′	102.41 m	FM958	June 21st and August 1st
Idalou, TX	Organic	N33°40.698′,W101° 38.115′	106.68 m	FM975	June 21st and August 1st
South Barrier, TX	Organic	N33°18.035′,W102° 25.050′	101.19 m	FM958	June 21st and August 1st
North Tucker, TX	Organic	N33°17.750′,W102° 22.820′	103.32 m	FM958	June 21st and August 1st
Lubbock-RACE, TX	Conventional	N33°35.707′,W101° 33.877′	103.02 m	AT81220B2RF NG4010B2RF FM2484B2F PHY499WRF CG3787B2RF	June 22nd and August 2nd
Dawson-RACE, TX	Conventional	N32°46.516′, W101°56.547′	102.41 m	AT81220B2RF NG4012B2RF FM2484B2F PHY367WRF CG3787B2RF	June 22nd and August 2nd

Table 2. Identification of fungal endophyte taxa isolated from cotton cultivated in Texas in June-August 2011.

Sequence accession Genbank number	Fungal taxa	Isolates	Sequence accession Genbank number	Fungal taxa	Isolates
KC800871	*Acremonium alternatum*	1	KC800875	*Colletotrichum capsici*	1
KC800839	*Alternaria alternata*	1	KC800892	*Coniolariella gamsii*	1
KC800833	*Alternaria brassicae*	6	KC800870	*Coniothyrium aleuritis*	1
KC800836	*Alternaria compacta*	2	KC800862	*Coniothyrium* sp.	2
KC800844	*Alternaria dianthi*	1	KC800885	*Corynespora cassiicola*	1
KC800829	*Alternaria longipes*	5	KC800865	*Diaporthe* sp.	1
KC800837	*Alternaria mali*	1	KC800859	*Diatrype* sp.	1
KC800896	*Alternaria sesami*	1	KC800834	*Drechslerella dactyloides*	1
KC800888	*Alternaria solani*	2	KC800863	*Embellisia indefessa*	1
KC800895	*Alternaria* sp.	3140	KC800886	*Epicoccum nigrum*	23
KC800894	*Alternaria tenuissima*	78	KC800831	*Epicoccum* sp.	2
KC800842	*Ascomycota* sp.	4	KC800830	*Exserohilum rostratum*	2
KC800889	*Bipolaris spicifera*	4	KC800873	*Fusarium chlamydosporum*	3
KC800881	*Cercospora canescens*	1	KC800890	*Fusarium* sp.	1
KC800866	*Cercospora capsici*	3	KC800880	*Gibellulopsis nigrescens*	4
KC800878	*Cercospora kikuchii*	1	KC800855	*Gnomoniopsis* sp.	1
KC800877	*Cercospora zinnia*	1	KC800848	*Lewia infectoria*	16
KC800876	*Chaetomium globosum*	9	KC800869	*Mycosphaerella coffeicola*	5
KC800846	*Chaetomium piluliferum*	1	KC800840	*Mycosphaerellaceae* sp.	1
KC800851	*Chaetomium* sp.	8	KC800864	*Nigrospora oryzae*	2
KC800874	*Cladosporium cladosporioides*	7	KC800891	*Nigrospora* sp.	2
KC800849	*Cladosporium* sp.	9	KC800893	*Nigrospora sphaerica*	1
KC800872	*Cladosporium uredinicola*	5	KC800867	*Paecilomyces* sp.	1
KC800838	*Cochliobolus* sp	5	KC800883	*Penicillium citrinum*	1
KC800850	*Phanerochaete crassa*	1	KC800861	*Retroconis* sp.	2
KC800857	*Phoma americana*	1	KC800832	*Rhizopycnis* sp.	1
KC800879	*Phoma subherbarum*	1	KC800860	*Schizothecium inaequale*	9
KC800882	*Phomopsis liquidambari*	1	KC800841	*Stagonospora* sp.	6
KC800853	*Phomopsis* sp.	2	KC800884	*Stemphylium lancipes*	3
KC800856	*Pleospora* sp.	3	KC800843	*Thielavia hyrcaniae*	2
KC800835	*Pleosporaceae* sp.	5	KC800845	*Thielavia* sp.	5
KC800858	*Polyporales* sp.	1	KC800852	*Ulocladium chartarum*	2
KC800854	*Preussia africana*	2	KC800868-	*Verticillium* sp.	4
KC800887	*Preussia* sp.	4	-	Unknown	24
KC816535	*Pseudozyma* sp.	1	-	Uncultured	9
KC800847	*Pyrenophora teres*	1			

examining the factors underlying variation in fungal endophyte communities have been relatively rare (*e.g.*, [29] [35]).

The study of endophytes is a methods-dependent process [54–55] with the identity and range of isolates obtained potentially influenced by a number of experimental variables that, in turn, can affect the comparability of endophyte datasets. Highly stringent surface sterilization protocols can potentially kill fungal endophytes [56], thereby reducing the ability to detect viable isolates grown on media. Our surface sterilization method was stringent enough to eliminate viable fungal pathogens and epiphytes living on the surface of surveyed tissues [57–58], as indicated by the lack of microbial growth on control PDA and V8 media following tissue imprints of sterilized plant fragments. However, the high number of endophytic fungal isolates cultured from many samples (Figure 2) suggests that the surface sterilization procedure did not systematically kill endophytic fungi. Our use of PDA and V8 media may also have affected the number of isolates obtained if they were unsuitable for certain taxa. Given that the same sterilization procedure, growth media and incubation conditions were applied to all samples collected for this study, our comparisons of communities across varieties, times of the season and tissue type should be unaffected by any bias related to the specific fungal isolation methods.

The 69 fungal endophyte OTUs isolated in this survey (Table 2), largely corresponds to the numbers obtained in similar studies of cotton and many other crops [29–31] [35] [52] indicating relative

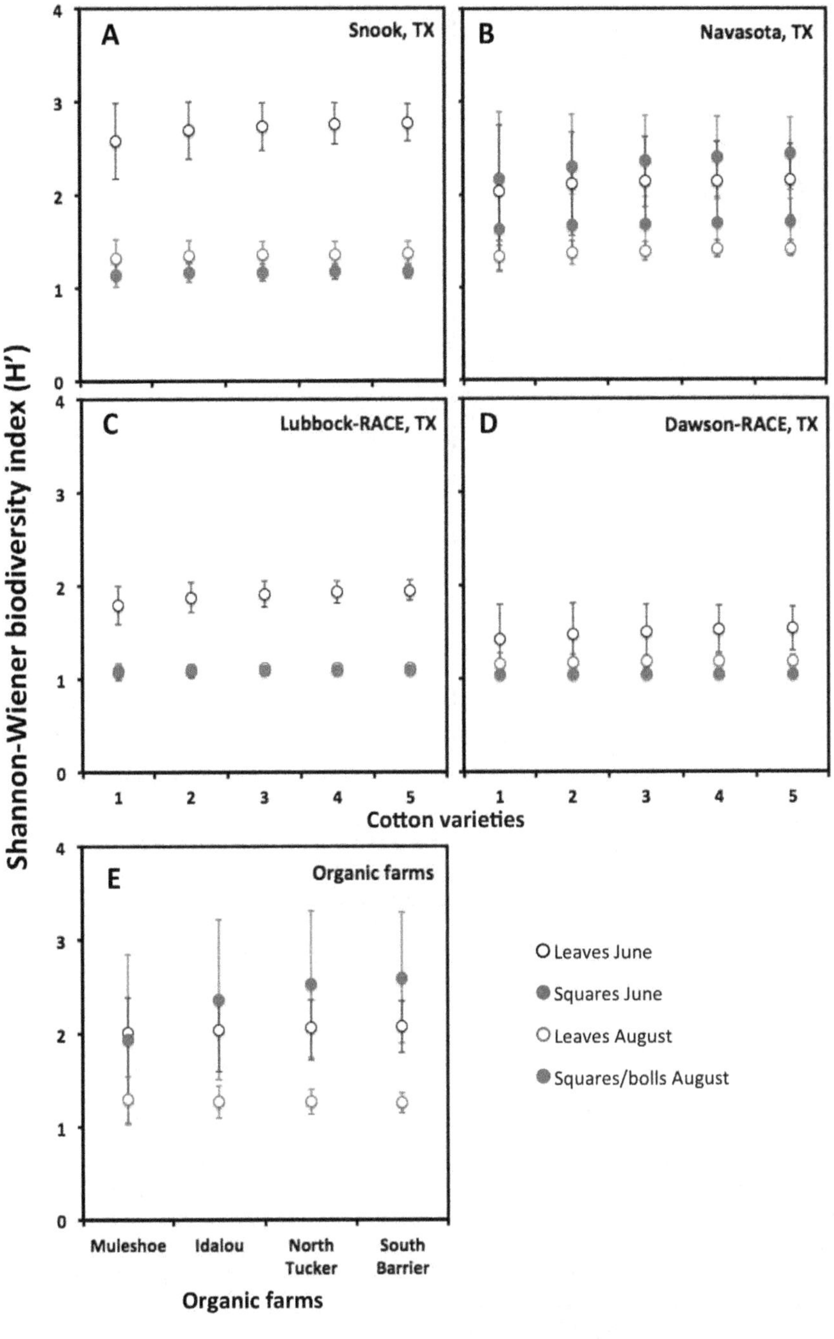

Figure 3. Fungal endophyte biodiversity analysis. The effects of variety and cultivation practices on fungal endophyte Shannon-Wiener biodiversity index (H'). Multiple varieties were sampled at four different sampling locations using conventional farming practices: (A) Snook, TX, (B) Navasota, TX, (C) Lubbock-RACE and (D) Dawson-RACE. Only a single variety was grown at each of the four sampled organic farms (E). Refer to Table 1 for the specific varieties sampled at each site.

consistency in endophyte isolation efficiency across studies. The percentage of plant fragments yielding fungal endophytes was high in the leaves and squares/bolls surveyed in conventional farms, and in the leaves surveyed in two of the organic cultivated farms (Muleshoe and Idalou) (Figure 2). We did not detect significant differences in isolation percentage depending on culture media used. However, there were significant differences in isolation percentage depending on time of the year and tissue surveyed. In the organic farms, fewer plant fragments yielded endophytic isolates overall relative to those from conventional farms. This suggests that specific organic farm practices may influence the prevalence of fungal endophytes. However, since different cotton varieties were grown in the organic and conventional farms, an effect of plant genotype cannot be ruled out. Future studies of endophytes from the same cotton variety grown under both organic and conventional cultivation practices will be required to address this possibility.

It is of interest to note that an overwhelming majority (93.5%) of the isolates recovered in this survey were members of the genus *Alternaria*. Isolates of some species such as *A. alternata* and *A.*

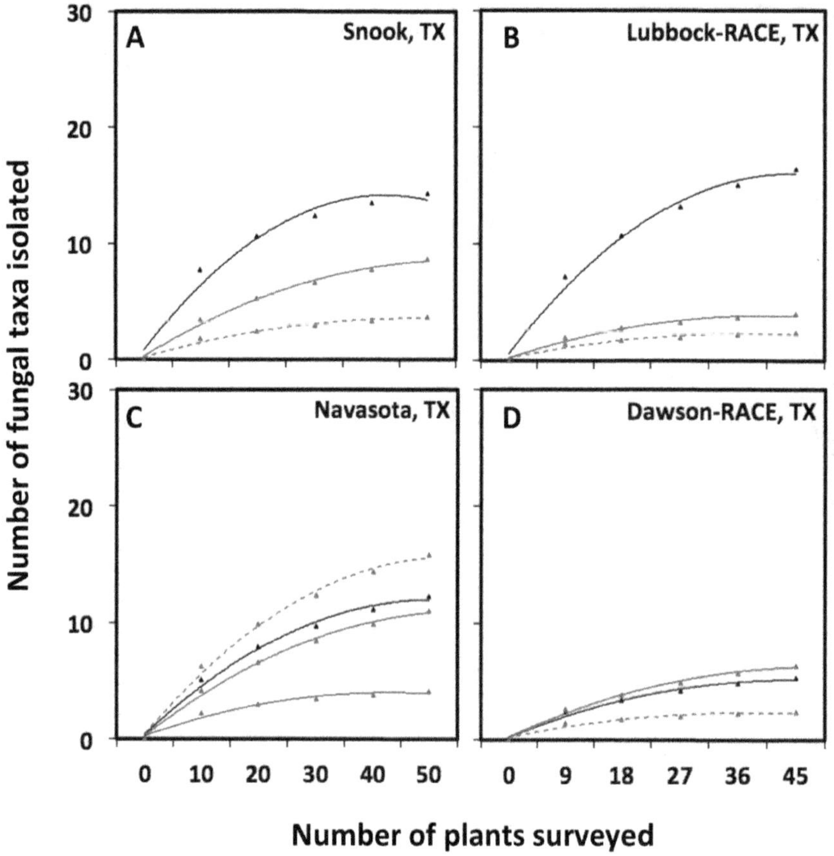

Figure 4. Accumulation curves of fungal taxa isolated from leaves, squares and squares/bolls. The graphs show the relationship between the number of fungal taxa isolated and the number of plants surveyed in (A) Snook, TX, (B) Lubbock-RACE, (C) Navasota, TX and (D) Dawson-RACE. Black continuous line = Leaves surveyed in June 2011, Blue continuous line = Squares surveyed in June 2011 (Navasota, TX), Red continuous line = Leaves surveyed in August 2011 and Red dashed lines = Squares/Bolls surveyed in August 2011. Whole endophyte communities were considered per tissue and location, as cotton variety did not appear to have effect on Shannon-Wiener biodiversity indexes in the farms managed under conventional practices as shown in Figure 3.

macrospora are cotton pathogens [59–60]. We recovered only one positively identified isolate of *A. alternata* and no *A. macrospora.* However, we were not able to identify most of the *Alternaria* isolates to the species level and cannot rule out that they may be pathogens. Previous studies have shown symptomless *Alternaria* infections in cotton [60] and variation among cultivars in susceptibility exists [59]. Thus, the abundance of *Alternaria* sp. that we recovered may reflect either the presence of many non-virulent species or strains, or latent pathogens and the absence of the environmental conditions that induce symptoms [60].

The diversity of endophytic fungi in a single host species can vary both temporally and geographically. In addition, fungal taxa composition can also depend on multiple factors such as plant density, nutrient availability, local environmental conditions and interaction with soil fungi and bacteria [61–65]. The Shannon-Wiener biodiversity index (H′) values, which take into account the number of species (richness) and relative abundance (evenness) of the individuals present in any given sample [37–38], were not significantly different among varieties (Figure 3). However, we did observe differences in H′ biodiversity values when they were compared within locations over time. At Snook, Lubbock-RACE and Dawson-RACE, the H′ values of fungal endophytes isolated earlier in the growing season in June were significantly higher than those of endophyte communities isolated later in August (Figure 3). However, this temporal effect was not consistent across all locations. In Navasota, Idalou, South Barrier and North Tucker, the biodiversity indexes of fungal endophytes isolated from leaves surveyed in August were significantly lower that the leaves surveyed earlier in June and the squares/bolls isolated in August (Figure 3). Interestingly, in Navasota, the H′ values were not different between squares and leaves surveyed in June. The H' biodiversity indexes of fungal endophytes isolated in this study exhibited spatial and temporal differences consistent with previous reports [12], but there were no consistent spatial or temporal patterns of variation in α-diversity.

Within a given location, we sought to test for an effect of variety, tissue or time of the season on fungal endophyte community composition by comparing three different measures of community similarity (β-diversity), each of which incorporates different information. Jaccard's index uses only binary presence-absence data while the Bray-Curtis coefficient incorporates quantitative species abundance data. Both measures exclude joint absences. Euclidean distance incorporates both quantitative abundance data and joint absences [39] [40]. Regardless of the community similarity measure employed, we did not observe any obvious clustering of endophyte communities associated with different cotton varieties at any of the sampled localities (Figure 5). We did observe some clustering at two sites indicating endophyte community similarity due to tissue type and time of the season, but this effect was notably inconsistent across locations (Figure 6).

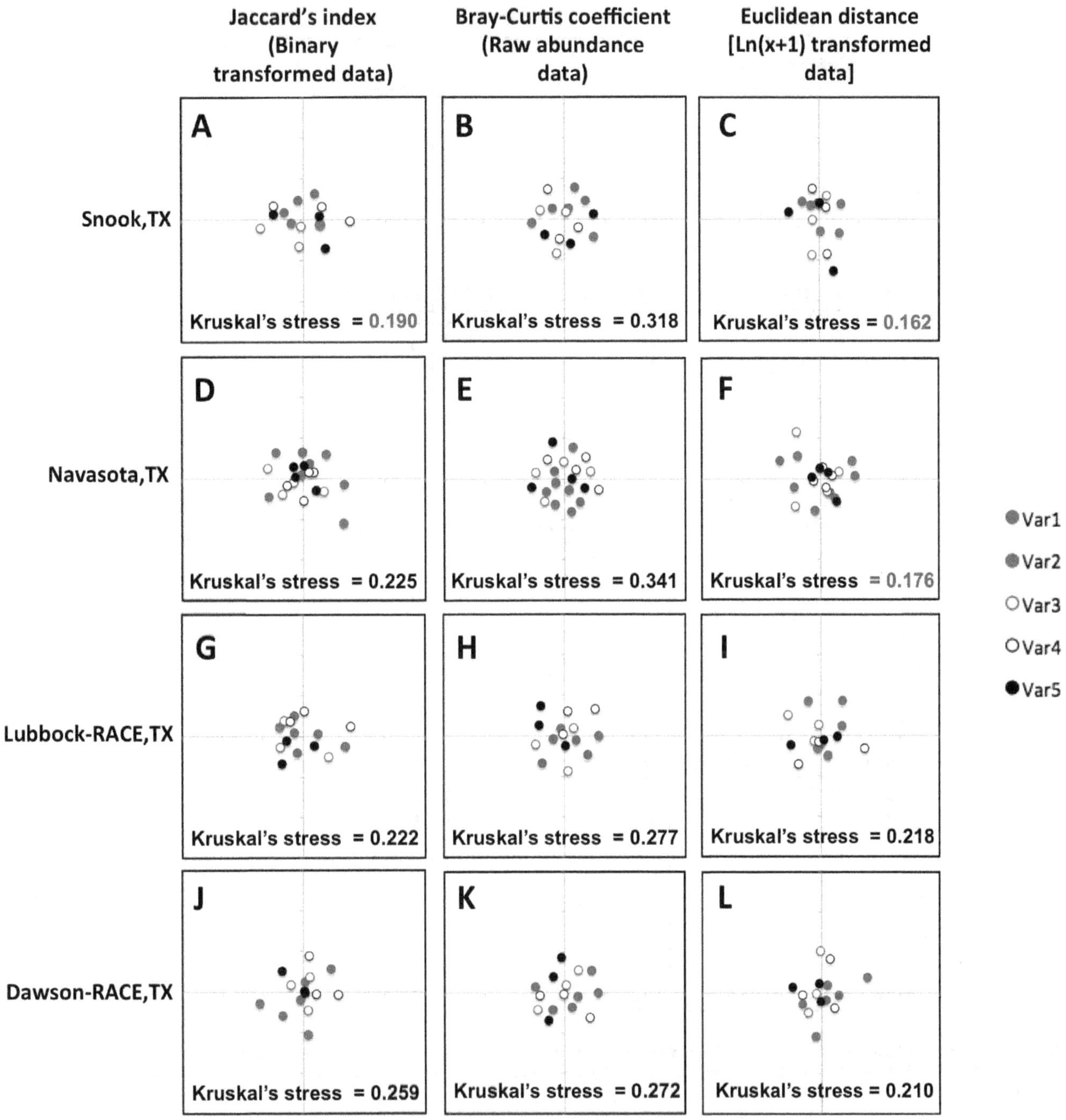

Figure 5. Effect of cotton variety (genotype) on fungal endophyte communities. Non-metric multidimensional scale (NMDS) plots corresponding to the clustering of endophyte communities isolated from five different commercial varieties per location indicated by different colors (see Table 1 for variety information). Each point represents a single endophyte community from a particular tissue and time of the season. For clustering analysis, three different community similarity measures were calculated: (A,D,G,J) Jaccard's index comparing fungal taxa presence or absence among samples from each variety within a site (Binary transformed data); (B,E,H,K) Bray-Curtis coefficient which compares fungal taxa presence or absence as well as abundance among samples from each variety within a site (Raw abundance data); and (C,F,I,L) Euclidean distance using total number of fungal taxa isolated per variety [Ln (x+1) transformed data]. Kruskal's stress values <0.2 indicating more confidence in the observed groupings are indicated in red.

We also sought to test for regional variation in fungal endophyte community composition across all locations including both conventional and organic farming practices, but could not make these comparisons within varieties because the same varieties were not grown at all sites. Given that neither the *H'* biodiversity values (Figure 3) nor the community similarity analyses (Figure 5) were affected by cotton variety, we pooled OTUs across varieties within each site to form a single endophyte community for subsequent comparisons across locations. The fungal taxa accumulation curves obtained to evaluate sampling intensity (Figure 4) indicated that our sampling of cotton at the variety trial sites [a total of 45–50 plants per location (9–10 individuals from 5 varieties)] was

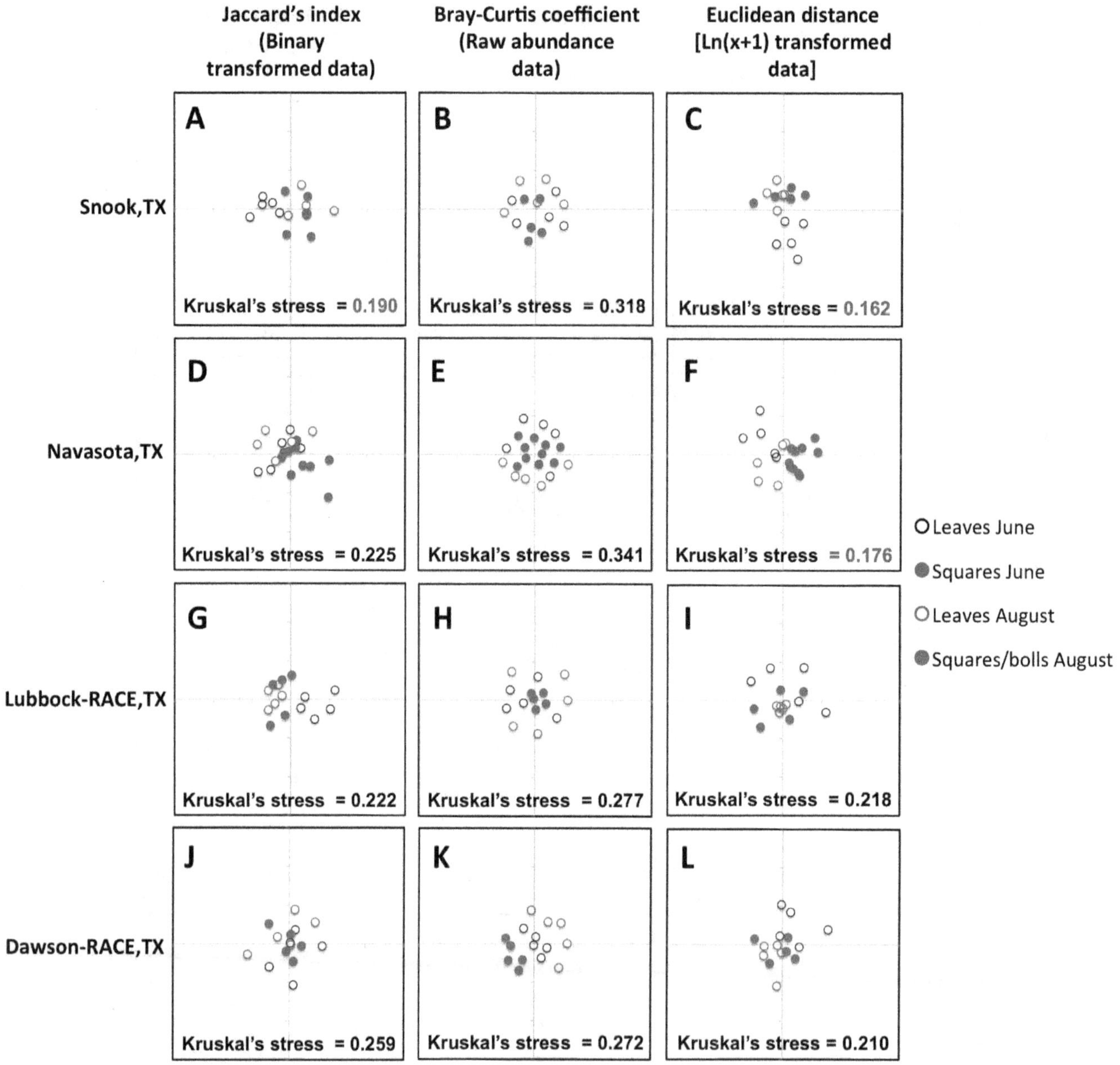

Figure 6. Effect of plant tissue and time of the season on fungal endophyte communities. Non-metric multidimensional scale (NMDS) plots as described for Figure 5, but labeled with different colors by tissue (Leaves, Squares and Squares/bolls) and time of the season surveyed (June and August).

sufficient to isolate rare fungal endophyte taxa regardless of the tissue being assessed, but that we likely under sampled the fungal endophyte communities at the organic farm sites were only 10 plants of a single variety were sampled (Figure 4). Thus, we did not include the organic farms in our regional analysis. With respect to the effect of organic farming practices on cotton fungal endophyte communities, it will be necessary to increase the intensity of sampling and compare different cotton varieties cultivated in those sites to know more about the ecology and distribution of any fungal endophytes that might be specifically affected by organic farming practices.

Evaluating whole fungal endophyte community composition across all conventional farming sites where multiple varieties were grown did not reveal any strong pattern of similarity due to either region, time of the season or tissue type regardless of the community similarity measure used (Figure 7). The only slight clustering effect of endophyte communities observed among these farms was due to time of the season when compared using Jaccard's similarity index (Figure 7B), indicating that the presence/absence of particular taxa played a more important role in differentiating the communities than either their relative abundances or the total number of taxa, as evidenced by the lack of clustering in the Bray-Curtis coefficient and Euclidean distance analyses (Figure 7E,H). However, confidence in the observed pattern was marginal (Kruskal's stress = 0.208, Figure 7B), thereby moderating this conclusion. Importantly, within each of the sites surveyed, varying "site-specific" management practices were followed with the common goal of obtaining high fiber yield, (*e.g.*, irrigation, insecticide, fungicide, herbicide and fertilizer treatments). Site-specific variation in fungal endophyte communi-

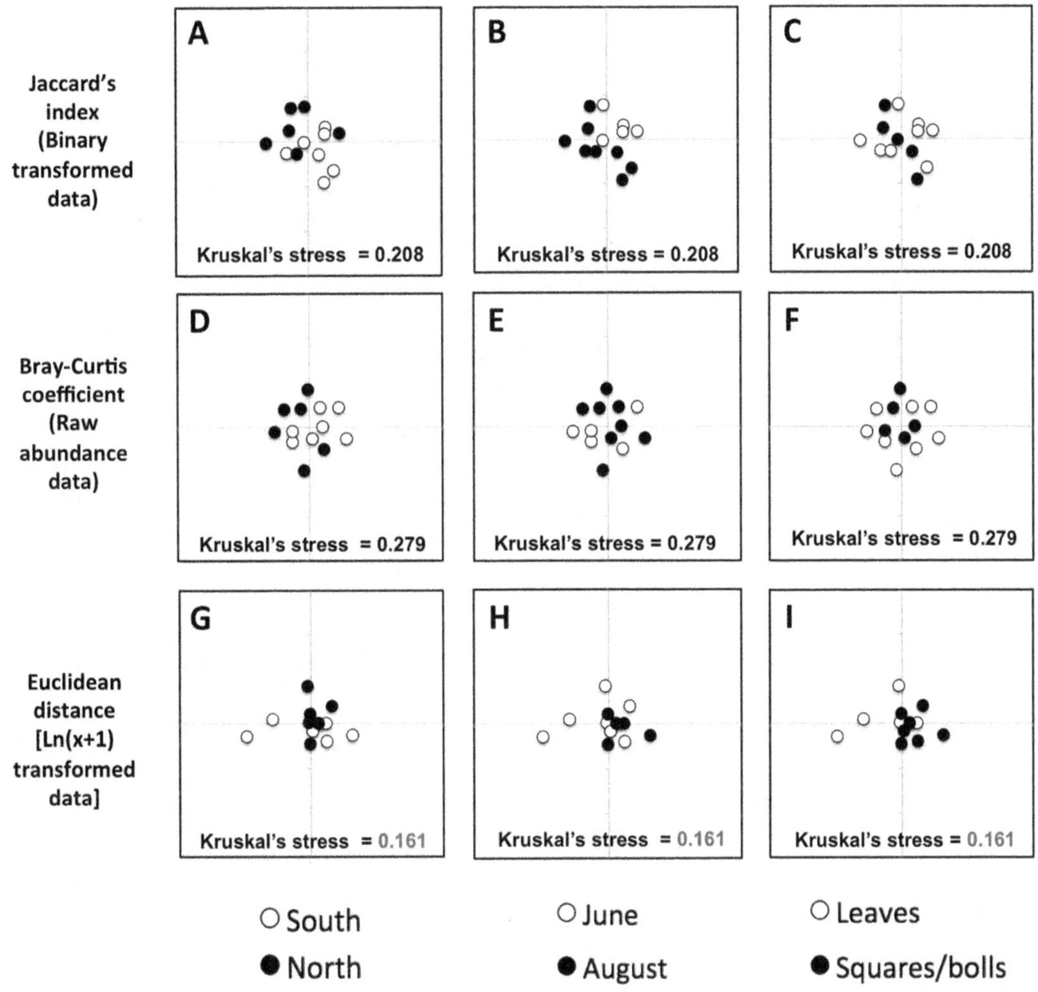

Figure 7. Effect of region, time of season, and tissue on whole fungal endophyte communities. Non-metric multidimensional scale (NMDS) plots corresponding to the clustering of endophyte communities grouped by location (North: Lubbock-RACE and Dawson-RACE; South: Snook and Navasota), time of the season (June and August) and tissues surveyed (Leaves, Squares and Squares/bolls). Analysis was done as explained in Figure 5, but whole endophyte communities per site were obtained by grouping taxa from all five varieties sampled at each location.

ties mediated by these factors was not addressed here and will require specific manipulative experiments. We cannot rule out the possibility that variation in site-specific treatment effects may have obscured our ability to detect broad patterns in endophyte community composition.

Among the fungal endophytes that we isolated, several are candidates for evaluation for use as beneficial endophytes in cotton based on their known effects, either as endophytes or in interactions outside of the plant, against a range of insect pests, nematodes and plant pathogens (Table 2). For example, an isolate of *Acremonium alternatum* is known to have a negative effect on the moth *Plutella xylostella*, when present as an endophyte in bean [66–67] and to induce resistance against *Leveillula taurica*, a causal agent of powdery mildew in tomato [68]. *Chaetomium* sp. isolates are known to produce compounds displaying a wide range of antimicrobial and antitumor activities [69] and along with *Acremonium* sp. and *Paecilomyces* sp., has been shown to have a negative effect on nematodes when present as an endophyte in cucumber [70]. *Cladosporium cladosporioides* is an entomopathogenic fungus of the two spotted spider mite *Tetranychus urticae* Koch [71], but effects on arthropods as an endophyte have not been explored to date to our knowledge. Similarly, *Drechslerella dactyloides* (*Arthrobotrys dactyloides*) is known as nematode-trapping fungus in the soil and has been used as inundative biocontrol agent for nematode control in mushroom culture and against root-knot nematode on tomato [72–73]. However, its functional significance as an endophyte remains unexplored.

Our study highlights the potential for surveys of fungal endophtyes in cotton and other plants to reveal a rich diversity of taxa. The ecological significance and potential for use in biological control of many of them is largely unknown. Further research is required to identify the functional and ecological significance of specific endophytic fungi within the plant under different conditions or at different sites. Should any of these isolates be further developed for beneficial applications, our analysis of spatial and temporal variation in community composition provides *a priori* reason to suspect that their presence in the plant will not be restricted to specific varieties of cotton or limited to any particular locations tested here, given that we observed no effects of variety or region on fungal endophyte biodiversity or community composition. Our results indicate that the presence of fungal endophytes in cotton can be affected by both time of the year and tissue, but in the absence of any consistent patterns across sites, the factors mediating specific interactions between the fungi and plant will need to be evaluated locally.

Materials and Methods

Ethics Statement

Texas A&M AgriLife Extension Service officers, located in College Station, TX and Lubbock, TX provided permission/ access to all eight cotton variety trial farms.

This work did not involve endangered or protected species.

Plant Sampling

Plant tissues [asymptomatic leaves, squares (developing flowers) and bolls (fruits)] were sampled at two different times of the growing season (June and late-July/early-August) in 2011 from multiple commercial cotton varieties grown during variety trials at eight sites distributed across two distinct growing regions in north and central Texas (Figure 1; Table 1). Texas in general suffered a severe drought in 2011 with a total annual precipitation of 385.57 mm, which was critically below the normal annual precipitation of 709.17 mm (Source: National Climatic Data Center/National Oceanic and Atmospheric Administration website: http://www.ncdc.noaa.gov/temp-and-precip/ranks.php?periods%5B%5D = 12&parameter = pcp&state = 41&div = 0&year = 2011&month = 12#ranks-form Accessed 2013 May 10). Although cotton is widely grown without irrigation in Texas, only cotton grown at irrigated sites was sampled for this study.

Plants (N = 9–10 individuals per variety) were randomly sampled across multiple replicate plots (in variety trials) to account for within-field spatial variation. Asymptotic tissues from apparently healthy plants were collected to reduce the chance of sampling pathogenic fungi (one leaf, and when present, one square and one boll per plant). For leaf samples, the 5th true leaves of young plants early in the season and leaves at the top of the canopy of mature plants later in the season were collected. The first developing flower buds (referred to as squares) were collected from young early season plants when present, and the squares at the top were collected from mature plants. Immature fruits (bolls) were collected from the middle of mature plants later in the season. Samples were stored in individual sealed plastic bags and kept refrigerated until processed in the lab.

Endophyte Isolation

Using a laminar flow cabinet as a sterile workspace, plant samples were rinsed in tap water and surface sterilized by immersion in 70% ethanol for 5 min, 10% bleach solution for 3 min, and rinsed twice with autoclaved distilled water. Samples were blotted dry using autoclaved paper towels. Five individual surface sterilized leaves, squares and bolls (N = 15 total samples) were randomly selected and imprinted onto fresh potato dextrose agar (PDA) and V8 media as a way to monitor surface sterilization efficiency. For endophyte isolation, leaves were cut in small fragments of approximately 1 cm^2. Squares and bolls were cut in six pieces. Any fiber present was removed and cut into six smaller pieces. Leaf fragments were placed upside down on PDA and V8 medium plates in triplicate. Each plate contained 3 leaf fragments for a total of 9 fragments assayed per plant. For squares collected early in the season, 3 slices per square were plated on PDA and V8 media as with the leaf fragments. Because of similarity in size and location within a plant, when collected later in the season, squares and bolls from a given plant were plated together on petri dishes containing two square slices, two boll slices and two pieces of fiber. Antibiotics Penicillin G (100 Units/mL) and Streptomycin (100 μg/mL) (Sigma, St Louis, MO, USA) were added to the media to suppress bacterial growth. All plates were incubated in the dark at room temperature for, in average, two weeks until growth of fungal endophyte hyphae from plant tissues was detected.

Endophyte Identification

We used an inclusive combination of morphological and molecular fungal endophyte identification. Once fungal hyphae were detected growing from the plant material, samples were taken to obtain pure fungal isolates. For identification by PCR, genomic DNA was extracted from mycelium of each isolated fungal strain, following a chloroform:isoamyl alcohol 24:1 protocol [74] and fungal specific primers were used to amplify the ITS (Internal Transcribed Spacer) region of nuclear ribosomal DNA [75–78]. This region is the primary barcoding marker for fungi [77] and includes the ITS1 and ITS2 regions, separated by the 5.8S ribosomal gene. In order to avoid introducing biases during PCR (taxonomy bias and introduction of mismatches), it has been suggested to amplify the ITS1 region only [77], therefore the primers ITS1 (5′ TCC GTA GGT GAA CCT GCG G 3′) and ITS2 (5′ GCT GCG TTC TTC ATC GAT GC 3′) were used to amplify and sequence the ~240 bp ITS1 region of each one of our isolated fungal strains. Sequencing was conducted at the Texas A & M University College of Veterinary Medicine & Biomedical Sciences sequencing facility and Macrogen Corp., Maryland, USA. The resulting sequences were aligned as query sequences with the publicly available databases GenBank nucleotide, UNITE [79] and PlutoF [80]. The last two are specifically compiled and used for fungi identification. In all the cases, the strains were identified to species level if their sequences were more than 95% similar to any identified accession from all three databases analyzed [81]. When the similarity percentage was between 90–95%, the strain was classified at genera, family, order, class, subdivision or phylum level depending on the information displayed in databases used. In addition, some of the isolates had lower similarity values (from 30–90%) and were classified as unknown or uncultured depending on the information displayed after BLAST analysis. In total, sequences from 69 unique fungal endophyte taxa were identified (Table 2). To support the molecular identification, fungal endophyte taxa were confirmed by inducing sporulation on PDA or V8 plates and using reported morphological criteria for identification of fruiting bodies structure and shape (*e.g.*, [82–83]). The specific accession numbers of our set of new fungal endophyte isolates obtained from *Gossypium hirsutum* are also shown in Table 2.

Endophyte Community Analyses

To quantify fungal endophyte species diversity within samples or sites (α−diversity), we calculated the Shannon-Wiener biodiversity index (H′) using the frequency of isolation of fungal taxa per variety, tissue, time of the season and location using EstimateS software [37–38]. In order to statistically compare H′ values across samples we used bootstrapping with replacement (1000 iterations) to generate 95% confidence intervals for each H′ value [37–38]. In addition, to assess species richness and determine if our sampling intensity was sufficient, EstimateS was used to calculate fungal endophyte taxa accumulation curves using 1000 randomizations separately for leaves and squares/bolls. The curves were plotted with data from all fungal taxa represented by one or more isolates obtained from each plant tissue per location [37–38].

We compared variation in community composition and structure among varieties, times of the season, tissues and locations (β-diversity) using three different ecological community similarity measures. Multiple different pairwise similarity measures were examined because they consider different kinds of information that can lead to different insights when comparing communities [40]. We calculated (i) the Jaccard's index comparing fungal taxa presence or absence among samples, (ii) the Bray-Curtis coefficient, which compares fungal taxa presence or absence as well as abundance among samples, and (iii) the Euclidean distance, which incorporates both quantitative abundance data and joint absences using total number of fungal taxa isolated. These parameters were calculated and cluster analyses performed using BOOTCLUS software to identify patterns and objectively determine groupings of multivariate data [39].

Matrices obtained by the cluster analyses of pairwise similarity measures of fungal endophyte communities were represented using non-metric multidimensional scale (NMDS) plots. Multidimensional scaling is designed to graphically represent relationships between objects in multidimensional space. The Kruskal's stress value is used to decide which grouping of the data, depending on the number of dimensions used, is the most accurate (commonly acceptable when it is <0.2) [40]. NMDS is a robust visual analysis method applicable to a range of data types, it is amenable to several user-defined standardizations and transformations of the data, flexible in terms of which dissimilarity or similarity measure is used, and can be used for describing patterns and testing *a priori* hypotheses [40].

Supporting Information

Figure S1 Three dimensional NMDS plots of the effects of region, time of season, and tissue on whole fungal endophyte communities. Three dimensional plots and associated Kruskal's stress values of endophyte community comparisons as shown in the two dimensional plots in Figure 7. Results indicate no major change in observed clustering patterns with increased dimensionality despite increased confidence based on reduced Kruskal's stress values (<0.2).

Acknowledgments

Mark Kelley (Texas A&M AgriLife Extension Service) assisted in providing access to sampling sites. Ryan Easterling provided valuable help preparing materials and reagents during sample processing. Dr. Juan Pedro Steibel from Department of Animal Sciences and Fisheries and Wildlife at Michigan State University provided statistical advice on bootstrapping analysis.

Author Contributions

Conceived and designed the experiments: MJER GAS. Performed the experiments: MJER WZ CUV JBA LLK. Analyzed the data: MJER WZ GAS. Contributed reagents/materials/analysis tools: GDM DLK. Wrote the paper: MJER GAS.

References

1. Porras-Alfaro A, Bayman P (2011) Hidden fungi, Emergent Properties: Endophytes and Microbes. Annu Rev Phytopathol 49: 291–315.
2. Schardl CL, Clay L (1997) Evolution of mutualistic endophytes from plant pathogens. In: Carroll GC, Tudzynski P (eds.) The Mycota. Vol V Part B, Plant Relationships. Springer Verlag, Berlin, Germany. 221–238.
3. Saikkonen K, Faeth SH, Helander M, Sullivan TJ (1998) Fungal endophytes: a continuum of interactions with host plants. Annu Rev Ecol Evol Syst 29: 319–343.
4. Johnson JM, Oelmüller R (2009) Mutualism or parasitism: life in an unstable continuum. What can we learn from the mutualistic interaction between *Piriformospora indica* and *Arabidopsis thaliana*? Endocytobiosis Cell Res 19: 81–111.
5. Petrini O, Stone J, Carroll FE (1992) Endophytic fungi in evergreen shrubs in western Oregon: a preliminary study. Can J Bot 60: 789–796.
6. Azevedo JL, Maccheroni W Jr., Pereira JO, Araujo WL (2000) Endophytic microorganisms: a review on insect control and recent advances in tropical plants. Electron J Biotechnol 3: 40–65.
7. Stone JK, Bacon CW, White JF (2000) An overview of endophytic microbes: endophytism defined. In: Bacon CW, White JF (eds.) Microbial endophytes. Marcel Dekker Inc., New York, New York, USA. 3–30.
8. Redman RS, Sheehan KB, Stout TG, Rodriguez RJ, Henson JM (2002) Thermotolerance generated by plant/fungal symbiosis. Science 298: 1581.
9. Arnold AE, Mejia LC, Kyllo D, Rojas EI, Maynard Z et al. (2003) Fungal endophytes limit pathogen damage in a tropical tree. Proc Natl Acad Sci U S A 100: 15649–15654.
10. Sikora RA, Pocasangre L, Felde AZ, Niere B, Vu TT et al. (2008) Mutualistic endophytic fungi and *in-planta* suppressiveness to plant parasitic nematodes. Biological Control 46: 15–23.
11. Bae H, Sicher RC, Kim MS, Kim SH, Strem MD et al. (2009) The beneficial endophyte, *Trichoderma hamatum* isolate DIS219b, promotes growth and delays the onset of the drought response in *Theobroma cacao*. J Exp Bot 60: 3279–3295.
12. Rodriguez RJ, White JF, Arnold AE, Redman RS (2009) Fungal endophytes: diversity and functional roles. New Phytol 182: 314–330.
13. Ownley BH, Gwinn KD, Vega FE (2010) Endophytic fungal entomopathogens with activity against plant pathogens: ecology and evolution. Biocontrol 55: 113–128.
14. Bing LA, Lewis LC (1992) Endophytic *Beauveria bassiana* (balsamo) Vuillemin in corn: the influence of the plant growth stage and *Ostrinia nubiladis* (Hubner). Biocontrol Sci Technol 2: 39–47.
15. Schulz B, Guske S, Dammann U, Boyle C (1998) Endophyte-host interactions. II. Defining symbiosis of the endophyte-host interactions. Symbiosis 25: 213–377.
16. Pinto LSRC, Azevedo JL, Pereira JO, Vieira MLC, Labate CA (2000) Symptomless infection of banana and maize by endophytic fungi impairs photosynthetic efficiency. New Phytol 147: 609–615.
17. Boyle C, Gotz M, Dammann-Tugend U, Schulz B (2001) Endophyte-host interactions III. Local *versus* systemic colonization. Symbiosis 31: 259–281.
18. Schulz B, Boyle C, Draeger S, Rommert AK, Krohn K (2002) Endophytic fungi: a source of novel biologically active secondary metabolites. Mycol Res 106: 996–1004.
19. Gurulingappa PG, Sword GA, Murdoch G, McGee PA (2010) Colonization of crop plants by fungal entomopathogens and their effects on two insect pests when *in planta*. Biological Control 55: 34–41.
20. Schardl CL, Phillips TD (1997) Protective grass endophytes: Where are they from and where are they going? Plant Disease 81: 430–438.
21. Liu CI, Zhou WX, Lu I, Tan RX (2001) Antifungal activity of *Artemisia annua* endophyte cultures against phytopathogenic fungi. J Biotechnol 88: 277–282.
22. Faeth SH (2002) Are endophytic fungi defensive mutualists? Oikos 98: 25–36.
23. Selosse MA, Baudoin E, Vandenkoornhuyse P (2004) Symbiotic microorganisms, a key for ecological success and protection of plants. C R Biol 327: 639–648.
24. Azevedo JL, Araujo WL (2007) Endophytic fungi of tropical plants: diversity and biotechnological aspects. In: Ganguli BN, Deshmukh SK (eds.) Fungi: multifaceted microbes. Anamaya Publishers, New Delhi, India. 189–207.
25. Crawford RF (1923) Fungi isolated from the interior of cotton seed. Phytopathology 13: 501–503.
26. Roy KW, Bourland FM (1982) Epidemiological and mycofloral relationships in cotton seedling disease in Mississippi. Phytopathology 72: 868–872.
27. Klich MA (1986) Mycoflora of cotton seed from the southern United States: A three year study of distribution and frequency. Mycologia 78: 706–712.
28. Baird R, Carling D (1998) Survival of parasitic and saprophytic fungi on intact senescent cotton roots. J Cotton Sci 2: 27–34.
29. Palmateer AJ, McLean KS, Morgan-Jones G, van Santen E (2004) Frequency and diversity of fungi colonizing tissues of upland cotton. Mycopathologia 157: 303–316.
30. Vega FE (2008) Insect pathology and fungal endophytes. J Invertebr Pathol 98: 277–279.
31. McGee PA (2002) Reduced growth and deterrence from feeding of the insect pest *Helicoverpa armigera* associated with fungal endophytes from cotton. Aust J Exp Agric 42: 995–999.
32. Farr DF, Bills GF, Chamuris GP, Rossman AY (1989) Fungi on plants and plant products in the USA. APS Press, St. Paul, Minnesota, USA.
33. Ragazzi A, Moricca S, Dellavalle I (1997) Fungi associated with cotton decline in Angola. Zeitschrift für Pflanzenkrankheiten und Pflanzenschutz 104: 133–139.
34. Wang B, Priest MJ, Davidson A, Brubaker CL, Woods MJ et al. (2007) Fungal endophytes of native *Gossypium* species in Australia. Mycol Res 3: 347–354.
35. de Souza Vieira PD, de Souza Motta CM, Lima D, Braz Torrez J, Quecine MC et al. (2011) Endophytic fungi associated with transgenic and non-transgenic cotton. Mycology 2: 91–97.
36. Freeman MF, Tukey JW (1950) Transformations related to the angular and the square root. The Annals of Mathematical Statistics 21: 607–611.
37. Colwell RK (2009) EstimateS: Statistical estimation of species richness and shared species from samples. Version 8.2. User's Guide and application published at: http://purl.oclc.org/estimates accessed 2013 January 16.

38. Colwell RK, Chao A, Gotelli NJ, Lin S-Y, Mao CX et al. (2012) Models and estimators linking individual-based and sample-based rarefaction, extrapolation, and comparison of assemblages. Journal of Plant Ecology 5: 3–21.
39. McKenna JE Jr. (2003) An enhanced cluster analysis program with bootstrap significance testing for ecological community analysis. Environmental Modeling & Software 18: 205–220.
40. Quinn GP, Keough MJ (2002) Multidimensional scaling and cluster analysis. In: Quinn GP and Keough MJ (eds.) Experimental design and data analysis for biologists. Cambridge University Press, Cambridge, UK. 473–493.
41. Anderson MJ, Crist TO, Chase JM, Vellend M, Inouye BD et al. (2011) Navigating the multiple meanings of β diversity: a roadmap for the practicing ecologist. Ecol Lett 14: 19–28.
42. Gonthier P, Gennaro M, Nicolotti G (2006) Effects on water stress on the endophytic mycota of *Quercus robur*. Fungal Divers 21: 69–80.
43. Oses R, Valenzuela S, Freer J, Sanfuentes E, Rodriguez J (2008) Fungal endophytes in xylem of healthy Chilean trees and their possible role in early wood decay. Fungal Divers 33: 77–86.
44. Taylor JE, Hyde KD, Jones EBG (1999) Endophytic fungi associated with the temperate palm *Trachycarpus fortune* within and outside its natural geographic range. New Phytol 142: 335–346.
45. Frohlich J, Hyde KD, Petrini O (2000) Endophytic fungi associated with palms. Mycol Res 104: 1202–1212.
46. Cheplick GP, Faeth SH (2009) In:Oxford University Press (eds.) Ecology and evolution of the grass-endophyte symbiosis. Oxford University Press Inc., New York, New York, USA. 1–233.
47. Alva P, McKenzie EHC, Pointing SB, Pena-Muralla R, Hyde KD (2002) Do sea grasses harbor endophytes? Fungal Diversity Research Series 7: 167–178.
48. Li WC, Zhou J, Guo SY, Guo LD (2007) Endophytic fungi associated with lichens in Baihua mountain of Beijing, China. Fungal Divers 25: 69–80.
49. Arnold AE, Maynard Z, Gilbert G, Coley PD, Kursar TA (2000) Are tropical fungal endophytes hyperdiverse? Ecol Lett 3: 267–274.
50. Arnold AE (2007) Understanding the diversity of foliar endophytic fungi: progress, challenges and frontiers. Fungal Biol Rev 21: 51–56.
51. Lodge DJ, Fisher PJ, Sutton BC (1996) Endophytic fungi of *Manilkara bidentata* leaves in Puerto Rico. Mycologia 88: 733–738.
52. Bayman P (2006) Diversity, scale and variation of endophytic fungi in leaves of tropical plants. In: Bailey MJ, Lilley AK, Timms-Wilson TM (eds.) Microbial Ecology of Aerial Plant Surfaces. CABI, Oxfordshire, UK. 37–50.
53. Porras-Alfaro A, Herrera J, Odenbach K, Lowrey T, Sinsabaugh RL et al. (2008) A novel root fungal consortium associated with a dominant grass in a semiarid grassland. Plant Soil 296: 65–75.
54. Hyde KD, Soytong K (2008) The fungal endophyte dilemma. Fungal Divers 33: 163–173.
55. Guo LD, Hyde KD, Liew ECY (2001) Detection and taxonomic placement of endophytic fungi within frond tissues of *Livistona chinensis* based on rDNA sequences. Mol Phylogenet Evol 19: 1–13.
56. Ownley BH, Griffin MR, Klingeman WE, Gwinn KD, Moulton JK et al. (2008) *Beauveria bassiana*: endophytic colonization and plant disease control. J Invertebr Pathol 98: 267–270.
57. Santamaria J, Bayman P (2005) Fungal epiphytes and endophytes of coffee leaves (*Coffea arabica*) Microb Ecol 50: 1–8.
58. Arnold AE, Lutzoni F (2007) Diversity and host range of foliar fungal endophytes: Are tropical leaves biodiversity hotspots? Ecology 88: 541–549.
59. Cotty PJ (1987) Evaluation of cotton cultivar susceptibility to *Alternaria* leaf spot. Plant Disease 71: 1082–1084.
60. Bashan Y (1994) Symptomless infections in alternaria leaf blight of cotton. Can J Bot 72: 1580–1585.
61. Rodrigues KT (1994) The foliar fungal endophytes of the Amazonian palm *Euterpe oleracea*. Mycologia 86: 376–385.
62. Carroll GC (1995) Forest endophytes: pattern and process. Can J Bot 73: S1316–S1324.
63. Wilson D (2000) Ecology of woody plant endophytes. In: Bacon CW, White JF Jr. (eds.) Microbial Endophytes. Mercel Dekker Inc., New York, New York, USA. 389–420.
64. Gamboa MA, Laureano S, Bayman P (2002) Measuring diversity of endophytic fungi in leaf fragments: Does size matter? Mycopathologia 156: 41–45.
65. Lingfei L, Anna Y, Zhiwei Z (2005) Seasonality of arbuscular mycorrhizal symbiosis and dark septate endophytes in a grassland site in southwest China. Mycorrhiza 17: 103–109.
66. Raps A, Vidal S (1998) Indirect effects of an unspecialized endophytic fungus on specialized plant-herbivorous insect interactions. Oecologia 114: 541–547.
67. Pineda A, Zheng SJ, van Loon JJA, Pieterse CMJ, Dicke M (2010) Helping plants to deal with insects: the role of beneficial soil-borne microbes. Trends Plant Sci 15: 507–514.
68. Kasselaki AM, Shaw MW, Malathrakis NE, Haralambous J (2006) Control of *Leveillula taurica* in tomato by *Acremonium alternatum* is by induction of resistance, not hyperparasitism. Eur J Plant Pathol 115: 263–267.
69. Mao BZ, Huang C, Yang GM, Chen YZ, Chen SY (2010) Separation and determination of bioactivity of oosporein from *Chaetomium cupreum*. Afr J Biotechnol 9: 5955–5961.
70. Yan X, Sikora RA, Zheng J (2011) Potential use of cucumber (*Cucumis sativus* L.) endophytic fungi as seed treatment agents against root-knot nematode *Meloidogyne incognita*. J Zhejiang Univ-Sci B (Biomedicine and Biotechnology). 12: 219–225.
71. Jeyarani S, Gulsar Banu J, Ramaraju K (2011) First record of natural occurrence of *Cladosporium cladosporioides* (Fresenius) de Vries and *Beauveria bassiana* (Bals.-Criv.) Vuill on two spotted spider mite, *Tetranychus urticae* Koch from India. Journal of Entomology 8: 274–279.
72. Cayrol JC (1983) Biological control of *Meloidogyne* by *Arthrobotrys irregularis*. Revue de Nematologie 6: 265–273.
73. Sayre RM (1986) Pathogens for biological control of nematodes. Crop Prot 5: 268–276.
74. Sambrook J, Fritsch E, Maniatis T (1989) In:Molecular Cloning: A Laboratory Manual. Cold Spring Harbor Laboratory Press, New York, New York, USA, Vol. 1, 2, 3.
75. Nilsson RH, Kristiansson E, Ryberg M, Hallenberg N, Larsson KH (2008) Intraspecific ITS variability in the kingdom fungi as expressed in the International Sequence Databases and its implications for molecular species identification. Evolutionary Bioinformatics 4: 193–201.
76. Seifert KA (2009) Progress towards DNA barcoding fungi. Mol Ecol Resour 9 (Suppl.1): 83–89.
77. Bellemain E, Carlsen T, Brochmann C, Coissac E, Taberlet P et al. (2010) ITS as an environmental DNA barcode for fungi: an in silico approach reveals potential PCR biases. BMC Microbiol 10: 189–197.
78. Vralstad T (2011) ITS, OTUs and beyond-fungal hyperdiversity calls for supplementary solutions. Mol Ecol 20: 2873–2875.
79. Abarenkov K, Nilsson RH, Larsson K-H, Alexander IJ, Eberhardt U et al. (2010) The UNITE database for molecular identification of fungi - recent updates and future perspectives. New Phytol 186: 281–285.
80. Abarenkov K, Tedersoo L, Nilsson RH, Vellak K, Saar I et al. (2010) PlutoF - a Web Based Workbench for Ecological and Taxonomic Research, with an Online Implementation for Fungal ITS Sequences. Evolutionary Bioinformatics 6: 189–196.
81. Zimmerman NB, Vitousek PM (2011) Fungal endophyte communities reflect environmental structuring across a Hawaiian landscape. Proc Natl Acad Sci U S A 109: 13022–13027.
82. Barnett HL, Hunter BB (1998) In:The American Phytopathological Society (eds.) Illustrated genera of imperfect fungi. APS Press, St. Paul, Minnesota, USA. 1–215.
83. Dugan FM (2008) In:The American Phytopathological Society (eds.) The identification of fungi. An illustrated introduction with keys, glossary and guide to literature. APS Press, St. Paul, Minnesota, USA. 1–176.

20

Organic Farming and Landscape Structure: Effects on Insect-Pollinated Plant Diversity in Intensively Managed Grasslands

Eileen F. Power[1]*, Daniel L. Kelly[1,2], Jane C. Stout[1,2]

1 Botany Department, School of Natural Sciences, Trinity College Dublin, Dublin, Republic of Ireland, 2 Trinity Centre for Biodiversity Research, Trinity College Dublin, Dublin, Republic of Ireland

Abstract

Parallel declines in insect-pollinated plants and their pollinators have been reported as a result of agricultural intensification. Intensive arable plant communities have previously been shown to contain higher proportions of self-pollinated plants compared to natural or semi-natural plant communities. Though intensive grasslands are widespread, it is not known whether they show similar patterns to arable systems nor whether local and/or landscape factors are influential. We investigated plant community composition in 10 pairs of organic and conventional dairy farms across Ireland in relation to the local and landscape context. Relationships between plant groups and local factors (farming system, position in field and soil parameters) and landscape factors (e.g. landscape complexity) were investigated. The percentage cover of unimproved grassland was used as an inverse predictor of landscape complexity, as it was negatively correlated with habitat-type diversity. Intensive grasslands (organic and conventional) contained more insect-pollinated forbs than non-insect pollinated forbs. Organic field centres contained more insect-pollinated forbs than conventional field centres. Insect-pollinated forb richness in field edges (but not field centres) increased with increasing landscape complexity (% unimproved grassland) within 1, 3, 4 and 5 km radii around sites, whereas non-insect pollinated forb richness was unrelated to landscape complexity. Pollination systems within intensive grassland communities may be different from those in arable systems. Our results indicate that organic management increases plant richness in field centres, but that landscape complexity exerts strong influences in both organic and conventional field edges. Insect-pollinated forb richness, unlike that for non-insect pollinated forbs, showed positive relationships to landscape complexity reflecting what has been documented for bees and other pollinators. The insect-pollinated forbs, their pollinators and landscape context are clearly linked. This needs to be taken into account when managing and conserving insect-pollinated plant and pollinator communities.

Editor: Jeff Ollerton, University of Northampton, United Kingdom

Funding: This project is part of a collaborative project led by Michael O'Donovan, Teagasc, and was funded by the Irish Department of Agriculture, Fisheries and Food (http://www.agriculture.gov.ie) Research Stimulus Fund (RSF), funded under the National Development Plan (2007–2013). The funders had no role in study design, data collection and analysis, decision to publish, or preparation of the manuscript.

* E-mail: eileen.power@newcastle.ac.uk

Introduction

Animal-mediated pollination is required for successful reproduction in many angiosperms [1]. However, there are concerns for the future of many pollinator species due to agricultural intensification [2]. Parallel declines in insect-pollinated plants and their pollinators have been reported [3]. Changes in plant communities in intensively managed arable systems are evident, with a dominance of self-pollinated plants that can better withstand disturbance (frequent soil cultivation, crop harvesting and crop rotations) over plants that depend on animals for pollination and/or other plants as pollen donors [4,5]. In comparison, natural or semi-natural systems tend to have lower proportions of self-pollinated plants [6].

It is not clear whether intensive grasslands also have reduced proportions of insect-pollinated plants. Most western European lowland grasslands - covering millions of hectares - are intensively managed [7]. Intensive grasslands receive high fertilizer application rates and frequent defoliation [8]. This results in degraded species pools and structurally homogenous swards. The majority of agricultural land in the Republic of Ireland is intensive grassland [9]; these intensive grasslands support considerably fewer plant [10] and pollinator [11,12] species compared to the communities of plants [13,14] and pollinators [15,16] found in semi-natural grasslands. Declines in plant species richness are likely to be reflected in declines in insect-pollinated plant numbers within intensive grassland plant communities. Declines in insect-pollinated plant numbers may, in turn, have further knock-on impacts on pollinators and plant-pollinator interaction networks.

Local factors such as abiotic conditions (e.g. nutrient availability and soil acidity) and field management (e.g. herbicide application and defoliation) can affect the distribution and species richness of arable plants [17,18]. In particular, organic farming has been found to mitigate the decline of plants and pollinators in arable systems [19,20], most likely through the prohibition of pesticides and chemical fertilizers (European Union Regulation 2092/91/

EEC). There is also evidence of a shift to more insect-pollinated plants in organically-farmed arable weed communities compared to their conventional counterparts, probably as a result of increases in both plant and pollinator diversity on organic farms [5,21]. This has important implications for organic intensive grasslands (characterised by lower stocking densities and no chemical fertilisers or pesticides), where increased plant and pollinator diversities have also been observed [12,22].

Landscape structure, calibrated in terms of landscape complexity and the proportion of the landscape that is organically farmed, has been found to affect the influence of organic farming on local plant and pollinator diversity [23,24,25]. Therefore, as plants and pollinators can be affected by landscape structure and pollinator diversity can decline in parallel with insect-pollinated plants [3], one might expect insect-pollinated plant diversity to be more affected by landscape structure than non-insect pollinated plant diversity. Landscape structure, calibrated in terms of hedgerow structure and distribution, may also influence insect-pollinated plant richness, as it is clear that linear features influence pollinator movements [26,27] and are important as food and nesting resources [16,28,29,30]. Similarly, the proximity of suitable habitats/resources nearby may have a strong influence on the local distribution of pollinators and the plants they pollinate [31]. Thus, unravelling the effects of farming system and landscape structure on insect-pollinated plants may help to inform management decisions pertaining to plant and pollinator conservation.

We hypothesised that, in landscapes dominated by conventional farming, (1) plant species richness is higher in organic than in conventional dairy farms. Specifically, we expected there to be (2) a greater number of insect-pollinated forb species in organic farms than in conventional. We also expected (3) that plant community composition would differ depending on position within the field (edge vs. centre). We further hypothesised that (4) the numbers of insect-pollinated and non-insect pollinated forbs would differ from each other depending on local factors (position in field and soil parameters) and also (5) on landscape factors (landscape complexity, presence of linear landscape features and distance metrics, i.e. proximity to different habitats). We expected that plant species richness (especially of insect-pollinated forbs) would be affected by (6) the interactions between local factors (especially local farm management) and landscape complexity. Overall, we aimed to evaluate how far grouping plants according to their principal pollen vector is useful in developing our understanding of plant relationships with local and landscape factors.

Methods

Ethics Statement

Permission to access sites, survey vegetation and take soil samples was obtained from all landowners.

Study Sites

Ten matched pairs of organic (managed according to the European Union Regulation 2092/91/EEC) and conventional (not managed according to organic regulations) dairy farms were selected in lowland permanent grassland (not ploughed or reseeded for at least 7 years) in the Republic of Ireland. Only dairy farms were chosen as effects of dairy versus drystock farming on biodiversity can differ substantially [10]. Organic farms had been certified for 10.5 years on average (range: 6–19 years), following a 2 year conversion period. Pairs were matched on geology, soil type and climatic similarity. Farms within a pair were 1–4 km apart, in landscapes dominated by conventional farming in central and southern Ireland (Fig. 1). All were in areas characterised by well-drained, fertile soils used predominantly for beef/dairy farming. The majority of organic and conventional farmers (except one conventional farmer) participated in an Irish agri-environmental scheme (the Rural Environmental Protection Scheme, REPS). Organic and conventional farms differed in terms of stocking density (average Livestock Units per hectare (LU/ha) on organic farms was 1.5 LU/ha compared with 2.5 LU/ha on conventional farms) and farm inputs. All conventional farmers in the study applied chemical fertilisers and herbicides in their fields, whilst organic farmers did not [12]. The dominant land-uses within 5 km radii around all study sites were: unimproved grassland (mean: 41%, range: 24–58%); improved grassland (mean: 35%, range: 18–46%); arable land (mean: 10%, range: 0–23%); woodland (mean: 5%, range: 0–19%); wetland (mean: 2%, range: 0–12%) and urban areas (mean: 1%, range: 0–9%).

Plant Surveys

We surveyed forbs (broadleaved herbaceous vascular plants) and graminoids (grasses, sedges and rushes) once in three fields from each farm between July and September 2008. Fields within each farm were sampled on the same day and farms within a pair were sampled on consecutive days. Similar field sizes were surveyed between farm types (mean organic field size = 3.2 hectares; mean conventional field size = 3.1 hectares). Only fields that were grazed within one week of surveying were sampled. Plant surveys were conducted in two 100 m transects per field: one transect located 1 m from and parallel to the base of an internal field boundary ("field edge") and the second transect located parallel to the first, 25 m from the same field boundary ("field centre"). The start of all edge transects were placed along stock-proof hedgerows, at least 100 m from field corners. The percentage cover of each species was recorded in five quadrats (0.5×0.5 m) placed 20 m apart in each transect, giving a total of 15 edge and 15 centre quadrats per farm. Quadrat number and size were deemed sufficient to capture the majority of plant species present by plotting species accumulation curves. For identification and nomenclature we follow Stace [32]. The forbs were classified into two groups based on their pollination types: insect and non-insect (self and/or wind-pollinated) by selecting the most frequent pollen vector according to Grime et al. [33] and the BIOFLOR database [34]. Graminoids at our sites were all wind pollinated [35] but were placed in a separate group to non-insect pollinated forbs in light of the unknown origins of many of the species (natural colonisation or planted as part of a seed mix).

Soil Analysis

Five ca 500 g soil samples were taken along each transect, 20 m apart and from approximately 0–10 cm under the soil surface. The five soil samples from each transect were then homogenised into one sample and, for each transect sample, the pH and the available macro-nutrients Phosphorus (P), Potassium (K) and Magnesium (Mg) were analysed following Alexander et al. [36] (Table S1).

As some soil parameters were collinear, they could not all be used in the same statistical model. Therefore, we determined the relationships among the soil parameters using principal components analysis (PCA) in Primer 6.1.13 [37]. K was correlated with the first principal component (PC1), P and Mg with PC2 and to a lesser extent PC3 and pH with PC4 (Table 1). Therefore, PC1, PC2 (both representing different plant macronutrients) and PC4 (representing soil acidity) were used as covariates in further analysis.

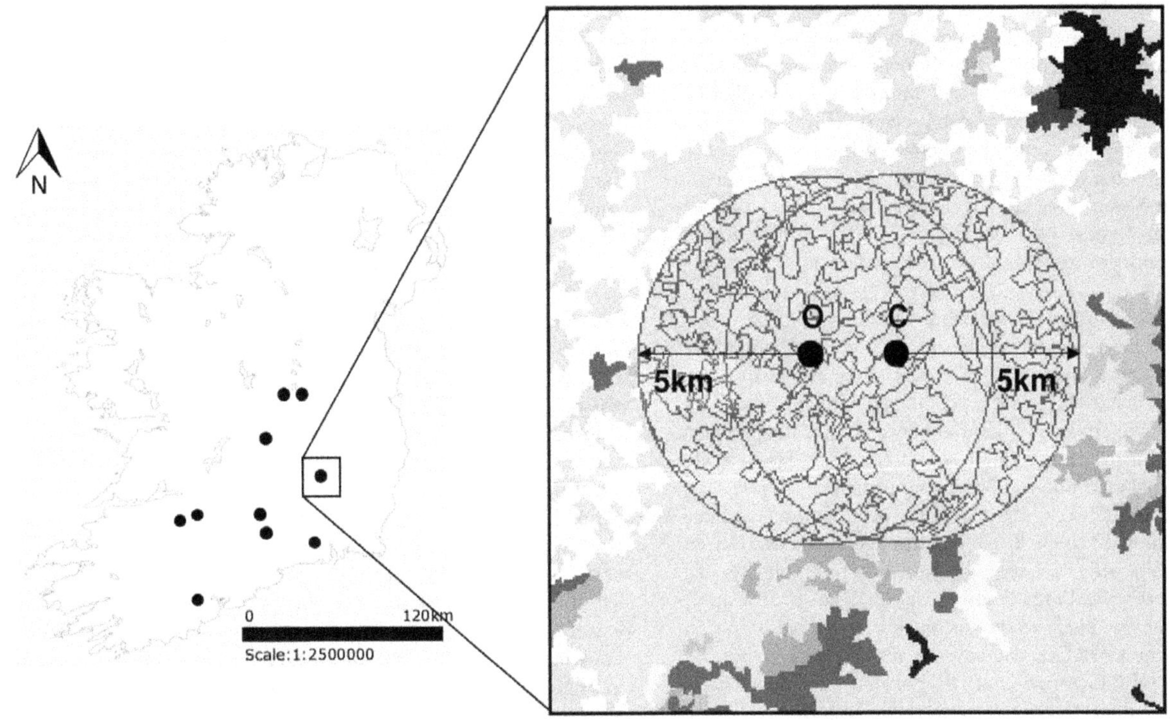

Figure 1. The distribution of the ten farm pairs within Ireland. For each farm, the percentage of different land-use types within 1, 2, 3, 4 and 5 km radii around farms was determined (the extent of the 5 km radii is shown as an example in this figure). Landscapes were similar and overlapping within a farm pair only. Note: CORINE Level 4 classification is displayed within the circular sectors while CORINE Level 3 classification is displayed outside them. O = organic farm, C = conventional farm.

Landscape Characterisation

We used the CORINE Land Cover 2000 [38] database (100 m^2 resolution) to characterise the landscape context of each site. This database was chosen over a more up to date version (CORINE Land Cover 2006 [39]) as it contained the level of detail required for this study (specifically, Level 4 classification which characterises pastures further into unimproved and improved grassland). The CORINE land cover information was visually compared for each site with aerial photographs (year 2005; Irish National Biodiversity Data Centre) and the CORINE Land Cover 2006 [39] database so that major changes in land cover could be identified. We discerned no major differences in Level 1 and 3 habitat classifications [38] between the databases and the aerial photographs. Difficulty in the visual identification of Level 4 classifications (unimproved and improved grassland) in the aerial photographs and the lack of this data in the CORINE Land Cover 2006 [39] database meant that the CORINE 2000 database was relied upon, in this respect.

Table 1. Eigenvalues for the first four principal components (PC1-4) from a principal component analysis (PCA) on soil nutrients and pH and the Pearson correlation coefficient (r) between soil parameters and the four principal components.

		Nutrient			
Principal component	**Eigenvalue**	**P**	**K**	**Mg**	**pH**
PC1	15.700	−0.120	−0.991***	−0.066	−0.003
PC2	0.470	−0.910***	0.136	−0.385***	−0.076
PC3	0.114	−0.386***	−0.015	0.920***	−0.069
PC4	0.009	−0.096	0.007	0.034	0.995***

Note: ***P<0.001.

Fifteen land use types were defined [using CORINE land cover classifications (Levels 1, 3 and 4)]: arable, unimproved grassland (medium management intensity pasture), improved grassland (high management intensity pasture), natural grassland (a low management intensity semi-natural grassland - of which there were negligible amounts recorded in the study landscapes), broadleaved forest, coniferous forest, mixed forest, transitional woodland scrub, urban areas, bogs, marshes, heaths, stream courses, water bodies and other habitats. Specific CORINE definitions of how unimproved and improved grassland are characterised are unavailable but these categories roughly correspond to semi-improved and improved grassland, respectively, in the classification of Sullivan et al. [40]. The unimproved grasslands referred to in our study deviate markedly from semi-natural grassland conditions.

The percent land cover of each of the 15 land use types was measured at five spatial scales, using 1–5 km radii around each study site (Fig. 1). Landscapes were similar around each organic and conventional farm within a pair (Figure S1) (as the spatial radii overlapped within pairs only) but not between pairs. The 1–5 km scales were chosen to reflect the potential dispersal distances of pollinators upon whom insect-pollinated forbs depend. Studies on bumblebees and honeybees have found that they can be related to the landscape at scales of up to 3 km radii [41,42] while hoverflies can be related to the landscape up to 4 km around study sites [43] and disperse up to 400 m in a day [26]. Some hoverfly species migrate across great distances, even across seas [16,44,45].

Percent cover of each of the 15 land use types was used to calculate landscape diversity (*H*) on each of the five spatial scales, using the Shannon index [46]. Only two out of the thirteen land-use types (unimproved grassland and improved grassland) were present at all sites and in all landscape sectors. However, the two were collinear (variance inflation factors (VIF) above 3 [47]). Unimproved grassland was the dominant land-use type in the study landscapes and was negatively correlated with landscape diversity (5 km radius: $r = -0.6893$, $P = 0.001$), and so unimproved grassland was used as an inverse indicator of landscape complexity.

Linear landscape features contribute to the complexity of a landscape and thus impact on biodiversity. In our study landscape, hedgerows are the dominant linear feature and consequently may affect plant richness. We estimated the percentage cover and length (km) of hedgerows and the number of connections between hedgerows within a 1 km radius around each site (larger scales could not be included due to budgetary restrictions). These three hedgerow measures were not found to be collinear (variance inflation factors (VIF) below 3 [47]). The hedgerow measures were also not significantly related to landscape complexity (i.e. percentage unimproved grassland) within the 1 km scale (unpublished data). This is likely to be due to differences between data sources used: the CORINE database (used to calculate landscape complexity) has a low resolution that does not include linear habitats smaller than 100 m^2 whereas visual inspection of aerial photographs (used to calculate hedgerow measures) can pinpoint smaller linear features and their connections. These hedgerow measures may give an indication of hedgerow structure within the landscape which is known to impact on plant diversity [48,49].

The isolation of sites, i.e. the distance (in metres) from each study site to the nearest edges of 10 habitat types, was calculated in relation to: broadleaved forest, coniferous forest, mixed forest, transitional woodland scrub, urban areas, bogs, marshes, heaths, stream courses and water bodies. Most of the 10 habitat types were found to be collinear (variance inflation factors (VIF) above 3 [47]) so were pooled into four categories: forest, urban, wetland and open water. For example: broadleaved forest, coniferous forest, mixed forest and transitional woodland scrub were pooled into the category named forest. All landscape analyses were done using ArcGIS 9.3.1, ESRI, Redlands, CA, USA.

Statistical Analysis

Plant community similarity (in terms of percentage cover of each species) between organic and conventional field edges and centres was graphically analysed using Non-Metric Multidimensional Scaling (NMDS) and significant differences were tested for using PERMANOVA+ (fixed factors were transect position (edge/centre), farm type (organic/conventional) and their interaction while random factors were field (1–3) and farm pair (1–10)), using PRIMER 6.1.13 [37]. To determine which plant species were primarily providing the discrimination (i.e. highest percentage contribution to the plant community) between organic and conventional field edges and centres, we performed SIMPER analysis using PRIMER 6.1.13 [37].

Plant richness was calculated for all plants ("total plants"), for insect-pollinated forbs, non-insect pollinated forbs and graminoids as the cumulative species richness of each transect. Each measure of richness was analysed in relation to: (1) local factors (farming system, position in the field and soil parameters (PC1, PC2 and PC4)) and landscape complexity for each of the five spatial scales; (2) local factors (farming system and position in the field) and linear landscape features and (3) local factors (farming system and position in the field) and distance metrics using Linear Mixed Effects Models. As individual soil parameters (K, P, Mg and pH) were collinear and so could not be analysed in the same model without using PCA scores (see Soil Analysis section), each parameter was also analysed using Linear Mixed Effects Models to elucidate individual soil parameter differences between farming system and position in the field. Landscape complexity and linear feature data were analysed in separate models because landscape complexity data was available for each of 1–5 km scales but only data for the 1 km scale was available for linear features. We accounted for the hierarchical structure of the data by including random terms: farm pair (1–10), farm (1–20) and field (1–3). In each local and landscape model, the fixed effects included: farm type (organic/conventional), location-within-field (edge/centre), soil parameters (PC1, PC2, PC4) and percent cover of unimproved grassland. In the linear landscape features models, summed hedgerow length, hedgerow percent cover, number of hedgerow connections, farm type and location-within-field were included as fixed effects. In the distance metrics models, farm type, location-within-field, distances to forest, urban, wetland and open water were included as fixed effects. In the individual soil parameter model, the fixed effects included: farm type and location-within-field. Biologically relevant two-way and three-way interactions between fixed effects were included. Models were simplified by removing, first, non-significant interactions ($P>0.05$) and then any non-significant main effects (that were not constituent within a significant interaction). Models were validated by: plotting standardised residuals against fitted values and each explanatory variable to assess homogeneity and independence, verifying normality of residuals using Normal QQ-plots and assessing the models for influential observations using Cook distances [50]. Mixed modelling was carried out using the nlme [51] package in *R* [52].

Results

A total of 69 plant species were found, with 61 in organic and 41 in conventional farms (Table S2). There were 31 insect-pollinated forbs (26 in organic, 20 in conventional), 18 non-insect pollinated forbs (17 in organic, 14 in conventional) and 20 graminoids (18 in organic, 14 in conventional). Of the non-insect pollinated species, 12 were normally self-pollinated and 6 were wind pollinated.

Plant Community Composition

Graminoids dominated the community in organic and conventional field edges and centres, followed by insect-pollinated forbs (Fig. 2). Non-insect pollinated forbs were far less frequent on all farms (Fig. 2). Multivariate analysis showed significant differences in plant community composition between organic and conventional farms (Pseudo-$F = 4.895$, $P = 0.009$) with conventional farm plant communities representing a subset of those in organic farms. There was a significant interaction between farm type and edge/centre (Pseudo-$F = 1.320$, $P = 0.042$) (Fig. 3) with the centres of conventional fields mainly characterised by graminoids (89% contribution to community (in terms of cover and frequency), particularly *Lolium perenne* and *Agrostis stolonifera*) while organic centres were characterised by a large percentage of insect-pollinated forbs (42% contribution to community, *Trifolium repens* and *Ranunculus repens*) and much fewer non-insect-pollinated forbs (2% contribution to community, *Taraxacum* spp.) as well as graminoids (46% contribution to community, *Lolium perenne, Holcus lanatus* and *Agrostis spp.*). Community composition in organic and conventional field edges was similar but all edges were significantly different from all field centres (Pseudo-$F = 20.899$, $P = 0.001$) (Fig. 3) as they contained varying mixtures of insect-pollinated

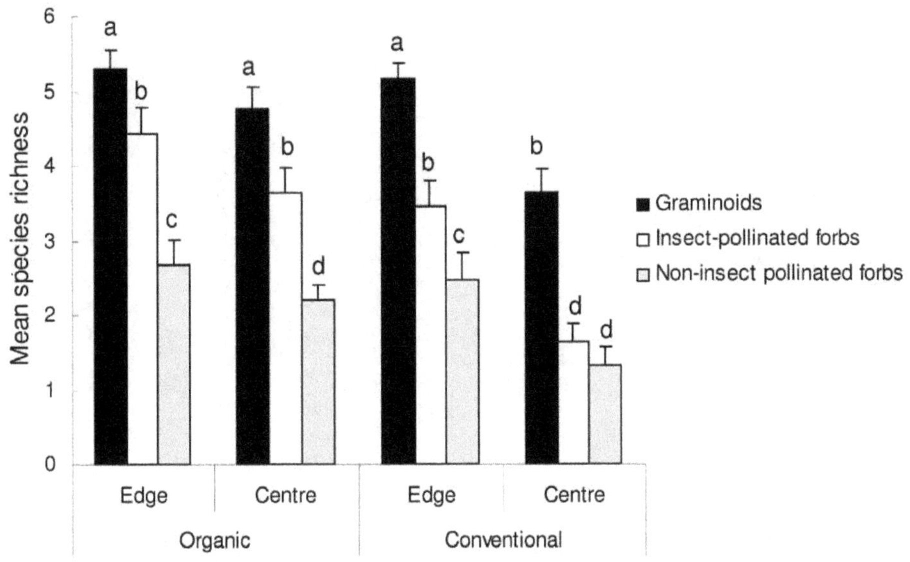

Figure 2. Mean (± standard error) graminoid/forb richness per transect in organic/conventional field edges and centres. Letters above the bars indicate significant differences among plant groups, farm type and edge/centre, using Linear Mixed Effects Models.

forbs (10–14% contribution to community), non-insect pollinated forbs (6–7% contribution to community) and graminoids (70–75% contribution to community).

Plant Richness: Local and Landscape Factors

Total plant richness, and that of insect-pollinated forbs and graminoids, were related to local factors and landscape factors, while non-insect pollinated forb richness was related to local factors only (Table 2). Total plant richness was significantly higher on organic farms than conventional and also higher in field edges than centres. Conventional field centres were particularly depauperate in terms of species richness (Fig. 2). Total plant richness was also significantly positively related to the PC1 scores of the principal components analysis (K soil parameter) but unrelated to PC2 or PC4 (see Table S3 for more detailed soil analysis results). The richness of insect-pollinated forbs and graminoids followed the same patterns as total plant richness in terms of local site characteristics but non-insect pollinated forbs did not. Non-insect pollinated forb richness was similar in both farming systems but was higher in all field edges compared to all field centres (Fig. 2). Non-insect pollinated forbs were also significantly positively related to PC1 scores only.

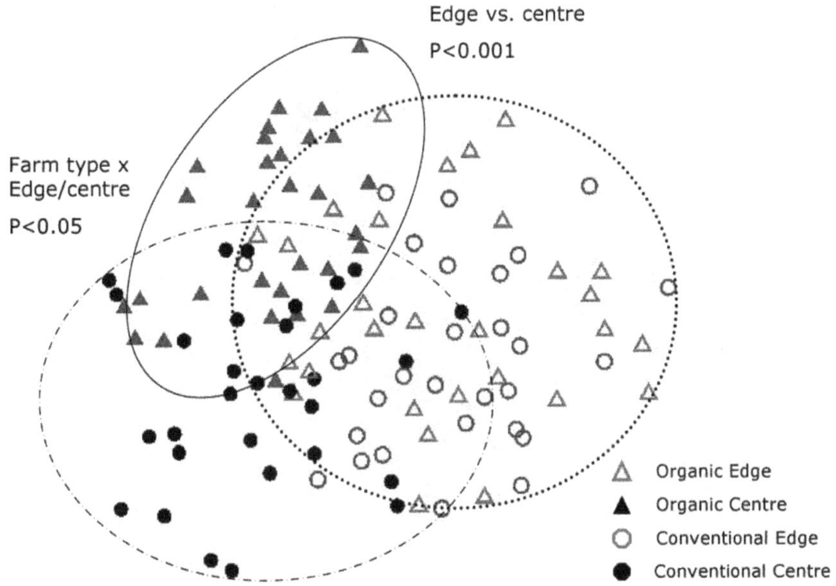

Figure 3. Plant community similarity between organic and conventional edge and centre transects. Illustrated using Non-metric Multidimensional Scaling (MNS with Bray - Curtis index, 2 Axes, 2D Stress =0.21), with significance values obtained using PERMANOVA+. Three distinct plant community groups emerged: organic and conventional edges (dotted line); organic centres (continuous line) and conventional centres (dashed and dotted line).

Table 2. Interacting relationships among local and landscape factors and plant richness (total plants, insect-pollinated forbs, non-insect pollinated forbs and graminoids) at five different spatial scales (1–5 km radii around sites).

		Local												Landscape		Local x landscape					
	Scale	FT		E/C		FT x E/C		PC1		PC2		PC4		%UG		FT x %UG		E/C x %UG		FT x E/C x %UG	
Richness	km	$t_{d.f.}$	P	$t_{d.f.}$	P	$t_{d.f.}$	P	$t_{d.f.}$	P	$t_{d.f.}$	P	$t_{d.f.}$	P	$t_{d.f.}$	P	$t_{d.f.}$	P	$t_{d.f.}$	P	$t_{d.f.}$	P
Total plants	1	3.092_{8}	0.015	5.612_{56}	0.000	-3.007_{56}	0.004	4.032_{56}	0.000	–	N.S.	–	N.S.	–	N.S.	–	N.S.	-2.537_{56}	0.014	–	N.S.
	2	2.964_{9}	0.016	7.290_{57}	0.000	-2.485_{57}	0.016	4.192_{57}	0.000	–	N.S.	–	N.S.	–	N.S.	–	N.S.	–	N.S.	–	N.S.
	3	3.005_{8}	0.017	4.903_{56}	0.000	-2.795_{56}	0.007	2.267_{56}	0.000	–	N.S.	–	N.S.	–	N.S.	–	N.S.	-2.227_{56}	0.030	–	N.S.
	4	2.960_{8}	0.018	5.251_{56}	0.000	-2.634_{56}	0.011	4.228_{56}	0.000	–	N.S.	–	N.S.	–	N.S.	–	N.S.	-2.674_{56}	0.010	–	N.S.
	5	2.918_{8}	0.019	5.255_{56}	0.000	-2.528_{56}	0.014	4.207_{56}	0.000	–	N.S.	–	N.S.	–	N.S.	–	N.S.	-2.805_{56}	0.007	–	N.S.
Insect-pollinated forbs	1	4.314_{8}	0.003	4.895_{56}	0.000	-2.647_{56}	0.010	2.455_{56}	0.017	–	N.S.	–	N.S.	–	N.S.	–	N.S.	-2.632_{56}	0.011	–	N.S.
	2	4.075_{9}	0.003	5.557_{57}	0.000	-2.102_{57}	0.040	2.728_{57}	0.008	–	N.S.	–	N.S.	–	N.S.	–	N.S.	–	N.S.	–	N.S.
	3	4.284_{8}	0.003	4.044_{56}	0.000	-2.379_{56}	0.021	2.697_{56}	0.009	–	N.S.	–	N.S.	–	N.S.	–	N.S.	-2.050_{56}	0.045	–	N.S.
	4	4.186_{8}	0.003	4.273_{56}	0.000	-2.214_{56}	0.031	2.661_{56}	0.010	–	N.S.	–	N.S.	–	N.S.	–	N.S.	-2.356_{56}	0.022	–	N.S.
	5	4.137_{8}	0.003	4.179_{56}	0.000	-2.115_{56}	0.039	2.630_{56}	0.011	–	N.S.	–	N.S.	–	N.S.	–	N.S.	-2.115_{56}	0.039	–	N.S.
Non-insect pollinated forbs	1	–	N.S.	3.996_{58}	0.000	–	N.S.	2.741_{58}	0.008	–	N.S.	–	N.S.	–	N.S.	–	N.S.	–	N.S.	–	N.S.
	2	–	N.S.	3.996_{58}	0.000	–	N.S.	2.741_{58}	0.008	–	N.S.	–	N.S.	–	N.S.	–	N.S.	–	N.S.	–	N.S.
	3	–	N.S.	3.996_{58}	0.000	–	N.S.	2.741_{58}	0.008	–	N.S.	–	N.S.	–	N.S.	–	N.S.	–	N.S.	–	N.S.
	4	–	N.S.	3.996_{58}	0.000	–	N.S.	2.741_{58}	0.008	–	N.S.	–	N.S.	–	N.S.	–	N.S.	–	N.S.	–	N.S.
	5	–	N.S.	3.996_{58}	0.000	–	N.S.	2.741_{58}	0.008	–	N.S.	–	N.S.	–	N.S.	–	N.S.	–	N.S.	–	N.S.
Graminoids	1	2.944_{8}	0.017	5.833_{53}	0.000	-3.560_{53}	0.001	3.874_{53}	0.000	–	N.S.	–	N.S.	–	N.S.	–	N.S.	-3.089_{53}	0.003	–	N.S.
	2	2.891_{8}	0.020	4.892_{53}	0.000	-3.473_{53}	0.001	4.007_{53}	0.000	–	N.S.	–	N.S.	–	N.S.	–	N.S.	-2.265_{53}	0.028	–	N.S.
	3	2.732_{9}	0.023	6.420_{54}	0.000	-3.067_{54}	0.003	3.801_{54}	0.000	–	N.S.	–	N.S.	–	N.S.	–	N.S.	–	N.S.	–	N.S.
	4	2.732_{9}	0.023	6.420_{54}	0.000	-3.067_{54}	0.003	3.801_{54}	0.000	–	N.S.	–	N.S.	–	N.S.	–	N.S.	–	N.S.	–	N.S.
	5	2.732_{9}	0.023	6.420_{54}	0.000	-3.067_{54}	0.003	3.801_{54}	0.000	–	N.S.	–	N.S.	–	N.S.	–	N.S.	–	N.S.	–	N.S.

Analysed using Linear Mixed Effects Models. Note: FT = Farm type (organic or conventional); E/C = Edge/Centre (position in the field); %UG = Percentage unimproved grassland (which reflects an inverse measure of landscape complexity); PC1 and PC2 = first and second principal components from a principal component analysis on soil parameters; N.S. = not significant.

Table 3. Relationships between plant richness (total plants, insect-pollinated forbs, non-insect pollinated forbs and graminoids) and hedgerow area, hedgerow length and number of hedgerow connections (and various interactions between factors) within 1 km radii around sites.

	HA		HL		Con		FT x E/C x HA		FT x E/C x HL		E/C x HA		HA x Con		HL x Con	
Richness	$t_{d.f.}$	P	$t_{d.f.}$	P	$t_{d.f.}$	P	$t_{d.f.}$	P	$t_{d.f.}$	P	$t_{d.f.}$	P	$t_{d.f.}$	P	$t_{d.f.}$	P
Total plants	–	N.S.	–	N.S.	–	N.S.	–	N.S.	2.242_{56}	0.029	–	N.S.	–	N.S.	–	N.S.
Insect-pollinated forbs	–	N.S.	–	N.S.	–	N.S.	2.603_{56}	0.012	–	N.S.	–	N.S.	–	N.S.	–	N.S.
Non-insect pollinated forbs	3.57_{5}	0.016	–	N.S.	–	N.S.	–	N.S.	–	N.S.	–	N.S.	-3.041_{5}	0.029	2.648_{5}	0.046
Graminoids	–	N.S.	–	N.S.	–	N.S.	–	N.S.	–	N.S.	2.039_{57}	0.046	–	N.S.	–	N.S.

Explanatory variables that were not significantly related with any response variable are not included in this table. Analysed using Linear Mixed Effects Models. Note: HA = Hedge area (%); HL = Hedge length (km); Con = number of connections between hedges; FT = Farm type (organic or conventional); E/C = Edge/centre (position in the field); N.S. = not significant.

The plant categories varied in terms of their relationships with landscape structure at different spatial scales (Table 2). Total plant and insect-pollinated forb richness in the field edges was significantly negatively related to the percent cover of unimproved grassland in the landscape between the scales of 1 and 5 km (though with no relationship evident at the 2 km scale). In contrast, graminoid richness in the field edges was negatively related to the percent cover of unimproved grassland in the 1 km and 2 km scales only, while non-insect pollinated forbs were unrelated to landscape structure at any spatial scale. The richness of all plants, insect-pollinated forbs, non-insect pollinated forbs and graminoids in all field centres were not related to the percent cover of unimproved grassland at any spatial scale. No interaction effect was found between farm type and the percent cover of unimproved grassland in any spatial scale on plant richness in general.

Linear features. Hedgerows were the dominant linear landscape feature around sites and covered a substantial percentage of the landscape within 1 km radii (Table S4). There were on average 5–6 km of hedgerows and approximately 30 hedgerow connections within these 1 km radii (with more hedgerow connections surrounding conventional farms than organic). Plant richness was significantly related to the various hedgerow measures. Total plant richness and insect-pollinated forb richness in organic field edges increased with increasing hedgerow length and area respectively, i.e. total plant richness was positively related to the interaction between farm type, edge/centre and hedgerow length while insect-pollinated forb richness was positively related to the interaction between farm type, edge/centre and hedgerow area (Table 3). Non-insect pollinated forb richness was: positively related to hedgerow area; negatively related to the interaction between hedgerow area and number of hedgerow connections; and positively related to the interaction between hedgerow length and number of connections. Graminoid richness in the edges of organic and conventional fields was positively related to hedgerow area.

Distance metrics. Insect-pollinated forb richness significantly increased with increasing proximity to wetlands ($t_{d.f.} = -2.405_{8}$, $P = 0.043$). There was no relationship between insect-pollinated plant richness and proximity to woodland, urban areas or open water and no interaction effects between farm type, edge/centre and distance metrics on this plant group. Total plant, non-insect-pollinated forbs and graminoid richness were not related to distance metrics.

Discussion

Local Factor Influences on Plant Communities

Total plant richness was higher in organic dairy farms compared to conventional farms because there was a higher plant richness in organic field centres than conventional centres; no difference was found between organic and conventional field edges. Similar patterns were evident in mixed arable and grassland systems in the UK [23] but not in Danish dairy systems [22].

It is not clear, which organic dairy farm management activities (lower fertilisation levels, lower stocking rates or lack of chemical herbicide applications) benefit plant diversity most or in which combination. In our study, macronutrients did not seem to be the most significant driver of plant diversity differences between organic and conventional systems. Though the macronutrient potassium was significantly related to plant richness, there was no interaction effect between any soil factor and farm type on plant richness and there were no significant differences among potassium levels in organic and conventional field edges and centres (Table S3). We did not include all macro-nutrients (e.g. nitrogen) in our soil analyses and this may lead to an underestimation of the influence of soil parameters on plant species richness in our study, but nitrogen enrichment of the soil (compared to other macronutrients such as phosphorus) has not been found to be the most important driver of species loss in semi-natural grasslands [53]. Though nutrient enrichment of the soil [53], can be responsible for species loss from semi-natural grasslands, in species-poor habitats, such as the commercially productive organic and conventional grasslands in our study, the effects of nutrient enrichment may become less important. This may be because most of the plant species that remain are likely to be relatively tolerant of high nutrient conditions. Previous studies in organic arable systems suggest that the exclusion of chemical herbicides in organic farming is important in explaining increased plant species richness in field centres [see: 24]. It seems likely that this is also the case in organic dairy field centres – chemical herbicides were applied annually in all the conventional farms in our study. However intensive grazing can also be responsible for species loss from semi-natural grasslands [54]. The conventional farms in our study had, on average, higher stocking densities. Therefore, more experimental research into the effects of organic and conventional farm management activities on species diversity is needed to elucidate the drivers of plant community composition in intensive dairy systems.

Unlike the finding for arable systems [5], there were more insect-pollinated forbs than non-insect pollinated forbs in both organic and conventional grasslands. Thus, intensive grassland plant communities appear to differ fundamentally from those of arable systems in terms of their relationships with pollinators and their plant-pollinator interaction networks, mutual dependence between plant and pollinator communities being more important in grassland than arable systems. If this is indeed the case, then changes in pollinator communities within grassland systems may have greater ramifications for plant-pollinator network stability than similar changes in arable systems.

Within the grassland systems in our study, plant community composition varied with farming system and position in the field. Considerably more forbs were present in organic field centres than conventional field centres. The forbs present in organic field centres were mainly insect-pollinated e.g. *Trifolium repens* and *Ranunculus repens*. While *T. repens* is a common component in organic seed mixtures [55], other insect-pollinated forbs are not and their increased presence in organic field centres could be beneficial to pollinators, attracting more insects and thus improving pollination services [12] (of course, certain forbs are potentially harmful to livestock, as is the case with *R. repens* and other buttercups [56,57]).

Local Versus Landscape Factor Influences on Plant Richness

Our results indicate that farm management activities influenced insect-pollinated forb richness in field centres, while increasing landscape complexity (calibrated by decreasing percentage of unimproved grassland within 1, 3, 4 and 5 km radii around sites) positively influenced insect-pollinated forb richness in field edges. Field edges tend to be less affected than field centres by farm management activities and are thus closer to a semi-natural condition [49]. Wild bees are important pollinators of wild plants [1] and they too can be influenced by landscape structure at scales of up to 3 km [41]. Thus, insect-pollinated forb richness at the field scale may be influenced by landscape complexity at large scales, mirroring the situation with some of the pollinators on which they depend. Parallel patterns have been recorded with insect-pollinated plant versus pollinator diversity in Great Britain and the Netherlands [3].

It is not clear why landscape complexity within the 2 km scale had no effect on insect-pollinated forb richness in field edges but may reflect the fact that relationships between local diversity, local factors and landscape structure change depending on the scale one is focusing on [58]. This illustrates the importance of choosing the correct scale in order to manage plants and pollinators in the landscape and, indeed, our study indicates that some scales may be more important for some plant groups than for others.

In contrast to the findings for insect-pollinated forbs, non-insect pollinated forb richness was unrelated to landscape complexity at any spatial scale – it was only related to position in the field and soil parameters. This may be because the majority of non-insect pollinated forbs in our study were self-pollinated. Graminoids in field edges were positively related to landscape complexity at lower scales than insect-pollinated forbs (as well as to farm type, position in the field and soil parameters). The reasons for this relationship between field edge graminoids and the landscape are unclear, but may be related to the fact that the graminoids in our study are wind pollinated [35] and so pollen transfer within the landscape could be dependent on climatic variables and landscape structure at smaller scales than for the factors important for insect-pollinated forbs. Our findings show the importance of using information on the pollination system of plants in order to understand their relationships with landscape structure. This is further illustrated by our findings for 'total plants' vs. landscape complexity which mirror the findings for insect-pollinated plants (the most species-rich group over all sites) but contrast greatly with findings for non-insect pollinated plants. Most landscape studies focus on total plant richness but our findings show that this may obscure the highly variable impacts of landscape structure on plant groups with different pollination systems.

Plant richness was also strongly related to linear features (hedgerows) within 1 km around sites. There was variation between plant groups in terms of which hedgerow measure was important, but hedgerow area, length or the interaction between these and the number of hedgerow connections were all significantly positively correlated with plant richness. Hedgerow structure is known to play an important role in determining hedgerow plant species distributions [48] and our results indicate that different plant groups (based on mode of pollination and ecology, i.e. graminoid vs. forb) respond differently to different aspects of hedgerow structure. There are many possible reasons for this. Hedgerows are known to act as corridors for movement of many plant species [49] and can act as barriers to or facilitators of pollen flow. Pollinators interact with linear landscape features [26,27,59] and hedgerows have been found to facilitate pollen dispersal between insect-pollinated forb populations through pollinator movements [60]. Hedgerows can also interrupt or slow down air fluxes [61] thus possibly affecting pollen dispersal in wind pollinated species. Maintenance or restoration of hedgerow networks in the landscape favours biodiversity but little is known of how hedgerow structure, pollen dispersal and plant community composition interact in the landscape.

Habitat fragmentation is known to affect different plant species in different ways [31] and so we tested whether the richness of insect-pollinated plants, non-insect pollinated plants and graminoids were related to proximity to different fragmented habitats, such as woodlands and wetlands. We found that insect-pollinated forb richness increased with proximity to wetlands only but found no relation between distance metrics and non-insect pollinated forbs or graminoids. The proximity of wetlands may boost plant richness as they will provide habitat for a range of species and may form a source from which colonisation can take place. The reasons for specific benefits to insect-pollinated plant richness are unclear, but may derive from a benefit to pollinator species from the proximity of wetlands: many pollinators (including hoverfly and solitary bee species) utilise wetlands as larval and forage habitats [16,30].

Conclusion

Plant community composition within intensive dairy grasslands, though species poor, differs from arable plant communities in having much higher proportions of insect-pollinated forbs in the community. This is likely to have implications for insect pollinator communities and plant-pollinator interaction networks. Organic farming was found to support greater plant richness in field centres than did conventional farming. The plant communities within field edges (organic or conventional) exhibit relationships with landscape structure that varied depending on their principal pollen vector. Our results indicate that insect-pollinated forbs in field edges differ from non-insect pollinated plant groups in terms of the extent to which landscape complexity at large spatial scales determines local richness. Such a relationship has not been demonstrated previously but comparisons can be drawn from the literature between this relationship and that found for pollinators, particularly bees, and landscape structure [41]. Our study also indicates that linear landscape features (hedgerows) and proximity to certain habitats influence the species richness of plant groups in grasslands in different ways, depending on their pollination system. The relationships between insect-pollinated forbs, their pollinators and landscape structure at different spatial scales need to be explored further, as all are inextricably linked.

Supporting Information

Figure S1 Percentage cover of unimproved grassland around each study site (farm pairs 1–10) for each of five spatial scales (1–5 km radii around sites).

Table S1 Mean ± standard error soil parameters measured in organic and conventional field edges and centres.

Table S2 List of all insect-pollinated forbs, non-insect pollinated forbs and graminoids and associated mean percentage cover per (0.5×0.5 m) quadrat (n = 600) recorded in the edges and centres of organic and conventional dairy fields.

Table S3 Relationships between soil parameters and farm type (organic/conventional), position in the field (edge/centre) and the interaction between farm type and edge/centre.

Table S4 Mean ± standard error hedgerow area, hedgerow length and number of connections between hedgerows within 1 km radii of organic and conventional farms.

Acknowledgments

We thank all the landowners who allowed us to sample in their fields; David Rockett for help with fieldwork; the Teagasc Analytical Laboratory, Johnstown Castle, Wexford, Ireland for soil analysis; and three anonymous journal editors and two anonymous reviewers for their comments on the manuscript.

Author Contributions

Conceived and designed the experiments: EFP JCS. Performed the experiments: EFP. Analyzed the data: EFP. Contributed reagents/materials/analysis tools: EFP JCS DLK. Wrote the paper: EFP. Secured funding: JCS DLK. Edited manuscript: JCS DLK EFP.

References

1. Buchmann SL, Nabhan GP (1996) The Forgotten Pollinators. Washington, DC: Island Press.
2. Steffan-Dewenter I, Potts SG, Packer L (2005) Pollinator diversity and crop pollination services are at risk. Trends in Ecology & Evolution 20: 651–652.
3. Biesmeijer JC, Roberts SPM, Reemer M, Ohlemuller R, Edwards M, et al. (2006) Parallel declines in pollinators and insect-pollinated plants in Britain and the Netherlands. Science 313: 351–354.
4. Baker HG (1974) The Evolution of Weeds. Annual Review of Ecology and Systematics 5: 1–24.
5. Gabriel D, Tscharntke T (2007) Insect pollinated plants benefit from organic farming. Agriculture, Ecosystems & Environment 118: 43–48.
6. Regal PJ (1982) Pollination by Wind and Animals: Ecology of Geographic Patterns. Annual Review of Ecology and Systematics 13: 497–524.
7. Plantureux S, Peeters A, McCracken D (2005) Biodiversity in intensive grasslands: Effect of management, improvement and challenges. Agronomy Research 3: 152–164.
8. Vickery JA, Tallowin JR, Feber RE, Asteraki EJ, Atkinson PW, et al. (2001) The management of lowland neutral grasslands in Britain: effects of agricultural practices on birds and their food resources. Journal of Applied Ecology 38: 647–664.
9. Department of Agriculture (2009) Fact sheet on Irish Agriculture.
10. McMahon BJ, Helden A, Anderson A, Sheridan H, Kinsella A, et al. (2010) Interactions between livestock systems and biodiversity in South-East Ireland. Agriculture, Ecosystems & Environment 139: 232–238.
11. Santorum V, Breen J (2005) Bumblebee diversity on Irish farmland. Irish journal of agri-environmental research 4: 79–90.
12. Power EF, Stout JC (2011) Organic dairy farming: impacts on insect–flower interaction networks and pollination. Journal of Applied Ecology 48: 561–569.
13. Martin JR, Gabbett M, Perrin PM, Delaney A (2007) Semi-natural grassland survey of counties Roscommon and Offaly. Dublin, Ireland: National Parks and Wildlife Service.
14. Ivimey-Cook RB, Proctor MCF (1966) Plant Communities of the Burren, Co. Clare. Proceedings of The Royal Irish Academy Dublin 64B: 211–301.
15. Fitzpatrick Ú, Murray TE, Paxton RJ, Breen J, Cotton D, et al. (2007) Rarity and decline in bumblebees - A test of causes and correlates in the Irish fauna. Biological Conservation 136: 185–194.
16. Speight MCD (2008) Database of Irish Syrphidae (Diptera). Irish Wildlife Manuals, No 36: National Parks and Wildlife Service, Department of Environment, Heritage and Local Government, Dublin, Ireland.
17. Kleijn D, Verbeek M (2000) Factors affecting the species composition of arable field boundary vegetation. Journal of Applied Ecology 37: 256–266.
18. Schippers P, Joenje W (2002) Modelling the effect of fertiliser, mowing, disturbance and width on the biodiversity of plant communities of field boundaries. Agriculture, Ecosystems & Environment 93: 351–365.
19. Bengtsson J, Ahnstrom J, Weibull AC (2005) The effects of organic agriculture on biodiversity and abundance: a meta-analysis. Journal of Applied Ecology 42: 261–269.
20. Morandin LA, Winston ML (2005) Wild bee abundance and seed production in conventional, organic, and genetically modified canola. Ecological Applications 15: 871–881.
21. Romero A, Chamorro L, Sans FX (2008) Weed diversity in crop edges and inner fields of organic and conventional dryland winter cereal crops in NE Spain. Agriculture, Ecosystems & Environment 124: 97–104.
22. Petersen S, Axelsen JA, Tybirk K, Aude E, Vestergaard P (2006) Effects of organic farming on field boundary vegetation in Denmark. Agriculture, Ecosystems & Environment 113: 302–306.
23. Gabriel D, Sait SM, Hodgson JA, Schmutz U, Kunin WE, et al. (2010) Scale matters: the impact of organic farming on biodiversity at different spatial scales. Ecology Letters 13: 858–869.
24. Rundlöf M, Edlund M, Smith HG (2010) Organic farming at local and landscape scales benefits plant diversity. Ecography 33: 514–522.

25. Roschewitz I, Gabriel D, Tscharntke T, Thies C (2005) The effects of landscape complexity on arable weed species diversity in organic and conventional farming. Journal of Applied Ecology 42: 873–882.
26. Wratten SD, Bowie MH, Hickman JM, Alison ME, Sedcole JR, et al. (2003) Field Boundaries as Barriers to Movement of Hover Flies (Diptera: Syrphidae) in Cultivated Land. Oecologia 134: 605–611.
27. Cranmer L, McCollin D, Ollerton J (2012) Landscape structure influences pollinator movements and directly affects plant reproductive success. Oikos 121: 562–568.
28. Jacobs JH, Clark SJ, Denholm I, Goulson D, Stoate C, et al. (2009) Pollination biology of fruit-bearing hedgerow plants and the role of flower-visiting insects in fruit-set. Annals of Botany 104: 1397–1404.
29. Ouin A, Burel F (2002) Influence of herbaceous elements on butterfly diversity in hedgerow agricultural landscapes. Agriculture Ecosystems & Environment, 93: 45–53.
30. Michener CD (2007) The Bees of the World. Baltimore, ed. Md. : Johns Hopkins University Press. 953 p.
31. Dauber J, Biesmeijer JC, Gabriel D, Kunin WE, Lamborn E, et al. (2010) Effects of patch size and density on flower visitation and seed set of wild plants: a pan-European approach. Journal of Ecology 98: 188–196.
32. Stace CA (2010) New flora of the British Isles. Cambridge: Cambridge University Press. 1232 p.
33. Grime JP, Hodgson JG, Hunt R (2007) Comparative Plant Ecology: A Functional Approach to Common British Species: Castlepoint Press, London.
34. Klotz S, Kühn I, Durka W (2002) BIOLFLOR - Eine Datenbank zu biologisch-ökologischen Merkmalen der Gefäßpflanzen in Deutschland. Schriftenreihe für Vegetationskunde 38. Bonn: Bundesamt für Naturschutz.
35. Cope T, Gray A, eds (2009) Grasses of the British Isles BSBI Handbook no 13. Botanical Society of the British Isles.
36. Alexander S, Black A, Boland A, Burke J, Carton OT, et al. (2008) Major and micro nutient advice for productive agricultural crops. In: Coulter BS, Lalor S, eds. 50th anniversary edition ed: Teagasc, Johnstown Castle, Co Wexford. 116 p.
37. Clarke KR, Gorley RN (2006) PRIMER v6: User Manual/Tutorial. Plymouth: PRIMER-E.
38. Bossard M, Feranec J, Otahel J (2000) CORINE land cover technical guide - Addendum 2000. Copenhagen: European Environment Agency.
39. European Environment Agency (2007) CLC2006 Technical guidelines. Copenhagen.
40. Sullivan CA, Skeffington MS, Gormally MJ, Finn JA (2010) The ecological status of grasslands on lowland farmlands in western Ireland and implications for grassland classification and nature value assessment. Biological Conservation 143: 1529–1539.
41. Steffan-Dewenter I, Munzenberg U, Burger C, Thies C, Tscharntke T (2002) Scale-dependent effects of landscape context on three pollinator guilds. Ecology 83: 1421–1432.
42. Westphal C, Steffan-Dewenter I, Tscharntke T (2006) Bumblebees experience landscapes at different spatial scales: possible implications for coexistence. Oecologia 149: 289–300.
43. Haenke S, Scheid B, Schaefer M, Tscharntke T, Thies C (2009) Increasing syrphid fly diversity and density in sown flower strips within simple vs. complex landscapes. Journal of Applied Ecology 46: 1106–1114.
44. Conn DLT (1976) Estimates of Population Size and Longevity of Adult Narcissus Bulb Fly Merodon equestris Fab. (Diptera: syrphidae). Journal of Applied Ecology 13: 429–434.
45. Dziock F (2006) Life-History Data in Bioindication Procedures, Using the Example of Hoverflies (Diptera, Syrphidae) in the Elbe Floodplain. International Review of Hydrobiology 91: 341–363.
46. Krebs JR, Wilson JD, Bradbury RB, Siriwardena GM (1999) The second silent spring? Nature 400: 611–612.
47. Zuur A, Ieno EN, Smith GM (2007) Analysing Ecological Data. New York: Springer.
48. Deckers B, Hermy M, Muys B (2004) Factors affecting plant species composition of hedgerows: relative importance and hierarchy. Acta Oecologica 26: 23–37.
49. Le Coeur D, Baudry J, Burel F, Thenail C (2002) Why and how we should study field boundary biodiversity in an agrarian landscape context. Agriculture, Ecosystems & Environment 89: 23–40.
50. Zuur AF, Ieno EN, Walker NJ, Saveliev AA, Smith GM (2009) Mixed Effects Models and Extensions in Ecology with R New York: Springer.
51. Pinheiro J, Bates D, DebRoy S, Sarkar D, the R Core team (2009) nlme: Linear and Nonlinear Mixed Effects Models. R package version 3.1-96. Available: http://CRAN.R-project.org/package=nlme. Accessed 2012 Feb 6.
52. R Development Core Team (2007) R: A language and environment for statistical computing. R Foundation for Statistical Computing Vienna, Austria. ISBN 3-900051-07-0. Available: http://www.R-project.org. Accessed 2012 Feb 6.
53. Ceulemans T, Merckx R, Hens M, Honnay O (2011) A trait-based analysis of the role of phosphorus vs. nitrogen enrichment in plant species loss across North-west European grasslands. Journal of Applied Ecology 48: 1155–1163.
54. Olff H, Ritchie ME (1998) Effects of herbivores on grassland plant diversity. Trends in Ecology & Evolution 13: 261–265.
55. Irish Organic Farmers and Growers Association (2006) The IOFGA Standards For Organic Food and Farming in Ireland - incorporating amendments A1 to A49.
56. James LF, Keeler RF, Bailey EMJ, Cheeke PR, Hegarty MP. Poisonous Plants; (1992) Iowa State University Press, Ames. 678 p.
57. Parton K, Bruere AN (2002) Plant poisoning of livestock in New Zealand. New Zealand Veterinary Journal 50: 22–27.
58. Dauber J, Purtauf T, Allspach A, Frisch J, Voigtländer K, et al. (2005) Local vs. landscape controls on diversity: a test using surface-dwelling soil macroinvertebrates of differing mobility. Global Ecology & Biogeography 14: 213–221.
59. Ekroos J, Piha M, Tiainen J (2008) Role of organic and conventional field boundaries on boreal bumblebees and butterflies. Agriculture, Ecosystems & Environment 124: 155–159.
60. Van Geert A, Van Rossum F, Triest L (2010) Do linear landscape elements in farmland act as biological corridors for pollen dispersal? Journal of Ecology 98: 178–187.
61. Burel F (1996) Hedgerows and their role in agricultural landscapes. Critical Reviews in Plant Sciences 15: 169–190.

21

Organic Production Enhances Milk Nutritional Quality by Shifting Fatty Acid Composition: A United States–Wide, 18-Month Study

Charles M. Benbrook[1]*, Gillian Butler[2], Maged A. Latif[3], Carlo Leifert[2], Donald R. Davis[1]

1 Center for Sustaining Agriculture and Natural Resources, Washington State University, Pullman, Washington, United States of America, **2** School of Agriculture, Food and Rural Development, Newcastle University, Northumberland NE, United Kingdom, **3** Organic Valley/CROPP Cooperative/Organic Prairie, Lafarge, Wisconsin, United States of America

Abstract

Over the last century, intakes of omega-6 (ω-6) fatty acids in Western diets have dramatically increased, while omega-3 (ω-3) intakes have fallen. Resulting ω-6/ω-3 intake ratios have risen to nutritionally undesirable levels, generally 10 to 15, compared to a possible optimal ratio near 2.3. We report results of the first large-scale, nationwide study of fatty acids in U.S. organic and conventional milk. Averaged over 12 months, organic milk contained 25% less ω-6 fatty acids and 62% more ω-3 fatty acids than conventional milk, yielding a 2.5-fold higher ω-6/ω-3 ratio in conventional compared to organic milk (5.77 vs. 2.28). All individual ω-3 fatty acid concentrations were higher in organic milk—α-linolenic acid (by 60%), eicosapentaenoic acid (32%), and docosapentaenoic acid (19%)—as was the concentration of conjugated linoleic acid (18%). We report mostly moderate regional and seasonal variability in milk fatty acid profiles. Hypothetical diets of adult women were modeled to assess milk fatty-acid-driven differences in overall dietary ω-6/ω-3 ratios. Diets varied according to three choices: high instead of moderate dairy consumption; organic vs. conventional dairy products; and reduced vs. typical consumption of ω-6 fatty acids. The three choices together would decrease the ω-6/ω-3 ratio among adult women by ~80% of the total decrease needed to reach a target ratio of 2.3, with relative impact "switch to low ω-6 foods" > "switch to organic dairy products" ≈ "increase consumption of conventional dairy products." Based on recommended servings of dairy products and seafoods, dairy products supply far more α-linolenic acid than seafoods, about one-third as much eicosapentaenoic acid, and slightly more docosapentaenoic acid, but negligible docosahexaenoic acid. We conclude that consumers have viable options to reduce average ω-6/ω-3 intake ratios, thereby reducing or eliminating probable risk factors for a wide range of developmental and chronic health problems.

Editor: Andrea S. Wiley, Indiana University, United States of America

Funding: Support for CMB and DRD came from the "Measure to Manage Program — Farm and Food Diagnostics for Sustainability and Health," Center for Sustaining Agriculture and Natural Resources at Washington State University. Support for MAL and milk sample testing came from CROPP Cooperative, La Farge, Wisconsin (http://www.farmers.coop/). Support for CL and GB came from Newcastle University, Northumberland NE, United Kingdom. The funders had no role in study design, data collection and analysis, decision to publish, or preparation of the manuscript.

Competing Interests: CROPP Cooperative is among the core funders of the "Measure to Manage Program" at Washington State University. MAL is the Director of Research & Development and Quality Assurance at CROPP Cooperative.

* E-mail: cbenbrook@wsu.edu

Introduction

Dairy products contribute significantly to dietary intakes of saturated fat in the United States and Europe, which has led to widely endorsed recommendations to limit consumption of whole milk and other high-fat dairy products, in favor of low- and non-fat dairy products [1]. However, these recommendations are based primarily on the serum-LDL ("bad")-cholesterol-raising effect of dairy fat, a single marker of risk for cardiovascular disease (CVD). They give little or no consideration to the CVD-risk reducing components in milk fat, especially omega-3 (ω-3) fatty acids (FAs), conjugated linoleic acid (CLA), the possibly beneficial *trans* FAs, *trans*-18:1 [2] and *trans*-16:1 [3], protective minerals, and a beneficial effect on serum HDL ("good") cholesterol [4].

Two recent reviews of epidemiological evidence question common beliefs about the health effects of dairy fat. One finds a contradiction between the evidence from long-term prospective studies and perceptions of harm from the consumption of dairy products [5]. The other review highlights inconsistent evidence of harm [4]. Most of the reviewed studies began before low-fat dairy products became widely used. These reviews conclude that high consumption of milk and milk fat may be overall neutral [4] or beneficial [5] regarding all-cause mortality, ischemic heart disease, stroke, and diabetes. Most recently, Ludwig and Willett have questioned the scientific basis for recommending reduced-fat dairy products [6]. Additional studies have linked dairy fat consumption to diminished weight gain [7], attenuated markers of metabolic syndrome, including waist circumference [8], and reduced risk of CVD [9] and colorectal cancer [10].

Milk products are good sources of many nutrients, including several of concern in at least some U.S. population cohorts—calcium, potassium, vitamin D (in fortified milk products), vitamin B_{12}, and protein [1,11]. Alpha-linolenic acid (ALA) and other ω-3 FA are also of concern, and are well recognized in milk products

[12], but some major reviews do not mention dairy sources [1,11]. There is increasing evidence that the dietary balance of ω-3 and ω-6 FA is perhaps as important as the dietary proportions of saturated, monounsaturated, and total fat [13,14].

Health concerns stemming from increasing dietary ω-6/ω-3 ratios have stimulated research on ways to improve the FA profile of common foods, including milk and dairy products [15–23]. Changes in dietary FA intakes during the last century have been brought about largely by: (a) increased consumption of major vegetable oils [13,24], and (b) generally low consumption of oily fish, vegetables, fruits, and beans [25]. Average dietary ratios of these two classes of polyunsaturated FA in the U.S. have increased from about 5 to about 10, with some ratios in excess of 20 [13,24–27]. The ω-6/ω-3 ratio in human breast milk has also increased dramatically in this time period, driven by changes in maternal dietary ratios [28].

Although the optimal ω-6/ω-3 ratio depends on the health measure in question and genetic factors [13], some authors have suggested a target ratio of 2.3. At this ratio, the conversion of ALA to long-chain ω-3 docosahexaenoic acid (DHA) is thought to be maximized [26,27]. In addition, epidemiological studies have reported no further CVD-prevention benefits from lowering the ω-6/ω-3 ratio below ~2.3 [24,25,29–31].

Milk from cows consuming significant amounts of grass and legume-based forages contains higher concentrations of ω-3 FAs and CLA than milk from cows lacking routine access to pasture and fed substantial quantities of grains, especially corn [15–23,32]. In turn, lactating women consuming such milk have an increased CLA concentration in their breast milk [33]. The balance of FAs in animal-derived foods like milk depends on the animal's dietary lipid intake and on its digestive physiology. The relationship between diet composition and lipid transfer into milk, meat and eggs has been reviewed by Woods and Fearon [34].

The rumen in dairy cattle influences the suitability of different feeds and also has a major impact on the nature of FAs absorbed and ultimately secreted into milk. Pigs and poultry, like humans, have a relatively simple digestive system and absorb FAs in approximately the same proportions as found in their diet. Lipid absorption by cattle and sheep is heavily influenced by rumen microbial activity that hydrogenates (saturates) up to 95% of dietary polyunsaturated fatty acids (PUFAs), making it challenging to increase the PUFA content of ruminant milk or meat. However, increased reliance on fresh herbage in dairy cow diets does elevate the ω-3 content of milk produced [15–23,32].

The U.S. National Organic Program (NOP) requires that lactating cows on certified organic farms receive at least 30% of daily Dry Matter Intake (DMI) from pasture during that portion of the year when pasture grasses and legumes are actively growing, with a minimum of 120 days per year [35]. Pasture and conserved, forage-based feeds account for most of the DMI year-round on a growing portion of organic dairy operations in the U.S. [36].

Although several European studies have compared the composition of organic and conventional milk [18–23,32], there is limited comparative data from the U.S. Also, published U.S. studies are based on relatively limited sampling and reflect milk production during only a portion of a year, hence providing no insight into seasonal changes in milk quality [16,17]. The two main objectives of the present study were first, to quantify average, annual FA differences between organic and conventional milk in an extensive, 18-month cross-U.S. survey, with attention to regional and seasonal variations, and second, to address the degree to which consumption of predominantly organic dairy products may enhance public health by decreasing dietary ω-6/ω-3 ratios from today's generally unhealthy levels [5,7–10,14,28–31,37–48].

Methods

We selected 14 commercial milk processors from 7 regions throughout the U.S. that produce organic milk products, and usually also conventional milk, pasteurized by either the high-temperature-short-time method (HTST) or by the ultra-high-temperature method (UHT, also known as ultra-pasteurization). The processors were located in the Northwest region (2 HTST, 1 UHT); California (1 HTST, 1 UHT—organic only); Rocky Mountains (1 HTST), Texas (1 HTST); Midwest (1 HTST, 1 UHT—organic only); Mid-Atlantic (1 HTST, 1 UHT—organic only); and Northeast (1 HTST, 2 UHT). These processors receive and market organic milk through the Organic Valley brand, the largest U.S. cooperative of organic farmers, based in La Farge, Wisconsin. Because three of the UHT processors produce only organic milk, we obtained more organic than conventional samples, and more UHT organic samples than UHT conventional samples. From each processor we obtained one fresh, whole-milk sample nominally every month for 18 months, January 2011 through June 2012. A total of 220 organic and 164 conventional samples were taken from either 1-gallon or half-gallon retail containers, transferred to sterile plastic bottles, refrigerated, and shipped with frozen ice packs by overnight courier to Silliker, Inc., an ISO/IEC 17025 accredited lab in Chicago Heights, Illinois. Analyses for FA and total fat used method AOAC 996.06, revised 2001 [49], with capillary column Supelco SP-2560, 100 m×0.25 mm, 0.2 μm film. The lab did not report non-quantifiable amounts defined as < 0.001 g/100 g.

Statistical Analysis

Digital laboratory data were transferred automatically to an Excel spreadsheet and spot-verified by one of us against printed lab reports. The data are available from authors MAL and DRD. Statistical analyses used NCSS 2004 software (Kaysville, Utah), plus Excel 2003 for some descriptive statistics and t distributions. Excel results were verified with NCSS [50]. Primary data were inspected with scatter plots and normal probability plots, and small numbers of clear outliers with large deviations from normality were removed as follows: First, 6 milk samples with total fat ≤ 2.52 and ≥4.17 g/100 g, leaving 218 organic and 160 conventional samples; then 7 extreme outlier values (g/100 g) for individual FA 12:1 (0.046), 18:2 linoleic acid (LA) (0.006), 20:1 (0.026, 0.026, 0.032), and 20:5 (eicosapentaenoic acid, EPA) (0.041, 0.084). The removed outliers deviated from the respective FA means by 5 to 70 SDs, and represent 0.07% of all values examined. Differences between means were evaluated by 2-tailed t test for approximately normal distributions or by 2-tailed Mann-Whitney test, as indicated. Mann-Whitney tests were used for non-normal distributions and those with insufficient numbers of samples to evaluate normality.

Ideally our data would have been amenable to ANOVA-based methods, but we chose alternative methods because of complex data that could not be consistently transformed to normality with equal variances, requiring both parametric and non-parametric methods. There are also some 2- and 3-fold differences in the numbers of organic and conventional samples, and in the number of regional samples. Because of multiple comparisons of organic vs. conventional FA concentrations in 2 tables, about 2 findings of $P=0.05$ and 0.5 finding of $P=0.01$ can be expected there by chance alone, but our key findings are highly significant ($P<0.0001$) and not vulnerable to Type-I errors (incorrectly rejecting the null hypothesis). To prevent or minimize potentially biased statistical comparisons caused by possible correlation between monthly FA concentrations, we limited statistical comparisons of monthly data between organic and conventional samples to individual months or groups of 6 months. To compare coefficients of variation we used the method by Forkman [51].

We calculated U.S. average fatty acid concentrations and ratios as means of samples from all 12 months in 2011 and all 7 geographical regions, calculated separately for organic and conventional samples. For regional concentrations and ratios, we averaged all 12 monthly samples separately for each region and for both production methods. Similarly, we calculated seasonal concentrations and ratios by averaging samples from all 7 geographical regions, separately for 18 months in 2011–2012 and for both production methods. The numbers of samples are shown in the resulting tables, figures, or figure caption.

Three ratios of FAs in conventional versus organic milk are reported. LA/ALA is the simplest ratio and represents the major ω-6 and ω-3 FAs in milk and human diets. The most commonly encountered ratio of ω-6 and ω-3 FAs is total ω-6/total ω-3, where the totals may include, besides LA and ALA, arachidonic acid (AA, ω-6), EPA (ω-3), DHA (ω-3) and possibly docosapentaenoic acid (DPA) (ω-3) and other minor FAs. Our inclusive ratio ω-6/ω-3 includes primarily LA (ω-6), 8,11,14-eicosatrienoic acid (20:3, ω-6), AA (ω-6), ALA (ω-3), EPA (ω-3), and DPA (ω-3), plus infrequently reported, small amounts of 18:3 γ-linolenic acid (ω-6), 20:2 eicosadienoic acid (ω-6), 22:2 docosadienoic acid (ω-6), 22:4 docosatetraenoic acid (ω-6), 18:4 moroctic acid (ω-3), 20:3 11,14,17-eicosatrienoic acid (ω-3), and DHA (ω-3). To more fully reflect the variations in levels of health-promoting dairy FAs, we include a third ratio in which we add to the ω-3 total the amount of CLA (18:2 conjugated). This FA is widely accepted as beneficial for heart health and prevention of cancer [52–54].

Diet Scenarios and LA/ALA Ratios

To estimate the impact on dietary ω-6/ω-3 ratios of replacing conventional with organic dairy products, we calculated LA/ALA ratios for hypothetical diets of moderately active women, age 19–30, selected in part because of the importance of ω-3 FA during women's child-bearing years. We limited this estimate to LA and ALA, because of the lack of reliable U.S. Department of Agriculture (USDA) composition data for EPA, DPA, and DHA in non-dairy foods. In addition, LA accounts for 90% of total ω-6 in both organic and conventional milk, and ALA accounts for 79% to 80% of total ω-3. Our estimated impacts on LA/ALA of a switch from conventional to organic dairy foods underestimates the full array of health benefits following such a switch, because CLA and long-chain, ω-3 FAs fall outside the LA/ALA ratio.

Table 1 shows model diets containing hypothetical servings of 4 common dairy foods in 3 diet scenarios—low-fat, average-fat, and high-fat (respectively 20%, 33%, and 45% of energy from fat). Except for the low-fat scenario, the "Moderate dairy intakes" follow the recommended 3 daily servings of milk and milk products in the Dietary Guidelines for Americans [1]. The moderate-dairy, low-fat scenario uses reduced amounts of 3 of the 4 dairy foods, thereby reducing its content of fat and other nutrients from dairy products, e.g., for calcium, from 1040 mg to 850 mg. (However, the reduced dairy intake adds nutrients from 128 kcal of additional non-dairy foods needed to keep total energy at 2,100 kcal.) To quantify the likely maximum benefit possible from consuming milk products with enhanced FA profiles, we use whole milk and full-fat cheese in the diet scenarios, instead of reduced-fat versions. For the recommended 3 cups fluid milk equivalent we use 1.5 cups milk + 1.5 ounces cheese (1 cup fluid milk equivalent) + 6 ounces low-fat yogurt with fruit (about 0.5 cup fluid milk equivalent). Ice cream is categorized separately as "dairy dessert" [1].

In Table 1 "Energy from fat" = 2100 kcal for a moderately active woman [1] × 20%, 33% or 45%; "Energy from dairy fat" = serving weight × fat content/100 g×8.79 kcal/g, summed over the 4 foods, using the footnoted values from USDA's National Nutrient Database for Standard Reference [55]; "Energy from other fat" = "Energy from fat" − "Energy from dairy fat."

We converted the dairy and non-dairy fat-energy values from Table 1 into fat weights using USDA's 8.79 kcal/g for dairy fat and 8.9 kcal/g for typical non-dairy fat, and then calculated the amounts of LA and ALA contained in these dairy and non-dairy sources. For dairy fat, we used the average amounts of LA, ALA, and total FAs found in this research, plus the conversion from milk FA weight to milk fat weight (0.933 g FA/g fat) [56]. For non-dairy fat, we used USDA data to calculate the amounts of LA and ALA per 100 kcal of fat in 8 foods—McDonald's French fries, plain tortilla chips, higher-fat chocolate chip cookies, soy oil (salad or cooking), regular stick margarine, ground chicken, composite pork cuts, and ground beef (15% fat).

We selected these 8 foods because they are commonly consumed and USDA reports amounts of their fully differentiated LA (*cis,cis*-18:2 n-6) and ALA (*cis,cis,cis*-18:3 n-3) [55]. On average they contain 23.23 g LA and 1.841 g ALA per 100 kcal fat, which we used to calculate the amount of LA and ALA in "other" (non-dairy) fat in Table 1.

To estimate the effect of reducing LA intakes, we made substitutions for 3 of the above 8 foods—pita chips instead of tortilla chips (LA/ALA = 20, instead of 40 for tortilla chips), canola oil instead of soy oil, and canola oil margarine instead of regular margarine, all with available fully differentiated data for LA and ALA [55]. These 3 substitutions yield average contents for non-dairy fat of 13.84 g LA (a 40% reduction) and 2.731 g ALA (a 48% increase) per 100 kcal of fat.

Fatty Acids in Fish Compared to Other Sources

The Dietary Guidelines for Americans [1] recommend consumption of about 8 ounces per week of a variety of fish (about twice the average U.S. consumption), with up to 12 ounces per week for women who are pregnant or breastfeeding. The fish varieties may include those with higher and lower amounts of EPA and DHA, but should include some with higher amounts, to achieve an average intake of 250 mg/day of EPA + DHA. Because fish also contain LA, ALA, and DPA, we used USDA data [55] to estimate the amounts of these 5 FAs in fish and compared them with amounts in the dairy products and non-dairy sources used in our dietary scenarios. We selected 7 representative fish with available data for fully differentiated ALA and (in most cases) LA.

Results

Table 2 shows FA concentrations in conventional and organic milk, averaged over a full calendar year of milk production (2011). The complete 18-month test period through June 2012 is used to assess seasonal variations. Table 2 includes sums of saturated, monounsaturated, and polyunsaturated FA, as well as ω-3, ω-6, and *trans* FA, plus CLA, CLA + ALA, and various ratios of FA levels, such as LA/ALA and ω-6/ω-3. FA names include the number of carbon atoms followed by a colon and the number of double bonds. Table 2 also shows descriptive statistics, including coefficients of variation (CV = SD/mean), a measure of nationwide percentage variation during 2011. Supplemental Table S1 reports the same values expressed as a percentage of total FAs. As expected, percentage variations (CVs) tend to be smaller in Table S1 than in Table 2, especially for the dominant fractions, total saturated FA (~67%) and monounsaturated FA (~25%).

Table 1. Hypothetical scenarios of dairy fat intake for a moderately active woman, age 19 to 30.

	Total Energy, kcal	Milk, cups	Cheese, oz.	Ice Cream, 0.5 cup	Yogurt, Low fat 6 oz.	Energy from Fat, kcal	Energy From Dairy Fat, kcal*	Energy from Other Fat, kcal
Low-Fat Diet (20% of Energy)								
Moderate dairy intake	2,100	1.25	1.00	0.75	1.00	420	239	181
High dairy intake	2,100	1.50	1.50	1.00	1.00	420	313	107
Average-Fat Diet (33% of Energy)								
Moderate dairy intake	2,100	1.50	1.50	1.00	1.00	693	313	380
High dairy intake	2,100	2.00	3.00	1.00	1.00	693	472	221
High-Fat Diet (45% of Energy)								
Moderate dairy intake	2,100	1.50	1.50	1.00	1.00	945	313	632
High dairy intake	2,100	2.00	3.00	2.00	1.00	945	536	409

*Based on the following serving sizes and USDA data:
Milk 1 cup (244 g), 3.25 g fat/100 g.
Cheddar cheese 1 oz. (28.35 g), 33.14 g fat/100 g.
Vanilla ice cream 0.5 cup (66 g), 11.0 g fat/100 g.
Low-fat yogurt with fruit 6 oz. (170.1 g), 1.41 g fat/100 g.
8.79 kcal/g dairy fat.

Supplemental Figures S1 and S2 illustrate key concentrations and ratios.

There were small or negligible differences between organic and conventional milk in total FAs (< 1%, not statistically significant), and in total saturated (4% higher in organic), monounsaturated (7% lower in organic), and *trans* FAs (2% lower in organic, not statistically significant). However, 12-month-average concentrations of total PUFAs, total ω-6, and LA were significantly higher in conventional compared to organic milk (by 11%, 33%, and 34% respectively, all $P<0.0001$), while concentrations of total ω-3, ALA and CLA were significantly higher in organic compared to conventional milk (by 62%, 60%, and 18% respectively, $P<0.0001$, <0.0001, and < 0.001).

Similarly, 12-month average concentrations of the more active, longer-chain ω-3 FAs EPA and DPA were significantly higher in organic milk compared to conventional milk (33% and 18% respectively, both $P<0.001$). Data for DHA are not shown, because over the 18 months of testing, quantifiable levels (≥ 0.001 g/100 g) were found in only 4 of 218 organic and 2 of 160 conventional samples. Our results are similar to other studies that report significant levels of EPA and DPA in dairy products, but little or no DHA [17,18,21]. Circulating EPA and DHA in cow blood are believed to have extremely low uptake in the udder. It is speculated that EPA, but very little DHA, is synthesized de novo in the udder, and that this synthesis is the primary source of EPA in cow milk [57].

As a result of the observed composition differences, the 2011 average ratios LA/ALA, ω-6/ω-3, and ω-6/(ω-3 + CLA) are much higher in conventional compared to organic milk—2.4-fold higher for LA/ALA, 2.5-fold higher for ω-6/ω-3, and 2.0-fold higher for ω-6/(ω-3 + CLA) (Table 2, all $P<0.0001$).

Regional Differences

National 12-month averages show a few clear differences between regions in both the concentrations of LA, ALA, and CLA (Figure 1) and in the FA ratios LA/ALA, ω-6/ω-3, and ω-6/(ω-3 + CLA) (Figure 2). Most notably, conventional milk from California (CA) (represented in this study by its far northern Humboldt County) is unusually low in LA and high in ALA and CLA, and it has a FA profile similar to nationwide organic milk. Conversely, conventional milk from the Mid-Atlantic region has unusually high ratios of LA/ALA and ω-6/ω-3. One-way analyses of variance among regions for conventional milk do not consistently conform to assumptions of normality and equal variance, but they suggest that values for the CA region differ from all other regions with high reliability ($P<0.000001$, ANOVA of log-transformed data, followed by Tukey-Kramer pair-wise comparisons). For other comparisons of possible interest in Figures 1 and 2, two values are likely reliably different ($P<0.05$) if their SE error bars are well separated in vertical position.

For some measures in Figures 1 and 2, variability among regions is notably larger for conventional compared to organic samples, as measured by CVs of the regional means. Even excluding the most variable values for conventional milk from CA, CVs are about 3-fold larger in conventional compared to organic samples for ALA (23% vs. 7.6%, $P=0.02$), LA/ALA (25% vs. 8.8%, $P=0.03$), and ω-6/ω-3 (29% vs. 8.9%, $P=0.015$). For CLA, there is an opposite trend toward greater regional variability in organic milk (CV = 18% vs. 8.2% without CA, $P=0.08$). There is negligible difference in regional variability for LA (organic 6.5% vs. conventional 7.2% without CA, $P=0.8$).

Seasonal Differences

Figure 3 shows monthly variations in the national average composition of organic and conventional milk over the full 18 months of the study. Concentrations of LA and ALA, and the differences between them, are similar across time (Figures 3A and 3B). In contrast, there is substantial seasonal variation in CLA, especially in organic milk (Figure 3C). We compared the national average CLA concentrations during the "summer" calendar months of May–October (May–Oct. 2011 and May–June 2012) with those during the "winter" months of November–April (Jan.–Apr. and Nov.–Dec. 2011, plus Jan.–Apr. 2012). In organic milk the CLA concentration was 55% higher in summer than in winter, with averages ± SE (in mg/100 g) of 0.0283 ± 0.0008 in summer vs. 0.0183 ± 0.0005 in winter ($P=0.000000$ by Mann-Whitney test, $n=96$ and 122 respectively). In conventional milk the

Table 2. Fatty acids in retail whole milk (g/100 g), 12 months ending December 2011.

	Organic					Conventional					Org/Conv	*P*(differ-ence)*
	Mean	**n**	**SD**	**CV**	**SE**	**Mean**	**n**	**SD**	**CV**	**SE**		
Total triglyceride (calculated)	**3.276**	**143**	**0.158**	**4.8%**	**0.013**	**3.265**	**108**	**0.133**	**4.1%**	**0.013**	**1.00**	**0.56**
Total fatty acids	**3.108**	**143**	**0.150**	**4.8%**	**0.013**	**3.098**	**108**	**0.127**	**4.1%**	**0.012**	**1.00**	**0.58**
Saturated fatty acids												
4:0 butyric	0.0751	143	0.0062	8.3%	0.0005	0.0738	107	0.0062	8.4%	0.0006	1.02	0.10
6:0 caproic	0.0607	143	0.0066	11%	0.0006	0.0577	108	0.0055	9.5%	0.0005	1.05	0.00014
8:0 caprylic	0.0407	143	0.0050	12%	0.0004	0.0384	108	0.0035	9.2%	0.0003	1.06	0.00007
10:0 capric	0.1288	143	0.0316	25%	0.0026	0.1241	108	0.0283	23%	0.0027	1.04	0.23
11:0 undecylic	0.0026	85	0.0009	36%	0.0001	0.0026	84	0.0008	32%	0.0001	0.98	0.69
12:0 lauric	0.1059	143	0.0131	12%	0.0011	0.0990	108	0.0100	10%	0.0010	1.07	0.00001
14:0 myristic	0.3490	143	0.0251	7.2%	0.0021	0.3274	108	0.0239	7.3%	0.0023	1.07	0.00000
15:0 pentadecanoic	0.0402	143	0.0036	9.0%	0.0003	0.0355	108	0.0039	11%	0.0004	1.13	0.00000
16:0 palmitic	0.9344	143	0.0798	8.5%	0.0067	0.8995	108	0.0519	5.8%	0.0050	1.04	0.00010
17:0 margaric	0.0243	143	0.0022	8.9%	0.0002	0.0216	108	0.0025	12%	0.0002	1.12	0.00000
18:0 stearic	0.3434	143	0.0390	11%	0.0033	0.3559	108	0.0362	10%	0.0035	0.96	0.010
20:0 arachidic	0.0064	141	0.0014	21%	0.0001	0.0055	107	0.0010	19%	0.0001	1.17	0.00000
22:0 behenic	0.0040	105	0.0010	24%	0.0001	0.0032	61	0.0013	41%	0.0002	1.25	0.00001
24:0 lignoceric	0.0025	73	0.0010	40%	0.0001	0.0024	23	0.0009	39%	0.0002	1.03	0.75
Total saturated†	2.116	143	0.120	5.7%	0.010	2.043	108	0.095	4.7%	0.009	1.04	0.00000
Monounsaturated fatty acids												
14:1 myristoleic	0.0290	143	0.0038	13%	0.0003	0.0269	108	0.0038	14%	0.0004	1.08	0.00002
16:1 palmitoleic	0.0468	143	0.0067	14%	0.0006	0.0467	108	0.0069	15%	0.0007	1.00	0.83
17:1 margaroleic	0.0080	137	0.0014	17%	0.0001	0.0070	106	0.0015	21%	0.0001	1.14	0.00000
18:1 incl. oleic	0.6505	143	0.0533	8.2%	0.0045	0.7074	108	0.0486	6.9%	0.0047	0.92	0.00000
20:1 incl. gadoleic	0.0071	126	0.0026	37%	0.0002	0.0067	96	0.0025	38%	0.0003	1.05	0.34
Total monounsaturated†	0.7410	143	0.0547	7.4%	0.0046	0.7944	108	0.0491	6.2%	0.0047	0.93	0.00000
ω-3 fatty acids												
18:3 α-linolenic, ALA	0.0255	143	0.0040	16%	0.0003	0.0159	108	0.0059	37%	0.0006	1.60	0.00000
20:5 eicosapentaenoic, EPA	0.0033	104	0.0012	35%	0.0001	0.0025	43	0.0010	41%	0.0002	1.33	0.00009
22:5 docosapentaenoic, DPA	0.0044	120	0.0012	28%	0.0001	0.0037	70	0.0010	26%	0.0001	1.18	0.00008
Total ω-3†	0.0321	143	0.0061	19%	0.0005	0.0198	108	0.0084	43%	0.0008	1.62	0.00000
ω-6 fatty acids												
18:2 linoleic, LA	0.0639	143	0.0079	12%	0.0007	0.0856	107	0.0146	17%	0.0014	0.75	0.00000
20:3 8,11,14-eicosatrienoic (γ)	0.0032	110	0.0010	32%	0.0001	0.0043	92	0.0012	27%	0.0001	0.75	0.00000
20:4 arachidonic, AA	0.0048	118	0.0014	29%	0.0001	0.0058	91	0.0016	28%	0.0002	0.83	0.00001
Total ω-6†	0.0711	143	0.0093	13%	0.0008	0.0948	107	0.0166	18%	0.0016	0.75	0.00000
Total Polyunsaturated†	0.1037	143	0.0126	12%	0.0011	0.1147	107	0.0165	14%	0.0016	0.90	0.00000
***trans* fatty acids**												
trans-16:1 *trans*-palmitoleic	0.0131	143	0.0023	17%	0.0002	0.0117	108	0.0019	17%	0.0002	1.12	0.00000
trans-18:1 incl. elaidic	0.0846	143	0.0235	28%	0.0020	0.0906	108	0.0143	16%	0.0014	0.93	0.00022‡
trans-18:2 octadecadienoic	0.0257	143	0.0084	33%	0.0007	0.0243	107	0.0061	25%	0.0006	1.05	0.17
Total *trans*§	0.1254	143	0.0292	23%	0.0024	0.1281	108	0.0190	15%	0.0018	0.98	0.40
Conjugated linoleic acid, CLA												
18:2 conjugated	0.0227	143	0.0084	37%	0.0007	0.0192	106	0.0049	25%	0.0005	1.18	0.00019
Sum												
ALA + CLA	0.0481	143	0.0105	22%	0.0009	0.0347	108	0.0098	28%	0.0009	1.39	0.00000
Ratios												
LA/ALA	2.568	143	0.544	21%	0.046	6.272	107	2.485	40%	0.240	0.41	0.00000
ω-6/ω-3	2.276	143	0.469	21%	0.039	5.774	107	2.520	44%	0.244	0.39	0.00000

Table 2. Cont.

	Organic					Conventional					Org/Conv	P(differ-ence)*
	Mean	n	SD	CV	SE	Mean	n	SD	CV	SE		
Total triglyceride (calculated)	**3.276**	**143**	**0.158**	**4.8%**	**0.013**	**3.265**	**108**	**0.133**	**4.1%**	**0.013**	**1.00**	**0.56**
Total fatty acids	**3.108**	**143**	**0.150**	**4.8%**	**0.013**	**3.098**	**108**	**0.127**	**4.1%**	**0.012**	**1.00**	**0.58**
ω-3/ω-6	0.456	143	0.083	18%	0.007	0.219	107	0.124	57%	0.012	2.08	0.00000
ω-6/(ω-3 + CLA)	1.353	143	0.345	26%	0.029	2.742	107	1.223	45%	0.118	0.49	0.00000

*Calculated by *t* test except as noted. Because of multiple comparisons, about 2 findings of $P = 0.05$ and 0.5 finding of $P = 0.01$ can be expected by chance.
†These group means (means of sums of saturated, monounsaturated, or polyunsaturated FA) are biased slightly low, because they Include some sums containing unreported small values (< 0.001) treated as zero. In contrast, when n is less than the number of samples (organic n<143, conventional n<108), means of individual FA are biased slightly high by omission of unreported small values. Thus these group means are slightly less than the sum of means of the individual FA.
‡Calculated by Mann-Whitney test due to non-normal distributions with medians 0.080 (organic) and 0.091 (conventional). ($P = 0.020$ calculated by *t* test).
§The *trans* FA group mean exceeds the sum of individual *trans* FA, because it includes small amounts of *trans*-14:1 omitted from the table due to small numbers of reported values (42 organic, 27 conventional).

summer increase was a much smaller 12% (0.0200±0.0006 vs. 0.0179±0.0003 mg/100, $P = 0.0074$, n = 67 and 90).

We also compared the above national average CLA concentrations between organic and conventional samples, separately by season. In the summer months the organic average was 42% higher (0.0283±0.0008 vs. 0.0200±0.0006 mg/100 g, $P = 0.000000$ by Mann-Whitney test, n = 96 and 67 respectively). But in the winter months the difference between organic and conventional samples was negligible (respectively 0.0183±0.0005 vs. 0.0179±0.0003 mg/100 g, $P = 0.86$, n = 122 and 90). Averaged over the full year of 2011, the CLA concentration was 18% higher in organic compared to conventional samples (Table 2).

Similarly for ALA in organic milk, Figure 3 shows a small (4%) but statistically significant increase in summer compared to winter (0.0262 ± SE 0.0003 vs. 0.0252 ± SE 0.0003 mg/100 g, $P = 0.022$, by Mann-Whitney test, n = 95 and 122). A corresponding 2.6% summer increase in conventional milk is not statistically significant ($P = 0.54$, n = 69 and 96). Differences between organic and conventional milk are highly reliable in both summer and winter ($P = 0.000000$) and reliable for each of the 18 months ($P \leq 0.020$).

For LA, we find no evidence for seasonal variability in national averages for either organic or conventional milk ($P = 0.13$ and 0.85, respectively, by Mann-Whitney tests). However, LA differences between organic and conventional milk are highly reliable in both summer and winter ($P = 0.000000$) and reliable for each of the 18 months ($P \leq 0.017$).

In Figure 3 the smaller variability of 18 monthly means in organic compared to conventional milk is statistically significant for ALA (CV = 5.2% vs. 9.8%, $P = 0.014$), but not for LA (CV = 3.7% vs. 4.2%, $P = 0.58$). For CLA monthly variability is larger in organic samples, 24% vs. 10%, $P = 0.0013$.

Diet Scenarios and LA/ALA

We analyzed a series of diet scenarios to assess the impact of conventional and organic dairy products on total dietary LA/ALA ratios (see Table 1). The top half of Table 3 shows the resulting intakes of LA and ALA in hypothetical diets for an adult woman consuming an average portion of energy from fat (33%) and typical non-dairy fat sources. The highest LA/ALA ratio of 11.3 occurred in the diet with moderate amounts (3 servings per day, fluid milk equivalent) of conventional whole milk and other mostly full-fat dairy products. Compared to this baseline value, the ratios decreased to 10.0 with either a switch to organic dairy products or a 50% increased consumption of conventional dairy products (4.5 servings per day, fluid milk equivalent). A combination of both changes (high intake of organic dairy products) reduced the LA/ALA ratio to 7.8. Expressed as a percentage of the reduction needed to reach a goal of LA/ALA = 2.3, either change alone (switch to organic or increase in conventional dairy consumption) achieves about 15% of the needed reduction, and a dual switch to organic and increased consumption of organic dairy products accomplishes about 40% of the necessary reduction.

The bottom half of Table 3 shows the effects of the same variations in types and amounts of milk in the context of a diet with reduced amounts of LA from non-dairy sources. In these low-LA scenarios, we substituted pita chips for tortilla chips, canola oil for soy oil, and canola-oil margarine for regular margarine, with no change in French fries, chocolate chip cookies, chicken, pork, or beef. These reductions in LA intake consistently reduced the overall dietary LA/ALA ratios to about 4 or 5, or 68% to 80% of the way toward 2.3, even with moderate consumption of conventional, mostly full-fat dairy products.

Figure 4 shows the above percentages of progress toward an LA/ALA ratio of 2.3, along with the results of additional scenarios in which we varied the baseline intake of total fat (to 20% and 45% of energy, in addition to 33%.

Fatty Acids in Fish Compared to Other Sources

Table 4 shows daily average FA contents and LA/ALA ratios of, (a) 8 ounces per week of 7 representative fish varieties (sorted by fat content) and their average, (b) conventional and organic dairy products, and (c) the two non-dairy fat sources used in our dietary scenarios. The serving amounts are based on the Dietary Guidelines for Americans, as previously described. The 7 fish range from poor to excellent sources of EPA and DHA. On average they contain 90 + 155 = 245 mg of EPA + DHA, close to the Dietary Guidelines goal of 250 mg/day from fish [1].

Discussion

This study is the first large-scale, national milk-fat composition survey in the U.S. comparing milk from organic and conventional farms. The results are based on 378 samples of organic and conventional milk from 7 regions collected over 18 months.

The sampling protocol allows assessment of the impacts of organic and conventional production systems on milk FA composition, as well as regional and seasonal differences.

A key finding is that the nationwide average ratios of LA/ALA and ω-6/ω-3 in 2011 were 2.6 and 2.3 respectively for organic milk samples, compared to 6.3 and 5.8 for the samples from cows

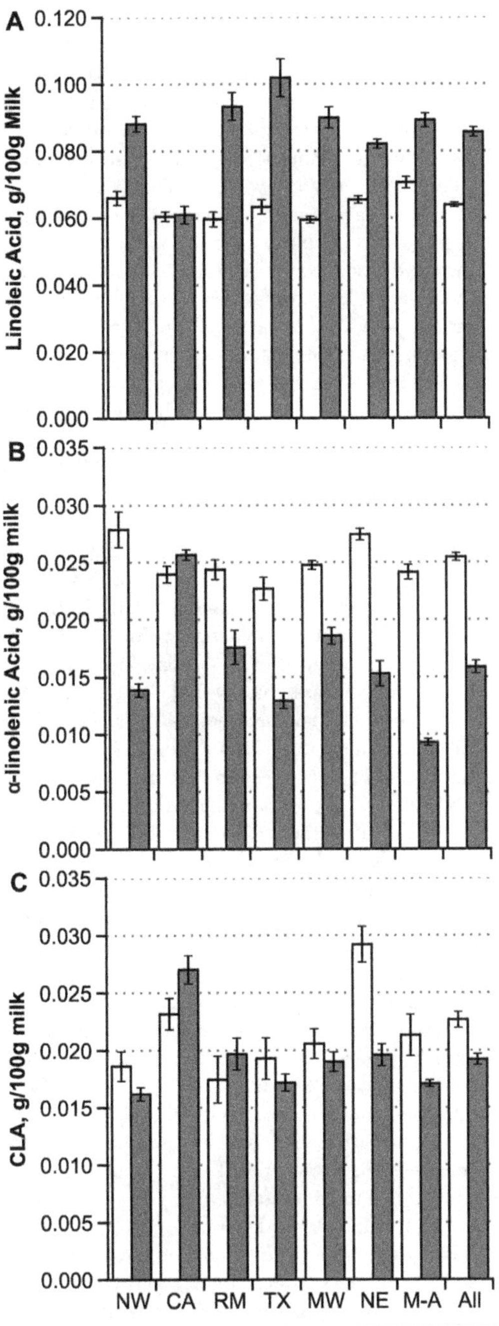

Number of Samples								
Org	19	22	11	10	24	36	21	143
Conv	19	11	10	11	12	35	10	108

Organic / Conventional

Figure 1. Regional variation in fatty acid content of retail whole milk, g/100 g (12-month average ± SE). A: Linoleic acid (LA, ω-6). **B:** α-linolenic acid (ALA, ω-3). **C:** Conjugated linoleic acid. Abbreviations: NW = Northwest, CA = California, RM = Rocky Mountain, TX = Texas, MW = Midwest, NE = Northeast, M-A = mid-Atlantic. Numbers of samples apply to panels B and C; for panel A conventional NE is 34 and All is 107. For LA and ALA, all differences between organic and conventional contents are statistically significant by Mann-Whitney test ($P<0.005$) except for the CA region ($P\geq0.10$). For CLA no such differences are statistically significant ($P>0.08$) except for the NE region and All regions ($P<0.001$).

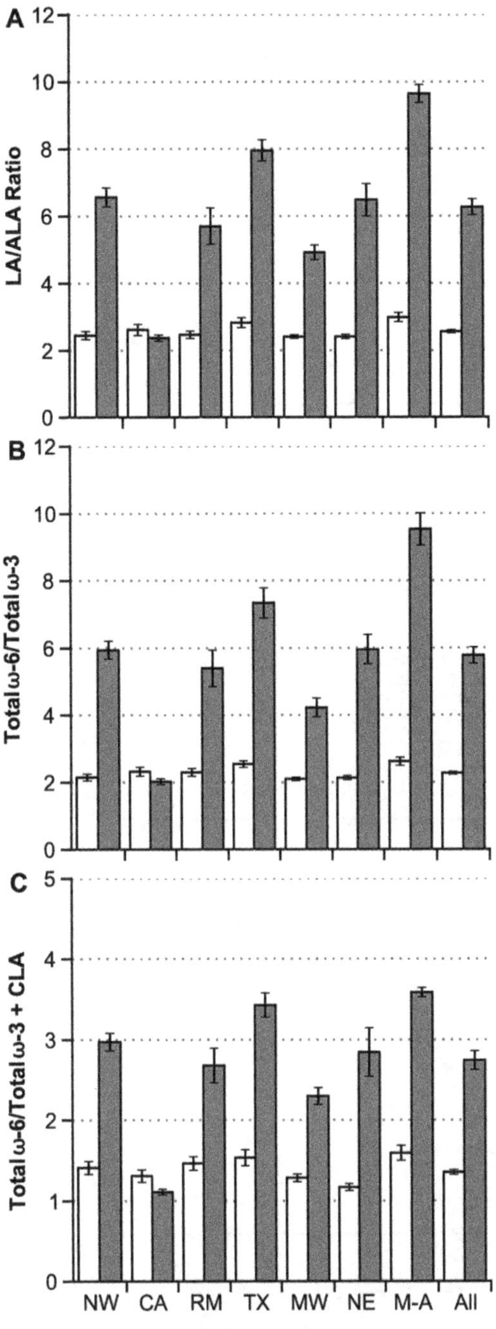

Number of Samples								
Org	19	22	11	10	24	36	21	143
Conv	19	11	10	11	12	34	10	107

Organic / Conventional

Figure 2. Regional variation in ratios involving ω-6 and ω-3 fatty acids (12-month average ± SE). A: Linoleic acid/α-linolenic acid (LA/ALA). **B:** Total ω-6/total ω-3. **C:** Total ω-6/(total ω-3 + CLA). Abbreviations: Same as Fig. 1. Numbers of samples apply to all panels.

on conventionally managed dairies ($P<0001$). Although the ω-6/ω-3 ratio of both conventional and organic dairy fat is healthier than the ratio of most other commonly consumed fat sources, full-fat organic dairy products offer clear advantages for individuals striving to reduce their overall dietary ω-6/ω-3 ratio.

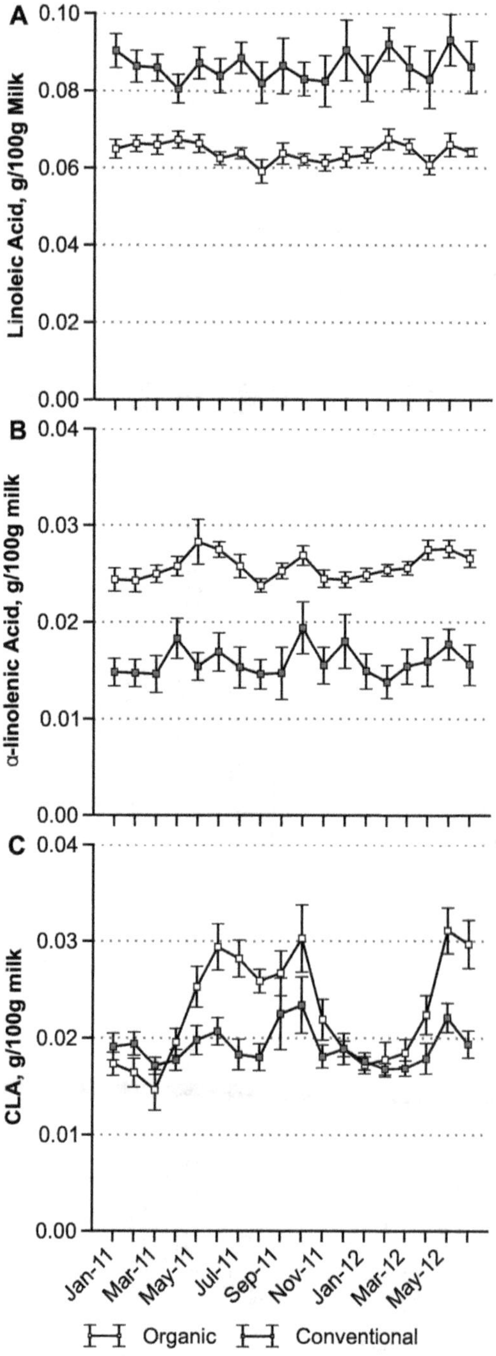

Figure 3. Seasonal variation in polyunsaturated fatty acid content of retail whole milk, g/100 g, 18 monthly averages ± SE. A: Linoleic acid (LA, ω-6). B: α-linolenic acid (ALA, ω-3). C: Conjugated linoleic acid (CLA). The number of monthly samples in all panels is 10 to 13 for organic and 7 to 10 for conventional milk, except 5 conventional in October 2011. For LA and ALA, all differences between organic and conventional contents are statistically significant by Mann-Whitney test (*P*<0.02).

This study confirms earlier findings that milk from cows consuming significant amounts of grass and legume-based forages contains less LA and other ω-6 FAs and higher concentrations of ALA, CLA, and the long-chain ω-3s EPA and DPA, compared to cows lacking routine access to pasture and fed substantial quantities of grains [17–23,32,58]. In most countries, lactating cows on organically managed farms receive a significant portion of daily DMI from pasture and conserved, forage-based feeds [18,19,21,23], while cows on conventional farms receive much less. In the most recent U.S. government dairy sector survey, only 22% of cows had access to pasture [59], and for most of these, access was very limited in terms of average daily DMI.

The greater regional variation in conventional compared to organic milk (Figs. 1 and 2) likely arises from large regional variations in the feed sources in lactating cow rations. For example, conventional dairy farms near vegetable oil, soy biodiesel, or ethanol plants are likely to feed byproducts from these plants [60]. Other farms might rely on brewers dried grain (from malting barley) or a wide range of food processing wastes. Organic dairy operations, in contrast, are much more dependent on relatively uniform pasture and forage-based feeds, in part because of the grazing requirement in the NOP rule [35]. Also, certified organic sources of most processing wastes and byproduct feeds are not available in substantial quantities.

The FA similarities between conventional and organic milk from our CA region (Figs. 1 and 2) were unexpected. These conventional and organic milk samples came from the Humboldt County area in far Northern CA, a coastal region where both types of dairy farms graze cattle for over 250 days per year. This heavy reliance on pasture contrasts sharply to the near-zero access to pasture on most conventional dairy farms throughout CA's central valley (the major dairy production region in CA) [61].

In the U.K. and much of Europe, cows on conventional farms have routine access to grazing, although less so than cows on organic farms in the U.S. In a study of organic and conventional dairy farms in North England, grazing accounted for 37% of average DMI on 29 organic farms, compared to 20% on conventional outdoor farms and 3% on conventional indoor operations (annual averages) [19]. During the cold season, indoor-period organic dairy diets had a higher ratio of conserved forage to concentrate compared to conventional dairy diets [18,22].

On most U.S. organic dairy farms, pasture and on-farm, forage-based feeds account for more than 30% to well over one-half of daily DMI for much of the year [36]. In contrast, the proportion of fresh forage in conventional dairy diets has decreased continuously over the last 40 years in the U.S. Grain-based "total mixed rations" now dominate the conventional U.S. dairy sector. In 2007, 94% of dairy farms milking 500 or more cows fed a total mixed ration, as did 71% of high-production dairies (herd average > 20,000 pounds annual milk production per cow) [59]. The differences in feeding regimes between organic and conventional dairy farms (and associated impacts on milk composition) would therefore be expected to be greater in the U.S. than in Europe.

We find little seasonal variability in milk concentrations of LA, ALA, and CLA, with the notable exception of CLA in organic milk (Fig. 3). In organic milk, CLA peaked during May through October and fell back in December through March to levels similar to conventional milk. Other studies show that CLA levels are especially dependent on pasture feeding with immature, nutrient-rich grasses and legumes, and that any form of mechanical harvest and storage leads to some loss of forage quality, affecting especially the CLA content of milk [15,19]. These findings likely explain why CLA levels in milk fall on most organic farms over the winter, and peak in the spring and summer, when pasture quality is optimized, and why there is little or no seasonal CLA variation in milk from conventional cows with little access to pasture.

Our results and others confirm that there are significant opportunities to improve the FA profile of milk and dairy

Table 3. LA and ALA contents of hypothetical average-fat diets with typical and low-LA non-dairy fat sources.

	LA from Dairy Fat, g*	ALA from Dairy Fat, G*	LA from Other Fat, g†	ALA from Other Fat, g†	Total LA, g	Total ALA, g	Total LA/ Total ALA	Decrease from Base-line LA/ALA	Decrease from Base-line as % of Decrease Needed to Reach LA/ALA = 2.3
With Typical Non-Dairy Fat Sources									
Conventional Dairy									
Moderate intake	0.92	0.17	9.91	0.79	10.83	0.96	11.33	--	--
High intake	1.38	0.26	5.77	0.46	7.15	0.71	10.01	1.31	15%
Organic Dairy									
Moderate intake	0.68	0.27	9.91	0.79	10.59	1.06	10.01	1.32	15%
High intake	1.03	0.41	5.77	0.46	6.80	0.87	7.83	3.50	39%
With Low-LA Non-Dairy Fat Sources									
Conventional Dairy									
Moderate intake	0.92	0.17	5.90	1.17	6.82	1.34	5.11	6.22	69%
High intake	1.38	0.26	3.44	0.68	4.82	0.94	5.15	6.17	68%
Organic Dairy									
Moderate intake	0.68	0.27	5.90	1.17	6.59	1.44	4.58	6.74	75%
High intake	1.03	0.41	3.44	0.68	4.47	1.09	4.10	7.23	80%

*Based on LA, ALA, and total FA from Table 2, 8.79 kcal/g dairy fat, and 0.933 g milk FA/g milk fat.
E.g., LA 0.92 = 313 (Table 1) × 0.0856×0.933/8.79/3.098.
†Based on 23.23 g LA and 1.841 g ALA per 100 kcal non-dairy fat, 8.9 kcal/g non-dairy fat.
E.g., LA 9.91 = 380 (Table 1) × 23.23/8.9/100.
Corresponding calculations for low-LA non-dairy fat use 13.84 g LA and 2.731 g ALA per 100 kcal non-dairy fat.

products. The potential human health benefits stemming from such improvements are less clear and must be evaluated in the context of overall dietary FA intakes and trends.

During the last century in the U.S. and other developed countries, increasing intakes of LA from vegetable oils, especially soy oil used by food processors, account for most of the shift in typical ω-6/ω-3 ratios from a relatively healthy ~5 in the early 1900s to ~15 in much of Europe [62] and around 10 to 15 in the U.S. [13,24,26,63]. LA and ALA are the main ω-6 and ω-3 PUFAs and account for respectively 84%–89% and 9%–11% of the total PUFA intake in Western diets [26]. In contrast, intakes of the more potent, longer-chain ω-3 FAs such as EPA, DPA, and DHA are relatively low in most Western diets, caused by low fatty fish intakes.

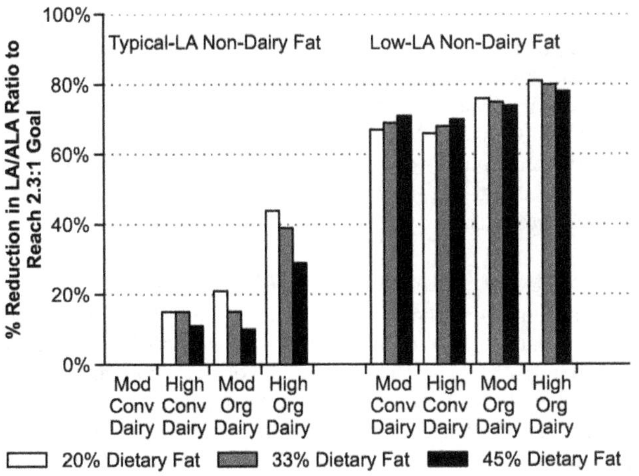

Figure 4. Percent progress toward a dietary LA/ALA ratio of 2.3 for hypothetical diets of an adult woman, relative to diets containing moderate amounts of conventional dairy products. The diets contain non-dairy fat sources with typical (left side) and low (right side) amounts of LA in the context of total dietary fat contributing 20%, 33%, or 45% of energy. Abbreviations: Mod = moderate, Conv = conventional, Org = organic.

Many studies and reviews have concluded that reducing dietary ω-6/ω-3 ratios during adulthood will lower risks of CVD [14,29–31], metabolic syndrome and diabetes [8,38,41,43,44,64], overweight [7,28,37,39,48,62], and violent behavior [40,41,65]. One study reported that a group of adults with the highest plasma ALA levels had ~30% lower incidence of diabetes [44], and a systematic review of lipid-lowering agents concluded that ω-3 FAs are as effective as statin drugs in lowering CVD risk [47].

Expected benefits from reduced dietary ω-6/ω-3 ratios, coupled with increased long-chain ω-3 intakes, are almost certainly greatest for women hoping to bear a child, for pregnant women and their babies, and for infants and children through adolescence [45–48]. High ω-6/ω-3 ratios and/or low long-chain ω-3 intakes predispose the developing fetus to a wide range of adverse neurological and immune system disorders, and can also impair the visual system [66]. Recent research shows that high LA/ALA dietary ratios depress long-chain ω-3 levels in the blood of pregnant women by two mechanisms—by depressing the conversion of ALA to long-chain ω-3s and by blocking incorporation of pre-formed, long-chain ω-3s into phospholipids [67–71].

Adults are able to convert a small fraction of ALA to EPA, DPA, and—mainly in women—to DHA [67,68,70,72–75]. However, excess dietary LA competes with ALA for the enzymes involved in these conversions [70,71]. One study found that an LA intake of 30 g/day reduces ALA conversion to DHA by ~40% [71], while

Table 4. Daily average contents of 8 oz. per week of cooked fish, with comparison to other daily fat sources.

Food	USDA	Fat	Fat Energy	LA	ALA	EPA	DPA	DHA	LA/ALA
	no.	g	kcal*	g	g	g	g	g	
Tuna, light, canned in water	15121	0.311	2.8	0.005	0.001	0.009	0.001	0.064	7.0
Tilapia, farmed	15262	0.859	7.8	0.092	0.015	0.002	0.019	0.042	6.3
Halibut with skin, Alaska	35188	0.885	8.0	0.006	0.003	0.079	0.017	0.118	2.0
Salmon, sockeye, Pacific	15273	1.508	13.6	0.035	0.018	0.111	0.029	0.211	1.9
Catfish, channel, farmed†	15234	1.926	17.4	0.294	0.021	0.007	0.006	0.022	14.0
Trout, rainbow, farmed†	15240	2.004	18.1	0.181	0.023	0.084	0.035	0.201	7.9
Salmon, Atlantic, farmed†	15236	4.352	39.3	0.350	0.058	0.335	0.153	0.430	6.1
Average of high- and low-fat fish		**1.692**	**15.3**	**0.138**	**0.020**	**0.090**	**0.037**	**0.155**	**6.5**
Milk products, conventional‡	--	--	313	0.92	0.170	0.027	0.040	--	5.4
Milk products, organic‡	--	--	313	0.68	0.273	0.035	0.047	--	2.5
Diet scenario non-dairy fat, typical LA§	--	--	380	9.91	0.79	--	--	--	12.6
Diet scenario non-dairy fat, low LA§	--	--	380	5.90	1.17	--	--	--	5.1

*Based on 9.02 kcal/g of fish fat (USDA).
†Cooked fish values, estimated from USDA's raw fish values × 1.2.
‡Fat energy from Table 1 (moderate dairy intake, average- or high-fat diet, 33% or 45% fat energy). FA amounts calculated from Table 2 values (sample calculation in Table 3).
§Fat energy from Table 1 (moderate dairy intake, average-fat diet, 33% fat energy). FA amounts calculated as shown in Table 3 sample calculation and footnote.

others have shown that certain diets allow pre-menopausal women to convert up to 3-fold more DPA to DHA than males [68]. Accordingly, the improved LA/ALA ratio in organic milk (2.6 organic vs. 6.3 conventional) secondarily benefits consumers by enhancing conversion of ALA to long-chain ω-3s.

Fatty Acids in Fish Compared to Other Sources

Table 4 shows that 8 ounces per week of a variety of fish (represented by the 7-fish average) contains small amounts of LA compared to dairy products, and only 12% and 7% as much ALA as conventional and organic dairy products, respectively. Compared to the more dominant non-dairy sources of PUFAs, fish contributes negligible LA and ALA, and its average LA/ALA ratio of 6.5 is not distinctive. An important implication is that recommended servings of fish cannot significantly alter U.S. dietary ratios of LA/ALA. Fish also cannot greatly alter ω-6/ω-3 ratios that are typically dominated by LA and ALA. However, our dietary scenarios show how LA/ALA ratios and presumably ω-6/ω-3 ratios can be improved by changing the types and amounts of dairy fat, and especially by reducing LA intake.

The most distinctive FA of fish in Table 4 is DHA, which does not occur in plant foods or significantly in cow's milk. Fish is less unique for EPA and not unique for DPA. Recommended amounts of dairy products, if mostly full-fat, contain about one-third as much EPA as a mixture of fish varieties, and contain as much DPA—or if organic, somewhat more DPA.

CLA and Other *Trans* Fatty Acids

Although industrially-produced *trans* FAs are recognized as generally harmful, CLA and the other major *trans* FA in cow's milk are probably beneficial or harmless to humans [76]. The dominant CLA in milk (75%–90%) [16] is *cis*-9,*trans*-11 18:2, known as rumenic acid and shown as "18:2 conjugated" in Table 2. Because CLAs have probable benefits in humans [52,53] and proven benefits in animals [16,54], the U.S. Food and Drug Administration does not count them as *trans* FA for food labeling purposes [77]. Conventional dairy products account for about 75% of U.S. CLA consumption [16], and organic production, especially spring pasture, is known to increase CLA levels [18–23,32]. We find an annual average 18% increase.

The major *trans* FA in dairy fat is *trans*-18:1 (included in "*trans*-18:1 incl. elaidic" in Table 2), of which the dominant isomer (25%–75%) is *trans*-11 18:1, vaccenic acid [16]. At the high range of human intakes, vaccenic acid has little or no effect on CVD risk factors [78]. Humans convert about 20% of it to the rumenic acid form of CLA [16]. In our samples, *trans*-18:1 is reduced by 7% in organic milk.

In human plasma, *trans*-16:1 comes almost exclusively from dairy fat and ruminant meats and thus serves as a marker for consumption of these foods. A recent study found that plasma levels are strongly associated with dairy fat consumption and also with broad health benefits—reduced incidence of new-onset diabetes, favorable CVD risk profile, reduced insulin resistance and inflammation (C-reactive protein), and slightly lower body fat [3]. We find that *trans*-16:1 is 12% higher in organic compared to conventional milk.

Organic Dairy Products and LA/ALA Ratios

In this study, organic production reduced the 12-month-average ω-6/ω-3 ratio of whole milk from the conventional milk level of 5.77 to only 2.28. The 18% higher level of CLA in organic milk is an additional health benefit [52–54], as are the generally higher levels of antioxidants in organic milk [18,19,22,23]. But a key question remains—is the shift in FA profiles in organic milk and dairy products sufficient to improve health outcomes? And if so, by how much, and for whom?

Our dietary scenarios show that organic dairy products can improve dietary LA/ALA and ω-6/ω-3 ratios in adults. Because both conventional and organic dairy fat have ω-6/ω-3 ratios superior to most other fat sources in typical Western diets, replacing non-dairy fat with full-fat dairy products, whether conventional or organic, will improve total dietary LA/ALA ratios. Without other changes, increasing dairy fat intake would increase overall dietary fat and calories, an unwelcome outcome for most people. But if coupled with reduced intakes of food products containing vegetable oils and/or other sources of saturated fat, overall fat content and energy intake can remain unchanged or even decline, while dramatically improving the diet's ALA content and ω-6/ω-3 ratio.

The impact—and importance—of selecting low-LA alternatives to high-LA foods is unmistakable in our dietary scenarios (Table 3 and Figure 4). They focus on women of childbearing age because of the heightened importance of adequate ω-3 intakes during pregnancy and lactation, as well as the need for efficient conversion of ALA to long-chain ω-3s. The scenarios suggest that the LA/ALA ratio can be reduced by ~30% to 45% of the way toward the target of 2.3 through high consumption of mostly full-fat organic dairy products, compared to moderate (Dietary Guidelines) consumption of corresponding conventional dairy products. But when coupled with partial reduction of high-LA foods, women can achieve ~80% of the reduction needed to reach a target ratio of LA/ALA ~2.3.

Our scenarios may be summarized as follows: For adult women consuming typical-LA non-dairy fat sources, an increase from moderate to 50% higher intakes of conventional dairy products alone reduces dietary LA/ALA ratios by about 10 to 15% of the way toward a target ratio of 2.3. Alternatively, a switch to only moderate amounts (3 servings per day) of mostly full-fat, organic dairy products achieves similar reductions. The switch to 50% higher amounts of organic dairy products adds a further roughly 25% increment toward the 2.3 goal (for a total increment near 40%). The additional step of partially choosing low-LA sources of non-dairy fats brings the overall reduction to ~75–80% of the way toward the 2.3 goal.

We conclude that increasing reliance on pasture and forage-based feeds on dairy farms has considerable potential to improve the FA profile of milk and dairy products. Although both conventional and organic dairies can benefit from grazing and forage-based feeds, it is far more common—and indeed mandatory on certified organic farms in the U.S.—for pasture and forage-based feeds to account for a significant share of a cow's daily DMI. Moreover, improvements in the nutritional quality of milk and dairy products should improve long-term health status and outcomes, especially for pregnant women, infants, children, and those with elevated CVD risk. The expected benefits are greatest for those who simultaneously avoid foods with relatively high levels of LA, increase intakes of fat-containing dairy products, and switch to predominantly organic dairy products.

Supporting Information

Figure S1 Fatty acid content of retail whole milk, g/100 g (12-month average ± SE). Some SE are too small to be visible. Abbreviations: Sat = saturated, Mono = monounsaturated, Poly = polyunsaturated, LA = linoleic acid, ALA = α-linolenic acid, CLA = conjugated linoleic acid, EPA = eicosapentaenoic acid, DPA = docosapentaenoic acid. Differences between organic and conventional contents are statistically significant by Mann-Whitney test ($P<0.001$) except for Total and *Trans* fatty acids ($P > 0.40$).

Figure S2 Ratios involving ω-6 and ω-3 fatty acids (12-month average ± SE). Low ratios denote increased amounts of ω-3 and other fatty acids that are commonly low in modern diets and are beneficial to heart, brain, eye, and other tissues and functions. Abbreviations: LA = linoleic acid, ALA = α-linolenic acid, CLA = conjugated linoleic acid. Differences between organic and conventional ratios are statistically significant ($P<0.0001$).

Acknowledgments

We thank the CROPP Cooperative milk-quality staff for managing the milk sampling process and compiling the results. We also thank the journal editor and reviewers for many helpful suggestions and comments. We appreciate the thoroughness and professionalism of Silliker, Inc. in carrying out and documenting the milk sample testing.

Author Contributions

Conceived and designed the experiments: CMB MAL. Performed the experiments: MAL. Analyzed the data: CMB DRD CL GB MAL. Wrote the paper: CMB DRD. Created the scenarios model, tables, and figures: DRD CMB.

References

1. U.S. Department of Agriculture (2010) Dietary guidelines for Americans, 7th Edition. Available: http://www.cnpp.usda.gov/DGAs2010-PolicyDocument.htm. Accessed 2013 Mar 15.
2. Chardigny J-M, Destaillats F, Malpuech-Brugère C, Moulin J, Bauman DE, et al. (2008) Do *trans* fatty acids from industrially produced sources and from natural sources have the same effect on cardiovascular diseases risk factors in healthy subjects? Results of the *trans* Fatty Acids Collaboration (TRANSFACT) study. Am J Clin Nutr 87: 558–566.
3. Mozaffarian D, Cao H, King IB, Lemaitre RN, Song X, et al. (2010) *Trans*-palmitoleic acid, metabolic risk factors, and new-onset diabetes in U.S. adults. Ann Intern Med 153: 790–799.
4. German JB, Gibson RA, Krauss RM, Nestel P, Lamarche B, et al. (2009) A reappraisal of the impact of dairy foods and milk. Eur J Nutr 48: 191–203.
5. Elwood PC, Pickering JE, Givens DI, Gallacher JE (2010) The consumption of milk and dairy foods and the incidence of vascular disease and diabetes: an overview. Lipids 45: 925–939.
6. Ludwig DS, Willett WC (2013) Three daily servings of reduced-fat milk: an evidence-based recommendation? JAMA Pediatr doi:10.1001/jamapediatrics.2013.2408. Accessed 3 July 2013.
7. Rosell M, Hakansson NN, Wolk A (2006) Association between dairy food consumption and weight change over 9 y in 19,352 perimenopausal women. Am J Clin Nutr 84: 1481–1488.
8. Stancliffe RA, Thorpe T, Zemel MB (2011) Dairy attenuates oxidative and inflammatory stress in metabolic syndrome. Am J Clin Nutr 94: 422–30.
9. Bonthuis M, Hughes MCB, Ibiebele TI, Green AC, van der Pols JC (2010) Dairy consumption and patterns of mortality of Australian adults. Eur J Clin Nutr 64: 569–577.
10. Larsson SC, Bergkvist L, Wolk A (2005) High-fat dairy food and conjugated linoleic acid intakes in relation to colorectal cancer incidence in the Swedish Mammography Cohort. Am J Clin Nutr 82: 894–900.
11. Rice BH, Quann EE, Miller GD (2013) Meeting and exceeding dairy recommendations: effects of dairy consumption on nutrient intakes and risk of chronic disease. Nutr Rev 71: 209–223.
12. Haug A, Høstmark AT, Harstad OM (2007) Bovine milk in human nutrition—a review. Lipids Health Dis 6: 25, doi: 10.1186/1476-511X-6-25 Available: http://www.ncbi.nlm.nih.gov/pmc/articles/PMC2039733. Accessed 2013 Aug 10.
13. Simopoulos AP (2006) Evolutionary aspects of diet, the omega-6/omega-3 ratio and genetic variation: nutritional implications for chronic diseases. Biomed Pharmacother 60: 502–507.
14. Russo GL (2009) Dietary n-6 and n-3 polyunsaturated fatty acids: From biochemistry to clinical implications in cardiovascular prevention. Biochem Pharma 77: 937–946.
15. Couvreur S, Hurtaud C, Lopez C, Delaby L, Peyraud JL (2006) The linear relationship between the proportion of fresh grass in the cow diet, milk fatty acid composition, and butter properties. J Dairy Sci 89: 1956–1969.
16. Lock AL, Bauman DE (2004) Modifying milk fat composition of dairy cows to enhance fatty acids beneficial to human health. Lipids 39: 1197–1206.

17. O'Donnell AM, Spatny KP, Vicini JL, Bauman DE (2010) Survey of the fatty acid composition of retail milk differing in label claims based on production management practices. J Dairy Sci 93: 1918–1925.
18. Butler G, Nielsen JH, Slots T, Seal C, Eyre MD, et al. (2008) Fatty acid and fat-soluble antioxidant concentrations in milk from high- and low-input conventional and organic systems: seasonal variation. J Sci Food Agric 88: 1431–1441.
19. Stergiadis S, Leifert C, Seal CJ, Eyre MD, Nielsen JH, et al. (2012) Effect of feeding intensity and milking system on nutritionally relevant milk components in dairy farming systems in North East of England. J Ag Food Chem 60: 7270–7281.
20. Krogh LM, Nielsen J, Leifert C, Slots T, Hald Kristensen G, et al. (2010). Milk quality as affected by feeding regimes in a country with climatic variation. J Dairy Sci 73: 2863–2873.
21. Butler G, Nielsen JH, Larsen ML, Rehberger B, Stergiadis S, et al. (2011) The effect of dairy management and processing on quality characteristics of milk and dairy products. NJAS – Wageningen J Life Sci 58: 97–102.
22. Butler G, Stergiadis S, Seal C, Eyre M, Leifert C (2011) Fat composition of organic and conventional retail milk in North East England. J Dairy Sci 94: 24–36.
23. Slots T, Butler G, Leifert C, Kristensen T, Skibsted LH, et al. (2009) Potentials to differentiate milk composition by different feeding strategies. J Dairy Sci 92: 2057–2066.
24. Blasbalg TL, Hibbeln JR, Ramsden CE, Majchrzak SF, Rawlings RR (2011) Changes in consumption of omega-3 and omega-6 fatty acids in the United States during the 20th century. Am J Clin Nutr 93: 950–962.
25. Hibbeln JR, Nieminen LRG, Blasbalg TL, Riggs JA, Lands WEM (2006) Healthy intakes of ω-3 and ω-6 fatty acids: estimations considering worldwide diversity. Am J Clin Nutr 83: 1483S–1493S.
26. Kris-Etherton PM, Taylor DS, Yu-Poth S, Huth P, Moriarty K, et al. (2000) Polyunsaturated fatty acids in the food chain in the United States. Am J Clin Nutr 71: 179S–188S.
27. Masters C (1996) Fatty acids and the peroxisome. Mol Cell Biochem 165: 83–93.
28. Ailhaud G, Massiera F, Alessandri J-M, Guesnet P (2007) Fatty acid composition as an early determinant of childhood obesity. Genes Nutr 2: 39–40.
29. Gebauer SK, Psota TL, Harris WS, Kris-Etherton PM (2006) n-3 fatty acid dietary recommendations and food sources to achieve essentiality and cardiovascular benefits. Am J Clin Nutr 83: S1526–15355.
30. Yamagishi K, Iso H, Date C, Fukui M, Wakai K, et al. (2008) Fish, omega-3 polyunsaturated fatty acids, and mortality from cardiovascular diseases in a nationwide community-based cohort of Japanese men and women. J Am Coll Cardio 52: 988–996.
31. De Caterina R (2011) n-3 fatty acids in cardiovascular disease. New Eng J Med 364: 2439–2450.
32. Molkentin J (2009) Authentication of organic milk using $\delta^{13}C$ and the α-linolenic acid content of milk fat. J Agric Food Chem 57: 785–790.
33. Rist L, Mueller A, Barthel C, Snijders B, Jansen M, et al. (2007) Influence of organic diet on the amount of conjugated linoleic acids in breast milk of lactating women in the Netherlands. Brit J Nutr 97: 735–743.
34. Woods VB, Fearon AM (2009) Dietary sources of unsaturated fatty acids for animals and their transfer into meat, milk and eggs: A review. Livestock Sci 126: 1–20.
35. Rinehart L, Baier A (2011) Pasture for organic ruminant livestock: understanding and implementing the national organic program (NOP) pasture rule. Agricultural Marketing Service, U.S. Department of Agriculture. Available: http://www.ams.usda.gov/AMSv1.0/getfile?dDocName = STELPRDC5091036. Accessed 2013May 1.
36. McBride W, Greene C (2009) Characteristics, costs, and issues for organic dairy farming. Economic research report no. 82, Washington, DC: U.S. Department of Agriculture. Available: http://www.ers.usda.gov/publications/err-economic-research-report/err82.aspx#.UZPDOUouffA. Accessed 2013 May 1.
37. Donahue SMA, Rifas-Shiman SL, Gold DR, Jouni ZE, Gillman MW, et al. (2011) Prenatal fatty acid status and child adiposity at age 3 y: results from a US pregnancy cohort. Am J Clin Nutr 93: 780–788.
38. Brostow DP, Odegaard AO, Koh WP, Duval S, Gross MD, et al. (2011) Omega-3 fatty acids and incident type 2 diabetes: the Singapore Chinese Health Study. Am J Clin Nutr 94: 520–526.
39. Moon RJ, Harvey NC, Robinson SM, Ntani G, Davies JH, et al. (2013) Maternal plasma polyunsaturated fatty acid status in late pregnancy is associated with offspring body composition in childhood. J Clin Endocrinol Metab 98: 299–307.
40. Schuchardt JP, Huss M, Strauss-Grabo M, Hahn A (2010) Significance of long-chain polyunsaturated fatty acids (PUFAs) for the development and behavior of children. Eur J Pediatr 169: 149–164.
41. Ryan AS, Astwood JD, Gautier S, Kuratko CN, Nelson EB, et al. (2010) Effects of long-chain polyunsaturated fatty acid supplementation on neurodevelopment in childhood: A review of human studies. Prostag Leukotr ESS 82: 305–314.
42. Kummeling I, Thijs C, Huber M, van de Vijver LPL, Snijders BEP, et al. (2007) Consumption of organic foods and risk of atopic disease during the first 2 years of life in the Netherlands. Brit J Nutr 99: 598–605.
43. Wang L, Folsom AR, Zheng ZJ, Pankow JS, Eckfeldt JH (2003) Plasma fatty acid composition and incidence of diabetes in middle-aged adults. Am J Clin Nutr 78: 91–98.
44. Djousse L, Biggs ML, Lemaitre RN, King IB, Song X, et al. (2011) Plasma omega-3 fatty acids and incident diabetes in older adults. Am J Clin Nutr 94: 527–533.
45. Safarinejad MR, Hosseini SY, Dadkhah F, Asgari MA (2009) Relationship of omega-3 and omega-6 fatty acids with semen characteristics, and antioxidant status of seminal plasma: A comparison between fertile and infertile men. Clin Nutr 29: 100–105.
46. Yehuda S, Rabinovitz-Shenkar S, Carasso RL (2011) Effects of essential fatty acids in iron deficient and sleep-disturbed attention deficit hyperactivity disorder (ADHD) children. Eur J Clin Nutr 65: 1167–1169.
47. Studer M, Briel M, Leimenstroll B, Glass TR, Bucher HC (2005) Effect of different antilipidemic agents and diets on mortality. Arch Int Med 165: 725–730.
48. Ailhaud G, Guesnet P (2004) Fatty acid composition of fats is an early determinant of childhood obesity: a short review and opinion. Obes Rev 5: 21–26.
49. AOAC International (2012) Official methods of analysis of AOAC International, 19th ed., Gaithersburg, MD, U.S.
50. McCullough BD, Heiser DA (2008) On the accuracy of statistical procedures in Microsoft Excel 2007. Comput Stat Data An 52: 4570–4578.
51. Forkman J (2009) Estimator and tests for common coefficients of variation in normal distributions. Commun Stat A-Theor 38: 233–251.
52. Smit LA, Baylin A, Campos H (2010) Conjugated linoleic acid in adipose tissue and risk of myocardial infraction. Am J Clin Nutr 92: 34–40.
53. Li G, Barnes D, Butz D, Bjorling D, Cook M (2005) 10t,12c-conjugated linoleic acid inhibits lipopolysaccharide-induced cyclooxygenase expression in vitro and in vivo. J Lipid Res 46: 2134–2142.
54. (2004) The role of conjugated linoleic acid in human health. Proceedings of a workshop. Winnipeg, Canada, March 13–15, 2003. Am J Clin Nutr 79: 1131S-1220S.
55. U.S. Department of Agriculture (2012) National nutrient database for standard reference, release 25. Available: http://ndb.nal.usda.gov. Accessed 2013 Mar 8.
56. Glasser F, Doreau M, Ferlay A, Chilliard Y (2007) Technical note: estimation of milk fatty acid yield from milk fat data. J Dairy Sci 90: 2302–2304.
57. Rymer C, Givens DI, Wahle KWJ (2003) Dietary strategies for increasing docosahexaenoic acid (DHA) and eicosapentaenoic acid (EPA) concentrations in bovine milk: a review. Nutr Abs Rev Series B, Livestock Feeds and Feeding 73: 9R-25R.
58. Palupi E, Jayanegara A, Ploegera A, Kahl J (2012) Comparison of nutritional quality between conventional and organic dairy products: a meta-analysis. J Sci Food Agric DOI: 10.1002/jsfa.5639.
59. U.S. Department of Agriculture (2008) Dairy 2007, Part II: Changes in the U.S. dairy cattle industry, 1991–2007, USDA-APHIS-VS, CEAH, Fort Collins, Co. #N481.0308. Available: http://www.aphis.usda.gov/animal_health/nahms/dairy/downloads/dairy07/Dairy07_dr_PartII.pdf. Accessed 2013 May 1.
60. Mathews KH, McConnell MJ (2009) Ethanol co-product use in U.S. cattle feeding: lessons learned and considerations. FDS-09D-01, Economic Research Service, U.S. Department of Agriculture. Available: http://usda01.library.cornell.edu/usda/ers/FDS//2000s/2009/FDS-04-17-2009_Special_Report.pdf. Accessed 2013 Oct 10.
61. Butler LJ (2002) The economics of organic milk production in California: a comparison with conventional costs. Am J Alt Agric 17: 83–91.
62. Massiera F, Barbry P, Guesnet P, Joly A, Luquet S, et al. (2010) A Western-like fat diet is sufficient to induce a gradual enhancement in fat mass over generations. J Lipid Res 51: 2352–2361.
63. Meyer BJ, Mann NJ, Lewis JL, Milligan GC, Sinclair AJ, et al. (2006) Dietary intakes and food sources of omega-6 and omega-3 polyunsaturated fatty acids. Lipids 38: 391–398.
64. Wood JAT, Williams JS, Pandarinathan L, Janero DR, Lammi-Keefe CJ, et al. (2010) Dietary docosahexaenoic acid supplementation alters select physiological endocannabinoid-system metabolites in brain and plasma. J Lipid Res 51: 1416–1423.
65. Amminger GP, Schafer MR, Papageorgiou K, Klier CM, Cotton SM, et al. (2010) Long-chain ω-3 fatty acids for indicated prevention of psychotic disorders: a randomized, placebo-controlled trial. Arch Gen Psychiatry 67: 146–154.
66. Ruxton CHS, Calder PC, Reed SC, Simpson MJA (2005) The impact of long-chain n-3 polyunsaturated fatty acids on human health. Nutr Res Rev 18: 113–129.
67. Childs CE, Romeu-Nadal M, Burdge GC, Calder PC (2008) Gender differences in the ω-3 fatty acid content of tissues, Proc Nutr Soc 67: 19–27.
68. Pawlosky R, Hibbeln RJ, Lin Y, Salem N (2003) n-3 fatty acid metabolism in women. Brit J Nutr 90: 993–994.
69. Williams CM, Burdge G (2006) Long-chain n-3 PUFA: plant v. marine sources. Proc Nutr Soc 65: 42–50.
70. Gibson RA, Muhlhausler B, Makrides M (2011) Conversion of linoleic acid and alpha-linolenic acid to long-chain polyunsaturated fatty acids (LCPUFAs), with a focus on pregnancy, lactation and the first 2 years of life. Matern Child Nutr 7 Suppl 2: 17–26.
71. Emkin EA, Adlof RO, Gulley RM (1994) Dietary linoleic acid influences desaturation and acylation of deuterium-labeled linoleic and linolenic acids in young adult males. Biochim Biophys Acta 1213: 277–288.
72. Arterburn LM, Hall EB, Oken H (2006) Distribution, interconversion, and dose response of n-3 fatty acids in humans. Am J Clin Nutr 83: 1467S–1476S.

73. Burdge GC, Calder PC (2005) α-Linolenic acid metabolism in adult humans: the effects of gender and age on conversion to longer-chain polyunsaturated fatty acids. Eur J Lipid Sci Technol 107: 426–439.
74. Brenna JT (2002) Efficiency of conversion of alpha-linolenic acid to long-chain n-3 fatty acids in man. Curr Opin Clin Nutr Metab Care 5: 127–132.
75. Burdge GC, Calder PC (2005) Conversion of alpha-linolenic acid to longer-chain polyunsaturated fatty acids in human adults. Reprod Nutr Dev 45: 581–597.
76. Gebauer SK, Chardigny J-M, Jakobsen MU, Lamarche B, Lock AL, et al. (2011) Effects of ruminant *trans* fatty acids on cardiovascular disease and cancer: a comprehensive review of epidemiological, clinical, and mechanistic studies. Adv Nutr 2: 332–354.
77. U.S. Food and Drug Administration (2003) Regulation 21 CFR 101.9(c)(2)(ii) "*Trans* fat" or "*Trans*." Federal Register 68: 41502. Available: http://web.archive.org/web/20070103035701/http://www.cfsan.fda.gov/acrobat/fr03711a.pdf. Accessed 2013 May 15.
78. Lacroix E, Charest A, Cyr A, Baril-Gravel L, Lebeuf Y, et al. (2012) Randomized controlled study of the effect of a butter naturally enriched in trans fatty acids on blood lipids in healthy women. Am J Clin Nutr 95: 318–325.

22

A Multi-Criteria Index for Ecological Evaluation of Tropical Agriculture in Southeastern Mexico

Esperanza Huerta[1]*, Christian Kampichler[2,3], Susana Ochoa-Gaona[4], Ben De Jong[4], Salvador Hernandez-Daumas[1], Violette Geissen[5,6]

1 El Colegio de la Frontera Sur, Unidad Campeche, Dpto. Agroecología, Campeche, México, **2** Universidad Juárez Autónoma de Tabasco, División de Ciencias Biológicas, Villahermosa, Tabasco, México, **3** Sovon Dutch Centre for Field Ornithology, Natuurplaza (Mercator 3), Nijmegen, The Netherlands, **4** El Colegio de la Frontera Sur, Unidad Campeche, Dpto. Sustainability Sciences, Campeche, México, **5** University of Bonn - INRES, Bonn, Germany, **6** Wageningen University and Research Center – Alterra, Wageningen, Gelderland, Netherlands

Abstract

The aim of this study was to generate an easy to use index to evaluate the ecological state of agricultural land from a sustainability perspective. We selected environmental indicators, such as the use of organic soil amendments (green manure) versus chemical fertilizers, plant biodiversity (including crop associations), variables which characterize soil conservation of conventional agricultural systems, pesticide use, method and frequency of tillage. We monitored the ecological state of 52 agricultural plots to test the performance of the index. The variables were hierarchically aggregated with simple mathematical algorithms, if-then rules, and rule-based fuzzy models, yielding the final multi-criteria index with values from 0 (worst) to 1 (best conditions). We validated the model through independent evaluation by experts, and we obtained a linear regression with an $r^2 = 0.61$ ($p = 2.4e\text{-}06$, $d.f. = 49$) between index output and the experts' evaluation.

Editor: Yong Deng, Southwest University, China

Funding: Financial support was obtained from the Conacyt-SEMARNAT Project "Uso sustentable de los recursos naturales en la frontera sur de México" (Sustainable use of natural resources on the southern border of Mexico) (code SEMARNAT-2002-C01-1109). El Colegio de la Frontera Sur provided infrastructural resources. The funders had no role in study design, data collection and analysis, decision to publish, or preparation of the manuscript.

Competing Interests: The authors have declared that no competing interests exist.

* Email: ehuerta@ecosur.mx

Introduction

In the past 60 years, degradation and deforestation of tropical forests worldwide have occurred much faster and more extensively than in any other period in history [1], [2]. Furthermore, countries like Mexico have been undergoing drastic land use changes. In the tropics of Mexico large parts of its lowland rainforest areas have been converted into pasture and cropland. In the state of Tabasco, for example, only 3.4% of the state is covered with original forest [3], whereas 76.4% of the surface was used for cattle production in 2000 [4] and 15.6% was used for agriculture, principally sugarcane and fruit plantations [3]. The ecological consequences of these land-use changes in Tabasco are well documented. Soil losses in hilly regions are very high up to 200 t ha^{-1} $year^{-1}$ [5]; high pesticide and fertilizer inputs to crops that have replaced forests have caused considerable environmental contamination [6], [7], and soil fertility is decreasing [8], [5], [9].

It is of the utmost importance to identify sustainable land use strategies which are economically attractive for the region's farmers and which may also reconcile the need for food production with that of soil conservation. In order to assist a variety of stakeholders at the local and regional level in making land use decisions, simple evaluation tools are needed. This is even more needed since a high percentage of the population consists of immigrants from other Mexican states, who are unfamiliar with the conditions of the humid tropics and use intensive techniques for farming the land. This is mostly due to large-scale agricultural development projects, such as "Plan Balancán-Tenosique," named after the two municipalities involved, in which, in the 1970 s, over 1100 km^2 of lowland rain forest was destroyed and converted into crop and pasture land. Ecological values of the 1970 s were very different from those of the present, and government representatives were willing to deforest in order to grant land to farmers [10].

The direct impact of farming is difficult to measure due to methodological difficulties (impossibility of measurement, complexity of the system) or practical reasons (time, costs) [11]. Therefore, the use of indicators appears to be an alternative way of guiding land use decisions [12], [13]. However, the "indicator explosion" [14], that is, the use of an exaggerated number of indicators aimed at assessing environmental impacts of agricultural activities, has been of little use to local decision-makers. Particularly in the tropics, land use decisions are still based on the informal opinion of local experts rather than on implementation of Decision Support Systems for environmentally sound resource management [15]. On the one hand, this is due to farmers' restricted access to modern communications and information technologies; on the other, application of indicators is often beyond the capability of local farmers. For example, the agricultural sustainability index proposed by Nambiar et al. [16], which aims to measure agricultural sustainability as a function of biophysical, chemical, economic, and social indicators, would require considerable training of government stakeholders and farmers in order to be applied, and such training is rarely available.

Thus, any tool which local farmers or regional decision makers may use to support their decision making must be as simple as possible. Agroecosystems (like any other ecosystem) are too complex to be precisely measured and evaluated [17]. We agree with the view of Darnhofer et al. [18] in favour of developing less precise rules of thumb which may be used by farmers as well as to guide local land-use decisions toward a more environmentally friendly system of agriculture.

This index may provide farmers and others involved with a tool to evaluate the sustainability of their management of crop land which is oriented toward diminishing soil damage and conserving soil fertility. The index, exclusively based on terms which describe the environmental conditions of the crop system, is accessible to most farmers, and may be calculated using a simple internet application. In this paper we present a simple, easy-to-use index in order to evaluate the ecological state of farms in south-eastern Mexico. We applied the indicator system to 52 crop production systems in south-eastern Mexico and compared the results of the indicator system with expert opinions.

Human knowledge of how to efficiently and sustainably manage complex systems (including agricultural systems) is incomplete and much of what is thought to be known about this topic is actually incorrect. Yet, decisions must be made by policy makers, agricultural extension agents, and farmers despite uncertainty and knowledge gaps [19]. Therefore, tools to support local decision-makers must be flexible, should not enter into too much detail or precision, and should allow for an adaptive strategy which promotes "learning through management" [19]. Consequently, our rationale for developing an index which aids farmers in making environmentally friendly land-use decisions is based on basic, simplified ecological concepts, i.e. the presence of trees, since trees within an agroecosystem enhance soil microclimate in terms of radiation partitioning (shading), evapotranspiration partitioning, and rain interception/redistribution [20]. These factors all help to retain soil moisture. Branches, bark, roots, and living and dead leaf surfaces provide shelter [21] for soil micro-, meso-, and macro-invertebrates. Tree cover for instance, enhances above- and below-ground diversity, serving to support agricultural sustainability [22].

Materials and Methods

1. Rationale of index composition

We define conventional agriculture as a cropping system, typically promoted by government development programs, that is "capital-intensive, large-scale, highly mechanized agriculture with monocropping and extensive use of synthetic fertilizers, and pesticides" [23], [24]. Furthermore, we acknowledge that farming systems are sustainable only if "they minimize the use of external inputs and maximize the use of internal inputs already present on the farm" [25], [26]. The strategy most frequently linked to sustainability is reduction or elimination of agrochemicals, particularly chemical fertilizers and pesticides [25], [27], [28], [29], [30], [31]. Another key to sustained productivity of agricultural systems is the maintenance of soil functions, such as organic matter and nutrient cycling [32], based on organic inputs [33], above-and below-ground biodiversity [22], and diversifying crop systems with nitrogen-fixing legumes [34]. The principal role of the index we propose is to characterize methods of tillage, external inputs, and crop structure.

2. Primary indicators

We chose 12 field variables as primary indicators related to the above mentioned aspects of ecologically sound agricultural land use based on farmer's practices. These are easy to evaluate in the field and characterize plot structure (primary indicators: tree cover, tree density, tree diversity), crop structure and crop conditions (primary indicators: crop type, crop rotation, crop density, crop colour), tillage (primary indicators: type of tillage, timing of tillage), the use of fertilizers (chemical versus organic) and pesticide application.

2.1 Tree cover. Tree cover is defined as the canopy of trees, measured in the field, and recorded as percentage classes of tree cover in three height classes (trees >15 m, 10–15 m, <10 m). Thus, this variable characterized one aspect of agroecosystem management: the farmers' decision to maintain the canopy of the trees in his or her agroecosystem.

2.2 Tree density. Tree density is defined as the number of trees per area. To measure this, we distinguished three categories of tree density: high density (abundant), medium density, and low density (isolated or no trees). A high number of trees per area guarantee carbon sequestration [35] while a stable microclimate is maintained. This variable, measured in the field, is one the variables that characterize the effect of trees in the agroecosystem.

2.3 Tree diversity. This variable was measured in the field by counting the number of trees species within the plot. In agroecosystems, biodiversity may; (i) contribute to constant biomass production and reduce the risk of crop failure in unpredictable environments, (ii) restore disturbed ecosystem services such as water and nutrient cycling, and (iii) reduce risks of pests and diseases through enhanced biological control or direct pest control [36], [20], [22].

2.4 Crop type. This variable indicates whether the crop is annual, seasonal, or perennial. We obtained this information by observing the type of crop. Annual crops in general have higher environmental impacts, ie: greenhouse emissions, and nutrient leaching, than perennial crops [37].

2.5 Crop rotation. In sustainable farm systems leguminous crops are increasingly used in crop rotations as a source of nutrients, particularly nitrogen for crop growth [38], [39], nitrogen-fixing legumes, contribute to maintaining biodiversity above and in the soil, contribute nitrogen to the soil/plant system, and help avoid the build-up of pest populations [34]. In this study, we asked farmers whether they planted another crop before planting the main crop and whether they practice crop rotation, as crop rotation may assist with weed and pest control [40], [34]. According to Bellon [41], an activity which leads into the maintenance or increase of renewable resources in agroecosystems, is considered as an ecological technology. This variable helped us to characterize the technology used in the agroecosystem.

2.6 Crop density. Crop density is defined as the number of plants (individuals) per area. Three categories were recorded in the field: abundant (high density: 3,000 plants/ha), medium density (1000–1600 plants/ha), and sparse (<1000 plants/ha). This variable also indicated the level of technology applied to the crop, as less intensive techniques typically yield lower densities [42].

2.7 Crop colour. The colour of a crop indicates the nutritional status of the plants; green plants generally have sufficient nutrients, while yellow plants lack nitrogen [43].

2.8 Type of tillage. Type of tillage was categorized into no tillage, manual tillage, and mechanical tillage using machinery, the latter of which generally indicates high disturbance of the soil surface and rapid loss of soil organic carbon and other nutrients

[44]. In the field, we asked the farmers how they prepared the land for crop planting.

2.9 Frequency of tillage. This variable indicates frequency of tillage - every year or every 2 years. With this information, it was possible to estimate the frequency of soil disturbance due to tillage.

2.10 Chemical input. Searching for sustainable productions, it is recommended no or low or use of inorganic fertilizers and pesticides [45], [46]. Long term use of some pesticides, as glifosate can decrease earthworm species number, density and biomass [47], [48]. This information allowed us to estimate the amount of chemical fertilizers and pesticides applied within a given area. These variables (chemical fertilizers and pesticides) are included in the indicator for chemical disturbance.

2.11 Green manure. Is defined as the presence or absence of leguminous crops mixed with the principal crop, generally used to increase total soil nitrogen content. Green manures should always be intercropped, as it has been proven that growing legumes with cereal crops decreases N_20 emissions [49], and therefore is a sustainable, environmentally friendly practice. Examples of green manure use have been observed in traditional Mesoamerican cultures, for example, intercropping beans, as well as other edible plants, within the *milpa* or traditional maize cropping system [50].

3. Index development

Primary indicators were hierarchically aggregated into higher levels, forming intermediate variables, which in turn are structured into a single index that evaluates the ecological condition of a given plot on a scale from 0 (worst) to 1 (best). An index close to zero either would mean that a more environmentally friendly farming techniques need to be implemented or that the plot should be subjected to a fundamental change in land-use, e.g., reforestation, in order to return to a more ecologically sound state. An index close to one, on the other hand, would indicate an ecologically sound land-use. Methods applied in indicator aggregation were (i) simple mathematical operations, ii) sets of if-then rules, and (iii) sets of if -then rules combined with fuzzy logic.

3.1. Aggregation through mathematical operations. Mathematical operations include calculating averages, weighed averages, minimum values, and maximum values. For example, if primary indicator A has the value a and primary indicator B has the value b, then the value x of the intermediate variable X is determined as $x = (a+b)/2$, $x = (w_1*a+w_2*b)$, where $w1$ and $w2$ are weights, or $x = \min(a, b)$ or $\max(a, b)$.

3.2. Aggregation by IF-THEN rules. If primary indicator A can have the discrete values a_1, a_2, and a_3, and primary indicator B can have the discrete values b_1, b_2, and b_3, then the value x of the intermediate variable X is determined by a set of nine (number of levels of A x number of levels of B) rules. For example, IF $A = a_1$ and $B = b_1$ THEN $X = x_1$.

3.3. Aggregation by IF-THEN rules and fuzzy logic. We used small fuzzy rule-based models for aggregation in the case of non-linear interactions among indicators using a continuous numerical scale or an ordinal scale with a large number of possible values. In classic set theory, an object can either be a member (membership = 1) or not (membership = 0) of a given set. The central idea of fuzzy set theory is that an object may have a partial membership of a set, which consequently may possess all possible values between 0 and 1. The closer an element is to 1, the more it belongs to the set; the closer the element is to 0, the less it belongs to the set. To apply the fuzzy set theory, three steps are involved in calculating the model's output. First, for any observed value of the primary indicators, its corresponding membership value in the fuzzy set domain is calculated (*fuzzification*); second, the memberships of the intermediate variable X are calculated, applying the rules in the fuzzy set theory (*fuzzy inference*); third, the fuzzy results are converted into a discrete numerical output (*defuzzification*; see Wieland 2008 for an introduction to fuzzy models). Fuzzy rule-based models have become popular in ecological modelling [51], [52], and several examples exist of its usefulness in the context of ecosystem evaluation, bioindication, and sustainable management [15], [53], [54]. Here, if both primary indicators A and B can have numerical values from 0 to 1, then the value x of the intermediate variable X is determined by a series of fuzzy set rules representing the linguistic variables "low A", "medium A", "high A", and "low B", "medium B", and "high B", as well as the output "low X", "medium X", and "high X". The value of the intermediate variable X is determined by nine levels of A x the number of levels of B; for example: If A = low and B = low Then X = [low, medium, or high].

To maintain the number of rules as well as their complexity as low as possible, we aggregated only two variables at a time. A simple example shows the reasoning behind this decision; if there are three primary variables, A, B, and C with three categories for each variable, a single rule node requires 3*3*3 = 27 If-Then rules of the type "If $A = a_1, a_2$, or a_3 and If $B = b_1, b_2$, or b_3 and if $C = c_1, c_2$, or c_3 Then...", whereas if an additional intermediate variable X is introduced to the model, only 18 rules are needed: 3*3 = 9 rules to aggregate A and B to X (If $A = a_1, a_2$, or a_3 and If $B = b_1, b_2$, or b_3), and 3*3 = 9 rules to aggregate X and C (If $X = x_1, x_2$, or x_3 and If $C = c_1, c_2$, or c_3).

4. Study area and application of the index

The state of Tabasco in south-eastern Mexico is characterized by a humid tropical climate with a mean annual rainfall between 1200 and 4000 mm and a mean annual temperature of 27°C [55]. Predominant soils are Gleysols and Fluvisols over alluvial sediments in the plains, Vertisols, Cambisols, Luvisols, and Acrisols over Miocene or Oligocene sediments, and Leptosols and Regosols over limestone mountains [9], [56]. We chose the municipalities Balancán and Tenosique in western Tabasco (17°81′50″–18°81′00″ N, 91°80′10″–91°84′60″ W) as a study area (Figure 1), we worked with private, *ejidal* (multipurpose land, where owners can or cannot sell the property according to their legal status in the National Agrarian File), and communal lands (coordinates of each plot are shown in Table 1). The region is mainly a plain towards the North (67% of area has an elevation <20 m. a. s. l.) with hills (29%, 20–200 m. a. s. l.), and mountains (4%, max. 640 m. a. s. l.) in the South, comprising a total area of 5474 km^2. These municipalities have undergone a high degree of land use change over the past 40 years. Until the early 1970 s, this region was still covered by lowland rain forest. The principal form of land use is pastureland, and covers 60% of the land [57]. An additional 30% is cropland, mainly cultivated under small-medium holder systems with seasonal conventional agriculture using high levels of agrochemical inputs (Table 2 & 3). Common crops are maize (*Zea mays*), a variety of hot peppers (*Capsicum* sp), cucumbers (*Cucurbita argyrosperma*), watermelon (*Citrullus lanatus*), perennial fruit crops such as papaya (*Carica papaya*), and biannual crops such as sugar cane (*Saccharum officinarum*) [58].

We chose 52 farms in the study area (Table 1), and selected those farms whose main economic activity (100%) is agriculture and chose one agricultural plot from each farm for evaluation. There were annual and biannual crops with or without trees (Table 2). Average plot size was 32.4±55.1 ha, and the average time that the plot had been used for a given crop system was 2.5±3.0 years.

Table 1. Characterization of plots.

Plot #	Municip	Lat	Long	Altitud (mls)	Plot size (ha)	Type of property	Original-vegetation
1	Bal	2007880	652445	43	18	Ejidal	A
2	Bal	1964368	642505	10	0.5	Private	TRF
3	Bal	1970936	647701	11	160	Private	TRF
4	Bal	1957459	667496	14	2	Ejidal	TRF
5	Bal	1964107	660656	14	7	Private	PT
6	Bal	1960297	658989	22	15	Private	TRF
7	Bal	1976300	676424	40	30	Private	TRF
8	Bal	1990326	665947	40	28	Private	TRF
9	Bal	1964590	667264	11	3	Private	RV
10	Bal	1972260	672382	28	2.5	Ejidal	TRF
11	Bal	1950339	666409	49	40	Private	TRF
12	Bal	1966112	682994	40	80	Private	PT
13	Ten	1931754	704474	65	14	Private	TRF
14	Ten	1912951	690381	130	2	Communal	TRF
15	Ten	1932644	665521	15	3	Ejidal	TRF
16	Ten	1926163	678608	56	68	Private	TRF
17	Ten	1933663	674227	59	35	Private	A
18	Ten	1926570	669486	35	8.5	Ejidal	TRF
19	Ten	1942431	663885	21	1.5	Ejidal	TRF
20	Ten	1942068	671156	31	7.25	Ejidal	TRF
21	Ten	1943167	650736	82	14	Ejidal	A
22	Ten	1923838	669841	118	0. 25	Private	TRF
23	Bal	1973301	695548	105	40	Ejidal	TRF
24	Bal	1982392	701383	74	20	Ejidal	A
25	Bal	1965354	709053	44	11	Ejidal	A
26	Bal	1960496	706464	51	44	Ejidal	TRF
27	Bal	1942484	709731	66	10	Ejidal	A
28	Ten	1944148	673673	16	320	Ejidal	TRF
29	Ten	1930350	659650	29	20	Private	TRF
30	Ten	1924549	656910	41	53	Ejidal	TRF
31	Ten	1926226	652978	65	3	Ejidal	TRF
32	Bal	1955194	688843	45	44	Private	TRF
33	Bal	1960980	679265	49	29	Ejidal	TRF
34	Bal	1942552	696222	48	10	Private	TRF
35	Ten	1938120	686470	49	5	Private	TRF

Table 1. Cont.

Plot #	Municip	Lat	Long	Altitud (mls)	Plot size (ha)	Type of property	Original-vegetation
36	Bal	1968750	695595	60	2	Private	A
37	Bal	1977546	699417	53	200	Ejidal	TRF
38	Bal	1949770	708854	53	23	Ejidal	TRF
39	Bal	1951216	697890	46	2	Ejidal	TRF
40	Ten	1931041	667544	31	17	Private	TRF
41	Ten	1927447	664223	3	3	Ejidal	TRF
42	Ten	1928226	664551	31	3	Ejidal	TRF
43	Ten	1930579	691785	54	15	Communal	TRF
44	Ten	1931219	677895	59	63	Private	TRF
45	Ten	1931525	676500	60	6.75	Ejidal	TRF
46	Ten	1908445	675040	140	40	Ejidal	TRF
47	Ten	1910875	669750	222	45	Ejidal	TRF
48	Ten	1924222	704702	42	30	Ejidal	A
49	Ten	1935076	712227	88	20	Ejidal	TRF
50	Ten	1914876	686285	113	4	Ejidal	TRF
51	Ten	1911365	684878	389	22	Ejidal	TRF
52	Ten	1911503	698606	139	10	Ejidal	TRF

Municip: municipality, Bal: Balancan, Ten: Tenosique. TRF: tropical rain forest, PA: pasture, A: acahual (secondary vegetation). Ejidal: multipurpose land, where owners can or cannot sell the property according to their legal status in the National Agrarian File.

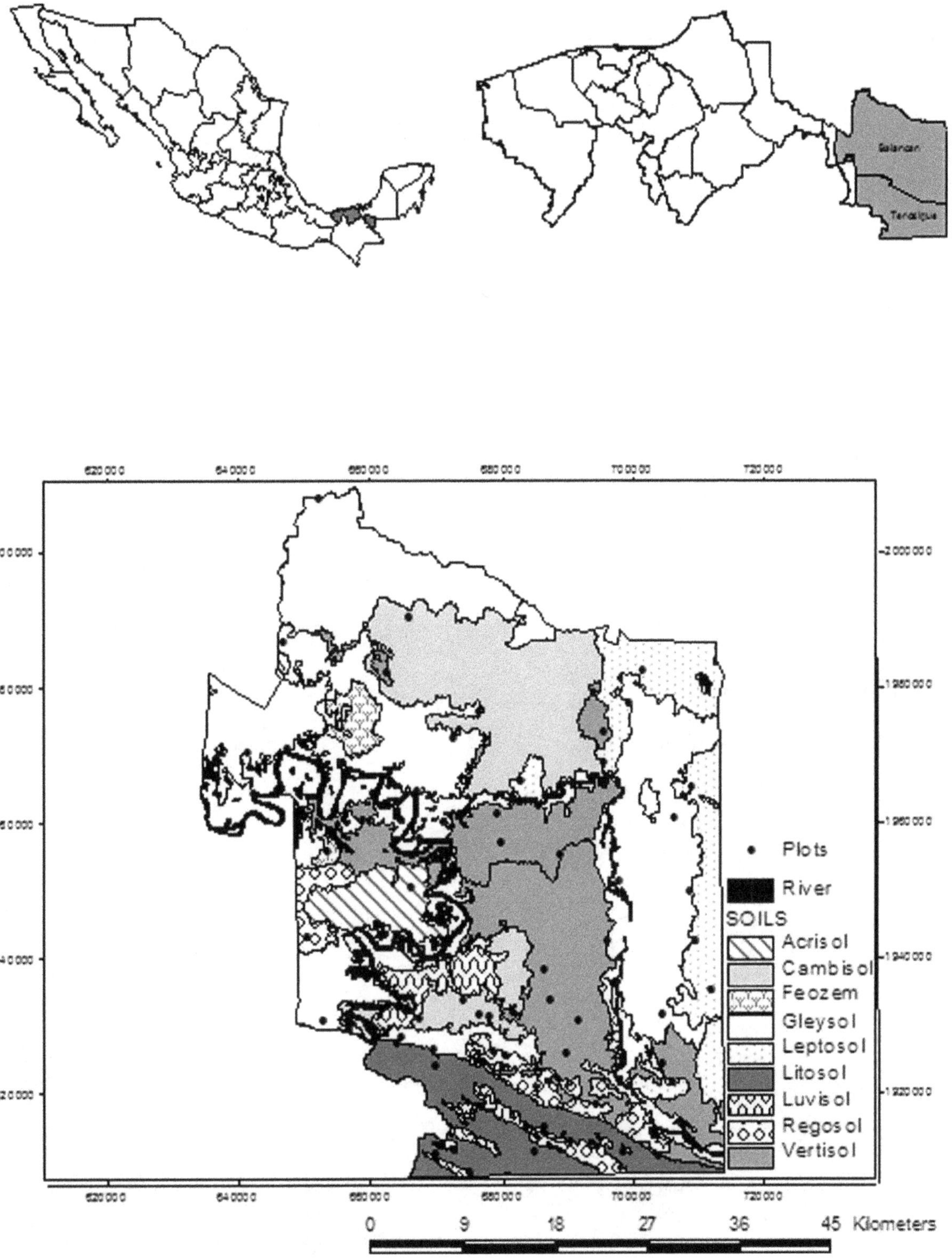

Figure 1. Distribution of 52 evaluated agricultural plots in tropical South-East Mexico.

We questioned farmers as to frequency, amount and type of chemical fertilizers and pesticides applied per area of cropland. After their verbal consent, farmers gave us their complete name and signed next to their name on the record sheet, validating all the information. Due to the fact that we only asked about the use of fertilizers and management of the land, the procedure approval from the Ethics Committee was not required.

Between March and October 2004, for each plot, the values of the primary indicators were determined and the index of ecological condition was calculated. Prior to index calculation, the plots were also evaluated by experts (2 scientists, each with

Table 2. Crop characterization.

Plot	Cycle	Main seasonal or perennial crop	Trees
1	s-s	*Zm, Pv, Cs, Cl*	*Tr, Pa, Cpa*
2	s-s	*Zm, Pv, Cs, Cl*	*Sh, Dg*
3	w-s	*Sv*	*Sa*
4	a-w	*Zm, Pv, Cpa*	*0*
5	Y	*Zm, Pv, Cl, Cm*	*0*
6	Y	*Zm*	*Cpa, Eg, Tr*
7	s-a	Ca	Tr
8	Y	*Cp*	*Cp*
9	Y	*Zm*	*0*
10	Y	*Zm*	*0*
11	w-s	*Zm, Pv, Cl*	*Sm, Cpa*
12	Y	*Co*	*Co*
13	s-s	*Zm*	*0*
14	s-s	*Le, Se*	*0*
		Cpa, Zm, Pv	
15	s-s	*Cl, Ta, Mp,*	*Fruit trees*
16	Y	*Jj*	*Gs, Hc, Cc, Tr*
17	Y	*So*	*0*
18	Y	*So*	*0*
19	Y	*So*	*0*
20	Y	*Zm*	*Mi*
21	s-s	*Zm*	*0*
22	Y	*Csi, Cli, Js*	*Co, Pa, Bc, Cn, Mi*
23	s-s	*Zm, Cpa*	*Cpe*
24	s-s	*Zm*	*Cpa*
25	s-s	*Zm*	*0*
26	s-s	*Zm, Cpa*	*0*
27	s-s	*Zm, Ca*	*Sm*
28	Y	*So*	*0*
29	s-s	*Zm, Pv, Cs, Cl*	*0*
30	Y	*Zm, Pv, Cs, Cl, Ta*	*0*
31	Y	*Zm, Pv, Cs*	*0*
		Cl, Os, Le	
32	s-s	*Zm, Pv, Cpa, Cl, Os*	*Tr*
33	Y	*Zm, Pv*	*Sm, Bg*
34	s-s	*Zm, Pv, Os*	*Cpa*
			Tr
35	Y	*Zm, Cl*	*Tr*
36	Y	*Zm, Pv, Cpa, Cl, Ca*	*0*
37	s-s	*Zm, Pv, Ca*	*Sm, Pa, Sh*
38	Y	*Zm, Pv, Cpa, Cl, Ta*	*Co, Sh*
39	Y	*Zm, Pv, Cpa, Cl, Ca*	
		Cs, Cm	
40	Y	*So*	*Tr, Cpe*
			Co, D a
41	Y	*So*	*Tr, Cpe*
			Co, D a
42	Y	*So*	*Tr, Cpe*
			Co, D a

Table 2. Cont.

Plot	Cycle	Main seasonal or perennial crop	Trees
43	Y	*So*	*Tr, Cpe*
			Co, Da
44	Y	*So*	*Tr, Cpe*
			Co, Da
45	Y	*So*	*Tr, Cpe*
			Co, D a
46	Y	*Zm, Pv, Cpa, Me*	*Sm*
47	Y	*Zm, Pv, Me, Ib*	*Bc, Mi, Dg, Cs, Ll*
48	Y	*Zm, Pv, Cpa, Cl*	*0*
49	Y	*Zm, Pv, Ca*	*Tr, Gs*
50	Y	*Zm, Pv, Cpa, Cl*	*Sm, Co, Ma*
		Cs, Ib, Me	0
51	Y	*Zm, Pv, Cpa, Cs*	*Fruit trees*
52	Y	*Zm, Pv, Cpa, Cl, Cs*	*Co, Sh, Bc*
		Mp, Ca, Me	0

Cycle: s–s: spring-summer, w-s: winter-summer, s-a: summer-autumn, Y: all the year. **Main seasonal or perennial crop**: Ca: *Cucurbita argyrosperma* (cushaw pumpkin); Cp: *Carica papaya* (papaya); Cl: *Citrullus lanatus* (watermelon); Cli: *Citrus limon* (lemon); Cm: *Cucumis melo* (muskmelon); Cpa: *Cucurbita pepo* (squash); Cs: *Capsicum* sp. (pepper); Csi: *Citrus sinensis* (orange tree); Ib: *Ipomoea batatas* (sweet potato); Jj: *Jarthropha jurcas* (oil palm); Js: *Jobo spondia* (plum); Le: *Lycopersicon esculentum* (tomato); Me: *Manihot esculenta* (cassava); Mp: *Musa paradisiaca* (banana); Os: *Oryza sativa* (rice); Pv: *Phaseolus vulgaris* (bean), So: *Saccharum officinarum* (sugarcane); Se: *Sechium edule* (pear squash); Sv: *Sorghum vulgare* (milo); Ta: *Triticum aestivum* (wheat); Zm: *Zea mays* (maize). **Trees**: *Bc: Byrsonima crassifolia* (nanche); *Cc: Crescentia cujete* (calabash tree); *Co: Cedrela odorata* (Mexican cedar); *Cpa: Carludovica palmate* (toquilla palm); *Cpe: Ceiba pentandra* (kapok); Cn: *Cocos nucifera* (Coconut); *Dg: Dialium guianense* (wild tamarind); Eg: *Eucalyptus grandis* (Eucalyptus); *Gs: Gliricidia sepium* (Cocoite, gliricidia, cacao de nance); *Ll: Leucaena leucocephala* (white lead tree, jumbay); *Ma: Mammea americana* (mamey apple); *Mi: Mangifera indica* (mango); *Pa: Persea americana* (avocado); *Sa: Sterculia apetala* (camoruco, manduvi tree); *Sh: Swietenia humilis* (small mahagoni); *Sm: Spondias mombin* (Yellow plum, bai makok); *Tr: Tabebuia rosea* (savannah oak, macuilis).

over 15 years of experience in agroecology) on a scale from 0 (poor condition) to 1 (good condition) in order to test the correlation between the experts' opinions and the index.

Data was normalized for carrying out multiple regression with plot size, altitude of plot site, or previous vegetation cover and the ecological index.

Finally we located the index in this public web address: http://201.116.78.102/~modelo/Index.html, where farmers in the near future can introduce independent data sets and evaluate or monitor their own agro ecosystems.

Results

1. Index structure

Ten nodes were used to aggregate primary indicators to the final index of ecological conditions of agricultural systems (Figure 2). Six of these used simple mathematical operations; two were based on rule sets, and two on fuzzy rule-based models (Table 4).

2. Index application

2.1. General characterization of sampled plots. A total of 67% of the plots were cultivated after cutting of primary lowland forest, 24% after cutting secondary forests of various ages, and 1.5% after cutting riparian vegetation, the rest conversion from pastureland (livestock production) to crop system. On 63% of the farms, maize, beans and pumpkins were cultivated, 13% sugar cane, 12% watermelon, 9% rice and 3% pepper. On 68.5% of the plots, trees were scattered among the crops. Conventional tillage was used on 55.5% of the plots, and pesticides were used on 79.4%, green manure was used on 21% of the plots and chemical fertilizers on 67%. All farmers understood the meaning of the variables evaluated for their plots (for a complete list of plots, see Table 5).

2.2. Plot evaluation with the index. Ecological condition of the plots ranges from 0.0 to 0.8125 (Figure 3; Table 5). One plot which was intercropped with timber and fruit trees presented the highest value. This site was characterized by an absence of agrochemical use, use of green manure, presence of annual crop rotation, and high tree diversity. We found 18 plots with index values of 0–0.25, 7 plots with values of 0.3–0.5, and 26 plots with values of 0.5–0.7. Thus, the majority of the plots evaluated in this study had an intermediate index value. 50% of the plots with this intermediate index are *ejidal* property, 38% private land, and 12% is under another type of ownership (smallholder or communal). 80% of these intermediate plots were lowland rain forest before being converted into agricultural land, 12% were secondary vegetation, and 8% were pastureland (Table 1). One might believe that a prior forest condition implies a more environmentally friendly agroecosystem that allows for preserving a more diverse system. However, plots with low index values were also lowland rain forest before being turned into agricultural land (Table 5). In this case, the land managers or owners decided to deforest the land to subsequently plant annual crops. 13 plots were larger than 40 ha (Table 1). All of these were previously covered with lowland rain forest; the ecological index ranges from 0.5–0.56 for 46% of these plots, and another 23% have an ecological index of 0.37–0.43. 10 plots fell under the smallest size category (0.25–3 ha) and had an ecological index of 0.5–0.68 (4 plots), 0 to 0.18 (5 plots), and 0.39 (1 plot).

Carrying out multiple regression with normalized data, we did not observe significant correlations between plot size, altitude of

Table 3. Crop land preparation and inputs characterization.

Plot	Tillage	Chemical Fertilization	Pesticides Use	Gm	Oa
1	Ma & Me	NPK 17:17:17 & U	Endosulfan**	0	0
2	Ma	0	Chlorpyrifos*	0	0
3	Me	Pholiar	0	0	c s m
4	Ma	0	0	0	0
5	Ma & Me	NPK 17:17:17	Chlorpyrifos**	Y	Y
6	Ma & Me	NPK 17:17:17	Chlorpyrifos*	0	0
7	Me	0	ID	0	0
8	Me	ID	ID	Y	0
9	Ma & Me	0	Carbofuran*	0	0
10	Me	NPK 18-46-0	ID	0	0
11	Me	NPK 18- 46- 0, U & P	ID	Y	0
12	Me	NPK 18-46-0	ID	0	0
13	Me	NPK 17:17:17 & U	Chlorothalonil**	0	0
			Chlorpyrifos*		
14	Ma	0	Endosulfan*	0	0
15	Me	0	0	Y	0
16	Ma	Urea	0	0	0
17	Me	NPK 19-19-19	ID	0	0
18	Me	NPK 19-19-19	ID	0	0
19	Me	Urea	ID	0	bg
20	Me	0	Methilic*	0	0
21	Me	NPK 17:17:17 & U	Zeta**	0	0
22	Ma	0	0	0	0
23	Me	NPK 17:17:17 & U	Chlorpyrifos**	0	lcf
24	Me	Urea & P	Zeta*	Y	lcf
25	Me	Urea	(2,4-D AMINA)*	Y	0
26	Me	NPK 17:17:17 & U	(Z)-(1R,3R)**	0	0
27	Me	0	Zeta*	N	0
28	Me	NPK 17:17:17	(RS)*	0	0
29	Me	NPK 17:17:17 & U	Chlorpyrifos**	0	0
30	Ma	0	Chlorpyrifos**	Y	0
31	Me	NPK 17:17:17 & U	Chlorpyrifos*	0	0
32	Me	0	0	Y	0
33	Me	0	0	0	0
34	Me	0	0	0	0
35	Me	NPK 17:17:17 & U	0	0	0
36	Me	NPK 17:17:17 & U	Chlorpyrifos*	Y	0
37	Me	0	Urea	0	0
38	Me	NPK 17:17:17 & U	Carbofuran*	Y	0
39	Me	NPK 17:17:17 & U	Chlorpyrifos**	0	0
40	Me	NPK 17:17:17	(Z)-(1R,3R)*	0	0
41	Me	NPK 17:17:17 & U	(Z)-(1R,3R)*	0	0
42	Me	NPK 17:17:17 & U	(Z)-(1R,3R)*	0	0
43	Me	NPK 17:17:17 & U	(Z)-(1R,3R)*	0	0
44	Me	NPK 17:17:17 & U	(Z)-(1R,3R)*	0	0
45	Me	NPK 17:17:17 & U	(Z)-(1R,3R)*	0	0
46	Ma	0	0	Y	0
47	Me	U	0	0	0
48	Ma &Sb	Urea	Chlorpyrifos**	0	0

Table 3. Cont.

Plot	Tillage	Chemical Fertilization	Pesticides Use	Gm	Oa
49	Me	Urea & P	Zeta*	0	0
50	Ma	Urea	Endosulfan*	0	0
51	Ma	0	ID	0	0
52	Me	NPK 17:17:17 & U	Chlorpyrifos*	0	0

Tillage: Ma: Manual; Me: Mechanized. **Pesticides**: Endosulfan: 6,7,8,9,10,10-Hexachlor- 1,5,5a,6,9,9a-Hexahidro-6,9-metane-2,4,3-benzodioxatiepin-3-oxide; Chlorpyrifos: 0,0-dimetil 0-(3,5,6-trichlore-2-piridinil) fosforotioate (33.8%), Permetrine: 3-fenoxibenzil (1RS)-cis, trans-3-(2,2 diclorovinil)-2,2 dimetil ciclopropane-carboxilate (4.8%); Carbofuran: 2,3 Dihidro-2,2-dimetil-7-benzofuranil metil carbamate; Chlorothalonil: Tetrachloroisoftalonitrile; Zeta: Zeta-cipermetrine a-ciano-3-(fenoxifenil) metil (±) cis-trans; (Z)-(1R,3R): (Z)-(1R,3R)-3-(2-chloro-3,3,3-trifluoroprop-1-enil)-2,2-dimetilcichlopropanecarboxilate de (S)-α-ciano-3-fenoxibencile & (Z)-(1S,3S)-3-(2-chloro-3,3,3-trifluoroprop-1-enil)-2,2-dimetilcichlopropanecarboxilate de (R)-α-ciano-3-fenoxibencile; (RS): (RS)- alfa- ciano-3-fenoxybencil(1RS)-cis-trans-3-(2,2-dichlorvinil)- 1,1-dimetilcichlopropanecasrboxilate. * Once per year, ** 2 per year. **Chemical Fertilization**: U: Urea, P: Phosphorus. **Gm**: Green manure, Y: Yes, 0: null application. **Oa**: Organic amendments: csm: cow, sheep manure, bg: burned grass, lcf: last crop fallow, 0: null application.

plot site, or previous vegetation cover and the ecological index (Kendall's Tau, $T = 0.016$ $p = 0.98$, $T = 0.11$ $p = 0.90$, $T = -0.31$ $p = 0.75$, respectively). Therefore, in this study, it seems that neither size nor location of the plot determines the type of plot management; rather, plot management is likely determined by government development programs and traditional farming techniques.

3. Correlation between index and expert opinion

We obtained a Pearson correlation coefficient of $r^2 = 0.61$ ($p = 2.4\text{e-}06$, $d.f. = 49$) between the index and the values determined by independent experts, indicating a satisfactory correspondence (Figure 4). However, the experts systematically awarded higher scores to the plots than did the index. Moreover, they suggested to include additional variables to the index which would yield better information regarding (i) type of organic inputs to the crop system, (ii) types of pest and disease control used, (iii) number of native plant species among the crop, (iv) origin of crop seeds, (v) vegetation surrounding the crop, (vi) presence of vertebrate fauna, and (vii) diversity of soil macroinvertebrates.

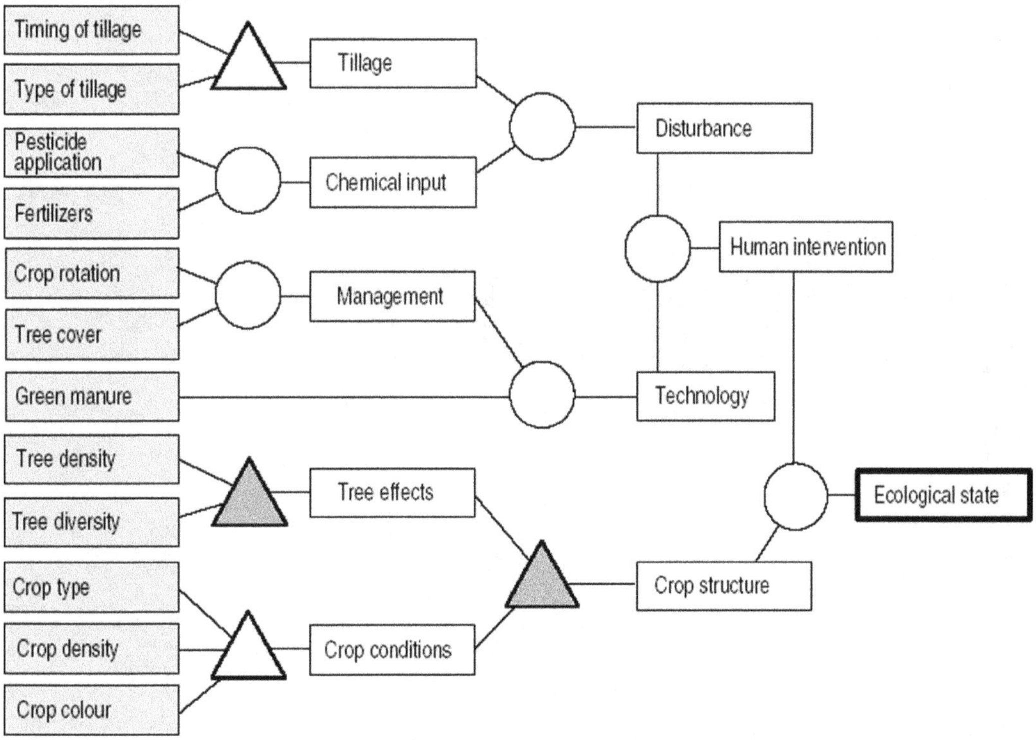

Figure 2. Structure of the Index of ecological condition of tropical agroecosystems. Primary indicators are shaded in grey. Circles represent simple mathematic algorithms, white triangles represent rule sets, and grey triangles represent rule sets, and grey triangles represent rule sets based on fuzzy logic. This index was presented together with other indexes within a frame of Indicators of environmentally sound land use in the humid tropics [15].

Table 4. Description of the measurement levels of the variables.

	Groupal variables	Field data variables	Best	Intermediate			Worst
				5 levels			
			1	0.75	0.5	0.25	0
	Management	Tree cover	all year	presence polyculture	present	rare	null
	Tree effect	Tree density	high	medium	low	isolated	null
				4 levels			
			1	0.66	0.33		0
		Tree diversity	>4 species	4 species	2 species		1 species
Crop structure	Crop conditions	Crop type	perennial polyculture	perennial monoculture	bianual		annual
				3 levels			
			1	0.33			0
	Plough (tillage)	Tillage time	null	frequent			constant
		Tillage form	null	manual			technical
Disturbance	Chemical inputs	Pesticides	null	frequent			constant
		Fertilizers	null	frequent			constant
	Management	Crop rotation	constant	frequent			null
Technology		Green manure	constant	frequent			null
				2 levels			
			1				0
	Crop conditions	Crop density	abundant				disperse
		Crop coleur	green				yellow

Table 5. Normalized Ecological Index for plots evaluated (52).

Plot	TT	TF	P	F	GM	RC	CA	DA	DivA	CT	CD	Ccol	Ev	Index
1	n	m	c	c	n	n	>1	l	>4	pm	a	g	0.24	0.56
2	n	n	c	n	n	n	>1	i	>4	a	a	g	0.35	0.5
3	c	m	n	c	c	n	>1	i	>4	a	a	g	0.26	0.56
4	n	n	n	n	n	n	n	n	1	a	a	g	0.01	0.37
5	n	n	c	c	c	c	>1	i	>4	a	a	g	0.09	0.62
6	n	m	c	c	n	n	>1	l	>4	a	a	g	0.22	0.5
7	c	t	c	n	n	n	>1	i	>4	a	d	b	0	0.15
8	c	t	c	c	c	c	n	n	1	pp	d	g	0.1	0.12
9	n	t	c	n	n	n	n	n	1	a	a	g	0.09	0.25
10	c	t	c	c	n	n	n	n	1	a	d	b	0	0
11	c	t	c	c	n	c	>1	i	>4	pm	ID	ID	0.03	0.21
12	c	t	c	c	n	n	>1	h	>4	ba	a	g	0.2	0.5
13	n	t	c	c	n	n	>1	i	>4	a	a	g	0.03	0.5
14	n	n	c	n	n	n	n	n	1	pp	a	g	0.29	0.37
15	n	t	n	n	n	c	>1	h	>4	pp	a	g	0.2	0.81
16	n	n	c	c	n	n	>1	l	>4	ba	a	g	0.25	0.5
17	c	t	c	c	n	n	n	n	1	a	a	g	0	0.25
18	c	t	c	c	n	n	n	n	1	a	d	g	0	0
19	c	t	c	c	c	n	n	n	1	a	d	g	0	0.06
20	n	t	c	n	n	n	>1	i	>4	a	a	g	0.1	0.5
21	c	t	c	c	n	n	n	n	1	a	a	g	0	0.25
22	n	n	n	n	n	n	>1	i	>4	pp	d	g	0.33	0.62
23	n	t	c	c	c	n	>1	i	>4	a	a	g	0.15	0.56
24	n	t	c	c	c	c	>1	l	>4	a	a	g	0.2	0.62
25	n	n	c	c	n	c	n	n	1	a	d	g	0.01	0.06
26	n	m	c	c	n	n	n	n	1	a	a	g	0.25	0.25
27	n	t	c	n	n	n	>1	i	>4	a	a	g	0.01	0.5
28	n	t	c	c	n	n	n	n	1	a	a	g	0	0.25
29	c	t	c	c	n	n	n	n	1	pm	a	g	0.01	0.37
30	n	m	c	n	n	c	n	n	1	pm	a	g	0.1	0.43
31	c	t	c	c	n	n	n	n	1	a	a	g	0.01	0.25
32	c	t	n	n	n	c	>1	i	>4	a	d	y	0.09	0.34
33	c	t	n	n	n	n	>1	h	>4	a	a	g	0.15	0.62
34	c	t	n	n	n	n	>1	i	>4	ba	a	g	0.13	0.62
35	c	t	n	c	n	n	>1	l	>4	a	a	g	0.01	0.5

Table 5. Cont.

Plot	TT	TF	P	F	GM	RC	CA	DA	DivA	CT	CD	Ccol	Ev	Index
36	c	t	c	c	n	c	n	n	1	a	a	g	0	0.31
37	c	t	c	n	n	n	>1	i	>4	pm	a	y	0.01	0.15
38	c	t	c	c	n	c	>1	h	>4	pm	a	g	0.2	0.68
39	c	t	c	c	n	n	>1	l	>4	a	a	b	0.06	0.18
40	c	t	c	c	n	n	>1	l	>4	a	a	g	0.19	0.5
41	c	t	c	c	n	n	>1	l	>4	a	a	g	0.19	0.5
42	c	t	c	c	n	n	>1	l	>4	a	a	g	0.17	0.5
43	c	t	c	c	n	n	>1	l	>4	a	a	g	0.2	0.5
44	c	t	c	c	n	n	>1	l	>4	a	a	g	0.2	0.5
45	c	t	c	c	n	n	>1	l	>4	a	a	g	0.2	0.5
46	n	n	n	n	n	c	>1	l	>4	pp	a	b	0.6	0.37
47	n	t	n	c	n	n	>1	l	>4	pp	d	g	0.18	0.5
48	c	n	c	c	n	n	n	n	1	a	a	b	0	0
49	c	t	c	c	n	n	>1	l	>4	a	a	b	0.07	0.18
50	c	n	c	c	n	n	>1	l	>4	pm	a	g	0.2	0.56
51	c	n	c	n	n	n	>1	l	>4	pm	a	g	0.15	0.56
52	c	t	c	c	n	n	>1	l	>4	pm	a	ID	0.24	0.18

TT: Tillage Time: c: >once per year; n: once per year; TF: Tillage Form: n: null, m: manual, t: technical; P. Pesticide use: c: constant, n: null, F: Use of chemical Fertilizers: c: constant, n: null; GM: Green Manure, c: constant, n: null; RC: Crop Rotation: c: constant, n: null; CA: tree cover: >1 year, n: null; DA: tree density: h: high, l: low, i: isolated, n:null; DivA: tree diversity: >4 species, 1 species, CT: crop type: a: annual, ba: biannual, pp: perennial polyculture, pm: perennial monoculture; CD: crop density: a: abundant, d: disperse; Ccol: crop colour: g: green, b: brown, y: yellow; EV: evaluation.

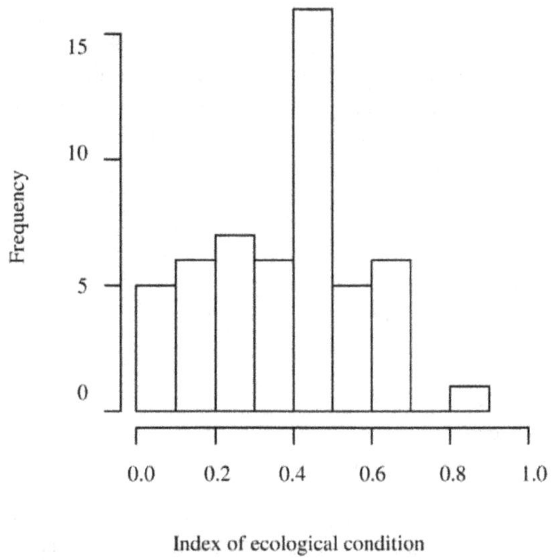

Figure 3. Frequency histogram of values of the index of ecological condition applied to 52 plots in South-East Mexico.

Discussion

Since the Rio Earth Summit, there has been a concerted effort to construct indicators to monitor progress toward sustainable development [59]. Most of these indicators have been developed in Europe (10) and Asia (2) [60]. In Latin America, sustainable indicators have been developed by Astier et al. [61]; this index mainly focuses on subsistence level agriculture, and evaluates the sustainability of a system.

Our index is geared toward small and medium scale producers who principally grow for market and have a fairly large crop area (32.4 ha on average). According to Bockstaller and Girardin [62], indicators must be elaborated according to a scientific approach, and one of the important steps in this elaboration is validation.

Our index was developed according to the consensus of a group of scientists, with knowledge in agroecology, and was validated independently by 2 scientists, each with over15 years of experience in agroecology. The evaluation included 3 important steps: design validation, output validation, and end use validation [62].

In previous studies, only seven indicators have been used to evaluate farm systems: crop diversity, crop succession, pesticide use, nitrogen level, phosphorus level, soil organic matter, and irrigation methods [63]. All of these practices depend on the farmer's decision, and to a large extent they impact the environment.

Our index is based on qualitative and quantitative concrete data and includes most of these 7 indicators, except nitrogen and phosphorus, both of which are observed indirectly via plant health, through the crop colour indicator (according to whether plants have a greenish or yellowish colour); organic matter, which is indirectly characterized by the technology applied in the system (green manure indicator, Figure 2); and irrigation, which in this study was not evaluated, given that all plots evaluated were only used for seasonal rainfed agriculture, according to local rainfall patterns.

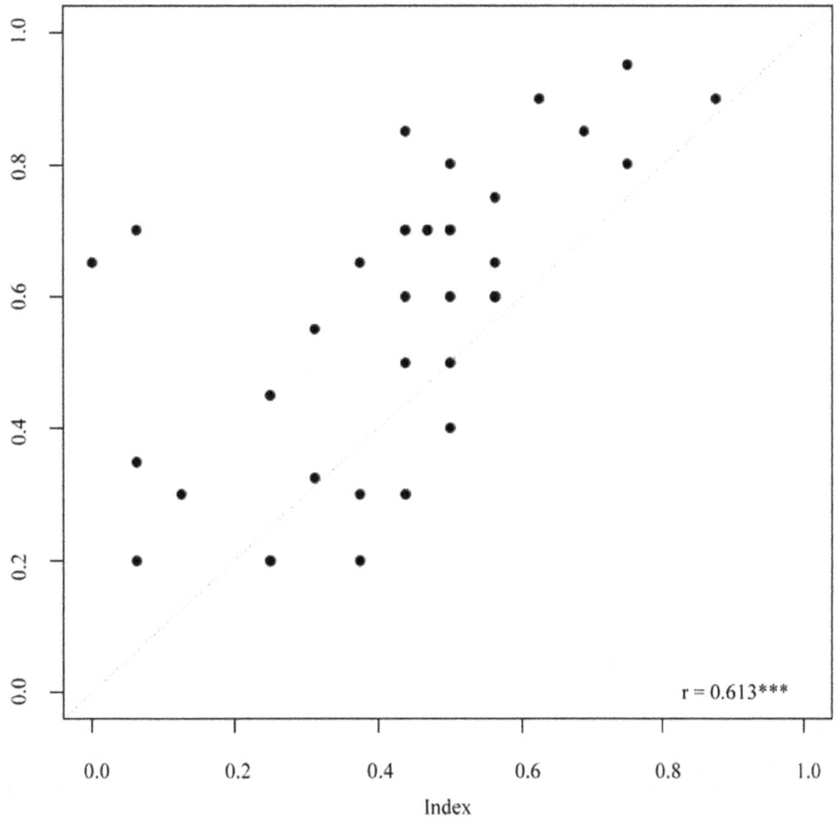

Figure 4. Scatterplot of the values of the index of ecological condition applied to 52 plots in South-East Mexico (x-axis) versus quality values between 0 (worst) and 1 (best) assigned by experts to the same plots (y-axis).

At the farm level there are indicators that evaluate the environmental impact of the agricultural practices and indicators that evaluate the effect of those practices at the local and global level [60].

Our index includes both types of indicators; evaluating agricultural practices: those variables taken in the field: type and frequency of tillage, pesticide application, fertilizer use, crop rotation, tree cover and density, green manure, crop density (see Table 2 & 3). Effect indicators used were disturbance (mainly soil disturbance) measured by tillage (frequency and type) and chemical inputs, technology used, and crop structure (see index, Figure 2). Some existing indexes focus on evaluation or evolution of environmental performance, thus encouraging environmentally sound practices [60], such as crop rotation, organic fertilizers, and no-tillage. Meanwhile, our index identifies the indicator that has the highest environmental impacts in each plot evaluated, with the idea that the farmer could potentially improve these with a given practice, ie. to monitor the soil ecological condition via the use of a soil macroinvertebrates index [64], where the lack of macroinvertebrates informs of a severe negative activity as pollution or conventional tillage.

In developing the variables to be included in our index, we reviewed bibliographic studies and carried out field work obtaining data which we hoped would reflect the negative and environmentally friendly practices commonly used in agro ecosystems of south-eastern Mexico. The index can be used by farmers, using the following web address: http://201.116.78.102/~modelo/Index.html.

However, we did not evaluate certain indicators such as water use, and water quality, as did - for example - the index (monitoring tool) of ecological indicators used on a Flemish dairy farm [65]. Nor did we evaluate environmental impacts due to energy consumption [66]. Within a tropical framework, in south-eastern Mexico, the priorities were to identify those practices that were soil perturbing and environment polluting, practices that can be modified by the farmers, by an attitude changing. Nambiar et al. [16], proposed an agricultural sustainability index (ASI) to measure sustainability as a function of biophysical, chemical, economic, and social indicators, our index only measures the ecological state of the agroecosystem, and provides easy to use tools for improving those practices which negatively impact the environment. Van der Werf and Petit [60], state that indicators based on farmer practices cost less in data collection but do not allow for an actual evaluation of environmental impact. In the case of our study, the experts' evaluation correlated satisfactorily with the index, although the index rendered more penalizing scores than did the experts. Some improvements must be made to our index, relating quality and quantity of the applied inputs for instance; the index should specify the kind of manures used and then to evaluate their effect when added to the systems. The consulted experts found important to integrate this information into the index, they also found that the possible relations among different crops and environmental effects of using cow manure, vermicompost, or traditional compost have to be considered. Another variable which the experts suggested should be added to the index is the presence of natural vegetation surrounding the crop. Farmers see advantages of having crops surrounded by secondary forest, diversity in the agricultural area can be increased, ie when different pollinators arrive.

The advantages of using this index is that the common agricultural practices (mechanized land preparation and use of common pesticides ie. Carbofuran, Chlorpyrifos), evaluated in this study as indicators are used throughout the world; over time, through practice, the index may be improved. Farmers and other land owners may realize which of the practices they use are disturbing the environment, due to the fact that these practices generate a value in each of the evaluated indicators. The variables obtained in the field contribute to the information of each of the indicators, and the index is the compendium of all the indicators. Our index doesn't give a sustainability measure, because it does not include socioeconomic indicators of the farms. Further studies are required in order to observe the acceptance of this index by farmers in a regional scale.

Acknowledgments

We thank Lorena Soto and Guillermo Jimenez for participating in the evaluation of the plots. We thank Simon Hernandez de la Cruz, Alejandra Sepulveda Lozada, Lauritania Ibarra Hernandez, and Marcelo Rodriguez Ricardes, who helped to collect data in the field. We are also thankful to Lorena Reyes for bibliographic support. El Colegio de la Frontera Sur provided infrastructural resources. Language revision was done by Ann Greenberg.

Author Contributions

Conceived and designed the experiments: EH CK SOG BDJ SHD VG. Performed the experiments: EH CK. Analyzed the data: EH CK. Contributed reagents/materials/analysis tools: EH CK SOG BDJ SHD VG. Contributed to the writing of the manuscript: EH CK SOG BDJ VG. Figures elaboration: CK SOG.

References

1. Houghton RA (1994) The world wide extent of land-use change. BioScience 44: 305–313.
2. Tilman D, Lehman C (2001) Human-cause environmental change: impacts on plant diversity and evolution. Proceedings of the National Academy of Sciences 98: 5433–5440.
3. INEGI (2008) Cuaderno Estadístico Municipal de Tenosique. Villahermosa, Tabasco: Gobierno del Estado de Tabasco.
4. Grande D, de Leon F, Nahed J, Perez-Gil F (2010) Importance and Function of Scattered Trees in Pastures in the Sierra Region of Tabasco, Mexico. Research Journal of Biological Sciences 5: 75–87.
5. Geissen V, Morales-Guzman G (2006) Fertility of tropical soils under different land use systems-a case study of soils in Tabasco, Mexico. Appl Soil Ecol 841: 1–10.
6. Melgar C, Geissen V, Cram S, Sokolov M, Bastidas P, et al. (2008) Pollutants in drainage channels following long-term application of Mancozeb to banana plantations in southeastern Mexico. Journal of Plant Nutrition and Soil Science 171: 597–604.
7. Aryal DR, Geissen V, Ponce-Mendoza A, Ramos-Reyes RR, Becker M (2012) Water quality under intensive banana production and extensive pastureland in tropical Mexico. Journal of Plant Nutrition and Soil Science 175(4): 553–559.
8. Ortiz SM, Anaya G, Estrada BW (1994) Evaluación, Cartografía y Políticas Preventivas de la Degradación de la Tierra. Chapingo, Mexico.
9. Geissen V, Sanchez-Hernandez R, Kampichler C, Ramos-Reyes R, Sepulveda-Lozada A, et al. (2009) Effects of land-use change on some properties of tropical soils – An example from Southeast Mexico. Geoderma 151: 87–97.
10. Moreno-Unda AA (2011) Environmental effects of the National Tree Clearing Program, Mexico, 1972–1982 Cologne: Cologne University of Applied Sciences 119 p.
11. Bockstaller CGL, Keichinger O, Girardin P, Galan MB, Gaillard G (2009) Comparison of methods to assess the sustainability of agricultural systems. A review. Agron Sustain Dev 29: 223–235.
12. Mitchell GMA, McDonald A (1995) PICABUE: a methodological framework for the development of indicators of sustainable development. Int J Sust Dev World 104–123.

13. Bockstaller C, Guichard L, Makowski D, Aveline A, Girardin P, et al. (2008) Agri-Environmental Indicators to Assess Cropping and Farming Systems: A Review Sustainable Agriculture. In: Lichtfouse E, Navarrete M, Debaeke P, Véronique S, Alberola C, editors. Springer Netherlands. 725–738.
14. Riley J (2001) The indicator explosion: local needs and International challenges. Agr Ecosyst Environ 87: 119–120.
15. Kampichler C, Hernández-Daumás S, Ochoa-Gaona S, Geissen V, Huerta-Lwanga E, et al. (2010) Indicators of environmentally sound land use in the humid tropics: The potential roles of expert opinion, knowledge engineering and knowledge discovery. Ecological Indicators 10: 320–329.
16. Nambiar KK, Gupta AP, Fu Qinglin, Li S (2001) Biophysical, chemical and socio-economic indicators for assessing agricultural sustainability in the Chinese coastal zone Agriculture Ecosystems and Environment 87: 209–214.
17. Cabell JF, Oelofse M (2012) An indicator framework for assessing agroecosystem resilience. Ecology and Society 17: 18 http://dx.doi.org/10.5751/ES-04666-170118.
18. Darnhofer I, Bellon S, Dedieu B, Milestad R (2010) Adaptiveness to enhance the sustainability of farming systems. A review. Agronomy for Sustainable Development 30: 545–555.
19. Allen CR, Fontaine JJ, Pope KL, Garmestani AS (2011) Adaptive management for a turbulent future. Journal of Environmental Management 92: 1339–1345.
20. Malezieux E, Crozat Y, Dupraz C, Laurans M, Makowski D, et al. (2009) Mixing plant species in cropping systems: concepts, tools and models. A review. Agron Sustain Dev 29: 43–62.
21. Jones CG, Lawton J, Shachak M (1997) Positive and negative effects of organisms as physical ecosystem engineers. Ecology and Society 78: 1946–1957.
22. Brussaard L, Caron P, Campbell B, Lipper L, Mainka S, et al. (2010) Reconciling biodiversity conservation and food security: scientific challenges for a new agriculture. Current opinion in Environmental sustainability 2: 34–42.
23. Knorr D, Watkins TR (1984) Alterations in Food Production Knorr DW, Watkins, T.R., editor. New York: Van Nostrand Reinhold.
24. Seufert V, Ramankutty N, Foley JA (2012) Comparing the yields of organic and conventional agriculture. Nature 485: 229–232.
25. Carter H (1989) Agricultural sustainability: an overview and research assessment. Calif Agric 43: 1618–1637.
26. Tellarini V, Caporali F (2000) An input/output methodology to evaluate farms as sustainable agroecosystems: an application of indicators to farms in central Italy. Agriculture, Ecosystems and Environment 77: 111–123.
27. Stinner BR, House GJ (1987) Role of ecology in lower-input, sustainable agriculture: an introduction. Am J Alternative Agric 2: 146–147.
28. Lockeretz W (1988) Open questions in sustainable agriculture. Am J Alternative Agric 3: 174–181.
29. Hauptli H, Katz D, Thomas BR, Goodman RM (1990) Biotechnology and crop breeding for sustainable agriculture. In: Edwards CA, Lal R, Madden P, Miller RH, House G, editors. Sustainable Agricultural Systems Soil and Water Conservation Society: Ankeny, Iowa. 141–156.
30. Madden P (1990) The economics of sustainable low-input farming systems. In: Francis CA, Flora CB, King LD, editors. Sustainable Agriculture in Temperate Zones. New York: John Wiley & Sons. 315–341.
31. Dobbs TL, Becker DL, Taylor DC (1991) Sustainable agriculture policy analyses: South Dakota on-farm case studies. J Farming Systems ResExt 2: 109–124.
32. Blair GJ, Lefroy RD, Lisle L (1995) Soil carbon fractions based on their degree of oxidation, and the development of a carbon management index for agricultural systems. Aust J Soil Res 46: 1459–1466.
33. Ouédraogo E, Mando A, Zombré NP (2001) Use of compost to improve soil properties and crop productivity under low input agricultural system in West Africa. Agriculture, Ecosystems and Environment 84: 259–266.
34. Pretty J, Toulmin C, Williams S (2011) Sustainable intensification in African agriculture. International Journal of Agricultural Sustainability 9: 5–24.
35. Lal R (2004) Soil carbon sequestration impacts on global climate change and food security. Science 304: 1623–1627.
36. Gurr GM, Wratten SD, Luna JM (2003) Multi-function agricultural biodiversity: pest management and other benefits. Basic Appl Ecol 4: 107–116.
37. Börjesson P (1999) Environmental effects of energy crop cultivation in Sweden I: Identification and quantification. Biomass and Bioenergy 16: 137–154.
38. Rommelse R (2001) Economic assessment of biomass transfer and improved fallow trials in western Kenya In: ICRAF Natural Resource Problems PaPP, editor. Natural Resource Problems. Nairobi (Kenya): International Centre for Research in Agroforestry (ICRAF).
39. Nyende P, Delve R (2004) Farmer participatory evaluation of legume cover crop and biomass transfer technologies for soil fertility improvement using farmer criteria, preference ranking and logit regression analysis. Exp Agric 40: 77–88.
40. Koocheki A, Nassiri M, Alimoradi L, Ghorbani R (2009) Effect of cropping systems and crop rotations on weeds. Agronomy for Sustainable Development 29: 401–408.
41. Bellon M (1995) Farmers 'Knowledge and Sustainable Agroecosystem Management: An Operational Definition and an Example from Chiapas, Mexico. Human Organization 54: 263–272.
42. Belalcázar S, Espinosa J (2000) Effect of Plant Density and Nutrient Management on Plantain Yield. Better Crops International 14: 12–15.
43. Shaahan MM, El-Sayed AA, Abou El-Nour EAA (1999) Predicting nitrogen, magnesium and iron nutritional status in some perennial crops using a portable chlorophyll meter. Scientia Horticulturae 82: 339–348.
44. Agbede TM (2008) Nutrient availability and cocoyam yield under different tillage practices. Soil Tillage Research 99: 49–57.
45. Edwards CA (1989) The Importance of Integration in Sustainable Agricultural Systems. Agriculture, Ecosystems and Environment 27: 25–35.
46. Edwards CA, Grove TL, Harwood RR, Pierce Colfer CJ (1993) The role of agroecology and integrated farming systems in agricultural sustainability. Agriculture, Ecosystems and Environment 46: 99–121.
47. García-Pérez JA, Alarcón-Gutiérrez E, Perroni Y, Barois I (2014) Earthworm communities and soil properties in shaded coffee plantations with and without application of glyphosate. Applied Soil Ecology 83: 230–237.
48. Correia FV, Moreira JC (2010) Effects of Glyphosate and 2,4-D on Earthworms (*Eisenia foetida*) in Laboratory Tests. Bull Environ Contam Toxicol 85: 264–268.
49. Dick J, Kaya B, Soutoura M, Skiba U, Smith R, et al. (2008) The contribution of agricultural practices to nitrous oxide emissions in semi-arid Mali Soil. Use and Management 24: 292–301.
50. Morales H, Perfecto I (2000) Traditional knowledge and pest management in the Guatemalan highlands. Agriculture and Human Values 17: 49–63.
51. Li BL, Rykiel EJ (1996) Introduction. Ecological Modelling 90: 109–110.
52. Salski A (1996) Introduction Fuzzy Logic in Ecological Modelling. Ecological Modelling 85: 1–2.
53. Mendoza GA, Prabhu R (2003) Fuzzy methods for assessing criteria and indicators of sustainable forest management. Ecological Indicators 3: 227–236.
54. Kampichler C, Platen R (2004) Ground beetle occurrence and moor degradation: modelling a bioindication system by automated decision-tree induction and fuzzy logic. Ecological Indicators 4: 99–109.
55. INEGI (2000) Cuaderno Estadístico Municipal de Tenosique. Gobierno del Estado de Tabasco; INEGI, editor. Villahermosa, Tabasco.
56. INEGI (1985) Carta Edafológica, Villahermosa In: E15-8, editor. 1: 250,000. Aguascalientes, México: Instituto Nacional de Estadística, Geografía e Informática.
57. Manjarrez-Muñoz B (2008) Ordenamiento territorial de la ganadería bovina en Balancán y Tenosique, Tabasco. Villahermosa Tabasco: El Colegio de la Frontera Sur. 105 p.
58. Isaac-Márquez R (2008) Análisis del Cambio de Uso y Cobertura del Suelo en los Municipios de Balancán y Tenosique, Tabasco, México. Villahermosa, Tabasco, México: El Colegio de la Frontera Sur.
59. Rigby D, Woodhouse P, Young T, Burton M (2001) Constructing a farm level indicator of sustainable agricultural practice. Ecological Economics 39: 463–478.
60. van der Werf HMG, Petit J (2002) Evaluation of the environmental impact of agriculture at the farm level: a comparison and analysis of 12 indicator-based methods. Agriculture, Ecosystems and Environment 93: 131–145.
61. Astier M, Speelman S, López-Ridaura S, Masera O, Gonzalez-Esquivel CE (2011) Sustainability indicators, alternative strategies and trade-offs in peasant agroecosystems: analysing 15 case studies from Latin America. International Journal of Agricultural Sustainability 9: 409–422.
62. Bockstaller C, Girardin P (2003) How to validate environmental indicators. Agr Syst 76: 639–653.
63. Bockstaller C, Girardin P, Van der Werf HGM (1997) Use of agroecological indicators for the evaluation of farming systems. Eur J Agron 7: 261–270.
64. Huerta E, Kampichler C, Geissen V, Ochoa-Gaona S, de Jong B, et al. (2009) Towards an ecological index for tropical soil quality based on soil macrofauna. Pesq agropec bras 44 (8): 1056–1062.
65. Meul M, Nevens F, Reheul D (2009) Validating sustainability indicators: Focus on ecological aspects of Flemish dairy farms. Ecological Indicators 9: 284–295.
66. Pervanchon F, Bockstaller C, Girardin P (2002) Assessment of energy use in arable farming systems by means of an agro-ecological indicator: the energy indicator. Agricultural Systems 72: 149–172.

Permissions

All chapters in this book were first published in PLOS ONE, by The Public Library of Science; hereby published with permission under the Creative Commons Attribution License or equivalent. Every chapter published in this book has been scrutinized by our experts. Their significance has been extensively debated. The topics covered herein carry significant findings which will fuel the growth of the discipline. They may even be implemented as practical applications or may be referred to as a beginning point for another development.

The contributors of this book come from diverse backgrounds, making this book a truly international effort. This book will bring forth new frontiers with its revolutionizing research information and detailed analysis of the nascent developments around the world.

We would like to thank all the contributing authors for lending their expertise to make the book truly unique. They have played a crucial role in the development of this book. Without their invaluable contributions this book wouldn't have been possible. They have made vital efforts to compile up to date information on the varied aspects of this subject to make this book a valuable addition to the collection of many professionals and students.

This book was conceptualized with the vision of imparting up-to-date information and advanced data in this field. To ensure the same, a matchless editorial board was set up. Every individual on the board went through rigorous rounds of assessment to prove their worth. After which they invested a large part of their time researching and compiling the most relevant data for our readers.

The editorial board has been involved in producing this book since its inception. They have spent rigorous hours researching and exploring the diverse topics which have resulted in the successful publishing of this book. They have passed on their knowledge of decades through this book. To expedite this challenging task, the publisher supported the team at every step. A small team of assistant editors was also appointed to further simplify the editing procedure and attain best results for the readers.

Apart from the editorial board, the designing team has also invested a significant amount of their time in understanding the subject and creating the most relevant covers. They scrutinized every image to scout for the most suitable representation of the subject and create an appropriate cover for the book.

The publishing team has been an ardent support to the editorial, designing and production team. Their endless efforts to recruit the best for this project, has resulted in the accomplishment of this book. They are a veteran in the field of academics and their pool of knowledge is as vast as their experience in printing. Their expertise and guidance has proved useful at every step. Their uncompromising quality standards have made this book an exceptional effort. Their encouragement from time to time has been an inspiration for everyone.

The publisher and the editorial board hope that this book will prove to be a valuable piece of knowledge for researchers, students, practitioners and scholars across the globe.

List of Contributors

Martin Lechenet, Sandrine Petit and Nicolas M. Munier-Jolain
Institut National de la Recherche Agronomique, Unité Mixte de Recherche 1347 Agroécologie, Dijon, Côte d'Or, France

Vincent Bretagnolle
Centre d'Etudes Biologiques de Chizé – Centre National de Recherche Scientifique, Beauvoir sur Niort, Deux-Sévres, France

Christian Bockstaller
Institut National de la Recherche Agronomique, Unité de Recherche 1121 Agronomie et Environnement, Colmar, Haut-Rhin, France
Université de Lorraine, Vandoeuvre-lés-Nancy, Meurthe-et-Moselle, France

Franc¸ois Boissinot
Chambre d'Agriculture des Pays de la Loire, Angers, Maine-et-Loire, France

Marie-Sophie Petit
Chambre Régionale d'Agriculture de Bourgogne, Quetigny, Côte d'Or, France

Jochen Krauss and Ingolf Steffan-Dewenter
Department of Animal Ecology and Tropical Biology, University of Würzburg, Biocentre, Würzburg, Germany
Population Ecology Group, Department of Animal Ecology I, University of Bayreuth, Bayreuth, Germany

Iris Gallenberger
Population Ecology Group, Department of Animal Ecology I, University of Bayreuth, Bayreuth, Germany

Mark F. Hulme and Philip W. Atkinson
British Trust for Ornithology, Thetford, Norfolk, United Kingdom

Juliet A. Vickery
The Royal Society for the Protection of Birds, Sandy, Bedfordshire, United Kingdom

Rhys E. Green
The Royal Society for the Protection of Birds, Sandy, Bedfordshire, United Kingdom
Department of Zoology, University of Cambridge, Cambridge, United Kingdom

Ben Phalan
Department of Zoology, University of Cambridge, Cambridge, United Kingdom

Dan E. Chamberlain
Dipartimento di Biologiá Animale e dell'Uomo, University of Turin, Turin, Italy

Derek E. Pomeroy and Raymond Katebaka
Department of Biological Sciences, Makerere University, Kampala, Uganda

Dianah Nalwanga and David Mushabe
NatureUganda, Kampala, Uganda

Simon Bolwig
Department of Management Engineering, Technical University of Denmark, Copenhagen, Denmark

Sarah J. Helyar
Molecular Ecology and Fisheries Genetics Laboratory, Bangor University, Bangor, Wales, United Kingdom

Food Safety, Environment & Genetics, Matís, Reykjavík, Iceland

Hywel ap D Lloyd
Molecular Ecology and Fisheries Genetics Laboratory, Bangor University, Bangor, Wales, United Kingdom
Food Safety, Environment & Genetics, Matís, Reykjavík, Iceland

Mark de Bruyn and Gary R. Carvalho
Molecular Ecology and Fisheries Genetics Laboratory, Bangor University, Bangor, Wales, United Kingdom

Jonathan Leake
Sunday Times, London, United Kingdom

Niall Bennett
Greenpeace UK, London, United Kingdom

Valentin H. Klaus, Norbert Hölzel and Till Kleinebecker
University of Münster, Institute of Landscape Ecology, Münster, Germany

Daniel Prati, Barbara Schmitt and Markus Fischer
University of Bern, Institute of Plant Sciences, Bern, Switzerland

Ingo Schöning and Marion Schrumpf
Max-Planck-Institute for Biogeochemistry, Jena, Germany

Stefano Casalegno, Richard Inger, Caitlin DeSilvey and Kevin J. Gaston
Environment and Sustainability Institute, University of Exeter, Penryn, Cornwall, United Kingdom

Aurelice B. Oliveira, Enéas Gomes-Filho and Maria Raquel A. Miranda
Universidade Federal do Ceará, Depto. Bioquímica e Biologia Molecular, Fortaleza-CE, Brazil

Carlos F. H. Moura
Embrapa Agroindustria Tropical, Fortaleza-CE, Brazil

Claudia A. Marco
Universidade Federal do Ceará, Campus do Cariri, Av. Tenente Raimundo Rocha s/n - Cidade Universitária, Juazeiro do Norte-CE, Brazil

Laurent Urban
Université d'Avignon et des Pays de Vaucluse, Campus Agroparc, Avignon, France

Péter Batáry
Agroecology, Georg-August University, Göttingen, Germany
MTA-ELTE-MTM Ecology Research Group, Budapest, Hungary

Laura Sutcliffe
Plant Ecology and Ecosystem Research, Georg-August University, Göttingen, Germany

Carsten F. Dormann
Computational Landscape Ecology, UFZ Centre for Environmental Research, Leipzig, Germany
Biometry and Environmental System Analysis, University of Freiburg, Freiburg, Germany

Teja Tscharntke
Agroecology, Georg-August University, Göttingen, Germany

Dionys Forster and Christian Andres
International Division, Research Institute of Organic Agriculture (FiBL), Frick, Switzerland

Rajeev Verma
Research Division, bioRe Association, Kasrawad, Madhya Pradesh, India

Christine Zundel
International Division, Research Institute of Organic Agriculture (FiBL), Frick, Switzerland
Ecology Group, Federal Office for Agriculture (FOAG), Bern, Switzerland

Monika M. Messmer and Paul Mäder
Soil Sciences Division, Research Institute of Organic Agriculture (FiBL), Frick, Switzerland

Basavaraj Basannagari and Chandra Prakash Kala
Ecosystem and Environment Management, Indian Institute of Forest Management, Bhopal, Madhya Pradesh, India

Ru Li, Martin H. Entz and W. G.Dilantha Fernando
Department of Plant Science, University of Manitoba, Winnipeg, Manitoba Canada

Ehsan Khafipour and Denis O. Krause
Department of Animal Science, University of Manitoba, Winnipeg, Manitoba, Canada
Department of Medical Microbiology and Infectious Diseases, Winnipeg, Manitoba, Canada

Teresa R. de Kievit
Department of Microbiology, University of Manitoba, Winnipeg, Manitoba, Canada

Xuelin Zhang, Qun Wang, Yilun Wang and Chaohai Li
The Incubation Base of the National Key Laboratory for Physiological Ecology and Genetic Improvement of Food Crops in Henan Province, Zhengzhou, China; Agronomy College of Henan Agricultural University, Zhengzhou, China

Frank S. Gilliam
Department of Biological Sciences, Marshall University, Huntington, West Virginia, United States of America

Feina Cha
Meteorological Bureau of Zhengzhou, Zhengzhou, China

Kosuke Takemura
Graduate School of Management, Kyoto University, Kyoto, Japan

Yukiko Uchida and Sakiko Yoshikawa
Kokoro Research Center, Kyoto University, Kyoto, Japan

América P. Durán, Stefano Casalegno and Kevin J. Gaston
Environment and Sustainability Institute, University of Exeter, Penryn, Cornwall, United Kingdom

Pablo A. Marquet
Departamento de Ecología, Facultad de Ciencias Biológicas, Pontificia Universidad Católica de Chile, Santiago, Chile
Instituto de Ecología y Biodiversidad (IEB), Santiago, Chile

Santa Fe Institute, Santa Fe, New Mexico, United States of America
Laboratorio Internacional en Cambio Global (LINCGlobal), Facultad de Ciencias Biológicas, Pontificia Universidad Católica de Chile, Santiago, Chile

Karen R. Siegel and K. M. Venkat Narayan
Nutrition and Health Sciences, Laney Graduate School, Emory University, Atlanta, Georgia, United States of America
Hubert Department of Global Health, Emory University, Atlanta, Georgia, United States of America

Mohammed K. Ali
Hubert Department of Global Health, Emory University, Atlanta, Georgia, United States of America

Adithi Srinivasiah
Emory College, Emory University, Atlanta, Georgia, United States of America

Rachel A. Nugent
Department of Global Health, University of Washington, Seattle, Washington, United States of America

Fanqiao Meng, Xiangping Sun and Wenliang Wu
College of Resources and Environmental Sciences, China Agricultural University, Beijing, China

Jørgen E. Olesen
Department of Agroecology and Environment, Faculty of Agricultural Sciences, Aarhus University, Tjele, Denmark

Georg K. S. Andersson and Henrik G. Smith
Centre for Environmental and Climate Research, Lund University, Lund, Sweden
Department of Biology, Lund University, Lund, Sweden

Maj Rundlöf
Department of Biology, Lund University, Lund, Sweden
Department of Ecology, Swedish University of Agricultural Sciences, Uppsala, Sweden

Nicolás L. Gutiérrez, David J. Agnew, Patricia L. Bianchi, Daniel D. Hoggarth, Chris Ninnes and Amanda Stern-Pirlot
Marine Stewardship Council, London, United Kingdom

Sarah R. Valencia and Christopher Costello
Bren School of Environmental Science and Management, University of California Santa Barbara, Santa Barbara, California, United States of America

Trevor A. Branch, Timothy E. Essington, Ray Hilborn and James T. Thorson
School of Aquatic and Fishery Sciences, University of Washington, Seattle, Washington, United States of America

Julia K. Baum
Department of Biology, University of Victoria, Victoria, British Columbia, Canada

Jorge Cornejo-Donoso
Interdepartmental Graduate Program in Marine Science, Marine Science Institute, University of California Santa Barbara, Santa Barbara, California, United States of America
Universidad Austral de Chile, Centro Trapananda, Coyhaique, Chile

Omar Defeo
UNDECIMAR, Facultad de Ciencias, Montevideo, Uruguay

Ashley E. Larsen, Rebecca L. Selden, Seeta Sistla and Sarah J. Teck
Department of Ecology, Evolution and Marine Biology, University of California Santa Barbara, Santa Barbara, California, United States of America

Keith Sainsbury
University of Tasmania, Tasmanian Aquaculture & Fisheries Inst, Taroona, Tasmania, Australia

Anthony D. M. Smith
Commonwealth Scientific and Industrial Research Organization, Wealth from Oceans Flagship, Hobart, Tasmania, Australia

Nicholas E. Williams
Department of Anthropology, University of California Santa Barbara, Santa Barbara, California, United States of America

María J. Ek-Ramos, Wenqing Zhou, César U. Valencia, Josephine B. Antwi, Lauren L. Kalns and Gregory A. Sword
Department of Entomology, Texas A & M University, College Station, Texas, United States of America

Gaylon D. Morgan
Department of Soil and Crop Sciences and Texas AgriLife Extension, Texas A & M University, College Station, Texas, United States of America

David L. Kerns
Department of Entomology, Texas A & M University, College Station, Texas, United States of America

AgriLife Extension Service, Texas A & M University, Lubbock, Texas, United States of America

Macon Ridge Research Station, LSU AgCenter, Winnsboro, Louisiana, United States of America

Eileen F. Power
Botany Department, School of Natural Sciences, Trinity College Dublin, Dublin, Republic of Ireland

Daniel L. Kelly and Jane C. Stout
Trinity Centre for Biodiversity Research, Trinity College Dublin Dublin, Republic of Ireland

Charles M. Benbrook and Donald R. Davis
Center for Sustaining Agriculture and Natural Resources, Washington State University, Pullman, Washington, United States of America

Gillian Butler and Carlo Leifert
School of Agriculture, Food and Rural Development, Newcastle University, Northumberland NE, United Kingdom

Maged A. Latif
Organic Valley/CROPP Cooperative/Organic Prairie, Lafarge, Wisconsin, United States of America

Esperanza Huerta and Salvador Hernandez-Daumas
El Colegio de la Frontera Sur, Unidad Campeche, Dpto. Agroecología, Campeche, México

Christian Kampichler
Universidad Juárez Autónoma de Tabasco, División de Ciencias Biológicas, Villahermosa, Tabasco, México

Sovon Dutch Centre for Field Ornithology, Natuurplaza (Mercator 3), Nijmegen, The Netherlands

Susana Ochoa-Gaona and Ben De Jong
El Colegio de la Frontera Sur, Unidad Campeche, Dpto. Sustainability Sciences, Campeche, México

Violette Geissen
University of Bonn - INRES, Bonn, Germany

Wageningen University and Research Center - Alterra, Wageningen, Gelderland, Netherlands

Index

Printed in the USA
CPSIA information can be obtained
at www.ICGtesting.com
JSHW052021301024
72690JS00004B/131

9 781632 397829